Lecture Notes in Artificial Intelligence 8918

Subseries of Lecture Notes in Computer Science

LNAI Series Editors

Randy Goebel
  University of Alberta, Edmonton, Canada
Yuzuru Tanaka
  Hokkaido University, Sapporo, Japan
Wolfgang Wahlster
  DFKI and Saarland University, Saarbrücken, Germany

LNAI Founding Series Editor

Joerg Siekmann
  DFKI and Saarland University, Saarbrücken, Germany

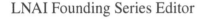

T0234593

Xianmin Zhang   Honghai Liu   Zhong Chen
Nianfeng Wang (Eds.)

Intelligent Robotics and Applications

7th International Conference, ICIRA 2014
Guangzhou, China, December 17-20, 2014
Proceedings, Part II

 Springer

Volume Editors

Xianmin Zhang
Zhong Chen
Nianfeng Wang
South China University of Technology
School of Mechanical and Automobile Engineering
Wushan Rd. 381, Tianhe District, Guangzhou 510641, China
E-mail: {zhangxm, mezhchen, menfwang}@scut.edu.cn

Honghai Liu
University of Portsmouth
School of Computing
Intelligent Systems and Biomedical Robotics Group
Buckingham Building, Lion Terrace
Portsmouth PO1 3HE, UK
E-mail: honghai.liu@icloud.com

ISSN 0302-9743   e-ISSN 1611-3349
ISBN 978-3-319-13962-3   e-ISBN 978-3-319-13963-0
DOI 10.1007/978-3-319-13963-0
Springer Cham Heidelberg New York Dordrecht London

Library of Congress Control Number: 2014956249

LNCS Sublibrary: SL 7 – Artificial Intelligence

Typesetting: Camera-ready by author, data conversion by Scientific Publishing Services, Chennai, India

Printed on acid-free paper

Springer is part of Springer Science+Business Media (www.springer.com)

Preface

The Organizing Committee of the 7th International Conference on Intelligent Robotics and Applications has been committed to facilitating interactions among those active in the field of intelligent robotics, automation, and mechatronics. Through this conference, the committee intends to enhance the sharing of individual experiences and expertise in intelligent robotics with particular emphasis on technical challenges associated with varied applications such as biomedical application, industrial automations, surveillance, and sustainable mobility.

The 7th International Conference on Intelligent Robotics and Applications was most successful in attracting 159 submissions on addressing the state-of-the art developments in robotics, automation, and mechatronics. Owing to the large number of submissions, the committee was faced with the difficult challenge of selecting of the most deserving papers for inclusion in these proceedings. For this purpose, the committee undertook a rigorous review process. Despite the high quality of most of the submissions, only 109 papers (68.5 % acceptance rate) were selected for publication in two volumes of Springer's *Lecture Notes in Artificial Intelligence* as subseries of *Lecture Notes in Computer Science*. The selected papers were presented during the 7th International Conference on Intelligent Robotics and Applications held in Guangzhou, China, during December 17 – 20, 2014.

The selected articles represent the contributions of researchers from 19 countries. The contributions of the technical Program Committee and the additional reviewers are deeply appreciated. Most of all, we would like to express our sincere thanks to the authors for submitting their most recent works and the Organizing Committee for their enormous efforts to turn this event into a smoothly running meeting. Special thanks go to South China University of Technology for the generosity and direct support. Our particular thanks are due to Alfred Hofmann and Anna Kramer of Springer for enthusiastically supporting the project.

We sincerely hope that these volumes will prove to be an important resource for the scientific community.

October 2014

Xianmin Zhang
Xiangyang Zhu
Jangmyung Lee
Huosheng Hu

Organization

International Advisory Committee

Tamio Arai	University of Tokyo, Japan
Hegao Cai	Harbin Institute of Technology, China
Tianyou Chai	Northeastern University, China
Toshio Fukuda	Nagoya University, Japan
Sabina Jesehke	RWTH Aachen University, Germany
Oussama Khatib	Stanford University, USA
Zhongqin Lin	Shanghai Jiao Tong University, China
Ming Li	National Natural Science Foundation of China, China
Nikhil R. Pal	Indian Statistical Institute, India
Grigory Panovko	Russian Academy of Science, Russia
Jinping Qu	South China University of Technology, China
Clarence De Silva	University of British Columbia, Canada
Bruno Siciliano	University of Naples, Italy
Shigeki Sugano	Waseda University, Japan
Michael Yu Wang	Chinese University of Hong Kong, China
Youlun Xiong	Huazhong University of Science and Technology, China
Ming Xie	Nanyang Technological University, Singapore
Lotfi A. Zadeh	University of California Berkeley, USA

General Chair

Xianmin Zhang	South China University of Technology, China

General Co-chairs

Xiangyang Zhu	Shanghai Jiao Tong University, China
Jangmyung Lee	Pusan National University, South Korea
Huosheng Hu	University of Essex, UK

Program Chair

Guoli Wang	Sun Yat-Sen University, China

Program Co-chairs

Youfu Li	City University of Hong Kong, SAR China
Honghai Liu	University of Portsmouth, UK

Organizing Committee Chairs

Han Ding   Huazhong University of Science and
                        Technology, China
Feng Gao   Shanghai Jiao Tong University, China
Zexiang Li   Hong Kong University of Science and
                        Technology, China
Tian Huang   Tianjin University, China
Hong Liu   Harbin Institute of Technology, China
Guobiao Wang   National Natural Science Foundation of China,
                        China
Tianmiao Wang   Beihang University, China
Hong Zhang   University of Alberta, Canada

Area Chairs

Suguru Arimoto   Ritsumeikan University, Japan
Jiansheng Dai   King's College of London University, UK
Christian Freksa   Bremen University, Germany
Shuzhi Sam Ge   National University of Singapore, Singapore
Peter B. Luh   Connecticut University, USA

Awards Chairs

Yangming Li   University of Macau, SAR China
Wei-Hsin Liao   Chinese University of Hong Kong, SAR China
Caihua Xiong   Huazhong University of Science and
                        Technology, China

Publication Chairs

Zhong Chen   South China University of Technology, China
Xinjun Sheng   Shanghai Jiao Tong University, China

Local Arrangements Chairs

Nianfeng Wang   South China University of Technology, China
Lianghui Huang   South China University of Technology, China

Finance Chair

Hongxia Yang   South China University of Technology, China
Wei Sun   South China University of Technology, China

General Affairs Chair

Zhiwei Tan   South China University of Technology, China

Additional Reviewers

We would like to acknowledge the support of the following people who peer reviewed articles for ICIRA 2014.

Xianmin Zhang	Guoli Wang	Nianfeng Wang
Guoying Gu	Zhong Chen	Tiemin Zhang
Yanjiang Huang	Xiangyang Zhu	Xinjun Liu
Caihua Xiong	Hegao Cai	Tianyou Chai
Tamio Arai	Yan Li	Clarence De Silva
Toshio Fukuda	Sabina Jesehke	Michael Yu Wan
Zhongqin Lin	Ming Li	Shigeki Sugano
Grigory Panovko	Jinping Qu	Jangmyung Lee
Bruno Siciliano	Lotfi A. Zadeh	Honghai Liu
Huosheng Hu	Youfu, Li	Zexiang Li
Han Ding	Feng Gao	Guobiao Wang
Tian Huang	Hong Liu	Suguru Arimoto
Tianmiao Wang	Hong Zhang	Shuzhi Sam Ge
Jiansheng Dai	Christian Freksa	Youlun Xiong
Peter B. Luh	Wei-Hsin Liao	
Xinjun Sheng	Yangming Li	

Table of Contents – Part II

Parallel Robotics

Robot Vision

Mechatronics

Industrial Robotics

System Optimization and Analysis

Mechanism Design

Table of Contents – Part I

Recent Advances in Research and Application of Modern Mechanisms

Rehabilitation Robotics

Underwater Robotics and Applications

Agricultural Robot

Bionic Robotics

Service Robotics

Kinematics Dexterity Analysis and Optimization of 4-UPS-UPU Parallel Robot Manipulator

Guohua Cui[*], Haiqiang Zhang, Feng Xu, and Chuanrong Sun

College of Equipment Manufacture,
Hebei University of Engineering,
Handan, Hebei Province, 056038, China
ghcui@hebeu.edu.cn, zhq19860905@126.com

Abstract. The development of a new parallel robot manipulator based on simulation analysis is a rapid approach to discover the unique features or advantage of a conceptual model. In this paper, a 5-DOF parallel robot manipulator which can generate three translations and two rotations was presented. The kinematics mathematical model and Jacobian matrix were derived analytically. The global conditions index (GCI) and the global gradient index (GGI) which represent the evaluation index of dexterity were introduced by considering the kinematics performance indices over the whole workspace. The workspace model of the mechanism was analyzed based on a simplified boundary searching method. The mathematical model of the global condition number was developed simultaneously. The multi-objective optimization model was deduced on the basis of the multidisciplinary design philosophy. The manipulator was optimized by using the design of experiment (DOE) and the multi-island genetic algorithm (MIGA). The optimal solution was chosen from the multi optimal solutions in a reasonable manner. Through the comparison of results before and after optimization, the kinematics performance of the mechanism was improved, which provide not only a guide to the multiple objectives optimal design but also an applicable method of dimensional synthesis for the optimal design of general parallel robot manipulator.

Keywords: Parallel Robot Manipulator, Kinematics Dexterity, Workspace, Multi-objective optimization.

1   Introduction

Parallel manipulator has the advantage of high rigidity, strong bearing capacity, and high precision and small error. Since the Stewart Parallel Manipulator, parallel manipulator has become an international research focus [1][2]. The dexterity and isotropy are of importance performance index to evaluate the mechanism. The kinematics dexterity can evaluate the transmission performance. Gosselin [3]

---

[*] This research was supported by the national natural science foundation of China under grant No. 51175143.

X. Zhang et al. (Eds.): ICIRA 2014, Part II, LNAI 8918, pp. 1–11, 2014.

introduced the concept of dexterity into parallel manipulator, and pointed out that the condition of Jacobian matrix can represent the dexterity; Zhang [4] adapted the kinematics condition index (KCI) to evaluated the dexterity, and draw the distribution atlas of spatial KCI; Moreno [5] regarded the condition of Jacobian matrix as the performance index of dexterity, calculated and analyzed the condition. Sergiu [6] proposed a large number of performance criteria dealing with workspace, quality transmission, manipulability, dexterity and stiffness, and the evaluation measures can be used for optimal synthesis; Chen [7] studied the dexterity of 4-UPS-UPU parallel manipulator focused on the seven performance indices; Qi [8] analyzed the structure of the five degree of freedom 4-UPS-UPU and proposed synthesis methods about the operation performance optimization on the orientation workspace.

In this paper, a five degree 4-UPS-UPU parallel manipulator was studied and its kinematics model and the dimensionless Jacobian matrix were established. Considering the kinematics performance on the workspace, the GCI and GGI were introduced as the evaluation index of the dexterity, which was optimized based on the multidisciplinary and multi-objective optimization software Isight. On the basis of workspace, the global condition index was developed and obtained the mathematical model of optimization. Last we obtained the Pareto solution by using design of experiment and multi-island genetic algorithm and selected the reasonable optimal solution. Compared the results before and after optimization, we can draw the conclusion that the new mechanism after optimization has excellent dexterity and transmissions performance, which provide a guide for design optimization and performance assessment. Therefore, it is necessary to seek an effective optimization procedure to improve the performance indices for achieving a higher score evaluation.

2   4-UPS-UPU Parallel Manipulator Model

As shown in Fig.1, 4-UPS-UPU parallel manipulator model and coordinate system were established, which consists of a moving platform, fixed platform and the legs connected the moving platform and the fixed platform, for four identifiable active chains UPS and one constraint active chain UPU, U stands for Hooke joint, P stand for Prismatic joint, S for Spherical joint, where the P joint is driven by a linear actuator.

Suppose that the platforms are circular and the connection points are distributed along the circumference of the moving platform and the fixed platform circles of radii r_a and r_b, respectively. The coordinate system $O-XYZ$ is fixed to the fixed platform and the coordinate $o-xyz$ is attached to the moving platform. x axis point to A_1, z axis perpendicular to the moving platform on the positive axis direction and y axis is given by the right hand. Similarly, X axis point to B_1 point, the Z axis is vertical. The points of intermediate branched Hooke joints are located on the point o and point O of the moving platform and the fixed platform, respectively.

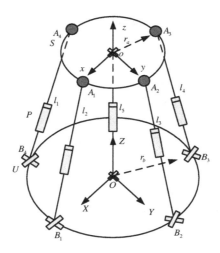

Fig. 1. The schematic diagram of 4-UPS-UPU parallel manipulator architecture

The number of degree of freedom for the parallel manipulator can be obtained by the general Kutahach-Grubler formula [9]

$$M = \mathrm{d}(n-g-1) + \sum_{i=1}^{g} f_i = 6 \times (12-15-1) + 29 = 5 \tag{1}$$

The 4-UPS-UPU parallel manipulator is a spatial 5-DOF, its moving platform can move in X, Y, Z and rotate around X and Y direction.

3   4-UPS-UPU Parallel Manipulator Kinematics Model

3.1   Inverse Kinematics Solution

As shown in Fig.1, position vector A_i, B_i and branched chains A_iB_i, for $i = 1,2,3,4$, the Cartesian coordinate of the moving platform is given by the position vector $^{o}A_i$ with respect to the moving coordinate system, and the position of the attachment point B_i with respect to the fixed coordinate system can be written as $^{O}B_i$. The Cartesian variables are chosen to be the relative position and orientation of $o-xyz$ frame with respect to $O-XYZ$ frame, where the position of o is specified by the position of its origin with respect to $O-XYZ$ frame, Furthermore, if vector $o = \begin{bmatrix} x & y & z \end{bmatrix}^T$ described the position of the attachment point o with respect to $O-XYZ$ frame. The coordinate can be represented as following,

$$^{o}A_i = \begin{bmatrix} A_{ix} & A_{iy} & A_{iz} \end{bmatrix}^T,$$

$$^{O}B_i = \begin{bmatrix} B_{iX} & B_{iY} & B_{iZ} \end{bmatrix}^T,$$

$$^{O}A_i = \begin{bmatrix} A_{iX} & A_{iY} & A_{iZ} \end{bmatrix}^T$$

$^{o}A_i$ is expressed with respect to the coordinate system $O-XYZ$ can be computed by

$$^{o}A_i = Q^{o}A_i + o \tag{2}$$

Q is a matrix describing the orientation of $o-xyz$ with respect to the $O-XYZ$, here RPY coordinate system representation is chosen to describe the pose, that is,

$$Q = \begin{bmatrix} c\beta & s\alpha s\beta & c\alpha s\beta \\ 0 & c\alpha & -c\alpha \\ -s\beta & s\alpha c\beta & c\alpha c\beta \end{bmatrix} \tag{3}$$

Where, s and c present sine and cosine, respectively
The length vector can be expressed as

$$L_i = A_iB_i = {}^{o}A_i - {}^{o}B_i \tag{4}$$

Then, the length of the active leg can be expressed by taking the norm of the vector of Eq.(4), we can get the

$$l_i = \|A_iB_i\| = \|{}^{o}A_i - {}^{o}B_i\| \tag{5}$$

3.2   Jacobian Matrix of the Parallel Manipulator

The relation between active joint velocities $\dot{i} = \begin{bmatrix} \dot{l}_1 & \dot{l}_2 & \dot{l}_3 & \dot{l}_4 & \dot{l}_5 \end{bmatrix}^{T}$ and twist of the end-effector $t = \begin{bmatrix} \dot{x} & \dot{y} & \dot{z} & \omega_x & \omega_y \end{bmatrix}^{T}$ can be described using a differential kinematics model, namely,

$$\dot{i} = \mathbf{J}t \tag{6}$$

Where $\mathbf{J}$ denote the Jacobian matrix

$$\mathbf{J} = \begin{bmatrix} q_1 & q_2 & q_3 & q_4 & q_5 \\ r_1 \times q_1 & r_2 \times q_2 & r_3 \times q_3 & r_4 \times q_4 & r_5 \times q_5 \end{bmatrix}^{T} = [\mathbf{J}_1 \vdots \mathbf{J}_2] \tag{7}$$

Where, the unit vector q_i, for $i = 1,2,3,4,5$ can be expressed in terms of position vectors, namely,

$$q_i = \frac{L_i}{l_i} \tag{8}$$

And vector r_i can be written as

$$r_i = Q^{o}A_i \tag{9}$$

$\mathbf{J}_1$ denoted Jacobian matrix of linear velocity and $\mathbf{J}_2$ denoted Jacobian matrix of angle velocity, whose dimension of $\mathbf{J}_1$ and $\mathbf{J}_2$ are both 6×3. Considering the unit difference of the Jacobian matrix was dimensionally inhomogeneous [10]. So we use a characteristic length, L_c, to homogenize the original Jacobian matrix in such a way that

$$\mathbf{J}_H = \mathbf{J} \cdot diag(1,1,1,\frac{1}{L_c},\frac{1}{L_c},\frac{1}{L_c}) \tag{10}$$

Where, $L_c = \sqrt{\dfrac{trace(J_2^T J_2)}{trace(J_1^T J_1)}}$ and $\mathbf{J}_H$ denotes the new homogeneous Jacobian matrix.

4   Performance Index of Kinematics Dexterity

In order to make the mechanism has good kinematics performance in the workspace, the kinematics dexterity optimization was studied and the GCI and the GGI were introduced as the evaluation index [11].

4.1   The Global Condition Index

The condition number of the Jacobian matrix changed along with the position and orientation of parallel robot manipulator, therefore, it cannot be measured the dexterity of the mechanism in the whole workspace. In order to obtain the kinematics performance in the whole workspace, Gosselin and Angeles [12] proposed the global condition index, which is a measure of its kinematics precision and control accuracy and which is defined as the ratio of the integral of the inverse condition numbers calculated in the whole workspace, dived by the volume of the workspace, i.e.,

$$GCI = \frac{\int_W v\,dW}{\int_W dW} \tag{11}$$

In which v is the local condition number defined as the reciprocal of the condition of the Jacobian matrix at a particular pose, and W is the workspace. It is noteworthy that the Global Condition number Index is bounded as (0, 1). If the *GCI* approach zero, the mechanism has a bad global performance and as the *GCI* approaches one the mechanism has a good global performance. Therefore, we should make the optimization objective *GCI* maximization.

4.2   The Global Gradient Index

The global gradient index reflected the average deviation level of the kinematics performance in the working space, and cannot reflect the fluctuation properties of mechanism in the working space. F.A.Lara-Molina [13] proposed the global gradient index, which represented the fluctuation information of the local performance index, and defined as

$$\nabla GGI = \max_W \left\| \nabla 1 / \kappa(J) \right\| \tag{12}$$

Where, the local gradient condition number can be expressed as

$$\nabla 1 / \kappa = [\frac{\partial 1 / \kappa(J)}{\partial x}, \frac{\partial 1 / \kappa(J)}{\partial y}, \frac{\partial 1 / \kappa(J)}{\partial z}, \frac{\partial 1 / \kappa(J)}{\partial \alpha}, \frac{\partial 1 / \kappa(J)}{\partial \beta}] \tag{13}$$

GGI is approximately equal to the maximum value of the local gradient throughout the workspace. As the gradient is bigger, so the fluctuation of the kinematics dexterity is greater. That means the kinematics performance of the parallel robot manipulator is up and down in the entire workspace. If the gradient is small, the kinematics performance of the mechanism in the working space is more stable. Therefore, the global gradient index should take the smallest value in the whole working space.

4.3   The Workspace Analysis

The workspace of parallel robot manipulator can be divided into constant orientation workspace and the dexterous workspace. Because of rotation around z axis constrained by the institution, so the mechanism doesn't have the dexterous workspace. In this paper, we established the GCI based on the constant orientation workspace, according the Section 4.1, we need to solve the workspace [14]. The workspace can be expressed as

$$W = \{(x, y, z) \in R \,|\, f(x, y, z) \leq 0\} \tag{14}$$

Where, $f(x, y, z) \leq 0$ denoted the constraint condition, namely,
(1) The active chains length constraints can be expressed by

$$l_{\min} \leq l_i \leq l_{\max} \tag{15}$$

Where $l_{\max}$ denoted the maximum link length, l_i denoted the link length of the i th link, and $l_{\min}$ denoted the minimum link length;
(2) The rotational angle of the spherical joint and the Hooke joint and their constraint can be computed by

$$\theta_u = arc \cos(l_i \cdot eb \,/\, \|l_i\|) \leq \theta_{u\,\max} \tag{16}$$

$$\theta_s = arc \cos(l_i \cdot ea \,/\, \|l_i\|) \leq \theta_{s\,\max} \tag{17}$$

Where, ea, eb represented the unit normal vector of the moving platform and fixed platform, respectively. $\theta_{u\,\max}$, $\theta_{s\,\max}$ represented the max angle limitation of the Hooke joint and Spherical joint, respectively.
(3) The mechanism was non-singular configuration, namely, the determinant of the Jacobian matrix was not equal to zero.

In order to obtain the position workspace of 4-UPS-UPU parallel manipulator quickly, we set the structural parameters of the parallel robot manipulator as, respectively: the circumcircle radius of the moving platform $r_a = 0.06$ m, the circumcircle radius of the fixed platform $r_b = 0.15$ m, the maximum shrinkage limit of the active chains are 0.05 m, the maximum elongation limit is 0.25 m, the maximum angle of the Hooke joint and the Spherical joint are both $\frac{\pi}{3}$. The translation ranges of the moving platform when $\alpha = \beta = 0°$ are $x \in [-0.1m, 0.1m]$, $y \in [-0.1m, 0.1m]$, $z \in [0.05m, 0.2m]$. The workspace is drawn using software MATLAB, as shown in Fig.2.

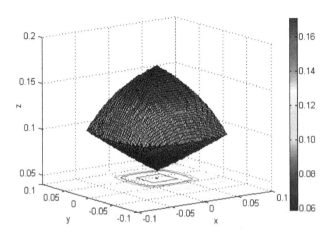

Fig. 2. Workspace of the 4-UPS-UPU parallel manipulator

5   The Multi-objective Optimization Problem of 4-UPS-UPU Parallel Robot Manipulator

5.1   The Optimization Model

The scale parameters of the parallel manipulator were radii of the moving platform r_a and the fixed platform r_b . Select the link length of the active chains l_i , and angle of the Hooke joint θ_u and Spherical joint θ_s as the constraint conditions, the optimization objective functions for *GCI* and the *GGI* , then the multi-objective optimization model can be expressed as

$$\begin{cases} \max f_1 = GCI(r_a, \; r_b) \\ \max f_2 = -GGI(r_a, \; r_b) \end{cases} \tag{18}$$

$$\text{s.t.} \begin{cases} 0.03 \le r_a \le 0.08 \\ 0.08 \le r_b \le 0.18 \\ 0.05 \le l_i \le 0.325 \\ \theta_s \le \dfrac{\pi}{3} \\ \theta_u \le \dfrac{\pi}{3} \end{cases} \tag{19}$$

5.2   The Optimization Results Analysis

Isight software integrated MATLAB, which adopted design of experiment and optimal algorithm to solve the maximum value of *GCI* and the minimum value of *GGI* [15]. The design of experiment (DOE) used the optimal Latin hypercube method and optimal algorithm used multi island genetic algorithm (MIGA).
Genetic algorithm parameters configuration are as follows:

Total population size: 100;
Sub population number: 10;
The number of the island: 20;
The total number: 100;
Cross probability: 0.8;
Migration rate: 0.45;
Interval algebra migration: 5;
The ratio of the individual to participate competition: 1;
The number of elite individuals of the next generation: 1;

In the multi-objective optimization process based on Isight, the samples points in the design of experiment were calculated and eliminated the values which were inconsistent with the constraints, and the values which were satisfied the constraints would access to the optimization part and conducted the multi-objective optimization solution. After several genetic iterations, we can obtain the Pareto frontier of the GCI and the GGI.

We can obtain the main effect diagram and Pareto diagram between design variables and objective functions from the design of experiment, as shown in Fig.3 to Fig.6. We can get the Pareto frontier between the GCI and the GGI and the feasibility of the design of optimization at the end of the MIGA optimization, in the following Fig.7 and Fig.8.

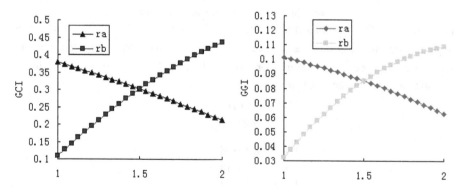

Fig. 3. The main effect between the design variables and GCI

Fig. 4. The main effect between the design variables and GGI

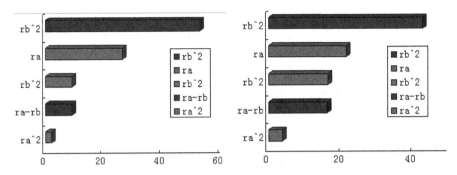

Fig. 5. Pareto diagram of the Global Condition Index

Fig. 6. Pareto diagram of the Global Gradient Index

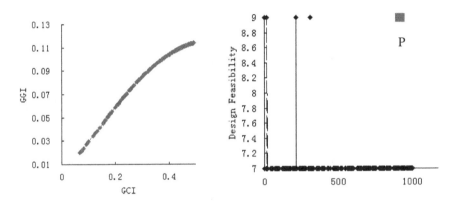

Fig. 7. Pareto frontier of the global performance indices

Fig. 8. The feasibility of the optimal design

As is shown in the Fig.3 and Fig.4 above, we can see that the design variables r_a and r_b have a large effect on the global performance indices GCI and GGI, and there is a linear relationship. From Fig.5 and Fig.6, r_b^2, the square of the design r_b, has a big contribution to GCI approximately sixty percent (the blue denoted the positive effect). Secondly, r_a has a big impact on the performance GGI (the red denoted the negative effect). The cross term $r_a - r_b$ has a small effect on the GCI. The influence trend of the design variables is substantially the same between GCI and GGI. Distribution from the Pareto solution in Fig.7, we can see that the Global Condition Index and the Global Gradient Index were the conflicting indices. If the GCI increased, simultaneously, the GGI would improve. Multi-objective optimization was different from the single objective optimization, not to obtain a solution of the function. Due to the conflicting of the multi-objective function, the Pareto solution may not be dominant. But if we simply optimize a target, we may make the other performance index poor.

As can be seen in Fig.8, the feasibility of the optimization design was more than seven, which indicated it was feasible to optimization design. What's more, the red box represented the recommended design point. In this paper, we choose P point as the optimal solution, and the best variables are revealed in Table 1 (Fig.8).

Table 1. Results comparison before and after optimization

	r_a	r_b	GCI	GGI
Before optimization	0.06	0.15	0.28	0.08
After optimization	0.044	0.198	0.46	0.1

The result shows that the kinematics dexterity increased, but loss the gradient index. So the designers need to weigh the results according to the specific application.

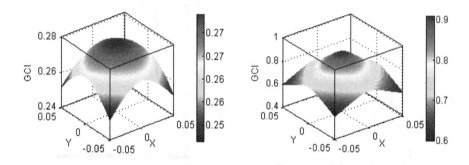

Fig. 9. Local dexterity before optimization when $z = 0.08$

Fig. 10. Local dexterity after optimization when $z = 0.08$

Through the comparisons between Fig.9 and Fig.10, we can see the kinematics dexterity increased obviously after optimization, in the position $x = 0$, $y = 0$, $z = 0.08$, has the best dexterity, and the value is close to one. Due to the symmetry of the mechanism, the dexterity was also symmetric distribution in the workspace. By weighing comprehensively, we can choose P point as the design optimization solution.

6   Conclusion

(1) In this paper, 4-UPS-UPU parallel manipulator with a five-degree of freedom was studied, and the kinematics model and the Jacobian matrix were established. Considering the kinematics performance in the workspace, we introduced the GCI and the GGI as the evaluation criterion of the kinematics dexterity.

(2) We established a mathematical model of the global index on the workspace; and we constructed the multi-objective optimization model of 4-UPS-UPU parallel manipulator. In order to obtain the global performance value, we must solve the workspace firstly.

(3) Multi-objective optimization research was conducted on the basis of the multidisciplinary design optimization software Isight, adopted the design of experiment and the multi-island genetic algorithm to optimize the 4-UPS-UPU parallel manipulator, and obtained the Pareto solutions.

(4) We choose the optimal solution from the number of the solutions in reasonable selection and determined the structural parameters and optimization parameters. The results between before and after optimization show that the kinematics performance improved highly. The methodology in this paper paves the way for providing not only the effective guidance but also a new approach of dimensional synthesis for the optimal design of general parallel mechanisms.

References

1. Gupta, A., O'Malley, K., Patoglu, V., et al.: Design, Control and performance of Rice Wrist: A Force Feedback Wrist Exoskeleton for Rehabilitation and Traning. The International Journal of Robotics Research 27(2), 233–251 (2008)
2. Refaat, S., Herve, J., Nahavandi, S.: Two-mode overatrained three-DOFs rotational translational linear motor based parallel-kinematics mechanism for machine tool application. Robotic 25, 461–466 (2007)
3. Gosselin, C.: Dexterity indices for planar and spatial robotic manipulators. In: 1990 IEEE International Conference Robotics and Automation, vol. 1, pp. 650–655 (1990)
4. Zhang, Y., Zhang, H.: Kinematics and Dexterity Analysis of a Novel Pure Translational Parallel Manipulator. Machine Tool and Hydraulics 08, 13–16 (2010)
5. Moreno, H.A., Pamanes, J.A., Wenger, P., et al.: Global optimization of performance of a 2PRR parallel manipulator for cooperative tasks. In: 3rd International Conference on Informatics in Control, Automation and Robotics (2006)
6. Stan, S., Manic, M., Szep, C., et al.: Performance analysis of 3DOF Delta parallel. In: 2011 4th International Conference on Human System Interaction (HSI), Yokohama, Japan, May 19-21 (2011)
7. Chen, X., Gao, Q., Zhao, Y.: Research on Dexterity Measures of 4-ups-upu Parallel Coordinate Measuring Machine. Computer Integrated Manufacturing System 18(6) (2012)
8. QI, M.: Dimensional synthesis of 4-UPS/UPU 5-DOF parallel mechanism. Journal of Harbin Institute of Technology 11(41), 160–164 (2009)
9. Yu, J., Liu, X., et al.: The robot mechanism mathematical foundation. Mechanical Industry Press (2008)
10. Xie, B., Zhao, J.: Advances in Robotic Kinematic Dexterity and Indices. Mechanical Science and Technology 08, 1386–1393 (2011)
11. Gosselin, C., Angeles, J.: A Global Performance Index for the Kinematic Optimization of Robotic Manipultors. Journal of Mechanical Design 113(3), 220–226 (1991)
12. Gosselin, C., Angeles, J.: The optimum kinematics design of a spherical three degree of freedom parallel manipulator. ASME Journal of Mechanisms, Transmissions, and Automation in Design 111(2), 202–207 (1989)
13. Lara-Molina, F.A., Rosario, J.M., et al.: Multi-Objective design of parallel manipulator using global indices. The Open Mechanical Engineering Journal 4, 37–47 (2010)
14. Cui, G., Zhou, H., Wang, N., et al.: Multi-objective Optimization of 3-UPS-S Parallel mechanism Based on Isight. Journal of Agricultural Machinery 09, 261–266 (2013)
15. Stan, S.D., Manic, M., Mătieş, M., Bălan, R.: Evolutionary Approach to Optimal Design of 3 DOF Translation Exoskeleton a Medical Parallel Robots. In: HSI 2008, IEEE Conference on Hum System Interaction, Krakow, Poland, May 25-27 (2008)

Inverse Dynamics Analysis
of a 6-PSS Parallel Manipulator

Weiyuan Xu[1], Yangmin Li[1,2,*], Song Lu[1], and Xiao Xiao[1]

[1] Department of Electromechanical Engineering, University of Macau,
Avenida da Universidade, Taipa, Macao SAR, China
ymli@umac.mo
[2] Tianjin Key Laboratory of the Design and Intelligent Control of the Advanced
Mechatronical System, Tianjin University of Technology, Tianjin 300384, China

Abstract. In this paper, a new six degrees of freedom (6-DOF) parallel manipulator with adjustable actuators is proposed. The kinematic model is firstly established and the kinematic analysis is performed afterward. Then the equations of motion are developed based on the concept of link Jacobian matrices. Finally, the principle of virtual work is applied to analyze the dynamics of this 6-PSS parallel manipulator. This methodology can be used on other types of parallel manipulators not only for 6-DOF but also with less than 6-DOF. To solve the inverse dynamics of the manipulator, a computational algorithm is developed and two trajectories of the moving platform are simulated.

Keywords: 6-PSS parallel manipulator, Kinematics, Dynamics, Virtual work.

1 Introduction

In the last decades, although the serial manipulators have been widely used in the industrial fields, the requirement for more efficient on the robotic operation is still increasing, which drives the engineers to design some typical parallel robots, such as Giddings & Lewis, Ingersoll and Hexel or even some micro parallel manipulators for the high precision application [1]. A parallel manipulator mostly consists of three parts: a moving platform, a fixed base and several limbs that connect the platform and the base. Because the actuators can be mounted on the fixed base of the manipulator, the weights of the moving components (limbs and moving platform) can be reduced, which will minimize the effect of the inertia of the limbs on the operation. Therefore, the parallel manipulator has some inherent advantages than traditional serial manipulator, such as: higher positioning accuracy, better rigidity and larger load capacity.

It is meaningful to develop the dynamical model of the robot because the dynamical analysis is essential for the computer simulation, control strategy development and physical prototype optimization [2]-[3]. Typically, there are two

---

* Corresponding author.

X. Zhang et al. (Eds.): ICIRA 2014, Part II, LNAI 8918, pp. 12–23, 2014.
© Springer International Publishing Switzerland 2014

problems for the dynamics analysis of parallel manipulator: forward and inverse dynamics [4]. The forward dynamics is about a situation that the input forces or the moments are given and we will calculate the position and orientation of the moving platform. On the other hand, the inverse dynamics is to gain the input forces or moments of the actuators with respect to the given motion trajectories of the moving platform. And this model later can be used to design the dynamic controller. Over the last three decades, several researchers have made contributions to the dynamic analysis of parallel manipulator. Some typical approaches that have been proposed include the Newton-Euler formulation [5]-[7], the Lagrangian formulation [8]-[10] and the principle of virtual work [2], [4], [11]. Other new approaches also have been studied such as the Kane method [12]-[14].

Because the kinematic model of the spatial parallel manipulator is complex, it is very normal to make some assumptions to simplify the expressions of the kinetic and potential energy when applying the Newton-Euler or Lagrangian methods [15]-[16]. Therefore, these approaches sometimes are not accurate and efficient enough for the dynamic analysis of parallel manipulator on some perspective. In this paper, we select the principle of virtual work to develop the dynamic modeling of this 6-PSS parallel manipulator. The method presented in this paper is similar to that used in Tsai [2] and Gosselin [11]. However, the process for developing the Jacobian matrices is different from that of [2], which makes it more easier and normal to form the motion equations. Moreover, this method is also suitable for other closed-loop structures dynamic analysis, such as other types of parallel manipulators.

In what follows, the structure of this 6-PSS parallel manipulator is illustrated with a three dimensional model. Then, the inverse kinematics are analyzed and a new method to define the link Jacobian is proposed. Thirdly, the dynamic equations of motion are formulated based on the principle of virtual work. Finally, a computational algorithm is developed to solve the inverse dynamic equations by MATLAB software and some simulations are made with respect to two given trajectories.

2 Kinematic Analysis

2.1 Illustration of the 6-PSS Parallel Manipulator

The architecture of the 6-PSS parallel mechanism is shown in Fig.1 that is composed of a fixed base, a moving platform, three triangle rail trusses and six identical limbs. The details of this manipulator has been described in [17].

As shown in Fig.1, the 3D prototype of the 6-PSS parallel manipulator, there are 14 links connected by 6 prismatic joints and 12 spherical joints. Hence, the number of the degrees of freedom of such mechanism is

$$F = \lambda(n - j - 1) + \sum_i f_i = 6(14 - 18 - 1) + (6 + 3 \times 12) = 12 \qquad (1)$$

However, there are 6 passive degrees of freedom associated with these six PSS limbs. Therefore, the moving platform possesses 6 degrees of freedom.

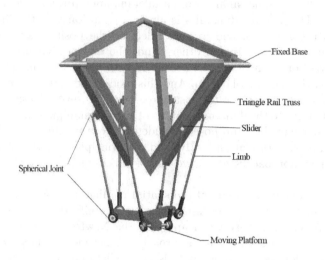

Fig. 1. A 3D prototype of the 6-PSS parallel manipulator

2.2 Kinematics Model

For the purpose of kinematic analysis, two Cartesian coordinate systems $O(x, y, z)$ and $B(u, v, w)$ are attached to the fixed base and the moving platform, respectively. As shown in Fig. 2, the $O(x, y, z)$ frame is attached at the center point O of the fixed congruent triangle base platform $\Delta M_1 M_2 M_3$ (M_1, M_2, M_3 are the cross sectional points of the central lines of the sides.). And the $B(u, v, w)$ frame is attached on the moving platform at point P that is the center of the hexagon $B_1 B_2 B_3 B_4 B_5 B_6$, which indicates the origin of frame $B(u, v, w)$ coincides with the center point P. The x-axis is along the direction of vector $M_2 O$, and the y-axis is parallel to vector $C_5 C_6$. And for the frame $B(u, v, w)$, the u-axis is perpendicular to the line $B_5 B_6$, same direction with x-axis and the v-axis is alongside the y-axis on origin. Both the z-axis and ω-axis are defined by the right-hand rule.

In this study, we assume that $OM_k = R$ ($k = 1, 2, 3$), $BB_i = r$, $C_i D_i = L$ and $A_i B_i = l$. The angle φ between planes $C_1 C_2 D_1$ and $C_1 C_2 C_4$ is defined as the angle layout of actuator, and θ is for the angle between PB_2 and the mid-perpendicular line of line segment $B_1 B_2$.

The coordinates transformation of the moving points B_i from the moving frame $B(u, v, w)$ to the fixed frame $O(x, y, z)$ can be described by the position vector $\boldsymbol{p} = \begin{bmatrix} p_x & p_y & p_z \end{bmatrix}^T$ of the centroid P and the rotation matrix $^O R_B$ in a $[3 \times 3]$ matrix. Let $\boldsymbol{u}, \boldsymbol{v}$ and $\boldsymbol{w}$ be the three unit vectors defined along with u, v and w axes of the frame $B(u, v, w)$, and the $^O R_B$ can be defined as a rotation of γ about the fixed x-axis, followed by a rotation of β about the fixed y-axis, and a rotation of α about the fixed z-axis, thus it yields $^O R_B$ to

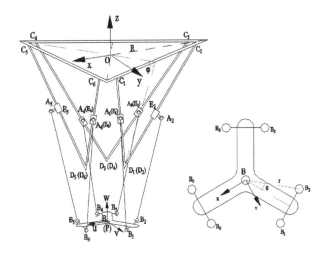

Fig. 2. Schematic representation of the 6-PSS parallel manipulator

$$
\begin{aligned}
^{O}R_B &= R_X(\gamma)R_Y(\beta)R_Z(\alpha) \\
&= \begin{bmatrix} c\alpha c\beta & c\alpha s\beta s\gamma - s\alpha c\gamma & c\alpha s\beta c\gamma + s\alpha s\beta \\ s\alpha c\beta & -s\alpha s\beta s\gamma + c\alpha c\gamma & -s\alpha s\beta c\gamma - c\alpha c\gamma \\ -s\beta & c\beta s\gamma & c\beta c\gamma \end{bmatrix}.
\end{aligned}
\tag{2}
$$

According to the structure of the model in Fig.2, the coordinates of the points B_i on the moving platform can be obtained with reference to the fixed frame O by using a closed-loop vector as follows:

$$
OC_i + d_i s_i + E_i A_i + l k_i = p + b_i .
\tag{3}
$$

where

d_i is the displacement of corresponding slider E_i;
k_i is the unit vector of limb i with respect to fixed frame O;
$b_i = {}^{O}R_B {}^{B}b_i$ and ${}^{B}b_i$ are the coordinates of B_i with respect to frame B;
s_i is the unit vector of the groove of the triangle truss i, respectively.

$$
s_i = \frac{C_i D_i}{L} .
\tag{4}
$$

Then by solving Eq.(3), we will find the vector k_i by

$$
k_i = \frac{p + b_i - OC_i - d_i s_i - E_i A_i}{l} .
\tag{5}
$$

2.3   Velocity Analysis

Before computing the motion equations of this manipulator, it is necessary to analyze the kinematic characteristics of each reference limb. According to the definition of the position and rotation matrix of the moving platform, we have the linear and angular velocities of it as follows

$$V_p = [\dot{p}_x \ \dot{p}_y \ \dot{p}_z]^T . \tag{6}$$

$$\boldsymbol{\omega}_p = \left[\dot{\gamma} \ \dot{\beta} \ \dot{\alpha}\right]^T . \tag{7}$$

The velocity of the center of a spherical joint B_i can be obtained by taking the derivative of the right-hand side of Eq.(3) with respect to time.

$$V_{bi} = V_p + \boldsymbol{\omega}_p \times \boldsymbol{b}_i . \tag{8}$$

Next, taking the derivative of the left-hand side of Eq.(3) with respect to time, we have another expression as follows:

$$V_{bi} = \dot{d}_i \boldsymbol{s_i} + l\boldsymbol{\omega}_i \times \boldsymbol{k}_i . \tag{9}$$

Dot multiplying both sides of Eq.(9) with $\boldsymbol{k}_i$ yields

$$\dot{d}_i = \frac{\boldsymbol{k}_i^T \cdot V_{bi}}{\boldsymbol{k}_i^T \cdot \boldsymbol{s}_i} . \tag{10}$$

Cross multiplying both sides of Eq.(9) with $\boldsymbol{k}_i$ yields to

$$\boldsymbol{\omega}_i = \frac{1}{l} \left[\boldsymbol{k}_i \times V_{bi} - \dot{d}\boldsymbol{k}_i \times \boldsymbol{s}_i\right] . \tag{11}$$

In this paper, we suppose that the center of mass of limb i is at the geometry center, then we have

$$C_{mi} = OC_i + d_i \boldsymbol{s}_i + E_i A_i + \frac{l}{2}\boldsymbol{k}_i . \tag{12}$$

Taking derivative of Eq.(12) with respect to time, we have the velocity of the center of mass of limb i as follows

$$V_{li} = \dot{d}_i \boldsymbol{s}_i + \frac{l}{2}\boldsymbol{\omega}_i \times \boldsymbol{k}_i . \tag{13}$$

2.4   Acceleration Analysis

The acceleration items of the moving platform can be obtained by taking the secondary derivative of the corresponding items as follows

$$\dot{V}_p = [\ddot{p}_x \ \ddot{p}_y \ \ddot{p}_z]^T . \tag{14}$$

$$\dot{\boldsymbol{\omega}}_p = \left[\ddot{\gamma} \ \ddot{\beta} \ \ddot{\alpha}\right]^T . \tag{15}$$

The acceleration of points B_i is obtained by taking the time derivative of Eq.(8).

$$\dot{V}_{bi} = \dot{V}_p + \dot{\omega}_p \times b_i + \omega_p \times (\omega_p \times b_i) . \tag{16}$$

By taking the derivative of Eq.(9) with respect to time, it yields another expression of the acceleration of point B_i as follows

$$\dot{V}_{bi} = \ddot{d}_i s_i + l \dot{\omega}_i \times k_i + l \omega_i \times (\omega_i \times k_i) . \tag{17}$$

Dot multiplying both sides of Eq.(17) with k_i yields to

$$\ddot{d}_i = \frac{1}{k_i^T \cdot s_i} \left(k_i^T \cdot \dot{V}_{bi} + l \omega_i^T \cdot \omega_i \right) . \tag{18}$$

To find the angular acceleration of limb i, we can cross multiply both sides of Eq.(17) with k_i.

$$\dot{\omega}_i = \frac{1}{l} \left[k_i \times \dot{V}_{bi} - \frac{k_i \times s_i}{k_i^T \cdot s_i} \left(k_i^T \cdot \dot{V}_{bi} + l \omega_i^T \cdot \omega_i \right) \right] . \tag{19}$$

The acceleration of the centre of mass of limb i can be obtained by taking derivative of Eq.(13) with respect to time.

$$\dot{V}_{li} = \ddot{d}_i s_i + \frac{l}{2} [\dot{\omega}_i \times k_i + \omega_i \times (\omega_i \times k_i)] . \tag{20}$$

3 Jacobian Matrices

3.1 Jacobian Matrix of the Moving Platform

The Jacobian matrices are necessary for formulating the equations of motion, while the derivatives of the components are essential for formulating the corresponding Jacobian matrices. Writing Eq.(8) in matrix form yields to

$$V_{bi} = J_{bi} \dot{X}_p . \tag{21}$$

where $\dot{X}_p = [V_p, \omega_p]$ is a $[6 \times 1]$ matrix representing the linear and angular velocities of the moving platform, and the Jacobian matrix J_{bi} is a $[3 \times 6]$ matrix.

$$J_{bi} = \begin{bmatrix} 1 & 0 & 0 & 0 & b_{iz} & -b_{iy} \\ 0 & 1 & 0 & -b_{iz} & 0 & b_{ix} \\ 0 & 0 & 1 & b_{iy} & -b_{ix} & 0 \end{bmatrix} . \tag{22}$$

The Eq.(10) can be expressed in the form of $y = ax$ as follows:

$$\dot{d}_i = J'_{inv\_i} V_{bi} . \tag{23}$$

where

$$J'_{inv\_i} = \frac{k_i^T}{k_i^T \cdot s_i} . \tag{24}$$

Substituting Eq.(21) (22) into Eq.(23) yields to

$$\dot{d}_i = J_{inv\_i}\dot{\boldsymbol{X}}_p \, , J_{inv\_i} = J'_{inv\_i}J_{bi} \, . \tag{25}$$

Rewriting the Eq.(25) for six times, we will find the inverse Jacobian matrix of the six actuators as follows

$$\dot{\boldsymbol{d}} = J_{inv}\dot{\boldsymbol{X}}_p \, . \tag{26}$$

where

$$J_{inv} = \begin{bmatrix} J_{inv\_1} \cdots J_{inv\_6} \end{bmatrix}^T_{1\times 6} \, . \tag{27}$$

3.2   Jacobian Matrix of the Sliders

According to the definition of $\dot{d}_i$, it can be seen that the value of the velocity of slider i is equal to $\dot{d}_i$.

$$\boldsymbol{V}_{si} = \dot{d}_i\boldsymbol{s}_i \, . \tag{28}$$

Substituting Eq. (25) into Eq.(28), we have

$$\boldsymbol{V}_{si} = J_{sVi}\dot{\boldsymbol{X}}_p \, . \tag{29}$$

where

$$J_{sVi} = \begin{bmatrix} s_{ix}J_{inv\_i} & s_{iy}J_{inv\_i} & s_{iz}J_{inv\_i} \end{bmatrix}^T_{1\times 3} . \tag{30}$$

Since the slider is constrained in the groove of the triangle truss, there is no rotation of the slider, i.e. $\boldsymbol{\omega}_{si} = 0$. Therefore, we can deduct the Jacobian matrix of the slider i as follows:

$$\dot{\boldsymbol{X}}_{si} = J_{si}\dot{\boldsymbol{X}}_p \, , J_{si} = \begin{bmatrix} J_{sVi} \\ \boldsymbol{0}_{3\times 6} \end{bmatrix} \, . \tag{31}$$

3.3   Jacobian Matrix of the Limbs

To find the Jacobian matrix of the limb i, we have to do some transformations on Eq. (13). Substituting Eq. (8), (10) and (11) into Eq.(13) yields to

$$\boldsymbol{V}_{li} = \boldsymbol{E}_i \cdot \boldsymbol{V}_{bi} \cdot \boldsymbol{F}_i + \frac{1}{2}\boldsymbol{k}_i \times \boldsymbol{V}_{bi} \times \boldsymbol{k}_i \, . \tag{32}$$

where

$$\boldsymbol{E}_i = J'_{inv\_i} \, , \boldsymbol{F}_i = \boldsymbol{s}_i - \frac{1}{2}\boldsymbol{k}_i \times \boldsymbol{s}_i \times \boldsymbol{k}_i \, . \tag{33}$$

By developing the vector $\boldsymbol{E}_i = \begin{bmatrix} E_{ix} & E_{iy} & E_{iz} \end{bmatrix}^T$, $\boldsymbol{F}_i = \begin{bmatrix} F_{ix} & F_{iy} & F_{iz} \end{bmatrix}^T$, $\boldsymbol{k}_i = \begin{bmatrix} k_{ix} & k_{iy} & k_{iz} \end{bmatrix}^T$, we can obtain the Jabobian matrix of linear velocity of limb i as follows

$$\boldsymbol{V}_{li} = J_{lVi}\boldsymbol{V}_{bi} \, . \tag{34}$$

where

$$J_{lVi} = J_{lV1\_i} + J_{lV2\_i} . \tag{35}$$

$$J_{lV1\_i} = \begin{bmatrix} E_{ix}F_{ix} & E_{iy}F_{ix} & E_{iz}F_{ix} \\ E_{ix}F_{iy} & E_{iy}F_{iy} & E_{iz}F_{iy} \\ E_{ix}F_{iz} & E_{iy}F_{iz} & E_{iz}F_{iz} \end{bmatrix} . \tag{36}$$

$$J_{lV2\_i} = \begin{bmatrix} k_{iy}^2 + k_{iz}^2 & -k_{ix}k_{iy} & -k_{ix}k_{iz} \\ -k_{ix}k_{iy} & k_{ix}^2 + k_{iz}^2 & -k_{iy}k_{iz} \\ -k_{ix}k_{iz} & -k_{iy}k_{iz} & k_{ix}^2 + k_{iy}^2 \end{bmatrix} . \tag{37}$$

Similarly, we can find the Jacobian matrix of angular velocity of limb i based on Eq. (10) and (11).

$$\boldsymbol{\omega}_i = \frac{1}{l} \left[\boldsymbol{k}_i \times \boldsymbol{V}_{bi} - \frac{\boldsymbol{k}_i \times \boldsymbol{s}_i}{\boldsymbol{k}_i^T \cdot \boldsymbol{s}_i} \left(\boldsymbol{k}_i^T \cdot \boldsymbol{V}_{bi} \right) \right] = \frac{1}{l} \left[\boldsymbol{k}_i \times \boldsymbol{V}_{bi} - \boldsymbol{Q}_i \left(\boldsymbol{k}_i^T \cdot \boldsymbol{V}_{bi} \right) \right] . \tag{38}$$

where

$$\boldsymbol{Q}_i = \frac{\boldsymbol{k}_i \times \boldsymbol{s}_i}{\boldsymbol{k}_i^T \cdot \boldsymbol{s}_i} = \begin{bmatrix} Q_{ix} & Q_{iy} & Q_{iz} \end{bmatrix}^T . \tag{39}$$

The two terms of Eq.(38) are as follows:

$$\boldsymbol{k}_i \times \boldsymbol{V}_{bi} = \begin{bmatrix} k_{iy}V_{biz} - k_{iz}V_{biy} \\ k_{iz}V_{bix} - k_{ix}V_{biz} \\ k_{ix}V_{biy} - k_{iy}V_{bix} \end{bmatrix} . \tag{40}$$

$$\boldsymbol{k}_i^T \cdot \boldsymbol{V}_{bi} = k_{ix}V_{bix} + k_{iy}V_{biy} + k_{iz}V_{biz} . \tag{41}$$

Substitute Eq.(39-41) into Eq.(38), it yields the angular velocity Jacobian matrix as follows

$$\boldsymbol{\omega}_i = J_{l\omega i} \boldsymbol{V}_{bi} . \tag{42}$$

where

$$J_{l\omega i} = \frac{1}{l} \begin{bmatrix} -k_{ix}Q_{ix} & -k_{iz} - k_{iy}Q_{ix} & k_{iy} - k_{iz}Q_{ix} \\ k_{iz} - k_{ix}Q_{iy} & -k_{iy}Q_{ix} & -k_{ix} - k_{iz}Q_{iy} \\ -k_{iy} - k_{ix}Q_{iz} & k_{ix} - k_{iy}Q_{iz} & -k_{iz}Q_{iz} \end{bmatrix} . \tag{43}$$

Therefore, the equation of motion of the limb i can be expressed as

$$\dot{\boldsymbol{X}}_{li} = \begin{bmatrix} \boldsymbol{V}_{li} \\ \boldsymbol{\omega}_{li} \end{bmatrix} . \tag{44}$$

By substituting Eq.(21), (34), (42) into Eq.(44), we get the Jacobian matrix of the limb i as follows

$$\dot{\boldsymbol{X}}_{li} = J_{li}\dot{\boldsymbol{X}}_p , J_{li} = \begin{bmatrix} J_{lvi} \\ J_{l\omega i} \end{bmatrix} J_{bi} . \tag{45}$$

4   Virtual Work

4.1   Applied and Inertia Wrenches

The resultant of the applied and inertia forces exerted at the center of mass of the moving platform is

$$F_p = \begin{bmatrix} \hat{f}_p \\ \hat{n}_p \end{bmatrix} = \begin{bmatrix} f_e + m_p g - m_p \dot{V}_p \\ n_e - {}^O I_p \dot{\omega}_p - \omega_p \times \left({}^O I_p \omega_p \right) \end{bmatrix} . \tag{46}$$

where f_e and n_e are the external force and moment exerted at the center of mass, and in this paper, we assume they are equal to zero. And ${}^O I_p$ is the inertia tensor of the moving platform taken about the center of mass and expressed in the fixed frame O.

In this paper, we assume that the external force exerted at the sliders and the limbs is only the gravitational force, and since there is no rotation for the slider i, i.e. $\omega_{si} = 0$, $\dot{\omega}_{si} = 0$, then the resultants of applied and inertia forces exerted at the center of mass of the slider i can be expressed as following equation.

$$F_{si} = \begin{bmatrix} \hat{f}_{si} \\ \hat{n}_{si} \end{bmatrix} = \begin{bmatrix} m_s g - m_s \dot{V}_{si} \\ 0 \end{bmatrix} . \tag{47}$$

In the Section 2 and 3, we have deducted the necessary items of the limbs, so it is straigtforward to find the force and moment of limb i.

$$F_{li} = \begin{bmatrix} \hat{f}_{li} \\ \hat{n}_{li} \end{bmatrix} = \begin{bmatrix} m_l g - m_l \dot{V}_{li} \\ -{}^O I_{li} \dot{\omega}_{li} - \omega_{li} \times \left({}^O I_{li} \omega_{li} \right) \end{bmatrix} . \tag{48}$$

4.2   Equations of Motion

In this section, the procedure for solving the inverse dynamics of this 6-PSS parallel manipulator is proposed. The principle of virtual work for implementation on this manipulator can be expressed as

$$\delta q_s{}^T \tau + \delta X_p^T F_p + \sum_1^6 \left(\delta X_{si}{}^T F_{si} + \delta X_{li}^T F_{li} \right) = 0 . \tag{49}$$

The virtual displacements δq_s, δX_{si}, δX_{li} in Eq.(49) should be compatible with the kinematic constraints imposed by the structure. Therefore, it is necessary to relate the above virtual displacements to a set of independent virtual displacements δX_p. Based on the d'Alembert's principle, the virtual displacement is equal to the derivative of the displacement with respect to time, hence we have

$$\delta q_s = J_{inv} \delta X_p , \delta X_{si} = J_{si} \delta X_p , \delta X_{li} = J_{li} \delta X_p . \tag{50}$$

Substituting Eq.(50) into Eq.(49) yields to

$$\delta X_p^T \left[J_{inv}^T \tau + F_p + \sum_{i=1}^{6} \left(J_{si}^T F_{si} + J_{li}^T F_{li} \right) \right] = 0 . \tag{51}$$

Since Eq.(51) is valid for any values of δX_p^T, the condition to satisfy it is

$$J_{inv}^T \tau + F_p + \sum_{i=1}^{6} \left(J_{si}^T F_{si} + J_{li}^T F_{li} \right) = 0 . \tag{52}$$

Equation (52) describes the dynamics of this 6-PSS parallel manipulator. Therefore, if J_{inv} is not singular, the input force of the six actuators can be determined by the solution of Eq.(52).

$$\tau = -J_{inv}^{-T} \left[F_p + \sum_{i=1}^{6} \left(J_{si}^T F_{si} + J_{li}^T F_{li} \right) \right] . \tag{53}$$

Because this analysis is based on the assumption of the inverse of the transpose of the manipulator Jacobian matrix, when the moving platform approaches a singular configuration, the computation of input forces may become numerically unstable.

5   Numerical Simulation

In this section, a simulation is preformed by the computer algorithm to verify this method. From the previous assumption, the external force acting on the items of the structure is only the gravitational force, and here, the gravity accelera-tion vector is $g = [0 \ 0 \ -9.807]^T \, m/s^2$. Some values of the relevant parameters of this program are listed as: $R = 400 \, mm$, $r = 120 \, mm$, $L = 450mm$, $\varphi = 65°$, $\theta = 24.13°$, and the others can be found in [16]. The mass properties of the relevant components are obtained by the Solidworks simulation function: $m_p = 829.3 \, g$, $m_l = 300.85 \, g$, $m_s = 73.93g$. Based on the dimensions of the components, the inertia tensors can be developed as follows:

$$^B I_p = \begin{bmatrix} 3.29 & 0 & 0 \\ 0 & 6.56 & 0 \\ 0 & 0 & 3.29 \end{bmatrix} \cdot 10^{-3} kg \cdot m^2, \ I_l = \begin{bmatrix} 0 & 0 & 0 \\ 0 & 2.04 & 0 \\ 0 & 0 & 2.04 \end{bmatrix} \cdot 10^{-2} kg \cdot m^2$$

For the simulation, there are two scenarios to perform it. The first scenario is that the orientation of the moving platform remains constant while the center of mass of it moves along with a given trajectory. Specifically, the trajectory of the moving platform is given as $\gamma=\beta=\alpha=0$, $p = [0 \ 0 \ -500 + 50 \sin t]^T \, mm$.

The input forces τ for the six linear actuators are calculated as functions of time t. The simulation result is plotted in Fig.3 (a), which shows that the six input forces coincide into a curve, i.e. they are equal to each other. This significance verifies the theoretical results due to the symmetrical arrangement of the six actuators.

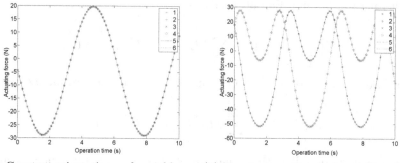

(a) Constant orientation and variable po- (b) Constant position and variable orien-
sition   tation

Fig. 3. Simulation results of the specific trajectories

For the second scenario, the trajectory of the moving platform is given as follows: the orientation of the moving platform varies by the rotation about the z-axis with a sinusoidal trajectory while the position remains constant. Specifically, the trajectory is specified as $\gamma=\beta=0$, $\alpha=\sin t$, $\boldsymbol{p}=\begin{bmatrix}0 & 0 & -400\end{bmatrix}^T mm$.

The results are plotted in Fig.3 (b) and similar to the first scenario, due to the symmetrical geometry, the input forces at actuators 1, 3 and 5 are equal to each other, and those at actuators 2, 4 and 6 are also equal to one another.

6   Conclusion

In this paper, a new 6-PSS parallel manipulator is investigated in 3D virtual environment and the kinematic model is built up. The inverse dynamic analysis for this parallel manipulator is performed based on the principle of virtual work. Based on the simulation results, the control strategies will be conducted for this parallel manipulator in our future research.

The implementation of the principle of virtual work leads to eliminating the constrained force at the outset. This makes it become more efficient than the conventional Newton-Euler approach on the dynamic analysis on this parallel manipulator. And the methodology of the link Jacobian matrices deduction is easy to understand, which can be also applied to the other types of parallel manipulators.

Acknowledgments. This work was supported in part by Macao Science and Technology Development Fund (108/2012/A3, 110/2013/A3), Research Committee of University of Macau (MYRG183(Y1-L3)FST11-LYM, MYRG203(Y1-L4)-FST11-LYM).

References

1. Yun, Y., Li, Y.M.: Design and Analysis of a Novel 6-DOF Redundant Actuated Parallel Robot with Compliant Hinges for High Precision Positioning. Nonlinear Dynamics 61(4), 829–845 (2010)
2. Tsai, L.W.: Sloving the Inverse Dynamics of a Stewart-Gough Manipulator by the Principle of Virtual Work. ASME Journal of Mechanical Design 122, 3–9 (2000)
3. Yun, Y., Li, Y.M.: A General Dynamics and Control Model of a Class of Multi-DOF Manipulators for Active Vibration Control. Mechanism and Machine Theory 46(10), 1549–1574 (2011)
4. Li, Y.M., Staicu, S.: Inverse Dynamics of a 3-PRC Parallel Kinematic Machine. Nonlinear Dynamics 67(2), 1031–1041 (2012)
5. Dasgupta, B., Choudhury, P.: A General Strategy Based on the Newton Euler Approach for the Dynamic Formulation of Parallel Maniplators. Mechanism and Machine Theory 34, 801–824 (1999)
6. Li, K.Q., Wen, R.: Closed-Form Dynamic Equations of the 6-RSS Parallel Mechanism Through the Newton-Euler Approach. In: Proceedings of the Third International Conference on Measuring Technology and Mechatronics Automation, vol. 01, pp. 712–715 (2011)
7. Harib, K., Srinivasan, K.: Kinematic and Dynamic Analysis of Stewart Platoform-Based Machine Tool Structures. Robotica 21, 541–554 (2003)
8. Nakamura, Y., Yamane, K.: Dynamics Computation of Structure-Varying Kinematic Chains and Its Application to Human Figures. IEEE Transactions on Robotics and Automation 16(2), 124–134 (2000)
9. Pang, H., Shahingpoor, M.: Inverse Dynamics of a Parallel Manipulator. Journal of Robotic System 11(8), 693–702 (1994)
10. Penda, H., Vakil, M., Zohoor, H.: Efficient Dynamic Equation of 3-PRS Parallel Manipulator Through Lagrange Method. In: Proceedings of the IEEE Conference on Robotics, Automation and Mehcatronics, vol. 2, pp. 1152–1157 (2004)
11. Wang, J.G., Gosselin, C.M.: A New Approach for the Dynamic Analysis of Parallel Manipulator. Multibody System Dynamics 2, 317–334 (1998)
12. Wang, Y.B., Zheng, S.T., Jin, J.: Dynamic Modeling of Spatial 6-DOF Parallel Manipulator Using Kane Method. In: International Conference on E-product E-service and E-entertainment, pp. 1–5 (2010)
13. Yun, Y., Li, Y.M.: A General Model of a Kind of Parallel Manipulator for Active Control based on Kane's Dynamics. In: IEEE Asia Pacific Conference on Circuits and Systems, Macao, China, pp. 1830–1833 (2008)
14. Liu, M.J., Li, C.X., Li, C.N.: Dynamics Analysis of the Gough-Stewart Platform Manipulator. IEEE Transactions on Robotics and Automation 16(1), 94–98 (2000)
15. Lee, K.M., Shah, D.K.: Dynamic Analysis of a Three-Degree-of-Freedom in Parallel Actuated Manipulator. IEEE Journal of Robotics and Automation 4(3), 361–367 (1988)
16. Lu, S., Li, Y.M.: Dimensional Synthesis of a 3-DOF Translational Parallel Manipulator Considering Kinematic Dexterity Property. In: IEEE International Conference on Information and Automation, Hailar, China, pp. 7–12 (2014)
17. Xu, W.Y., Li, Y.M., Xiao, X.: Kinematics and Workspace analysis for a novel 6-PSS parall manipulator. In: IEEE International Conference on Robotics and Bimimetics, pp. 1869–1874 (2013)

Elastodynamics of the Rigid-Flexible 3-RRR Mechanism Using ANCF Method

Xuchong Zhang and Xianmin Zhang[*]

Guangdong Provincial Key Laboratory of Precision Equipment
and Manufacturing Technology
South China University of Technology
510640, Guangzhou, Guangdong, China
zhangxuchong06@163.com, zhangxm@scut.edu.cn

Abstract. The elastodynamics of the 3-RRR mechanism is studied in this paper. The absolute nodal coordinate formulation (ANCF) is used to model the flexible links, the generalized α method with several efficient methods are adopted to solve the equations of motion of the system. A comparison is made between the rigid and flexible links of the mechanism. The results show that the flexibility of the link affects the displacement, velocity and acceleration of the moving platform significantly, and the flexure mechanism exhibits high frequencies vibrations

Keywords: Elastodynamics, 3-RRR mechanism, ANCF, generalized α method.

1   Introduction

The 3-RRR parallel mechanism is widely used in industry applications and laboratory researches for its high moving velocity and wide range of motion [1, 2]. The link flexibility causes vibration and noise of the mechanism, making it hard to model and control. The requirement of predicting the system dynamics accurately is increasingly necessary with the development of precision mechanical engineering. Therefore, it is important to establish a way to model the 3-RRR mechanism with the link flexibility.

Some researchers have studied the influence of the body flexibility in the dynamics performance of multibody systems. Tian [3, 4] proposed a new methodology for the dynamic analysis of rigid-flexible multibody systems with ElastoHydraDynamic lubricated cylindrical joints. It showed that the bearing flexibility plays a significant role in the system responses, extends the lubricant distribution space and, consequently, reduces the lubricant pressure. Zhao [5] proposed a numerical approach for the modeling and prediction of wear at revolute clearance joints in flexible multibody systems by integrating the procedures of wear prediction with multibody dynamics, which demonstrated that the wear result predicted is slightly reduced after taking the flexibility of components into account.

---

[*] Corresponding author.

X. Zhang et al. (Eds.): ICIRA 2014, Part II, LNAI 8918, pp. 24–35, 2014.
© Springer International Publishing Switzerland 2014

The ANCF is firstly proposed by Shabana [6, 7], which employs the mathematical definition of the slopes to define the element coordinates instead of the infinitesimal and finite rotations. This method is able to describe flexible multibody systems with large deformations accurately, and the constant mass matrices can be obtained without neglecting and projecting any nonlinear inertia terms. This method has been considered a benchmark in flexible multibody dynamics development [8].

The ANCF coupled deformation modes are associated with high frequencies, so that the explicit integration methods such as the Baumgarte stabilization method [9] can be very inefficient or even fail. For such a stiff system, implicit integration methods are much more efficient than explicit methods, as the implicit integrator does not capture the high frequency oscillations which can have a negligible effect on the solution accuracy. One of the most popular implicit integrators is the generalized-α method [10]. Among the attractive features of this method are the stability of the solution and an optimal combination of accuracy at low-frequency and numerical damping at high-frequency. The generalized-α algorithm results from successive contributions by Newmark [11], Hilber, Hughes and Taylor [12], and Chung and Hulbert [13].

In this paper, a general computational methodology for modeling and analysis of planar flexible multibody systems is presented, in which the ANCF is implemented to describe the large deformation, and the generalized α method with several efficient skills are used to solve the equations of motion of the system. This paper is organized as follows: in section 2, the planar ANCF-based beam element is briefly introduced. In section 3, the 3-RRR planar parallel mechanism with rigid and flexible links is described and modeled. In section 4, the computational strategy for the solution of the equations of the flexible 3-RRR system is described. After that, the simulation results are obtained and discussed. Finally, the main conclusions are drawn.

2   Planar ANCF Based Beam Element

This section will present the molding of rigid beam and flexible beam. For rigid beam, the coordinates of the link is described using natural coordinate system, see Fig. 1(a). The coordinates of the link in global coordinate system are defined as

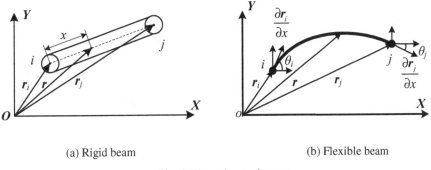

(a) Rigid beam   (b) Flexible beam

Fig. 1. Planar beam element

$$e = \left[r_i, r_j \right]^T = \left[x_i, y_i, x_j, y_j \right]^T \tag{1}$$

The location of generic point along the central axis of the link is

$$r = \begin{bmatrix} X \\ Y \end{bmatrix} = \begin{bmatrix} x_i + (x_j - x_i)x/l \\ y_i + (y_j - y_i)x/l \end{bmatrix} = Se \tag{2}$$

where X and Y denote the position coordinates defined in the global coordinate system, x denotes the nodal coordinate along the beam central line, l denotes the element length. S is the shape function

$$S = \begin{bmatrix} 1-x/l & 0 & x/l & 0 \\ 0 & 1-x/l & 0 & x/l \end{bmatrix} \tag{3}$$

The mass matrix of the rigid beam is expressed as

$$M^e = \int_V \rho S^T S dV \tag{4}$$

in which ρ is density and V is the volume.

As shown in Fig. 1(b), the planar ANCF-based deformable beam element is also described by two nodes. The coordinates of an arbitrary point along the central axis of the beam can be defined by the third-order polynomials in the global coordinates, that is

$$r = \begin{bmatrix} X \\ Y \end{bmatrix} = \begin{bmatrix} a_0 + a_1 x + a_2 x^2 + a_3 x^3 \\ b_0 + b_1 x + b_2 x^2 + b_3 x^3 \end{bmatrix} = Se \tag{5}$$

where X and Y denote the position coordinates defined in the global coordinate system, x denotes the nodal coordinate along the beam central line. S is the shape function. The absolute nodal coordinates e for node i and j can be expressed as

$$e = \left[e_1, e_2, e_3, e_4, e_5, e_6, e_7, e_8 \right]^T$$

$$= \left[r_{i1}, r_{i2}, \frac{\partial r_{i1}}{\partial x}, \frac{\partial r_{i2}}{\partial x}, r_{j1}, r_{j2}, \frac{\partial r_{j1}}{\partial x}, \frac{\partial r_{j2}}{\partial x} \right]^T \tag{6}$$

The vector $r_m = \left[r_{m1}, r_{m2} \right]^T$, $m=(i, j)$ indicates the position coordinates defined in the global coordinate system. The angle θ_m indicates the beam cross section orientation, which can be expressed by vector $\dfrac{\partial r_m}{\partial x} = \left[\dfrac{\partial r_{m1}}{\partial x}, \dfrac{\partial r_{m2}}{\partial x} \right]^T$. According to the conventional finite element method, the element shape function can be obtained

$$S = \left[S_1 I_2 \ S_2 I_2 \ S_3 I_2 \ S_4 I_2 \right] \tag{7}$$

where I_2 is the identity matrix of size two and

$$\begin{cases} S_1 = 1 - 3\xi^2 + 2\xi^3 \\ S_2 = l\left(\xi - 2\xi^2 + \xi^3\right) \\ S_3 = 3\xi^2 - 2\xi^3 \\ S_4 = l\left(-\xi^2 + \xi^3\right) \end{cases} \tag{8}$$

with $\xi = x / l$.

By using the Newton-Euler formulation, the element equations of motion are as follows

$$M^e \ddot{e} + F^e = Q^e \tag{9}$$

where Q^e represents the element generalized force vector, M^e denotes the element constant mass matrix

$$M^e = \int_V \rho S^T S dV \tag{10}$$

in which ρ is density and V is the volume. F^e denotes the element elastic force vector that can be deducted by the continuum mechanics approach [14], which is expressed as

$$\begin{aligned} F^e &= \left(\frac{\partial U_l}{\partial e}\right)^T + \left(\frac{\partial U_t}{\partial e}\right)^T \\ &= \int_0^l EA\varepsilon S_l e dx + \int_0^l EIS''^T S'' e dx \\ &= \left(\int_0^l EA\varepsilon S_l dx\right) e + \left(\int_0^l EIS''^T S'' e dx\right) e \\ &= K_l e + K_t e \end{aligned} \tag{11}$$

where

$$\begin{aligned} K_l &= \int_0^l EA\varepsilon S_l dx \\ &= EA \int_0^l \frac{1}{2}\left(e^T S_l e - 1\right) S_l dx \\ &= \frac{1}{2} EA \int_0^l S_l e e^T S_l dx - \frac{1}{2} EA \int_0^l S_l dx \\ &= K_{l1} + K_{l2} \end{aligned} \tag{12}$$

To improve the computation efficiency, Garcia-Vallejo et al. [15] firstly proposed an invariant matrix method to calculate the elastic force and the tangent matrix of the elastic force. Based on the following matrix transformation

$$\left(S_{l}ee^{T}S_{l}\right)_{ij} = e^{T}\left(S_{l}\right)_{,i}\left(S_{l}\right)_{j,}e \tag{13}$$

the element of K_{l1} can be expressed as

$$\left(K_{l1}\right)_{ij} = e^{T}\left(\frac{1}{2}EA\int_{0}^{l}\left(S_{l}\right)_{,i}\left(S_{l}\right)_{j,}dx\right)e = e^{T}C_{l1}^{ij}e \tag{14}$$

where C_{l1} is called the constant matrix of K_{l1}.

When solving the motion equations of the multibody system in an interactive way, that is, in an implicit way, the derivation of the elastic force respect to generalized coordinates will be used, which can be expressed as

$$\left(\frac{\partial F^{e}}{\partial e}\right)_{ij} = \frac{\partial\left(F^{e}\right)_{i}}{\partial e_{j}} = \left(K_{t} + K_{l2} + K_{l1}\right)_{ij} + \sum_{k}^{8}\sum_{s}^{8}e_{s}\left(C_{l1}^{ik} + \left(C_{l1}^{ik}\right)^{T}\right)_{sj}e_{k} \tag{15}$$

and it will be the most time consuming part in simulation.

3   Modeling of the 3-<u>R</u>RR Mechanism

This section will describe the configuration and motion equations of the 3-<u>R</u>RR (the underline means the driving joint) planar parallel robot, as is shown in Fig. 2. The mechanism is a trigonal symmetry structure which is composed by 7 parts, link A_iB_i (i=1~3) is the driving part, link B_iC_i is the passive part, and the plate part is the moving platform. Joints A_1, A_2 and A_3 are attached to the drivers, while joint B_1, B_2, B_3 and C_1, C_2, C_3 are passive revolute joints. All the driving links are treated as rigid, while all the passive links are treated as flexible, and each is described by one ANCF beam element. The moving platform is treated as rigid which can be described by three reference point coordinates. There are totally 27 generalized coordinates in the system, as labeled in the figure.

The kinematic constraints of the system can be written as

$$\boldsymbol{\Phi}(\boldsymbol{q}) = \begin{bmatrix} (q_{1}-x_{A1})^{2}+(q_{2}-y_{A1})^{2}-l_{1}^{2} \\ q_{5}-(q_{25}+r\cos(q_{27}+7\pi/6)) \\ q_{6}-(q_{26}+r\sin(q_{27}+7\pi/6)) \\ (q_{9}-x_{A2})^{2}+(q_{10}-y_{A2})^{2}-l_{1}^{2} \\ q_{13}-(q_{25}+r\cos(q_{27}+11\pi/6)) \\ q_{14}-(q_{26}+r\sin(q_{27}+11\pi/6)) \\ (q_{17}-x_{A3})^{2}+(q_{18}-y_{A3})^{2}-l_{1}^{2} \\ q_{21}-(q_{25}+r\cos(q_{27}+\pi/2)) \\ q_{22}-(q_{26}+r\sin(q_{27}+\pi/2)) \end{bmatrix} = \boldsymbol{0} \tag{16}$$

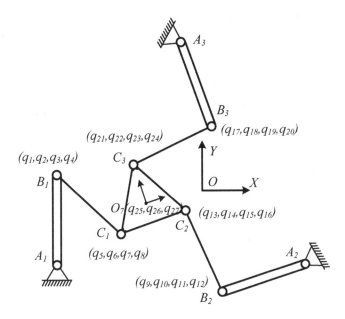

Fig. 2. Geometry and global coordinates of the 3-RRR mechanism

The first time derivative with respect to time of Eqn. (16) provides the velocity constraint equations

$$\boldsymbol{\Phi}_q \dot{q} = -\boldsymbol{\Phi}_t \tag{17}$$

where $\boldsymbol{\Phi}_q$ is the Jacobin matrix of the constraint equations, that is, $\boldsymbol{\Phi}_q = \partial \boldsymbol{\Phi} / \partial q$. In the same way, $\boldsymbol{\Phi}_t = \partial \boldsymbol{\Phi} / \partial t$.

Based on the ANCF, the assembly of the elements can be carried out by the conventional finite element method. The element nodal e can be easily transformed into the flexible multibody system generalized coordinates q. Without considering the damping of the system, the Newton-Euler equations for the constrained flexible multibody system in Cartesian coordinates can be written as

$$M\ddot{q} + \boldsymbol{\Phi}_q^T \lambda + F_e = F_d \tag{18}$$

where M is the system mass matrix, λ is the vector of Lagrange multipliers, F_e denotes the system elastic force vector, and F_d is the driving force vector.

4   Solving the Equations of Motion

The process of the integration of the equations of motion for a flexible multibody system is different from that of rigid body system. The high frequency responses are stimulated by the finite element discretization. Conversely, the high-frequency responses can produce very large inner forces, which in turn lead to the simulation divergence. The convergence for the solution of the high frequency mode is poor if it cannot be damped out. The generalized-α method allows an optimal combination of accuracy at low-frequency and numerical damping at high frequency. Thus it can eliminate the contribution of nonphysical high-frequency modes, which are generally present in finite element models.

According to the generalized-α method [10], the Eqn. (16) is appended with Eqn. (18) and rewritten as

$$\begin{cases} M\ddot{q}_{n+1} + (\Phi_q^T)_{n+1}\lambda + F_e(q_{n+1}) - F_d = 0 \\ \Phi(q_{n+1}, t_{n+1}) = 0 \end{cases} \tag{19}$$

The following equations exist:

$$\begin{cases} q_{n+1} = q_n + h\dot{q}_n + h^2\left((1/2-\beta)a_n + \beta a_{n+1}\right) \\ \dot{q}_{n+1} = \dot{q}_n + h\left((1-\gamma)a_n + \gamma a_{n+1}\right) \end{cases} \tag{20}$$

in which

$$\begin{cases} (1-\alpha_m)a_{n+1} + \alpha_m a_n = (1-\alpha_f)\ddot{q}_{n+1} + \alpha_f\ddot{q}_n \\ a_0 = \ddot{q}_0 \end{cases} \tag{21}$$

where n denotes the nth interaction, $\alpha_m, \alpha_f, \beta, \gamma$ are the algorithm parameters. For the fixed step size h, in order to preserve the low frequency responses and dissipate the spurious high frequency responses, Chung and Hulbert [13] have proposed optimal algorithmic parameters for second-order accurate

$$\alpha_m = \frac{2\rho-1}{\rho+1}, \alpha_f = \frac{\rho}{\rho+1}, \beta = \frac{1}{4}\left(1+\alpha_f-\alpha_m\right)^2, \gamma = \frac{1}{2}+\alpha_f-\alpha_m \tag{22}$$

where $\rho \in [0,1]$. When $\rho = 0$, the algorithm has the maximum energy dissipation, and $\rho = 1$, the energy will be preserved. The Broyden-Newton interaction approach [16] is used to solve the nonlinear equations.

The generalized constrains G and its Jacobian matrix J are defined as

$$G = \begin{bmatrix} M\ddot{q} + F_e(q) + \Phi_q^T\lambda - F_d \\ \Phi(q,t) \end{bmatrix} \tag{23}$$

$$J = \frac{\partial G}{\partial[\ddot{q}\ \lambda]} = \begin{bmatrix} M + \left(\dfrac{\partial F(q)}{\partial\ddot{q}} + \dfrac{\partial(\Phi_q^T\lambda)}{\partial\ddot{q}}\right)\bar{\beta} & \Phi_q^T \\ \Phi_q & 0 \end{bmatrix} \tag{24}$$

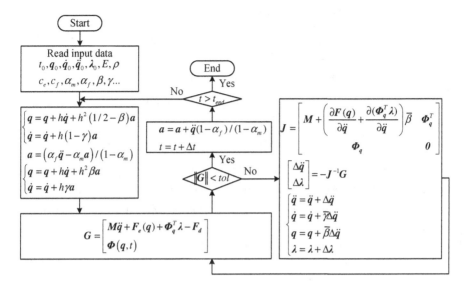

Fig. 3. Flowchart of the computational scheme

The whole computational scheme is illustrated in the flowchart of Fig. 3, in which the following relations are utilized

$$\begin{cases} \bar{\beta} = \dfrac{1-\alpha_f}{1-\alpha_m} \beta h^2 \\[4mm] \bar{\gamma} = \dfrac{1-\alpha_f}{1-\alpha_m} \gamma h \end{cases} \tag{25}$$

which can make

$$\frac{\partial q}{\partial \ddot{q}} = I\bar{\beta}, \frac{\partial \dot{q}}{\partial \ddot{q}} = I\bar{\gamma} \tag{26}$$

5   Results and Discussions

The geometric and material parameters of the 3-RRR mechanism are listed in Tab. 1-2. The Young's module of the passive links is set smaller than Aluminum in order to enlarge the influence of the link flexibility. In the simulation, the preset moving trajectory of 3-RRR platform is a circle with radius $0.1m$, which is expressed as

$$\begin{cases} q_{25} = 0.1\cos(\omega t) \ m \\ q_{26} = 0.1\sin(\omega t) \ m \\ q_{27} = 0 \ rad \end{cases} \tag{27}$$

Table 1. Geometry parameters of the mechanism

Member	Driving link	Passive link	Moving stage	Fixed stage
Length[m]	0.245	0.242	0.112	0.400

Table 2. Material parameters of the mechanism

Member	Young[Pa]	Density[kg/m^3]	Width[m]	Thickness[m]
Driving link	201e9	2.7e3	0.03	0.01
Passive link	7e8	2.7e3	0.01	0.005

in which ω is the angle frequency whose value is set to be $2\pi \ rad \ / \ s$. At the start of the simulation, the initial position of the platform is $q_{25} = 0.1 \ m$, $q_{26} = 0 \ m$, $q_{27} = 0 \ rad$, the initial velocity is $\dot{q}_{25} = 0 \ m/s$, $\dot{q}_{26} = 0.1\pi \ m/s$, $\dot{q}_{27} = 0 \ rad/s$. The initial positions and velocities necessary to start the dynamic analysis are obtained from the kinematic calculation of the 3-RRR mechanism. The simulation results are plotted in Fig. 4-8, the dash line represents the value with all rigid links, while the solid line represents the result with flexible links.

Fig. 4 shows the configuration of the mechanism at different times. It can be observed that the driving links keep undeformed all the time, while the passive links have obvious deformations once the platform is moving. These deformations will lead to the position deviation of the moving platform, as shown in Fig. 5. The position error of the platform along x direction and y direction are around +/-0.1mm, while the angle error is around 0.1mrad, and these values will be influenced by the stiffness of the flexible links. It indicates that the flexure of the link is an important influence factor of positioning accuracy, especially in precision mechanism. However, it should be taken into consideration that the numerical error will also cause deviations of the results.

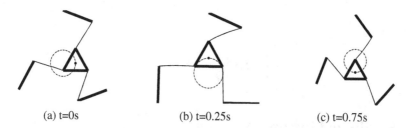

(a) t=0s   (b) t=0.25s   (c) t=0.75s

Fig. 4. Configuration of the mechanism at different times

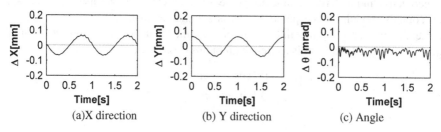

(a)X direction   (b) Y direction   (c) Angle

Fig. 5. Pose error of the platform

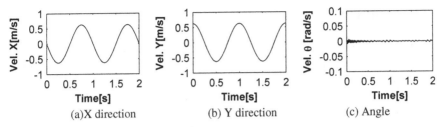

Fig. 6. Moving velocity of the platform

The velocities of the moving platform along different directions are plotted in Fig. 6. It shows that the velocity along axis x and along axis y are very smooth and approximant to their ideal values. But the rotational velocity of the moving platform is a little different from its ideal value, it varies around zero, and the amplitude is about 0.01rad/s.

In Fig. 7, the accelerations of the moving platform along different directions are plotted. The accelerations are very different from their ideal values, that is, the accelerations are enlarged when the link flexibility is taken into consideration. The curves also present high frequency responses which are caused by the vibrations of the flexure links, theses vibrations will make the system unstable and difficult to control. Moreover, the vibration amplitude is pretty large at the start of the simulation, and with the program running, the curves become smoother, as can be clearly seen in Fig. 7(c).

The same phenomena can also be observed in the curves of driving link moments represented in Fig. 8. The enlarged forces will decrease the service life of the components or even lead to failure of the mechanism. In Fig. 8, the driving moment for each crank is different, and these values are influenced by many factors, such as moving trajectory, moving velocity, or the loads applied on the moving platform.

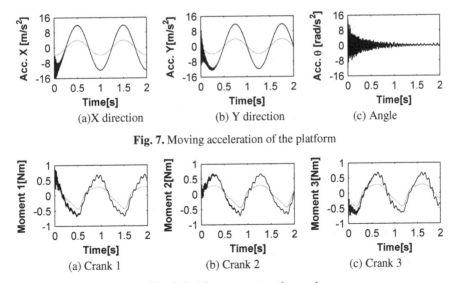

Fig. 7. Moving acceleration of the platform

Fig. 8. Driving moment on the crank

6   Conclusions

In this paper, a methodology is proposed to study the elastodynamics of the rigid-flexure 3-RRR mechanism, in which the rigid driving links are molded using natural coordinates, while the flexure passive links are molded using ANCF. The generalized α method with several efficient skills are adopted to solve the equations of motion of the system.

The simulation results show that the link flexibility affects the positioning accuracy of the moving platform apparently. It has great influence on the acceleration of the platform and driving moments of the cranks, the flexure mechanism also exhibits high frequencies vibrations. These analyses are the basis for the optimal design and control of the flexure multibody systems, in the future work, the effects of joint clearance will be taken into consideration.

Acknowledgments. This research was supported by the National Natural Science Foundation of China (Grant No.91223201), the Natural Science Foundation of Guangdong Province (Grant No.S2013030013355), Project GDUPS (2010) and the Fundamental Research Funds for the Central Universities (2012ZP0004). These supports are greatly acknowledged.

References

1. Zhang, X.C., Zhang, X.M., Chen, Z.: Dynamic analysis of a 3-RRR parallel mechanism with multiple clearance joints. Mechanism and Machine Theory 78, 105–115 (2014)
2. Zhang, X.C., Zhang, X.M.: Dynamic analysis of a 3-RRR parallel robot with Joint clearances using natural coordinates. In: ASME 2014 International Design Engineering Technical Conferences & Computers and Information in Engineering Conference (IDETC/CIE 2014), DETC2014-34609, New York, pp. 1–9 (2014)
3. Tian, Q., Zhang, Y.Q., Cheng, L.P., Yang, J.Z.: Simulation of planar flexible multibody systems with clearance and lubricated revolute joints. Nonlinear Dyns. 60, 489–511 (2010)
4. Tian, Q., Zhang, Y.Q., Cheng, L.P., Flores, P.: Dynamics of spatial flexible multibody systems with clearance and lubricated spherical joints. Comput. Struct. 87, 913–929 (2009)
5. Zhao, B., Zhang, Z.N., Dai, X.D.: Modeling and prediction of wear at revolute clearance joints in flexible multibody systems. Proceedings of the Institution of Mechanical Engineers, Part C: Journal of Mechanical Engineering Science 4, 1–13 (2013)
6. Shabana, A.A.: An absolute nodal coordinates formulation for the large rotation and deformation analysis of flexible bodies. Technical Report. No. MBS96-1-UIC, University of Illinois at Chicago (1996)
7. Shabana, A.A.: Definition of the slopes and absolute nodal coordinate formulation. Multibody Syst. Dyn. 1, 339–348 (1997)
8. Shabana, A.A.: Dynamics of Multibody Systems, 3rd edn. Cambridge University Press, New York (2005)
9. Flores, P., Machado, M., Seabra, E., Sliva, M.T.: A parametric study on the Baumgarte stabilization method for forward dynamics of constrained multibody systems. ASME J. Comput. Nonlinear Dyn. 6, 011019/1–9 (2011)

10. Arnold, M., Bruls, O.: Convergence of the generalized-a scheme for constrained mechanical systems. Multibody Syst. Dyn. 18, 185–202 (2007)
11. Newmark, N.: A method of computation for structural dynamics. ASCE J. Eng. Mech. Div. 85, 67–94
12. Hilber, H., Hughes, T., Taylor, R.: Improved numerical dissipation for time integration algorithms instructural dynamics. Earthq. Eng. Struct. Dyn. 5, 283–292 (1977)
13. Chung, J., Hulbert, G.: A time integration algorithm for structural dynamics with improved numerical dissipation: The generalized-α method. ASME J. Appl. Mech. 60, 371–375 (1993)
14. Berzeri, M., Shabana, A.A.: Development of simple models for the elastic forces in the absolute nodal coordinate formulation. Journal of Sound and Vibration 235(4), 539–565 (2000)
15. Garcia-Vallejo, D., Mayo, J., Escalona, J.L., Dominguez, J.: Efficient evaluation of the elastic forces and the Jacobian in the absolute nodal coordinate formulation. Nonlinear Dynamics 35, 313–329 (2004)
16. Broyden, C.G.: A class of methods for solving nonlinear simultaneous equations. Math. Compu. 19(92), 577–593 (1965)

Kinematic Analysis and Control
of a 3-DOF Parallel Mechanism

Hongyang Zhang and Xianmin Zhang

Guangdong Provincial Key Laboratory of Precision Equipment
and Manufacturing Technology
South China University of Technology
510640, Guangzhou, Guangdong, China
z.hongyang@mail.scut.edu.cn, zhangxm@scut.edu.cn

Abstract. Parallel mechanisms have advantages to carry out tasks requiring high speed and high accuracy. This paper focuses on the motion control of a 3-PRR planar parallel mechanism. In order to build the control system, kinematic analysis of the 3-PRR planar parallel mechanism is studied. The inverse and forward kinematic solutions and the speed Jacobian matrix are obtained. Control system based on the joint space and task space of the mechanism is built. Moreover, the iterative learning method is used to improve the PD controller's performance. Simulation result shows that the error of the platform is substantially reduced.

Keywords: 3-PRR, Kinematic, Iterative Learning Controller.

1   Introduction

Compared with the serial manipulators, potential advantages of the parallel architectures are higher kinematical precision, lighter weight, better stiffness, greater load bearing, stabile capacity and suitable positional actuator arrangements[1]. Parallel mechanisms have more advantages to carry out tasks requiring high speed and high accuracy[2]. Since 1830s, there have been plenty of researchers promoting the development of parallel mechanisms. More and more parallel mechanisms are used in practical work[3].

With the development of parallel mechanism research, lower-mobility parallel mechanisms have attracted much attention due to their simplicity in structure, low cost in design, manufacturing and control[4].

The performance of parallel mechanism depends on its structural parameters, drivers and controller. Control system is the core part of the parallel mechanism. In order to improve the accuracy and stability of the parallel mechanism, it is necessary to do in-depth research and design for its control system.

PD controller has simple structure and is easy to be built. Therefore, it is widely used in industry. However, because of the nonlinearity and coupling of the parallel mechanism, PD controller cannot achieve desired performance. Therefore, this paper does some research on kinematic of parallel mechanism and builds iterative learning

X. Zhang et al. (Eds.): ICIRA 2014, Part II, LNAI 8918, pp. 36–47, 2014.

control system of 3-PRR mechanism. Compare with PD control system, iterative control system can achieve better performance.

2   Kinematic Analysis of Mechanism

2.1   Mechanism Description

The architecture and coordinate system of the 3-PRR mechanism is shown in Fig.1[5].The origin of the fixed coordinate system is located at O , the positive direction of X-axis is parallel to the direction of vector $\mathbf{A_1A_2}$ and the positive direction of Y-axis is parallel to the direction of vector $\mathbf{OA_3}$. The origin of the moving coordinate system is located at P , the positive direction of X'-axis is parallel to the direction of vector $\mathbf{C_1C_2}$ and the positive direction of Y'-axis is parallel to the direction of vector $\mathbf{OC_3}$ .Each kinematic chain is of PRR-type and consists of one actuated prismatic joint, P ; two revolute joints, R ; and two links.

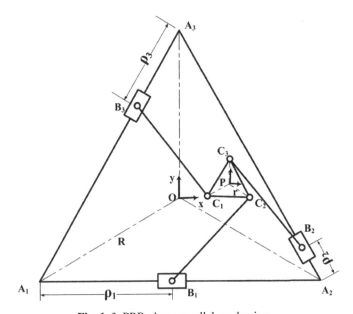

Fig. 1. 3–PRR planar parallel mechanism

The geometric center of platform $C_1C_2C_3$ denoted by P , is the operation point of the mechanism. The displacements of the three prismatic joints, i.e., ρ1, ρ2 and ρ3, are the input variables whereas the Cartesian coordinates of point P , i.e., x_p, y_p, and φ are the output variables. Point O is the base part's geometric center and the origin of the reference frame. The prismatic actuators are aligned to its sides and are attached to points A_i (i=1, 2, 3) with orientation angles α_1, α_2, and α_3 being equal to 0°, 120° and 240°, respectively. Here are the parameters describing the manipulator geometry:

- R :   radius of the circumscribed circle of triangle $A_1A_2A_3$ of center O, i.e., $R = OA_i$;
- r :   radius of the circumscribed circle of triangle $C_1C_2C_3$ of center P, i.e. $r = PC_i$;
- L : the length of the intermediate links, i.e., $L = B_iC_i$;

2.2   Inverse Kinematics

Define the input vector of the 3-PRR mechanism as $Q = [\rho_1 \quad \rho_2 \quad \rho_3]^T$ and the output vector as $P = [x_p \quad y_p \quad \varphi]^T$. Hence, the coordinates of the three endpoints of the moving platform, i.e., C_1, C_2 and C_3, in the moving coordinate system can be expressed as:

$$
[C_1', C_2', C_3'] = \begin{pmatrix} -\dfrac{\sqrt{3}}{2}r & \dfrac{\sqrt{3}}{2}r & 0 \\ -\dfrac{r}{2} & -\dfrac{r}{2} & r \\ 1 & 1 & 1 \end{pmatrix} \tag{1}
$$

The coordinates of the three endpoints of the base platform, i.e., A_1, A_2 and A_3, in the fixed coordinate system can be expressed as:

$$
[A_1, A_2, A_3] = \begin{pmatrix} -\dfrac{\sqrt{3}}{2}R & \dfrac{\sqrt{3}}{2}R & 0 \\ -\dfrac{R}{2} & -\dfrac{R}{2} & R \\ 1 & 1 & 1 \end{pmatrix} \tag{2}
$$

The coordinates of the three endpoints of the links, i.e., B_1, B_2 and B_3, in the fixed coordinate system can be expressed as:

$$
[B_1, B_2, B_3] = \begin{pmatrix} \dfrac{\sqrt{3}}{2}R + \rho_1 \cos\alpha_1 & -\dfrac{\sqrt{3}}{2}R + \rho_2 \cos\alpha_2 & \rho_3 \cos\alpha_3 \\ -\dfrac{R}{2} + \rho_1 \sin\alpha_1 & -\dfrac{1}{2}R + \rho_2 \sin\alpha_2 & R + \rho_3 \sin\alpha_3 \\ 1 & 1 & 1 \end{pmatrix} \tag{3}
$$

Based on the mechanism geometric feature, $\alpha_1 = 0°, \alpha_2 = 120°, \alpha_3 = 240°$. Transfer matrix from the moving coordinate system to the fixed coordinate system can be expressed as:

$$\,^{O}_{C}T = \begin{pmatrix} \cos\varphi & -\sin\varphi & x_p \\ \sin\varphi & \cos\varphi & y_p \\ 0 & 0 & 1 \end{pmatrix} \tag{4}$$

Hence, the coordinates of C_1, C_2 and C_3 in the moving coordinate system can be obtained:

$$[C_1, C_2, C_3] = \,^{O}_{C}T[C_1', C_2', C_3']$$

$$= \begin{pmatrix} x_p - \dfrac{\sqrt{3}}{2}r\cos\varphi + \dfrac{r}{2}\sin\varphi & x_p + \dfrac{\sqrt{3}}{2}r\cos\varphi + \dfrac{r}{2}\sin\varphi & x_p - r\sin\varphi \\ y_p - \dfrac{\sqrt{3}}{2}r\sin\varphi - \dfrac{r}{2}\cos\varphi & y_p + \dfrac{\sqrt{3}}{2}r\sin\varphi + \dfrac{r}{2}\cos\varphi & y_p + r\cos\varphi \\ 1 & 1 & 1 \end{pmatrix} \tag{5}$$

Since the lengths of the three links are equal and constant, i.e. $l^2 = B_iC_i^2$, it can be obtained:

$$\rho_1 = -a_1 \pm \sqrt{l^2 - b_1^2} \tag{6}$$

$$\rho_2 = \frac{a_2 - \sqrt{3}b_2 \pm \sqrt{(2l)^2 - (\sqrt{3}a_2 + b_2)^2}}{2} \tag{7}$$

$$\rho_3 = \frac{a_3 - \sqrt{3}b_3 \pm \sqrt{(2l)^2 + (\sqrt{3}a_3 - b_3)^2}}{2} \tag{8}$$

Where,

$$a_1 = -x_p + \frac{\sqrt{3}}{2}r\cos\varphi - \frac{r}{2}\sin\varphi - \frac{\sqrt{3}}{2}R, b_1 = -y_p + \frac{\sqrt{3}}{2}r\sin\varphi + \frac{r}{2}\cos\varphi - \frac{R}{2} \tag{9}$$

$$a_2 = -x_p - \frac{\sqrt{3}}{2}r\cos\varphi - \frac{r}{2}\sin\varphi + \frac{\sqrt{3}}{2}R, b_2 = -y_p - \frac{\sqrt{3}}{2}r\sin\varphi + \frac{r}{2}\cos\varphi - \frac{R}{2} \tag{10}$$

$$a_3 = -x_p + r\sin\varphi, b_3 = y_p + r\cos\varphi - R \tag{11}$$

2.3   Kinematic Jacobian Matrix

The Jacobian matrix defines the relationship between the actuators and mobile platform velocity vectors. The Jacobian matrix satisfies the following equation:

$$\dot{\rho} = J\dot{P} \tag{12}$$

Where $\dot{P}$ is the cartesian velocity vector, i.e. $\dot{P} = [\dot{x}_p, \dot{y}_p, \dot{\phi}]^T$, $\dot{\rho}$ is the prismatic joints' velocity vector, i.e. $\dot{\rho} = [\dot{\rho}_1, \dot{\rho}_2, \dot{\rho}_3]^T$.

According to the initial structure, the corresponding inverse kinematics result can be obtained:

$$\rho_1 = -a_1 - \sqrt{l^2 - b_1^2} \tag{13}$$

$$\rho_2 = \frac{a_2 - \sqrt{3}b_2 - \sqrt{(2l)^2 - (\sqrt{3}a_2 + b_2)^2}}{2} \tag{14}$$

$$\rho_3 = \frac{a_3 - \sqrt{3}b_3 - \sqrt{(2l)^2 - (\sqrt{3}a_2 + b_2)^2}}{2} \tag{15}$$

Differentiating,

$$\dot{\rho}_1 = -\dot{a}_1 + \frac{2b_1}{\sqrt{l^2 - b_1^2}}\dot{b}_1 \tag{16}$$

$$\dot{\rho}_3 = (1 - \frac{2\sqrt{3}(\sqrt{3}a_3 + b_3)}{\sqrt{(2l)^2 - (\sqrt{3}a_3 + b_3)^2}})\dot{a}_3 - (\sqrt{3} + \frac{2(\sqrt{3}a_3 + b_3)}{\sqrt{(2l)^2 - (\sqrt{3}a_3 + b_3)^2}})\dot{b}_3 \tag{17}$$

It can be obtained that:

$$\dot{\rho}_i = c_i \dot{a}_i + d_i \dot{b}_i \tag{18}$$

Where,

$$c_1 = -1, d_1 = \frac{2b_1}{\sqrt{l^2 - b_1^2}} \tag{19}$$

$$c_2 = (1 - \frac{2\sqrt{3}(\sqrt{3}a_2 + b_2)}{\sqrt{(2l)^2 - (\sqrt{3}a_2 + b_2)^2}}), d_2 = -(\sqrt{3} + \frac{2(\sqrt{3}a_2 + b_2)}{\sqrt{(2l)^2 - (\sqrt{3}a_2 + b_2)^2}}) \tag{20}$$

$$c_3 = (1 - \frac{2\sqrt{3}(\sqrt{3}a_3 + b_3)}{\sqrt{(2l)^2 - (\sqrt{3}a_3 + b_3)^2}}), d_3 = -(\sqrt{3} + \frac{2(\sqrt{3}a_3 + b_3)}{\sqrt{(2l)^2 - (\sqrt{3}a_3 + b_3)^2}}) \tag{21}$$

Hence, the kinematic Jacobian matrix can be obtained:

$$
J = \begin{bmatrix}
-c_1 & -d_1 & -(\dfrac{\sqrt{3}}{2}rc_1 + \dfrac{1}{2}rd_1)\sin\varphi + (\dfrac{1}{2}rc_1 - \dfrac{\sqrt{3}}{2}rd_1)\cos\varphi \\[2mm]
-c_2 & -d_2 & -(\dfrac{\sqrt{3}}{2}rc_2 + \dfrac{1}{2}rd_2)\sin\varphi + (\dfrac{1}{2}rc_2 + \dfrac{\sqrt{3}}{2}rd_2)\cos\varphi \\[2mm]
-c_3 & d_3 & -rd_3\sin\varphi + rc_3\cos\varphi
\end{bmatrix}
\tag{22}
$$

2.4   Forward Kinematics

Forward kinematic defines the relationship between the actuators and mobile platform displacement. Because of the specific properties of parallel mechanism, it is difficult to obtain the direct kinematics solution by analytical method. Many numerical methods have been proposed. Jacobian matrix method is one of the numerical methods that have high speed and precision[6].

Define P_0 as the initial output vector and Q_0 is the corresponding input vector. Assume the current input vector is P_1 and Q_1 is the corresponding input vector. Define J as the corresponding Jacobian matrix and set ε as the required accuracy.

As shown in Fig.2, the main steps of the method can be described as:

1. Set the initial output vector P_0.
2. Calculate the corresponding input vector Q_0 according to the inverse kinematic result.
3. Calculate the corresponding Jacobian matrix J.
4. Calculate the new output vector P_0 according to the equation:
   $P_0 = P_0 + J(Q_0 - Q_1)$
5. Calculate the new input vector Q_0.
6. If $\|P_0 - P_1\| > \varepsilon$, go to step 3.Else, $P_1 = P_0$ is the desired solution.

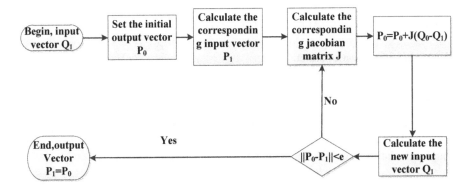

Fig. 2. Flowchart of forward kinematic

3   Control Method

3.1   Iterative Learning Controller

Iterative learning control (ILC) is based on the notion that the performance of a system that executes the same task multiple times can be improved by learning from previous executions[7]. The objective of ILC is to improve performance by incorporating error information into the control for subsequent iterations. It does not need precise mathematical model. Therefore, it is very useful for the tasks that are nonlinear, complex and difficult to build mathematical model. The control system of parallel mechanism has the character of high nonlinear and coupling. Iterative learning control is suitable and available to solve such problem.

Assume the desired control input is $u_d(t)$, the object of the algorithm is to reach the desired output, i.e. $y_d(t)$. Each time we run the program, the initial status $x_k(0)$ is recorded. After several rounds of repetition and correction, the desired input and output value is obtained, i.e. $u_k(t) \rightarrow u_d(t), y_k(t) \rightarrow y_d(t)$.

Consider the discrete-time system

$$\dot{x}_k(t) = f(x_k(t), u_k(t), t), \quad y_k(t) = g(x_k(t), u_k(t), t) \tag{23}$$

Where t is the time index, k is the iteration index, $y(t)$ is the output, $u_k(t)$ is the control input, and $x_k(t)$ is the status of the system. The performance or error signal is defined by $e_k(t) = y_d(t) - y_k(t)$.

In the open-loop learning control method, the relationship between each generation can be expressed as

$$u_{k+1}(t) = L(u_k(t), e_k(t)) \tag{24}$$

In the close-loop learning control method, the relationship between each generation can be expressed as

$$u_{k+1}(t) = L(u_k(t), e_{k+1}(t)) \tag{25}$$

Where, L is a linear operator.

3.2   Control System Based on the Task Space

Control system based on the task space considers the mechanism as a MIMO system[8]. The desired trajectory is given and the actual position is measured. Through calculating the error, the input control signal can be determined.

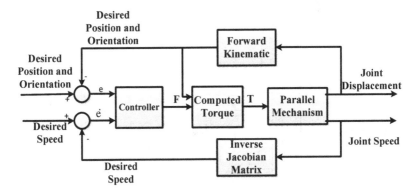

Fig. 3. Control system based on the task space

3.3 Control System Based on the Joint Space

Control System based on the joint space considers each joint of the parallel mechanism as a SISO system. The input control signals are given to each joint respectively, while the coupling between the joints is not considered[9].

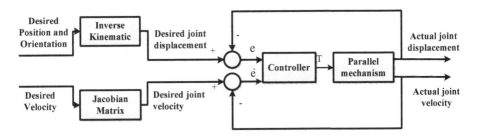

Fig. 4. Control system based on the joint space

4 Simulations and Results Analysis

As shown in Fig.5, the simulation model is built via MATLAB. The mechanism and joint are replaced by modules in the SimMechanic toolbox.

The joint actuators provide driving forces to the prismatic joints. The joint sensors measure the speed and position of the joints, which are part of the feedback signals of the controller[10].

4.1 PD Control Based on the Task Space

Fig. 6 shows the block diagram of the PD control system based on the task space. In the control system, we should transfer the force in the joint space to the task space. According to inverse kinematic, the relationship between the input vector i.e. Q and output vector P can be expressed as:

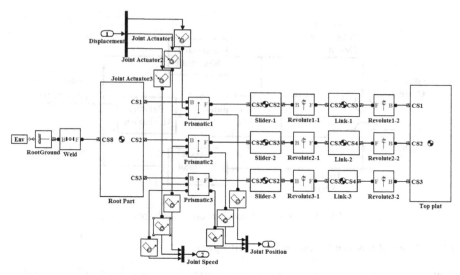

Fig. 5. Simulation model of the 3-PRR mechanism

$$Q = J \cdot P$$

Since the principle of conservation of energy, it can be obtained that

$$F \cdot Q = T \cdot P$$

Where, F is the generalized force acting on the joints and T is the generalized force acting on the platform.

Thus,

$$T = F \cdot J$$

The F2T module in the block diagram achieves the function of generalized force transformation.

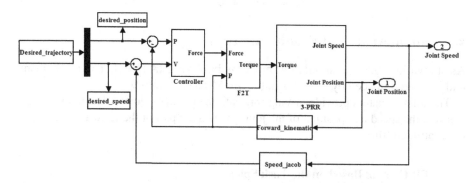

Fig. 6. PD control system based on the task space

As shown in Fig. 7, we can see the error of the platform. Compared with the iterative learning control, the error of PD controller remains constant and has larger value, whose average is 1mm. The trend of the error curve is related to the trajectory.

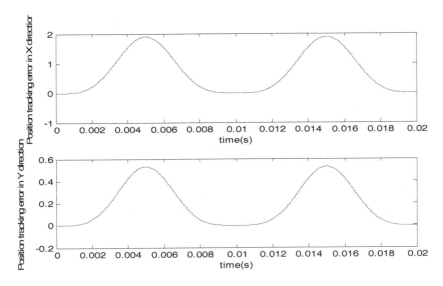

Fig. 7. Errors of the platform

4.2   Iterative Learning Control Based on the Joint Space

As shown in Fig.8, the control system of the 3-PRR mechanism is built via MATLAB simulation toolbox. The controller uses the iterative learning control algorithm. The 3-PRR mechanism is simulated as shown in Fig.5. The desired trajectory is given and the actuators' displacement and speed are calculated based on the result of inverse kinematic.

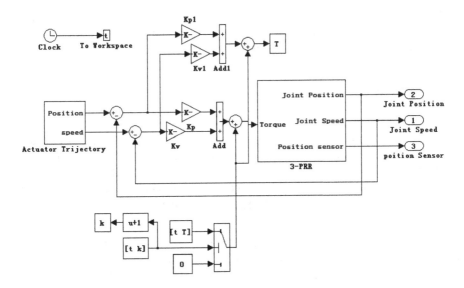

Fig. 8. Block diagram of the iterative learning control system

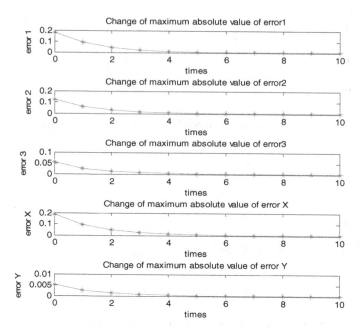

Fig. 9. Error of the actuators and platform

As shown in Fig.9 errors of the actuators and platform decrease to near zero after 10 times of iteration. Error1, error2 and error3 represent the displacement errors of the prismatic joints i.e. B_1, B_2 and B_3, respectively. Error X and error Y represent the error of the platform in the X and Y directions.

The simulation of the model shows that iterative learning control algorithm is applicable and effective. After enough times of iteration, we can always get the desired accuracy as long as it is convergent.

5   Conclusion

Parallel mechanism has high stiffness, strong bearing capacity and high accuracy. The motion of parallel mechanism is nonlinear and coupling.This paper obtains kinematic solutions of 3-PRR mechanism and builds control systems based on joint space and task space. PD control algorithm and iterative learning algorithm were used to build the controller respectively. Simulation results show that iterative learning control based on the joint space can achieve good effect.

Acknowledgements. This research was supported by the National Natural Science Foundation of China (Grant No.91223201), the Natural Science Foundation of Guangdong Province (Grant No.S2013030013355), Project GDUPS (2010) and the Fundamental Research Funds for the Central Universities (2012ZP0004). These supports are greatly acknowledged.

References

1. Merlet, J.P.: Parallel robots, 2nd edn., pp. 6–9. Springer, Netherlands (2006)
2. Gao, M., Zhang, X., Liu, H.: Experiment and Kinematic Design of 3-RRR Parallel Robot with High Speed. ROBOT 35(6) (2012)
3. Zhang, X., Zhang, X., Chen, Z.: Dynamic analysis of a 3-RRR parallel mechanism with multiple clearance joints. Mechanism and Machine Theory 78, 105–115 (2014)
4. Huang, Z., Li, Q.C.: General methodology for type synthesis of symmetrical lower-mobility parallel manipulators and several novel manipulators. The International Journal of Robotics Research 21(2), 131–145 (2002)
5. Ur-Rehman, R., Caro, S., Chablat, D., et al.: Multiobjective design optimization of 3-PRR planar parallel manipulators. Global Product Development, 1–10 (2010)
6. Yi, D.: Analysis and calibration of a 3-PRR micro-motion manipulator. Shanghai Jiao Tong University (2011)
7. Bristow, D.A., Tharayil, M., Alleyne, A.G.: A survey of iterative learning control. IEEE Control Systems 26(3), 96–114 (2006)
8. Cong, S., Shang, W.: Parallel robots-modeling, Control optimization and Applications. Publish House of Electronic Industry (2010)
9. Shang, W., Cong, S.: Nonlinear adaptive task space control for a 2-DOF redundantly actuated parallel manipulator. Nonlinear Dynamics 59(1-2), 61–72 (2010)
10. Liu, H., Zhang, X.: Control system design for a high speed and accuracy planar 3-RRR parallel robot basing on dSPACE. Electronic Design Engineering 22(11) (2014)

Experimental Study on Joint Positioning Control of an Ultrasonic Linear Motor Driven Planar Parallel Platform

Jiasi Mo, Zhicheng Qiu, Junyang Wei, and Xianmin Zhang*

Guangdong Province Key Laboratory of Precision and Manufacturing Technology,
South China University of Technology, Guangzhou 510641, China
{mo.jiasi,wei.jy01}@mail.scut.edu.cn,
{zhchqiu,zhangxm}@scut.edu.cn

Abstract. Planar parallel mechanism is commonly used in planar precision positioning platform. Its precision is influenced by the nonlinear factors such as friction, clearance and the elastic deformation, especially the accuracy of the driven joint, directly affects the positioning precision of the platform. The content of this study is proposed one kind of the positioning joint which directly driven by the ultrasonic linear motor and its control method. Because driven directly on the ultrasonic linear motor, which making the driven joint has a quick response, no clearance, and realize the submicron positioning accuracy. In this paper, the drive characteristics of ultrasonic linear motor have been studied, and the model of ultrasonic linear motor has been established, also identify the transfer function of the motor. This paper used the integral separation PID algorithm to control the joint, experiments shows that the accuracy of this system can satisfy the precision positioning.

Keywords: Positioning control, ultrasonic linear motor, planar parallel platform.

1 Introduction

Traditional planar parallel positioning platform usually used the rotary servo motor to drive the driven joint. If the joint driven by revolute pair, it must be installed reducer to increase torque; if driven by prismatic pair, it is not only need to install reducer, but also need to use screw rod and slider to transform rotary motion to linear motion [1]. The installation of screw rod and reducer, will introduce various nonlinear factors such as friction, clearance, elastic deformation and so on, which increase the uncertainty and increase the volume of the mechanism [2], lead to difficult to miniaturization. In order to solve the above problems, this paper proposes a joint positioning device which is driven by ultrasonic linear motor and its control method. This device drive directly by ultrasonic linear motor, have no need to install the reducer and screw rod. Ultrasonic motor driven by friction is a no clearance driven approach [3], which can reduce the clearance error. The design of the joint is shown in Fig. 1, which is makes up by the base plate, motor stand, linear slider, ultrasonic linear motor, linear encoder. This joint has been used for 3PRR parallel positioning platform for the driven joint.

---

*Corresponding author.

X. Zhang et al. (Eds.): ICIRA 2014, Part II, LNAI 8918, pp. 48–59, 2014.
© Springer International Publishing Switzerland 2014

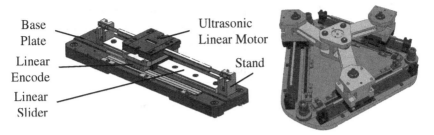

Fig. 1. Left: The design of the prismatic joint; Right: 3PRR positioning platform used this joint

Nonlinear factor is the main factor affecting the positioning accuracy [4], if driven joints exist too many nonlinear factors, the parallel mechanism which used that kind of joint to drive, the nonlinear phenomena will be more serious because of the coupling relationship with the other joints, even lead to the self-excited vibration [5]. Generally there are two ways to solve the nonlinear problems, the first way is avoid to introducing nonlinear factors on the design phase, the second way is to use nonlinear control algorithm [6]. In order to solve the nonlinear problem fundamentally, it need to introduce nonlinear factors as less as possible when design, and combining control algorithm, to compensate the nonlinear factors, so it can achieve high precision positioning.

The basis of the precision positioning is the precision way of drive and an effective control algorithm. In the study of driven way, Sang-chae Kim design a linear actuator using the piezoelectrically driven friction force [7], but it is low level of integration, so it is difficult to make the mechanism miniaturization; The ultrasonic motor is a good way to realize high accuracy of positioning, and modeling is the foundation of the control, but the ultrasonic motor is too complicated to modeling. Kenji Uchino make an overview of the traditional ultrasonic motors [8], it can found that the ultrasonic linear motor from PI is the new type of motor, there is no one has modeling it before. Lihua Lu using macro and micro two models to control the screw rod positioning platform, it can achieve the encoder accuracy[9], but if switching the model in an inappropriate timing will lead to a decline in system stability. In this paper, the ultrasonic linear motor is used for the driven joint; the model of the motor has been established. The control experiments of this joint show that it can satisfy the submicron positioning.

2   System Modeling and Identification

The closed-loop system diagram of the above prismatic joint positioning system is shown as Fig. 2, including the controller, motor drive and the driven joint. The signal flow of the system is input the target position to the controller, the controller generate a DC reference voltage as control signal, the control signal input the motor drive, make the motor approach the target position, in the meantime, the motor carry the linear encoder to detect the actual position and feed back to the controller, to realize closed-loop control. Hence, this system is an electro-mechanical coupling system, the mathematical model of this system including the motor drive model, the driven mechanism model, the friction model and the inverse piezoelectric effect model.

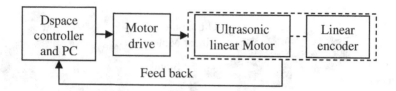

Fig. 2. The closed-loop system diagram

2.1   System Modeling

In order to establish the mathematical model of this system, it is necessary to analyze the structure of the ultrasonic linear motor and the drive properties. This system used the ultrasonic linear motor type is U-264, the motor drive type is C-872, which are from Physik Instrumente (PI) Germany. Through the signal lines from the motor drive to the motor body, there are two PZTs to drive one motor, and each PZT has two electrodes, respectively access to the voltage waveform phase1 and phasse2, and the two PZTs driven lines in parallel, it can be learned that the two PZTs move at the same time. Through the test of waveform as shown in Fig. 3, it can get the driven frequency is 160 kHz, phase1 and phase2 has no phase difference.

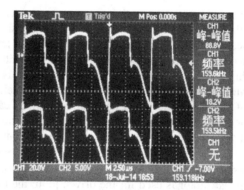

Fig. 3. Driven wave (top: phase1; bottom: phase2)

The peak-to-peak voltage (V_{pp}) of the waveform as shown in Fig. 3 has a relationship between the input DC reference voltage of the controller, the allowable input DC reference voltage range of the C-872 is -10~10VDC, the test curve of DC reference voltage and V_{pp} of the waveform as shown in Fig. 4. When the DC reference voltage is positive, the V_{pp} of phase1 is much bigger than the phase2, the phase1 work at this time, making the motor move forward; whereas the DC reference voltage is negative, the phase2 work at this time, making the motor move backward. The DC reference voltage and the V_{pp} of the waveform have the linear relation, for fitting the linear equation as below

$$V_{pp} = 21.08DC + 26.91 \tag{1}$$

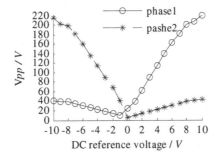

Fig. 4. Relationship between the DC reference voltage and output V_{pp}

The U-264 make up by the motor body and rod, the motor body contains two PZTs, each PZT connected a drive foot, and control by two phases voltage call phase1 and phase2, respectively control the forward and backward movement. Through the waveform of Fig. 3, phase1 and phase2 has no phase difference, and it is also not a sine wave, so it is not possible to form an elliptic motion, therefore, the U-264 has difference between the traditional ultrasonic motor [3]. Assuming that the right side is the forward direction, when the motor move forward, the V_{pp} of phase1 is much higher than the phase2, making the position of the electrode which connected to the pahse1 extend much longer than the position of phase2.The result of that is the drive foot turn to the phase2 side, two drive feet extend at the same way and friction the rod simultaneously, then the motor move forward in micro step, the principle diagram of the PI ultrasonic linear motor as shown in Fig. 5.

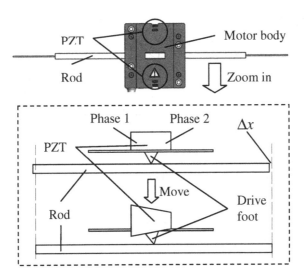

Fig. 5. The principle diagram of the PI ultrasonic linear motor

To establish the mathematical model of ultrasonic linear motor, as shown in Fig. 6, considering the motion state and force analysis of one PZT work in phase1 status, and establish the kinematics and dynamics model. Consider the junction between the PZT and the driven foot has the minimum stiffness, thus can be equivalent to a compliant revolute hinge, set to point A. When no voltage supply to the PZT, and if there is a gap between the driven foot and rod, the rod will separate from the motor body, it is not conform to the actual situation, so $\Delta d = 0$. At this time, the driven foot critical contact with the rod, the contact point is C, initial driven foot length is L. When phase1 electrify, the driven foot extend because the PZT extend, contact point move from point C to point B, the extended distance of PZT is dp. If the rod and the driven foot without slip, the micro step $\Delta x = |BC|$. The equation is shown as below

$$\Delta x = \sqrt{(L+dp)^2 - (L\sin\theta + \Delta d)^2} - \cos\theta(L + \Delta d / \sin\theta) \qquad (2)$$

If $\Delta d = 0$, there is

$$\Delta x^2 + 2L\cos\theta\Delta x + (-dp^2 - 2Ldp) = 0 \qquad (3)$$

So it can get Δx by solving quadratic equation(3). Assume that the motor frequency is f_m, so it move $f_m\Delta x$ distance in one second, so the motor moving velocity $\bar{v}_m = f_m\Delta x$, the acceleration $a_m = \dot{v}_m$.

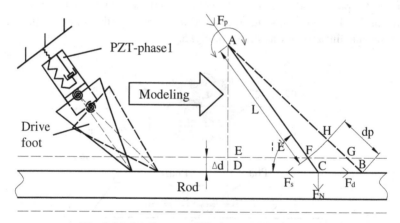

Fig. 6. Model diagram of PI ultrasonic linear motor

To analysis the force of Fig. 6, the PZT driven force F_p provides the lengthways driven force F_d and the positive pressure F_N, which generate the frictional force, assuming no relative slip, so the frictional force F_s provide the driven force, there is

$$\begin{cases} F_d = F_p \cos\theta \\ F_N = F_p \sin\theta \\ F_s = F_d \end{cases} \qquad (4)$$

Assume that the motor equivalent mass is m, when two PZTs drive at the same time, there is

$$2F_p \cos\theta = m\dot{v}_m \tag{5}$$

In the extended process of PZT, θ is not a constant value, but gradually become smaller, so the F_d is also not a constant value, but in a step cycle ($1/f_m$) changes from small to big.

The relationship between the DC reference voltage and the output V_{pp} is already established (equation(1)), and the relationship between dp and Δx has been got (equation(3)), so, if the relationship between the V_{pp} and dp can find, will be able to establish the system input (DC reference voltage) and output (Δx) model. The relationship between the V_{pp} and dp can use the inverse piezoelectric effect equation to describe, the equation is shown as below [10]

$$s_j = d_{ij}E_i, (i=1,2,3; j=1,2,3,4,5,6) \tag{6}$$

Where ——— s_j is the strain under the electric field;

d_{ij} is the piezoelectric strain constant;

E_i is the field strength;

1, 2, 3 means the direction of the x, y, z axis;

4, 5, 6 means the tangential effect around the x, y, z axis;

Because the PZT belongs to the 6mm point group, according to the symmetry and piezoelectric, the nonzero and independent piezoelectric strain constant is d_{31}, d_{33}, d_{15}, so the equation (6) changes as below [10]

$$\begin{bmatrix} s_1 \\ s_2 \\ s_3 \\ s_4 \\ s_5 \\ s_6 \end{bmatrix} = \begin{bmatrix} 0 & 0 & d_{31} \\ 0 & 0 & d_{31} \\ 0 & 0 & d_{33} \\ 0 & d_{15} & 0 \\ d_{15} & 0 & 0 \\ 0 & 0 & 0 \end{bmatrix} \begin{bmatrix} E_1 \\ E_2 \\ E_3 \end{bmatrix} \tag{7}$$

Assumes that the motor is using a longitudinal piezoelectric effect, the polarization direction along the thickness direction, when applying an electric field, according to equation(7), the relationship between the strain and electric field can be established, as below [10]

$$\Delta T = Td_{33}E_T = d_{33}U_T \tag{8}$$

Where ——— T is the thickness of the crystal;

d_{33} is the lengthways piezoelectric strain constant;

E_T is the lengthways field strength;

U_T is the lengthways voltage;

Replace the V_{pp} of U_T, and put it into the formula, so we can solve the deformation dp of the PZT, by considering the equation(1)、(3)and(8), mathematical model of this system can be obtained.

2.2   System Identification

The input of this system is DC reference voltage, and the output is linear displacement of the motor, the input and output model of this system can be obtained on above section. But the PZTs of the ultrasonic linear motor have been packed and installed inside the motor, so it is unable to get the parameters of the PZTs such as the thickness、 the piezoelectric strain constants and so on, which result in it cannot get mathematical model with the specific parameters. Accordingly, the system identification methods can be used to get the system model, determining the system structure and parameters by the input and output data.

This paper uses the sine frequency response method to identify system model, through the sine wave input with different frequency, the frequency response curve of this system can be got, and the fitting curve of magnitude frequency characteristic plot as below

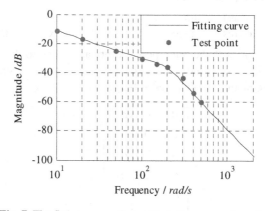

Fig. 7. The fitting curve of magnitude frequency characteristic

According to the magnitude frequency characteristic curve, System consists of an integration element and a second order oscillation element, making the system become a third order system, the fitting transfer function shown as below,

$$\frac{K}{s(\dfrac{s^2}{\omega^2}+2\xi\dfrac{s}{\omega}+1)} \tag{9}$$

Where   —— $K=2.82$ is the open-loop gain;

$\omega=200$ is the corner frequency;

$\xi=0.5$ is the damping factor;

3   Positioning Experiments

Experimental platform make up by the base plate, linear slider, motor stand and ultrasonic linear motor from PI (open-loop resolution $0.1\ \mu m$); use the Renishaw linear encoder (resolution $0.05\ \mu m$) to feedback the actual position, use the Omron photoelectric switch as the limit switch; the control system make up by the computer, Dspace DS1103 semi-physical simulation card, and the PI motor drive, the component of experiment system is shown in Fig. 8.

Ultrasonic
Motor Drive

Dspace
DS1103
Controller
& PC

DC Source

Limit
Switch

Ultrasonic
Linear Motor

Linear
Encoder

Fig. 8. The component of experiment system

This paper uses PID algorithm to realize positioning control, PID algorithm has strong commonality, and it is not based on the system model, control only through the deviation of input and output. The transfer function of the PID algorithm is shown as below

$$G_{PID}(s) = K_p(1 + \frac{1}{\tau_I s} + \tau_D s) \qquad (10)$$

The main design parameters of PID controller is the gain K_p, integration time τ_I and differential time τ_D, through the experiment, the parameters tuning as K_p =50, $1/\tau_I$ =10, τ_D =0.001, sample time is $0.0001s$. In order to prevent the excessive amount of overshoot which cause by the integral action, and improve the dynamic performance, so introduce the integral separation PID algorithm. When feedback deviation is less than $1mm$, then introduce the integral action, minimize the residual error; when feedback deviation is large than $1mm$, only use the PD control. Integral separation PID control experiment system diagram is shown as below:

Performance of the positioning system is including the absolute accuracy, repeatability, trajectory tracking ability and so on. The following experiments for test the various performances.

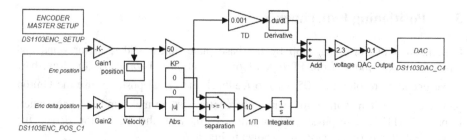

Fig. 9. Integral separation PID control diagram

3.1   Absolute Accuracy Test

This experiment in order to test the absolute accuracy of the motor under different scale of step, in this experiment, it is set 4 scales of step size respectively as $1\mu m$, $10\mu m$, $100\mu m$, $1000\mu m$, each scale step forward 12 times, repeat three times, the positioning errors of 4 scales of step are as below.

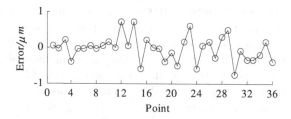

Fig. 10. $1\mu m$ step error curve

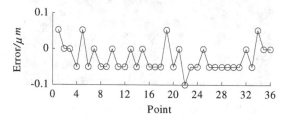

Fig. 11. $10\mu m$ step error curve

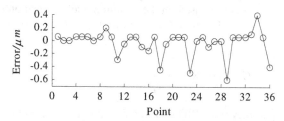

Fig. 12. $100\mu m$ step error curve

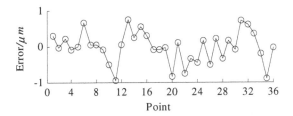

Fig. 13. $1000\mu m$ step error curve

Every scale proceeds three times of experiment, 36 times of step in total. The error statistics is shown as below

Table 1. The error statistics of 4 scales of step

Step Scale / μm	Average Error / μm	Variance	Maximum Error / μm
1.000	0.025	0.0368	0.100
10.000	0.032	0.1946	0.600
100.000	0.036	0.3484	0.750
1000.000	0.029	0.4361	0.950

3.2 Repeatability Test

The repeatability experiment in order to test the positioning ability that is in different direction and different location moves toward to the same point. This test picks up 12 points in two directions in random, making the motor move from the 12 points toward to the point 10mm, the path and the data are shown as below

Table 2. Repeatability experiment data

No.	Actual/*mm*	Error / μm	No.	Actual/*mm*	Error / μm
1	19.999950 10.000050	0.050	7	7.000050 10.000100	0.100
2	11.000050 10.000650	0.650	8	14.999700 9.999900	-0.100
3	8.999900 10.000500	0.500	9	13.999950 9.999900	-0.100
4	5.000350 9.999850	-0.150	10	12.999850 10.000050	0.050
5	12.999900 9.999500	-0.500	11	3.000600 9.999750	-0.250
6	11.999950 10.000100	0.100	12	0.000050 10.000150	0.150

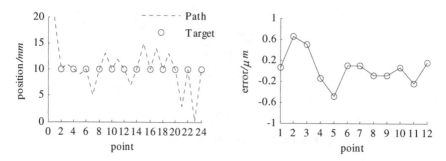

Fig. 14. Left: The path of the repeatability test; Right: The errors of the repeatability test

As shown in Table 2, the maximum error in 12times positioning to the 10mm point is $0.65 \mu m$. The path of the repeatability test is shown as the left of Fig. 14.

3.3   Sine Wave Trajectory Tracking

Straight lines and arcs appear most in the trajectory inverse kinematics about the planar 3 DOFs parallel platform, in order to apply this joint on planar parallel mechanism to achieve good trajectory tracking, the trajectory tracking of this system must meet the accuracy requirement. In order to test the trajectory tracking of straight lines and arcs, using sine wave to test the trajectory tracking, because the sine wave can be equivalent to the straight lines and arcs well, and it can realize the reverse movement. The sine wave trajectory tracking curve is shown as below

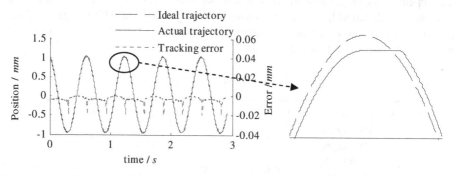

Fig. 15. Left: Sine wave trajectory tracking and the error; Right: largest tracking error appears in wave peak

According to the result of Fig. 15, this system can track the sine wave trajectory, which can meet the initial requirements of trajectory tracking. If zoom in the left figure of Fig. 15, it can find that the most tracking error appears in the wave peaks /troughs as the right figure of Fig. 15, the largest tracking error is $19.7 \mu m$, the other errors of the sine wave tracking is all less than this value.

4   Conclusion

This paper propose one kind of the positioning joint which directly drive by the ultrasonic linear motor, with integral separation PID algorithm to control the joint in some experiments. Through the results of the experiments, come to some conclusions as below.

1) The absolute accuracy experiment test four scales of step in total, The maximum error is $0.95\mu m$ which appears in the $1000\mu m$ step test, the total error average less than $0.036\mu m$, consider that absolute accuracy of this system is $0.95\mu m$;

2) In the experiment of repeatability test, the maximum error is $0.65\mu m$, so consider that the repeatability of the system is $0.65\mu m$;

3) Integral separation PID algorithm can suppress the overshoot and improve the dynamic performance; all the overshot of the point to point test is less than 20%;

4) Though this system minimize the nonlinear factors, the nonlinear factor from the frictional force is unable to eliminate, which is lead to the sine wave trajectory tracking error is bigger than $1\mu m$.

The experiment proved that this joint can meet the requirements of the submicron level on point to point positioning. But the PID algorithm is a linear algorithm, which make it appear a big error on trajectory tracking. It is need to design a nonlinear algorithm to improve the trajectory tracking accuracy in the subsequent study.

Acknowledgement. This research was supported by the National Natural Science Foundation of China (Grant No. 91223201) and the Natural Science Foundation of Guangdong Province (Grant No. S2013030013355).

References

1. Norton, R.L.: Design of Machinery-An Introduction to the Synthetic and Analysis of Mechanisms and Machines, vol. 2. McGraw-Hill Education, New York (1999)
2. Spong, M.W., Vidyasagar, M.: Robot Dynamics and Control. John Wiley & Sons, American (2008)
3. Zhao, C.: Ultrasonic Motor Technology and Application. Science Press, Beijing (2007)
4. Spong, M.W., Hutchinson, S., Vidyasagar, M.: Robot Modeling and Control, vol. 3. Wiley New York, American (2006)
5. Gao, M.: Design, Analysis and Control of Planar Parallel Robot. Doctoral Degree. South China University of Technology, Guangzhou, China (2011)
6. Castillo-Castaneda, E., Takeda, Y.: Improving Path Accuracy of a Crank-Type 6-Dof Parallel Mechanism by Stiction Compensation. Mechanism and Machine Theory 43(1), 104–114 (2008)
7. Kim, S., Kim, S.H.: A precision Linear Actuator Using Piezoelectrically Driven Friction Force. Mechatronics 11(8), 969–985 (2001)
8. Uchino, K.: Piezoelectric Ultrasonic Motors: Overview. Smart Materials and Structures 7(3), 273 (1998)
9. Lu, L.: Research on Modeling and Control of Long Range Nanometer Positioning System Based on BallScrew. Doctoral Degree. Harbin Institute of Technology, Harbin (2007)
10. Kang Sun, F.Z.: Piezoelectricity. National Defense Industry Press, Beijing (1984)

Stiffness Modeling and Optimization Analysis of a Novel 6-DOF Collaborative Parallel Manipulator[*]

Yao Liu, Bing Li, Peng Xu, and Hailin Huang

Shenzhen Graduate School, Harbin Institute of Technology,
Shenzhen, 518055, P.R. China
libing.sgs@hit.edu.cn

Abstract. This article presents a novel 6-DOF cooperative parallel manipulator assembled by two independent 3-DOF parallel manipulators that can complete tasks through relative motion. It provides broader workspace, higher stiffness and better controllability for high precision machining demands compared to traditional parallel kinematics manipulator. The architectures of the two 3-DOF parallel manipulators are firstly stated, followed by inverse kinematics analysis and stiffness analysis through establishing global Jacobian matrix. Finally the cooperative parallel manipulator's static stiffness and modal analysis are conducted. First order of natural frequency optimization is established so that optimal radius of drive rod can be achieved. Comparing the results before and after optimization, it can be seen that performances are obviously improved.

Keywords: cooperative parallel manipulator, kinematics, stiffness, parametric optimization.

1 Introduction

Over the past two decades, many efforts have been put on cooperative serial robot (CSR) and hybrid cooperative manipulator (HCM). CSR, inheriting the feature of serial mechanism, adds 1~3 DOFs on the work piece to generate cooperative machining between tool and work piece, which broaden the workspace and can undertake the machining of more complex surface. Attractive results have been achieved, such as Chang, K. S.'s analyses on cooperative manipulation [1], Liu, Y. H.'s cooperative control research [2]. HCM, combined of series robot and parallel manipulator, provides not only broader workspace and large mass/stiffness ratio but also low accumulative error, which gains an advantageous place in multi-DOF complex surface machining [3]. Waldron, K. J. designed a hybrid series/parallel 6-DOF manipulator and researched into the dualities of motion screw axes and wrenches with Ball [4, 5]. Shahinpoor, M. and Sklar, M. came up with a novel

---

[*] This work was financially supported by the National Natural Science Foundation of China (No. 51175105), the Shenzhen Key Lab Fund of Advanced Manufacturing Technology (No. CXB201203090033A) and the Shenzhen Research Fund (No. SGLH20131010144128266, JCYJ20140417172417129).

X. Zhang et al. (Eds.): ICIRA 2014, Part II, LNAI 8918, pp. 60–71, 2014.
© Springer International Publishing Switzerland 2014

parallel-serial robot and researched into its kinematics and dynamics [6, 7]. Yet the presence of serial mechanism in both CSR and HCM makes it unsuitable in high precision machining, which leads to the idea of cooperative parallel manipulator (CMP).

CMP that means that two parallel manipulators cooperate to accomplish assigned tasks, which obviously provides broader workspace, larger mass/stiffness ratio and lower accumulative errors, is proposed [8]. All these advantages reduce the cost and improve the practicability. Joint efforts have been made into realizing this wonderful idea [9, 10]. However, most work has been focused on structure analysis, control strategy and the scope of application [11, 12], very few of which systematically researched on its performances. Therefore, current researches on CPM machine tool are very limited, which hinders its application in manufacture.

This paper is organized as follows: in Section 2, the architecture of CPM is described and in Section 3 the inverse kinematic which is based on a closed-loop vector method is presented. In Section 4, the stiffness of parallel manipulator is derived. In Section 5, mathematic model is established between the first order natural frequency and parameters of drive rods, which is used for optimizing the size of the CPM. The conclusion is drawn in the last section.

2   Description of the CPM

A novel 6-DOF CPM machine tool that can cooperate to complete given tasks, is designed as shown in Fig. 1. These two manipulators, designed according to screw theory [13], are placed up and down. The upper manipulator, a 3-PSC/PU parallel manipulator, is the tool system, which can rotate around x, y axis and stretch out/draw back along z axis. The lower manipulator, a 3-PU*R parallel manipulator, is able of translation along x, y, z axis, which is named work piece system.

Fig. 1. 3-D schematic representation of a general CPM

Tool system, as shown in Fig. 2(a), consist a moving platform, a fixed base and three limbs of identical kinematic structure which are 120° symmetric distribution. Each limb connects the fixed base to the moving platform by a P joint, a S joint and a C joint in sequence, where the P joint is actuated by a linear actuator assembled on the fixed base. A central passive limb connects the fixed base to the moving platform by a

P joint and a U joint. Thus, the moving platform is attached to the base by three identical PSC linkages and one PU linkage. To achieve the desired movement, the axe of each C joint should satisfy some geometric conditions. That is, the angel of the axes of C joint is 60°, and they are perpendicular to the axis of the moving platform.

Meanwhile, work piece system as shown in Fig. 2(b) consist a moving platform, a fixed base, and three limbs of identical kinematic structure as well. Each limb connects the fixed base to the moving platform by a P joint, a 3-UU PM, and a R joint in sequence, where the P joint is actuated by a linear actuator assembled on the fixed base. Some geometric conditions should be met to make sure the PM achieve the desired movement, such as the axes of R joint are perpendicular to the axis of the moving platform which is a regular triangle. Moreover, the trajectories of the linear actuators should intersect in one point.

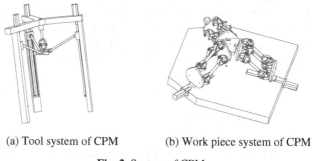

(a) Tool system of CPM   (b) Work piece system of CPM

Fig. 2. System of CPM

3   Inverse Kinematics Modeling

The inverse kinematics problem resolves the input variables from a given position of the moving platform. The schematic diagram of 3-PSC/PU PM and a typical kinematic chain of the structure are given in Fig. 3.

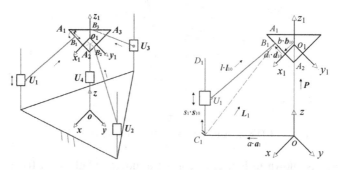

(a) Schematic diagram of 3-PSC/PU PM   (b) One typical kinematic chain

Fig. 3. Schematic diagram of cutter system

Inside the equilateral triangle sets a body-fixed coordinate $o_1\text{-}x_1y_1z_1$, of which the x_1 axis lies perpendicular to one side of the triangle, z_1 lies perpendicular to the platform. Meanwhile, a fixed Cartesian coordinate $o\text{-}xyz$, of which the x, y, z axis lie parallel to x_1, y_1, z_1 axis, is established at the platform center. The length of the rod connecting C joint and S joint is represented by l. B_i' ($i=$ 1, 2, and 3) expresses the original position of the moving platform on a circle of b-radius. Slideway C_iD_i ($i=$ 1, 2, and 3) lies parallel to Cartesian coordinate's z axis, with the other two slideways spread along the circle every $120°$ and intersect the x-y plane at points c_1, c_2 and c_3, which lie on a circle of a-radius. Slider U_i is restrained within slideway C_iD_i. Without losing generality, let the x axis point along $\overrightarrow{OC_1}$ and the y-axis direct along $\overrightarrow{O_1B_1'}$. Angle α is defined from the x_1 axis to $\overrightarrow{OC_2}$ in the fixed frame and also from the x-axis to $\overrightarrow{O_1B_2'}$ in the moving frame. Similarly, angle β is measured from the x axis to $\overrightarrow{OC_3}$ in the fixed frame and also from the x_1-axis to $\overrightarrow{O_1B_3'}$ in the moving frame. For simplicity, we assign that $\alpha=120°$ and $\beta=240°$, vector $^{O_1}b_i$ represents the position of B_i' in the $o_1\text{-}x_1y_1z_1$, while vector $^{O}a_i$ represents the position of C_i in the $o\text{-}xyz$. In most cases the position and direction of platform center can be denoted by position vector $\boldsymbol{P}=$ (0, 0, z). Since the moving platform of a 3-PSC/PU PM possesses two rotational and a translational motion, a 3×3 rotation matrix $^{O}R_{O_1}$ is shown as fellow:

$$^{O}R_{O_1} = \begin{bmatrix} \cos\beta & 0 & \sin\beta\sin\alpha \\ 0 & \cos\alpha & -\sin\alpha \\ -\sin\beta & \cos\beta\sin\alpha & \cos\beta\cos\alpha \end{bmatrix} . \tag{1}$$

Then, one obtains:

$$\boldsymbol{d}_{10} = {}^{O}R_{O_1}\,{}^{O_1}\boldsymbol{d}_1 \ , \tag{2}$$

where, $\boldsymbol{d}_{10}$ represents the unit direction vector of slideway A_1A_2 in fixed frame.

Referring to Fig. 3(b), a vector-loop equation can be written for i-th limb as follows:

$$ll_{10} = \boldsymbol{L}_1 - s_1\boldsymbol{s}_{10} \ , \tag{3}$$

where, $\boldsymbol{L}_1=\boldsymbol{p}+b\boldsymbol{b}_{10}+d_1\boldsymbol{d}_{10}-a\boldsymbol{a}_1$, l_{10} is direction vector along $\overline{U_iB_i}$, s_1 represents the displacement of motion of i-th slider, $\boldsymbol{s}_{10}$ is the unit direction vector of the slider's motion. d_1 denotes the displacement of C joint along slideway A_1A_2 from its original position. From the geometry relations, it can be concluded that l_{10} is perpendicular to $\boldsymbol{d}_{10}$, while $\boldsymbol{b}_{10}$ lies perpendicular to $\boldsymbol{d}_{10}$. Substituting (3) into (2) and dot-multiplying both sides of the equation by $\boldsymbol{s}_{10}$ allow the derivation of d_1:

$$s_1 = (d_1\boldsymbol{d}_{10} + a\boldsymbol{a}_1 - \boldsymbol{p} + b\boldsymbol{b}_{10})\cdot\boldsymbol{d}_{10}^{T} \ . \tag{4}$$

The geometry of kinematic chain of 3-PU*R PM is similar to 3-PSC/PU PM. Thus, following the same method, the inverse kinematics of 3-PU*R PM can be obtained.

4   Stiffness Analysis

4.1   Stiffness Modeling

In this section, the reciprocal screws theory is applied to get the Jacobian matrix of a parallel manipulator. With v and ω denoting the vectors of the linear and angular velocities respectively, the velocity of the center of moving platform can be expressed as $V=[v, \omega]^T$ in the generalized coordinate system, where $v=[v_x, v_y, v_z]^T$ and $\omega=[\omega_x, \omega_y, \omega_z]^T$. v_i donates the velocity of the points that contact the moving platform with C Joints. v_{ri} donates the scalar speed of slider.

The moving platform will be affected by external workload W, which can be simplified to a force f and torque m, so $W =[f^T, m^T]^T$. The balance of W are the internal driving force F_{ai}, constraint force F_{cj} and constraint moment $T_{\tau k}$. For 3-PSC/PU PM, there exist two constraint forces and one constraint moment. And constraint force F_{cj} and constraint moment $T_{\tau k}$ do not doing any work, so one obtains:

$$F_{cj} \cdot v + \left(d_i \times F_{cj}\right) \cdot \omega = 0 \quad , \tag{5}$$

where, $F_{cj} = c_{cj} F_{cj}$.

So:

$$\left[c_j^T \quad \left(d_i \times c_j\right)^T \right]\begin{bmatrix} v \\ \omega \end{bmatrix} = 0 \quad , \tag{6}$$

where, j donates the number of constraint force, and so:

$$T_{\tau k} \cdot e_i = 0 \quad . \tag{7}$$

The velocity of the mechanism can be obtained:

$$\begin{bmatrix} v_{ri} \\ 0 \end{bmatrix} = \begin{bmatrix} r_i^T & \left(e_i \times r_i\right)^T \\ c_j^T & \left(d_i \times c_j\right)^T \\ 0 & e_i \end{bmatrix}\begin{bmatrix} v \\ \omega \end{bmatrix} . \tag{8}$$

Hence, Jacobian matrix is derived as follows:

$$J = \begin{bmatrix} r_i^T & \left(e_i \times r_i\right)^T \\ c_j^T & \left(d_i \times c_j\right)^T \\ 0 & e_i \end{bmatrix}_{6\times 6} . \tag{9}$$

According to the principle of virtual work, external workload can be derived as:

$$W = -J^T F_r \ ,$$

(10)

where, $-J^T$ donates force Jacobi matrix. Velocity Jacobi matrix donates the mapping between input velocity and output velocity and force Jacobi matrix donates the mapping between the driving and constraining force and the external workload of moving platform.

The driven sliders and their links are considered as flexible. Therefore, the 3-PSC/PU PM includes three elastic deformations in every limb, for instance axial elastic and bending deformation of every limb and twist deformation of each slider. Under the effect of external workload **W**, the stiffness with respect to driven sliders and limbs can be derived as follows.

Firstly stiffness matrix of driven slider is established by taking the torsional deformation of linear driven slider into consideration. The axial stiffness of the i-th linear driven slider is obtained as:

$$k_{ai} = \frac{F_{ai}}{\delta l_{ai}} = \frac{2k_{\theta i}^{'}}{P_h \mu_c d_s} \ ,$$

(11)

where, μ_c is the friction coefficient, d_s represents the pitch diameter of legs. P_h donates the lead of screw

Secondly the stiffness matrix of limbs is established. k_{si} and $k_{\theta i}^{'}$ represent longitudinal and transverse compliance of the i-th limb respectively. Hence, one obtains:

$$k_{si} = \frac{E_i A}{l} \ , \quad k_{\theta i}^{''} = \frac{3E_i I}{l^3} \ ,$$

(12)

where, E_i and I represent the modulus of elasticity and the polar moment of inertia, l and A donate the length and the cross section area of each limb.

It is assumed that $f=[f_x, f_y, f_z]^T$ denotes a force and $m=[m_x, m_y, m_z]^T$ denotes a torque. Additionally, the reaction forces or torques of driven sliders and constraints can be assumed to be τ_a and τ_c, respectively. In the ignorance of gravity, the external workload is balanced by the reaction forces or torques exerted by the driven sliders and constraints:

$$W = J^T \begin{bmatrix} \tau_a \\ \tau_c \end{bmatrix} \ .$$

(13)

The reaction forces or torques can be expressed as:

$$\tau_a = \chi_a \Delta q_a \ , \ \tau_c = \chi_c \Delta q_c \ ,$$

(14)

where, Δq_a and Δq_c denote the displacements of actuations and constraints, respectively, and the diagonal matrices are $\chi_a=\mathrm{diag}[K_{a1},\ K_{a2},\ K_{a3}]$ and $\chi_c=\mathrm{diag}[K_{c1},\ K_{c2},\ K_{c3}]$ respectively.

Moreover, let $\Delta x=[x,\ y,\ z]^{\mathrm{T}}$ and $\Delta\theta=[\Delta\theta_x,\ \Delta\theta_y,\ \Delta\theta_z]^{\mathrm{T}}$ be the infinitesimal deformation of translation and rotation of the moving platform with resorting to three axes of the coordinate frame. And then, the principle of virtual work is applied and this allows the generation of:

$$W^T \Delta X = \tau_a^T \Delta q_a + \tau_c^T \Delta q_c \ , \tag{15}$$

where, $\Delta X=[\Delta x^{\mathrm{T}},\ \Delta\theta^{\mathrm{T}}]$ represents the moving platform's twist displacement in the axis coordinate.

By analyzing the equations above, it leads to the expression of

$$W = K\Delta X \ , \tag{16}$$

where, $K=J\chi J^{\mathrm{T}}$ is defined as the 6×6 global stiffness matrix of 3-PSC/PU PM taking the influence of actuations and constraints into consideration, with the 6×6 diagonal matrix $\chi=\mathrm{diag}[\chi_a, \chi_c]$.

Similarly, the stiffness matrix of 3-PU*R PM can be obtained by introducing a concept which is compliant virtual joints.

4.2 Numerical Simulation

By analyzing the stiffness matrix of 3-PSC/PU PM and 3-PU*R PM, eigenvalue of each matrix is obtained to evaluate the stiffness characteristic within workspace. Since each PM owns 3 DOFs, one DOF is fixed in order to express the stiffness characteristic in Cartesian coordinate system. Stiffness characteristic of these two PMs is shown in Fig. 4 through three-dimension search method.

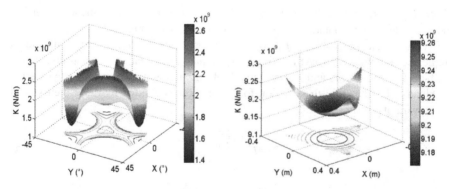

(a) The X-Y mapping of 3-PSC/PU PM   (b)The X-Y mapping of 3-PU*R PM

Fig. 4. The X-Y mapping of stiffness

In Fig. 4(a), Z axis donates stiffness characteristic and X, Y axis express rotation degree when the moving platform of 3-PSC/PU PM stays around 200mm. As shown in Fig. 4(b), X, Y axis mean displacement of moving platform of 3-PU*R PM which stays around 220mm. Seen from Fig. 4(a), minimum stiffness value occurs in the border of workspace while in the center of workspace for 3-PSC/PU PM has an central passive chain which makes the stiffness value bigger in the center. The maximum stiffness appears in the border of workspace for these two PMs are close to their singular configuration.

5   Optimization Study

In order to improve the performance of machine tool, optimization analysis of the size of the mechanism must be studied. This section mainly discusses the relationship between the first order of natural frequency (FNF) of the machine tool and the radius of cross section of drive rod.

5.1   Optimal Design of the Mechanism

The optimization will be accomplished within certain working space, containing structural parameters as design variables, which can be classified into two groups, rod lengths and sectional areas. The former affects the kinematic properties while the latter influences the stiffness, stress and strain. Sectional area influences natural frequency significantly. Therefore, radius ratio will be introduced for the math model concerning structural parameters and the FNF.

While keeping the total mass unchanged, aimed at the FNF, by changing the sectional area of drive rods, namely rod radius r_a (3-PSC/PU), r_b (3-PU*R), to optimize the machine tool's property. Here introduces the radius ratio to convert the problem from area to ratio:

$$h_i = \frac{r_{ai}}{r_{bi}} (i = 1, 2, 3) \quad . \tag{17}$$

After the shape and length of connecting rod is determined, optimization of the sectional area becomes prior to increasing natural frequency. Set $h(r_a, r_b)$ as the radius function of CPM's connecting rod and $f(t, h(r_a, r_b))$ as the function. Consider the total mass of the CPM unchanged and the model of CPM's optimization can be reached:

$$Max(\overline{f}) = \frac{\int_{t_0}^{t_1} f(t, h(r_a, r_b)) dt}{t_1 - t_0} \quad ,$$

$$s.t. \quad \pi(r_a^2 + r_b^2) = \cos nt_j \quad , \tag{18}$$

$$\sigma_{ia \, max} \leq [\sigma_{ia}], \ \sigma_{ib \, max} \leq [\sigma_{ib}]; \ \tau_{a \, max} \leq [\tau_{ia}], \ \tau_{ib \, max} \leq [\tau_{ib}]; \ \varepsilon_{max} \leq [\varepsilon] \quad ,$$

where, $\sigma_{ia\,max}$ stands for the maximum normal stress of drive rod and passive rods of 3-PSC/PU PM, $\sigma_{ib\,max}$ stands for the maximum normal stress of the drive rod of 3-PU*R CPM. $\tau_{ia\,max}$ stands for the maximum shear stress of motion rod and passive rods of 3-PSC/PU PM, $\tau_{ib\,max}$ stands for the maximum shear stress of the motion rod of 3-PU*R PM. $[\sigma_{ia}]$, $[\sigma_{ib}]$, $[\tau_{ia}]$ and $[\tau_{ib}]$ stand for the allowable value of the stresses mentioned above. ε_{max} stands for the maximum elastic error. t_0, t_1 stand for the starting/ending time.

To keep the CPM stable and economical, set the parameters of similar chains the same, that is to say, the sectional radius ratio h of 3-PSC/PU and 3-PU*R are set the same, $h_1=h_2=h_3=h$. Based on the model deliberated above, conduct parametric simulation in MATLAB. The results are listed in table 1.

Table 1. Optimization results of first order natural frequency

Parameters	Radius of cross section of drive rod (mm)			Total mass (kg)	FNF (Hz)
	r_a	r_b	$h= r_a/r_b$		
1	7.434	9.292	0.8		30.862
2	8.166	9.073	0.9		32.428
3	9	9	1		34.034
4	10.619	8.849	1.2	101.24	43.926
5	12.106	8.647	1.4		39.729
6	13.448	8.405	1.6		31.168
7	16.316	8.158	2		27.546

Making use of the MATLAB optimization toolbox, the optimization is accomplished and the curve of the FNF changing with radius ratio h is obtained as shown in Fig. 5. Seen from the figure, under the condition of total mass unchanged, the forth sectional area radius makes the FNF maximum, about 43.926Hz. Therefore, the optimal result occurs when r_a, radius of input rod of 3-PSC/PU CPM, equals to 10.619mm and r_b, radius of input rod of 3-PU*R CPM equals to 8.849mm.

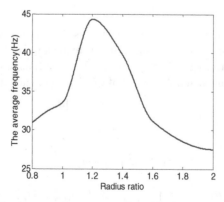

Fig. 5. Relation between the average value of FNF and radius of active legs

To conclude, when the total mass of the machine tool remain unchanged, through altering the sectional area of connecting rods, the stiffness of CPM can be increased so that stability and carrying capacity can be improved. Processing errors can be reduced and better precision can be reached.

5.2   Characteristics Analysis before and after Optimization

Static properties include maximum stress and strain in equilibrium location, which are important parameters to machine tools and directly influence capacity and precision. Therefore static properties need to be analyzed before and after dynamic optimization, which will be conducted in ANSYS Workbench and the results are shown in table 2.

Table 2. Data comparison before and after optimization

Parameters	Radius of cross section (mm)		Maximum stress (Mpa)	Maximum deformation (mm)
	r_a	r_b		
Before	9	9	1.04	7.491×10^{-3}
After	10.619	8.849	0.91	8.154×10^{-3}
Alteration	+17.99%	-1.68%	-12.5%	+8.8%

Seen from table 2, after optimization under the same load, the stress is reduced by 12.5% while strain increases but remains within 0.008mm in the safe range. The result is satisfactory.

The modal property of machine tool is one of the most important dynamic properties. After the optimization conducted in MATLAB, higher order frequencies should also be assessed. Thus modal analysis is conducted in ANSYS Workbench module aiming at the vibration frequency and modal of CPM. The first six order modal value of optimized CPM is obtained from modal analysis, as is shown in Fig.6.

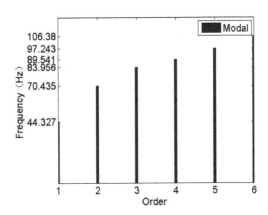

Fig. 6. Frequency of each order

As can be seen from Fig. 6, the first order natural frequency of CPM is about 44.327 Hz, which is much larger than design requirement, namely 20 Hz. Moreover, comparison of each set of frequencies before and after the optimization is made as shown in Fig. 7, from which we can see that after optimization all sets of frequencies have grown much.

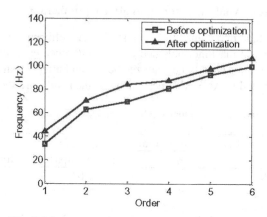

Fig. 7. Mode comparison before and after optimization

Seen from Fig. 7, frequency of each order has increased after optimization. The first order frequency increased from 33.534Hz to 44.327Hz by 32.2%, which proves the optimization result satisfactory. Therefore stiffness, capacity and precision of the machine tool are optimized.

6   Conclusion

A novel 6-DOF cooperative parallel manipulator for high precision machining is presented. The mechanism's kinematic architecture, 3-PSC/PU PM and 3-PU*R PM, is firstly deliberated, which lay the foundation for the stiffness analysis. Then through the establishment of global Jacobian, the stiffness of 3-PSC/PU PM and 3-PU*R PM has been calculated and both have been proved to possess satisfactory mass-stiffness ratio. With the concept of radius ratio, the math model of the CPM's first order of natural frequency optimization is established so that optimal radius of drive rod can be achieved. Comparing the results before and after optimization, it can be seen that results are obviously improved. Further research will be made into the mechanism's dynamic modeling and control, which remains a tough work to finish.

References

[1] Chang, K.S., Holmberg, R., Khatib, O.: The augmented object model: Cooperative manipulation and parallel mechanism dynamics. In: Proceedings of the IEEE International Conference on Robotics and Automation, ICRA 2000, vol. 1, pp. 470–475 (2000)

[2] Liu, Y.H., Xu, Y., Bergerman, M.: Cooperation control of multiple manipulators with passive joints. IEEE Transactions on Robotics and Automation 15(2), 258–267 (1999)

[3] Romdhane, L.: Design and analysis of a hybrid serial-parallel manipulator. Mechanism and Machine Theory 34(7), 1037–1055 (1999)

[4] Waldron, K.J., Raghavan, M., Roth, B.: Kinematics of a hybrid series-parallel manipulation system. Journal of Dynamic Systems, Measurement, and Control 111(2), 211–221 (1989)

[5] Waldron, K.J., Hunt, K.H.: Series-parallel dualities in actively coordinated mechanisms. The International Journal of Robotics Research 10(5), 473–480 (1991)

[6] Shahinpoor, M.: Kinematics of a parallel-serial (Hybrid) manipulator. Journal of Robotic Systems 9(1), 17–36 (1992)

[7] Sklar, M., Tesar, D.: Dynamic analysis of hybrid serial manipulator systems containing parallel modules. Journal of Mechanisms, Transmissions and Automation in Design 110(2), 109–115 (1988)

[8] Huang, H., Deng, Z., Li, B., Liu, R.: Analysis and synthesis of a kind of mobility reconfigurable robot with multi-task capability. Journal of Advanced Mechanical Design, Systems, and Manufacturing 5(2), 87–102 (2011)

[9] Schneider, S.A., Cannon, J. R.H.: Object impedance control for cooperative manipulation: Theory and experimental results. IEEE Transactions on Robotics and Automation 8(3), 383–394 (1992)

[10] Callegari, M., Suardi, A.: Hybrid kinematic machines for cooperative assembly tasks. In: Proc. Intl. Workshop: Multiagent Robotic Systems: Trends and Industrial Applications, Robocup (2003)

[11] Caccavale, F., Chiaverini, S., Natale, C.: Geometrically consistent impedance control for dual-robot manipulation. In: Proceedings of the IEEE International Conference on Robotics and Automation, ICRA 2000, vol. 4, pp. 3873–3878. IEEE (2000)

[12] Surdilovic, D., Grassini, F., De Bartolomei, M.: Synthesis of impedance control for complex co-operating robot assembly task. In: Proceedings of the 2001 IEEE/ASME International Conference on Advanced Intelligent Mechatronics, vol. 2, pp. 1181–1186. IEEE (2001)

[13] Huang, Z., Li, Q.: Type synthesis of symmetrical lower-mobility parallel mechanisms using the constraint-synthesis method. The International Journal of Robotics Research 22(1), 59–79 (2003)

Experimental Characterization of Self-excited Vibration of a 3-RRR Parallel Robot

Sheng Liu, Zhicheng Qiu, Jiasi Mo, and Xianmin Zhang[*]

Guangdong Provincial Key Laboratory of Precision Equipments and Manufacturing
Technology, South China University of Technology
510640, Guangzhou, Guangdong, China
liusheng20080818@163.com,
{zhchqiu,mo.jiasi,zhangxm}@scut.edu.cn

Abstract. A kind of 3-RRR parallel robot is described. The dynamics model is analyzed. Experiment is performed to characterize the self-excited vibration of a 3-RRR parallel robot. The measured data is analyzed to describe the dynamics characteristics of the parallel robot system. It is found that the hysteresis is obviously related to the displacements and driving torques of the corresponding actuated joints. The hysteresis, which is caused by the friction and backlash of the speed reducer gear existed in the mechanism, could describe the positioning nonlinearity.

Keywords: 3-RRR parallel robot, Electromechanical coupling, Self-excited vibration, Equivalent hysteresis nonlinear.

1 Introduction

Self-excited vibrations occur in numerous physical systems, such as chatter in metal cutting machine tools, galloping of transmission lines, bending vibrations of drill strings and the rotor motion. There are different reasons for the initiation of self-excited vibrations. Some literatures hold the view that the reason is either a negative damping or cross-stiffness.

The machine oscillation emerging at the tool centre point may influence the accuracy negatively. Brecher gives a new possible method to compensate the machine oscillation by using a controller-integrated compensation principle [1]. The active and adaptive compensation strategies are used to compensate the oscillations at the tool centre point of the machines with parallel kinematics. Belhaq and Mohamed study the interaction effect of a vertical fast harmonic parametric excitation and time-delay state feedback on self-excited vibration in a Van der Pol oscillator [2]. Friction-induced self-excited vibration is very common in mechanical and electromechanical systems [3]. Many active control techniques are developed to improve the control performances in terms of eliminating the limit cycle and the steady-state error in the friction-induced self-excited vibration systems [4]. Sinou investigates the effects of damping and nonlinearities on flutter instability and the associated limit cycles using a minimal two-degrees-of-freedom model [5]. The influence of nonlinear parameters

X. Zhang et al. (Eds.): ICIRA 2014, Part II, LNAI 8918, pp. 72–80, 2014.

could be analyzed by state trajectory, Poincare maps, frequency spectra and bifurcation diagrams for investigating the non-linear dynamic behaviors of self-excited vibration systems [6]. Recently, many researchers have investigated time-delay state feedback control of vibration [7]. Cai and Huang investigate instantaneous optimal control method for vibration control of linear sampled-data systems with time-delay [8]. The effects of time-delay on the self-excited oscillation of single and two degrees-of-freedom systems under nonlinear feedback have been studied [9].

However, few of researches are related to the self-excited vibration of parallel robots. The previous researches only focus on the field of kinematics and dynamics [10-12]. For 3-RRR parallel robot, the self-excited vibration is a complex and nonlinear physical phenomenon with some uncertain-ties.

The controlled positioning system is an integral part of 3-RRR parallel robot. A large number of nonlinear factors cause high level of undesirable positioning inaccuracies by inducing self-excited oscillations around the desired position. Self-excited vibrations generate additional dynamic loads degrading the system performances, sometimes even causing breakdown of the mechanical structure. It is meaningful to analyze the principle of self-excited vibration in 3-RRR parallel robot, which is beneficial to reducing or eliminating the vibration and improving the performance of the 3-RRR parallel robot.

The rest of this paper is structured as follows: In section 2, the kinematic model and the dynamic model of a 3-RRR parallel robot are discussed. In section 3, the experimental characterization of the self-excited vibration is performed. Through section 3, the principle of self-excited vibration in 3-RRR parallel robot is analyzed. Finally, the conclusions are presented in Section 4.

2   Model of the 3-RRR Planar Parallel Robot

The experiment platform is a planar parallel robot, which is composed by a mobile platform and three RRR serial chains that join it to a fixed base. Each RRR chain is a serial chain composed of three rotational joints. The geometrical configuration is shown in Fig.1.

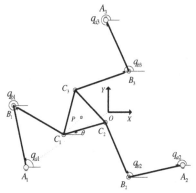

Fig. 1. Structure diagram of 3-RRR parallel robot

The parallel robot is actuated by three alternating current servo motors located at the base A_1, A_2, and A_3. P(x, y) is the position of the mobile platform in the plane and θ is the orientation. The rotational articulations of each limb are defined by A_i, B_i, C_i, (i=1, 2, 3). Points A_i are actuated by the actuators fixed to the base. The actuated joint vector is defined as $q_a=[q_{a1}\ q_{a2}\ q_{a3}]^T$. The unactuated joint vector is defined as $q_b=[q_{b1}\ q_{b2}\ q_{b3}]^T$, while $x=[x\ y\ \theta]^T$ defines the task coordinates vector.

Closure loop equation is the term used in mechanics referring to the geometric equation associated to a kinematic loop. Based on Fig.1, the two vectorial closure loop equations can be expressed as follows.

$$\begin{cases} \overrightarrow{OA_1} + \overrightarrow{A_1B_1} + \overrightarrow{B_1C_1} - \overrightarrow{C_1C_2} - \overrightarrow{C_2B_2} - \overrightarrow{B_2A_2} - \overrightarrow{A_2O} = 0 \\ \overrightarrow{OA_1} + \overrightarrow{A_1B_1} + \overrightarrow{B_1C_1} - \overrightarrow{C_1C_3} - \overrightarrow{C_3B_3} - \overrightarrow{B_3A_3} - \overrightarrow{A_3O} = 0 \end{cases} \tag{1}$$

Eq (1) could be divided into an x-axis component and a y-axis component, the closed-loop constraint equations of the parallel robot can be written as

$$W(q) = 0 \tag{2}$$

Differentiating the kinematic constraint equations with respect to time, the link Jacobian matrix can be calculated.

$$\frac{dW(q)}{dt} = \frac{\partial W(q)}{\partial q}\dot{q} = S(q)\dot{q} = 0 \tag{3}$$

where $S(q)$ is constraint matrix of the joint velocity.

The acceleration Jacobian matrix is determined by differentiating the velocity constraint equations with respect to time. The joint acceleration can be calculated as

$$\frac{d}{dt}(S\dot{q}) = S\ddot{q} + \dot{S}\dot{q} = 0 \tag{4}$$

The open chain dynamic equation can be obtained by applying the Lagrange Multiplier formulation, with which the dynamic model is expressed as combination of the model of the open-chain system with the closed-loop constrains, namely

$$M\ddot{q} + C\dot{q} = \tau + S^T\lambda \tag{5}$$

where M is inertia matrix in the joint space, C is Coriolis and centrifugal force matrix in the joint space, $\dot{q}$ and $\ddot{q}$ are the joint velocity and acceleration, τ are the actuator torque vector, $S^T\lambda$ represents the constraint force, λ is a multiplier representing the magnitude of the constraint force, matrix S is constraint matrix of the joint velocity. Differentiating the closed-loop constraint equations of the parallel robot yields

$$S\dot{q} = SJ\dot{q}_x = 0, \forall \dot{q}_x \in \mathbf{R}^3 \tag{6}$$

The constraint force $S^T\lambda$ is an unknown term and can be eliminated by the expression for the null-space of the matrix S, Considering the constraint equation

$SJ\dot{q}_x = 0$, the equation $SJ = 0$, or $J^T S^T = 0$ equivalently can be obtained. Thus, the term of $S^T \lambda$ can be eliminated, and the dynamic model in the joint space can be written as

$$J^T M J \ddot{q}_x + J^T \left(M \dot{J} + C J \right) \dot{q}_x = J^T \tau \tag{7}$$

Where $q_x = [x \ y \ \theta]^T$ is the position coordinates of the moving platform, the elements of vector τ represent the actuator torque vector of the actuated joints, $J^T M J$ is the inertia matrix in the task space, and $J^T \left(M \dot{J} + C J \right)$ is the Coriolis and centrifugal force matrix in the task space, J is the velocity Jacobian matrix between the moving platform and the actuated joints.

The above-mentioned dynamics model is a common dynamics model. It is not including the existed nonlinear factors in the practical parallel mechanism. In the investigated experimental setup, three speed reducers are connected the three AC servomotors respectively. Thus, the nonlinear factors exist in the dynamic model of the practical system, including nonlinear friction torque and the backlash of each speed reducer's gear transmission. The hysteresis phenomenon will be caused due to the mentioned nonlinear factors. In this paper, we mainly provide the experimental results of the caused self-excited vibration. The further theoretical analyses will be conducted subsequently, including building the dynamics model comprising the nonlinear factor and the mechanism analysis on self-excited vibration.

3   Experimental Characterization of Self-excited Vibration

The 3-<u>R</u>RR parallel robot is mainly composed of 3-<u>R</u>RR mechanism, AC servo system and Dspace1103 physical simulation platform. The overview of the experiment setup is shown in Fig.2.

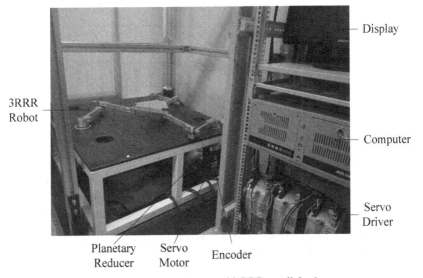

Fig. 2. Experimental setup of 3-<u>R</u>RR parallel robot

The host computer is an industrial computer for off-line programming. Three AC servomotors with rotary encoders provide the angular motion of the actuated joints. They have low ratio speed reducers. High-resolution encoders with 2^{20} counts per revolution, give the angular displacement of the actuated joints. Three amplifiers drive the servomotors. The software implemented in the Dspace1103 Controller is a multitasking application, which includes tasks such as data acquisition from encoders and computation of complex algorithm for trajectory generation. The control system structure is shown in Fig.3.

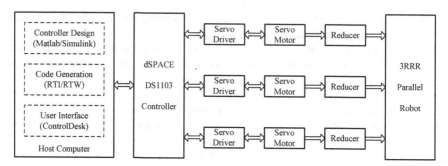

Fig. 3. Control system structure of 3-RRR parallel robot

3.1   The Positioning Self-oscillation

The phenomenon of self-excited vibration of 3-RRR parallel robot during point-to-point motion could be validated by the experiment. The velocity and acceleration are programmed by law of positive rotation during the point-to-point movement. When the 3-RRR parallel robot arrives at the target point, its acceleration is not zero and it saves energy. Then, it enters into a state of self-excited vibration.

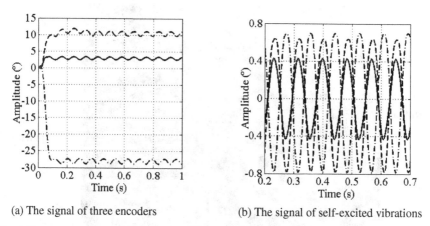

(a) The signal of three encoders   (b) The signal of self-excited vibrations

Fig. 4. The time-domain signal of self-excited vibration The joint1 is showed by the solid line, the joint2 is showed by the dashed line, the joint3 is showed by the chain-dotted line.

The target point (x, y) is (0.1m, 0.1m). The signal of three incremental rotary encoders could be seen in Fig.4. The whole moving process of three actuated joints could be seen in Fig.4 (a). The parallel robot is in a state of self-excited vibration after running the command signal, see Fig.4 (b).

A set of experimental data is presented. The desired trajectory is considering the following time dependent functions for the Cartesian coordinate, which is expressed as

$$\begin{cases} x(t) = x_d \cdot [0.5 - 0.5\cos(20\pi \cdot t)] & (0 \le t \le 0.05s) \\ y(t) = y_d \cdot [0.5 - 0.5\cos(20\pi \cdot t)] & (0 \le t \le 0.05s) \\ \theta(t) = 0 & (0 \le t \le 0.05s) \end{cases} \tag{8}$$

The phase plane method can be used to analyze second-order autonomous systems with nonlinearity and is appropriate to study the self-excited vibration occurring in second-order systems. Fig.5-7 depicts the limit cycles of three joints. Fast Fourier transform (FFT) with 1024 points and a sampling rate of 1000 is applied to the time signal. The frequency of self-excited vibration is 12.21Hz.

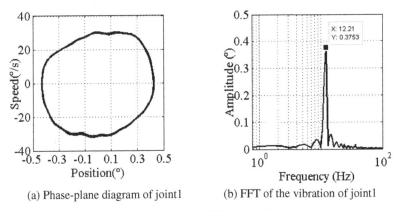

(a) Phase-plane diagram of joint1   (b) FFT of the vibration of joint1

Fig. 5. Phase-plane diagram and FFT of the vibration of joint1

The limit cycle of joint1 can be expressed as the following form.

$$\dot{x}^2 / 32.21^2 + x^2 / 0.42^2 = 1 \tag{9}$$

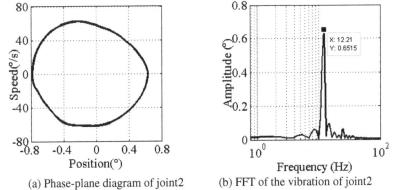

(a) Phase-plane diagram of joint2   (b) FFT of the vibration of joint2

Fig. 6. Phase-plane diagram and FFT of the vibration of joint2

The limit cycle of joint2 can be expressed as the following form.

$$\dot{x}^2 / 60.58^2 + x^2 / 0.79^2 = 1 \tag{10}$$

The limit cycle of joint3 can be expressed as the following form.

$$\dot{x}^2 / 59.81^2 + x^2 / 0.78^2 = 1 \tag{11}$$

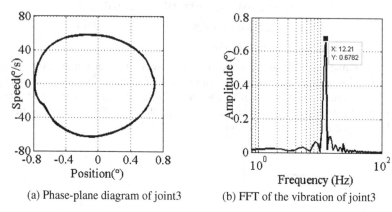

(a) Phase-plane diagram of joint3   (b) FFT of the vibration of joint3

Fig. 7. Phase-plane diagram and FFT of the vibration of joint3

Many target points are chosen in the experiment, and the experiment results show that the frequency of the self-excited vibration is constant and the amplitude of the self-excited vibration has nothing to do with the initial disturbance. The self-excited vibration of the 3-RRR parallel robot is a sympathetic resonant vibration of mechanical system and AC servo motor system.

3.2   The Characteristics of the Mechatronics Servo System

The closed-loop position control system of servo drivers, which is as shown in Fig.8, allows adjustment in response to changes of displacement during self-excited vibration.

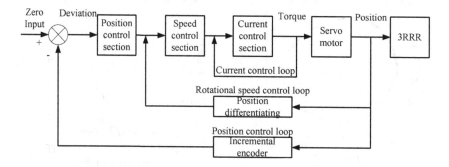

Fig. 8. Block diagram of the servo motor control system

For the parallel robot, the actuated joint angles can be measured directly with the incremental encoders, while the angular velocity and acceleration cannot be measured directly. Sometimes, the acceleration signals are calculated by signal differentiation method, in which the true acceleration signals may be submerged by large noise. In this paper, incremental encoders are installed to measure the actuated joint angles. The joint velocity can be estimated by the numerical difference algorithm with filter. Acceleration sensors are provided to measure the acceleration of links and the moving platform.

According to the displacement, velocity and acceleration, the actual torque actuated by servo motor could be calculated by the dynamic model established in section two. The dynamic characteristic of the self-excited vibration is obtained. The actuated joint angle and the actual torque are plotted in one figure, the input-output characteristic of the electromechanical position servo system could be obtained, as shown in Fig.9. From the measured torque and angle position, one can know that the hysteretic nonlinear existed in the practical system, which may cause the self-excited vibration.

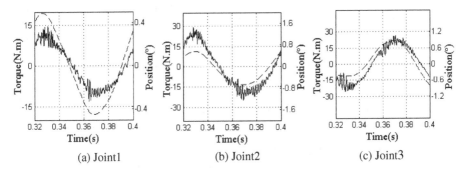

(a) Joint1   (b) Joint2   (c) Joint3

Fig. 9. Displacement and torque of three actuated joints The torque is showed by the solid line, the displacement is showed by the dashed line.

The friction and backlash in the planetary reducer, joint clearance and the pole pieces of the elastic deformation lead to the hysteresis in electromechanical position servo system, which may cause the self-excited vibration.

4   Conclusion

In this paper, laboratory experiment is performed to characterize the self-excited vibration of 3-RRR parallel robot. Experimental results show that the frequency of self-excited vibration is a constant in different configuration, and the joint driving torque lags behind joint displacement. Since the 3-RRR parallel robot isolated from the servo motor system should be classified into forced vibration category, the influence of nonlinear factors of the parallel mechanism could cause the self-excited vibration.

Acknowledgments. This research was supported by the National Natural Science Foundation of China (Grant No. 91223201), the Natural Science Foundation of Guangdong Province (Grant No. S2013030013355), Project GDUPS (2010), and the Fundamental Research Funds for the Central Universities (2012ZP0004). These supports are greatly acknowledged.

References

1. Brecher, C., Weck, M., Yamasaki, T.: Controller-integrated predictive oscillation compensation for machine tools with parallel kinematics. International Journal of Machine Tools and Manufacture 46(2), 142–150 (2006)
2. Belhaq, M., Mohamed Sah, S.: Fast parametrically excited van der Pol oscillator with time delay state feedback. Int. J. Nonlinear Mech. 43(2), 124–130 (2008)
3. Chatterjee, S.: Non-linear control of friction-induced self-excited vibration. Int. J. Nonlinear Mech. 42(3), 459–469 (2007)
4. Wang, Y.F., Wang, D.H., Chai, T.Y.: Active control of friction-induced self-excited vibration using adaptive fuzzy systems. J. Sound Vib. 330(17), 4201–4210 (2011)
5. Sinou, J.J., Jézéquel, L.: The influence of damping on the limit cycles for a self-exciting mechanism. J. Sound Vib. 304(3-5), 875–893 (2007)
6. Cheng, M., Meng, G., Jing, J.: Numerical and experimental study of a rotor–bearing–seal system. Mech. Mach. Theory 42(8), 1043–1057 (2007)
7. Chatterjee, S.: Time-delayed feedback control of friction-induced instability. Int. J. Nonlinear Mech. 42(9), 1127–1143 (2007)
8. Cai, G., Huang, J.: Instantaneous optimal method for vibration control of linear sampled-data systems with time delay in control. J. Sound Vib. 262(5), 1057–1071 (2003)
9. Chatterjee, S.: Self-excited oscillation under nonlinear feedback with time-delay. J. Sound Vib. 330(9), 1860–1876 (2011)
10. Zhang, X., Zhang, X., Chen, Z.: Dynamic analysis of a 3-RRR parallel mechanism with multiple clearance joints. Mech. Mach. Theory 78, 105–115 (2014)
11. Zubizarreta, A., Marcos, M., Cabanes, I., Pinto, C., Portillo, E.: Redundant sensor based control of the 3RRR parallel robot. Mech. Mach. Theory 54, 1–17 (2012)
12. Noshadi, A., Mailah, M., Zolfagharian, A.: Intelligent active force control of a 3-RRR parallel manipulator incorporating fuzzy resolved acceleration control. Appl. Math. Model. 36(6), 2370–2383 (2012)

Design of Less-Input More-Output Parallel Mechanisms

Huiping Shen, Xiaorong Zhu, Lifang Dai, and Jiaming Deng

School of Mechanical Engineering, Changzhou University, Changzhou 213016, China
shp65@126.com

Abstract. An important concept of less-input more-output parallel mechanisms (Li-Mo PMs) is proposed based on the research results achieved by the authors on less-input three-output parallel vibrating screen and considering the status of increasing application of PMs and development demands of the emerging technology. First, the correlations between the number of input (w), degrees of freedom (dof) and numbers of output position and orientation (N_{out}) of this kind of Li-Mo PMs are further described. Secondly, the research and application of the single-input three-output parallel vibrating screen developed by the authors are illustrated. Sequentially, conceptual designs of two kind rehabilitation training devices for shoulder joint based on single-input three-rotation output PMs are presented. Li-Mo PMs has important theoretical and extensive practical value for novel industrial equipment design and is expected to be an important research direction in the field of PMs.

Keywords: Parallel Mechanisms, Less Input, More Expected Output, Parallel Equipment.

1 Research Background

1.1 The Li-Mo PMs Existing in Industrial Fields

With the development of industrial technology, it is eager to develop more novel mechanisms, which possess more spatial motion under the least input, in order to achieve the target of simple structure, easy manufacture, effective and reliable run, convenient maintenance, and energy conservation, etc. Therefore, research and discovery of parallel mechanisms with less input and more dimensional expected output (Li-Mo PMs, for short) is one of the most effective technologies to achieve this target.

First, there has been a large number of such devices in industrial fields, which can achieve multi-output but just with one or two inputs. For example, the compound pendulum jaw crusher with single input is widely used in the industry of mining, building, road and so on. A schematic of the mechanism is shown in Figure 1. If link BC is selected as the output components (moving jaw), it has two planar compound motions (one translation and one rotation). That is , any point S on the moving jawt can translate along X-axis, Y-axis, and rotate around Z-axis when the crank AB is choosen as one input. The materials in the crushing cavity have widely range motion to roll from up to down, which contributes to uniform segments. The productivity is

X. Zhang et al. (Eds.): ICIRA 2014, Part II, LNAI 8918, pp. 81–88, 2014.

increased by 20%-30% compared with that of a simple pendulum jaw crusher (single output, and rocker CD as a moving jaw) with the same parameters. At the same time, the mechanism has the advantage of lower weight, less components, and more compact structure, etc. Therefore, the conventional simple pendulum jaw crushers have been basically replaced recently by the compound pendulum jaw crushers [1].

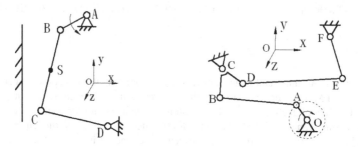

Fig. 1. 1-dof compound pendulum jaw crusher   **Fig. 2.** 1-dof linkage vibrating screen

Another example, the coupler type vibrating screen with single or double deck is widely used in the industry, e.g. metallurgic, coal, cereal and oil, chemical and so on. A schematic of the mechanism is shown in Figure 2. When the crank OA is selected as a one rotational input, the screen box mounted with link AB and link DE has two-planar compound motions (one translation and one rotation). Three output components of the screen box, including translate on the XY direction and rotate around the Z-axis, can provide the grains with larger tumbling up and down, and flow from the entrance to the discharge end. Therefore, it can improve the screening efficiency, and has the advantages of low power loss, easy operating, and small in size [2].

Further, for the multi-*dof* parallel mechanisms, research on the relationship between driving input and output motion has more important theorical and application value. For example, the 3-RPS parallel mechanism is a typical one with three degrees of freedom, where three prismatic joints(P) are actuaced. When the axes of three revolute joints attached to the base are parallel to a same plane, the moving platform will produce five output motions, including both translation and rotation, but there are four output motions if their axes are not parallel to a same plane. However, for both cases, there are just only 3-dimensional independent motion, the others are non-independent derived or parasitic ones [3].

1.2   The Needs of Emerging Industries

With the development of the strategic emerging industry technology, rehabilitation robot and rehabilitation exercises aids for the elderly and the disabled have received great attention. However, currently, the rehabilitation training equipments are made of whether parallel or serial structure, suffering some drawbacks such as: (a) complex structure, and (b) more actuators. These deficiencies make these equipments not suitable for some rehabilitating tasks.

In addition, some novel equipments, which are adapted to site disposal of nuclear contamination under complex dangerous enviroment, and to clean slag at high temperature and large sewage pool, as well as to swing for entertainment, all these are intended to obtain more spatial motion with the least inputs, to realize simple structure, convience for manufuring, green operation, energy-saving and emission-reduction, etc[4]

From the above, we can get the following research results: using less inputs can produce more spatial motion ($dof+\triangle$) than the number of input(dof), meanwhile can achieve the desired performance. Based on this kind of research strategy, PMs can be classified into following two types according to the different demands in application, which is helpful to expand PMs applications.

2   Classification of PMs and Definition of Li-Mo PMs

2.1   Classification of PMs and Description of the Design Problem

On one hand, many studies have been carried out extensively by numerous researchers both on theory and on application of PMs, with the focus on multi-axes NC machine tool, pick-and-place application and assembly automation. Each output motions of these equipments must be independent and controled precisely, which means that the number of dependent output motion equals the degrees of freedom . It is defined in this paper as: Type I PMs, namely, "multi input - equal dimensional accurate output ". There has been much research on this type of PMs at parallel mechanisms community for many years [5-8].

On the other hand, some traditional equipments, such as large vibrating screen, crusher, slag cleaning machine and motion simulator, etc, are characterized by high power consumption, high wastage, noise and poor operating environment. Nevertheless, some emerging industrial technology, such as rehabilitation training machine and nuclear contamination cleaning machine, have the features of special service enviroment, high performance-price ratio, energy-saving and environmental protection, etc, which require reducing the number of components, raising efficiency, reducing consumption and noise, and longer serving life. From the view of output, the end-effector of such equipments is mostly desired to produce more output motion, including both independent and non-independent motions, in order to ensure the demands of spatial multi-operating. Among these motions, the parasitic non-independent components is required to participate corresponding operability outside together with the independent motions. However, each output motion component is not independent necessarily. For such equipments, the requirements for the number of output motions is higher and more rigid than the requirement for accuracy. At the same time, the number of inputs should be as less as possible, which will facilitate simple structure and be convenient for production, processing and maintenance, in order to gain the desired working reliability and to achieve the purpose of energy consumption reduction. All these can be realized fully by utilizing the advantages of moving platform of PMs, which can produce spatial compound motion. Therefor, this kind of PMs is here defined as: Type II PMs, namely, "Less input – More output desired with non-independent motions ", which is abbreviated as Li-Mo PMs. However, the related systematic topological structure design method in this family have never been reported at parallel mechanisms community so far.

2.2   The Relationship among PMs Type and Number of Outputs and Inputs

Assumed respectively that the number of input is w, degrees of freedom is dof, and the number of output position and orientation is N_{out}. According to the relationship between w with dof, there are three cases to consider:

The first case arises when $w=dof=N_{out}$, named accurate input and accurate output. The first type of PMs mentioned above "multi input - equal dimensional accurate output " belongs to this category, which has been widely used in actual industrial application.

The second case occurs when $w>dof$, named redundant input. The redundant input is now a very effective method which is used to reduce or eliminate singularities and improve the performance of PMs[9]. However, the parallel redundant mechanism is much more complex than corresponding non-redundant PMs. Meanwhile, it is hard to control because of the multi input - multi output system with strong I-O coupling and nonlinear dynamic properties.

The third case, named as underactuated input, occurs when $w<dof$, e.g. the system has less input than the dof. A coupling between the actuated input joint and underactuated joint is available, thus it is possible to control the position of robots or equipments via the dynamics coupling[10]. Because of the decrease of actuated input joint, it has the merits of compact configuration, reducing weight and costs and low energy consumption and soon.

At present, almost all existing literatures on PMs show that researches have being done mainly on the relationship between w and dof. But few literatures have involved the relationship among w, dof and N_{out}.

Here needs to be emphasized: The topological structure design method and application of Type II PMs mentioned in this paper are executed under the condition of considering the relationship among w, dof and N_{out}, and satifying the relation of $w=dof < N_{out}$. Obviously, this type of PMs belongs to neither redundant input nor underactuated input PMs, but expansion of accurate input. That is to say, less dimensional input (dof) will be adopted to realize more output motions($dof+\Delta$), where, $N_{out}=dof+\Delta$, and $2\leq N_{out}\leq 6$, normally $1\leq w=dof\leq 5$, and here $\Delta=1\sim 5$. These mechanisms work under making full use of the non-independent parasitic motions. Thus, because of the less number of actuating input, the device has such advantages as the simple structure and reliable work, easy manufacturing, usage and maintenance. The 3-"dimension" parallel vibrating sieve with single input, which has been proposed by the authors and has been industrial trial runned[11] , is the best proof.

Here, it deserves to be specially noted that,

1) The "less input" in this paper implies that the number of dof of PMs is less than that number of output motions of the platform ($dof+\Delta$). However, the "lower-freedom parallel mechanisms" named by mechanism community is compared with 6-dof mechanisms, whose 2~5 dimensional inputs produce the same dimensional outputs.

2) The Type II PMs mentioned above, i.e., "Li-Mo PMs" does not refer to increasing the additional transmission lines installed the inside of the machine. That is to say, it does not divide one input into multi transmission output. Actually, it means that the mechanism has *dof* inputs and (*dof*+△) outputs, and it satifies the relation as follows: $2 \leqq dof+\triangle \leqq 6$, *dof*=1~ 5 and △=1~ 5. Therefore, such PMs can produce 2~6 output motions, including the non-independent output components, even though it is driven by only single input.

3   Design for 3-Output PMs with Single Input

3.1   Study on 3D Parallel Vibration Sieve with Single Input

We have studied several kinds of parallel vibration sieve with 1- or 2-*dof* input since we originally proposed the concept of parallel vibration sieve in 2006[12-13]. In the Figure 3, the single input parallel vibration sieve is illustrated, including the schematic representation, photo of prototype and application example.

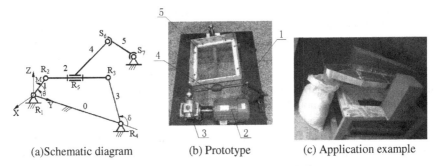

(a)Schematic diagram   (b) Prototype   (c) Application example

Fig. 3. 3D parallel vibration sieve with single input

As shown in Figure 3(a), when the crank 1 is chosed as input link, vibrating screen box fixed with link 4 will produce one-translation along Z-axis and two-rotation respectively around X-axis and link 2. Here, for easy understanding,we still call it 3D parallel vibration sieve even if the three motions are not all independent. All these three output motion components make the inner grains in the vibrating screen box vibrate intensely. Among these three output motion components, only one is independent and the other two are its parasitic motions. This device has the advantages of smooth operation and lower voice beause of novel design concept and rational topological configuration construction, and it improves the efficiency greatly. A lot of experiments show that the screening efficiency and the rate of pass sieves of this parallel vibrating screen are higher by 15% and 20% than the tradtional linear vibrating screen respectively, moreover with 35% power reduction[14]. It has been successfully applied in the sieving of sulphur for rubber products.

3.2   Study on the 3-Rotation Parallel Rehabilitation Training Device for Shoulder Joint with Single Input

Recently, we proposed several kinds of less dimensional input - more output PMs, including single input - three rotatation output PMs[15,16]. As depicted in

Figure 4(a), a parall mechanism consists of three branched chains $\underline{R}SS$, S and RS; and Figure 4(b) represents another parallel mechanism composed of three branched chains $\underline{R}RS$, S and SS (the underlined means the input joint). For these two parallel mechanisms, when the crank 1 rotates fully, the moving platform 3 has three rotational angles around the spherical joints S_{31}. The rotation angles may change as the mechanism parameters change, while one of them may be fixed or equal zero by special link parameters.

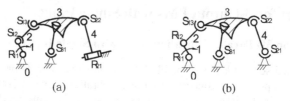

(a)   (b)

Fig. 4. Two kinds of 3-rotational parallel mechanisms with single input

These two kinds of single input 3-rotation output parallel mechanisms can be used as candidate mechanisms of rehabilitation training for shoulder. The design scheme of rehabilitation training for shoulder joint based on the type of Figure 4(b) has been put forward, and its installation instruction is shown in Figure 5(a). The moving platform 3 is connected with the upper arm, and the base is directly mounted on the upper shoulder. As indicated in Figure 5(a), given the crank 1 fully rotation, the upper arm can rotate around the shoulder (spherical joints S_0) with (α, β, γ) as three angles of rotation. Modification and optimizatoin of the mechanism parameters will control the output number and change the ranges of rotation angles to meet the rehabilitation requirements in different stage and ages. Furthermore, the mobility of the moving platform is analyzed and Figure 5(b) shows the variation of output angles of upper arm.

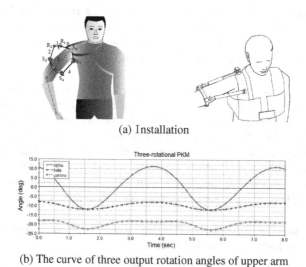

(a) Installation

(b) The curve of three output rotation angles of upper arm

Fig. 5. A rehabilitation training device for shoulder joint based on 1-*dof* 3-rotate output PMs

4   Conclusions

In this paper we tlak about the concept, definition, design examples of the Li-Mo PMs. The following conclusions have reached.

(1) A novel concept of less input-more output parallel mechanisms (Li-Mo PMs) is proposed based on the authors' previous research works. This kind of PMs has important theoretical and extensive practical value to the design of novel equipments serving for industrial development, and is expected to be an important research direction for expanding the design theory and application of PMs.

(2) This type of Li-Mo PMs involves a theoretical issue about PMs, e.g, pursuing the relationship among type and number of output motion, number of output motion and input. Essentially, Li-Mo PMs is a kind of multi outputs ($dof+\Delta$) mechanism with less inputs (dof), here, $N_{out}=dof+\Delta$, $2 \leqq N_{out} \leqq 6$, generally, $1 \leqq W=dof \leqq 5$, and $\Delta=1 \sim 5$. Because of the decrease of input joints, the device has such advantages as the simple structure and reliable working, easy manufacturing and maintenance and so on.

(3) Based on the concept of Li-Mo PMs, two kinds of novel PMs with three-output and single input have been presented. The novel 3D parallel vibrating sieve based on 1-dof 1-translation and 2-rotation output PMs has been industrial trial application. While a novel rehabilitation training device for shoulder joint based on 1-dof 3-rotate output PMs is under prototype manufacturing. They can be the best proof for the feasibility of Li-Mo PMs. Therefore, Li-Mo PMs is worthy to study for mechanism acadmic and industry community.

Acknowledgments. This project is supported by National Natural Science Foundation of China (Grant No. 51075045, 51375062, 51405039).

References

1. Luo, H.P.: Study on the Effective Space in the Crushing Cavity of the Compound Pendulum Jaw Crusher. Mining & Processing Equipment 38(5), 20–25 (2010)
2. Song, S.Z., Zhou, Z.D.: General Narration to the Development of Vibration Screening Machinery and Primary Probe to New Type Vibration Screen. Mining & Processing Equipment 34(4), 230–236 (2006)
3. Yang, T.L., Liu, A.X., Luo, Y.F., et al.: Theory and Application of Robot Mechanism Topology. Science Press, Beijing (2012)
4. China National Natural Science Fund Committee, Development Strategic Report on Mechanical Engineering (2011-2020), Beijing. Science Press (2010)
5. Huang, Z., Kong, L.F., Fang, Y.F., et al.: Theory and Control of Parallel Robot Mechanism. Machinery Industry Press, Beijing (1997)
6. Huang, T., Wang, P.F., Zhao, X.M., et al.: Design of a 4-DOF Hybrid PKM Module for Large Structural Component Assembly. CIRP Annals-Manufacturing Technology 59, 159–162 (2010)
7. Gao, F., Yang, J.L., Ge, Q.D.: Type Synthesis Parallel Mechanisms Based on G_f Set. Science Press, Beijing (2011)

8. Michael McCarthy, J.: 21st Century Kinematics (2012) 978-1-4471-4510-3
9. Qing, J.X., Li, J.F., Fang, B.: Drive Optimization of Tricept Parallel Mechanism with Redundant Actuation. Journal of Mechanical Engineering 46(5), 8–14 (2010)
10. Chen, W., Yu, Y.Q., Zhang, X.P.: A Survey on the Underactuated Robots. Machine Design and Research 21(4), 22–26 (2005)
11. Shen, H.P., Zhang, J.T., Li, J., et al.: A Novel Vibration Sieve Based on the Parallel Mechanism. In: Proceedings of 2009 IEEE 10th International Conference on Computer-aided Industrial Design and Conceptual Design, Wen Zhou, November 26-29, pp. 2328–2332 (2009)
12. Shen, H.P., Yang, T.L.: A Kind of 2-dof Spatial Parallel Mechanism for Parallel Equipment. China Patent, ZL 2006100881264 (March 2006)
13. Shen, H.P., Xue, C.Y., Zhang, J.T., et al.: A Novel PKM-Based Vibrating Sifter and its Screening Efficiency Experimental Study. In: Proceedings of 13th World Congress in Mechanism and Machine Science, Mexico, June 19-25, pp. A7–A300 (2011)
14. Wang, X.X.: Experimental study on screening performance of parallel vibrating screen. Changzhou University (2012)
15. Shen, H.P., Yu, T.Z., Huang, T., et al.: A kind of 3-dimension motion mechanism with single input. Patant: 201210307595.6 (2012)
16. Shen, H.P., Yu, T.Z., Yin, H.B., et al.: A kind of 3-dimension motion equipment with single input. Patant: 201210307592.2 (2012)

Type Synthesis Approach for 2-DOF 1T1R Parallel Mechanisms Based on POC

Ju Li, Hongbo Yin, Huiping Shen[*], Lifang Dai,
Xiaomeng Ma, and Jiaming Deng

School of Mechanical Engineering, Changzhou University, Jiangsu, 213016, China
wangju0209@163.com

Abstract. Type synthesis of 1-translational and 1-rotational (in short, 1T1R) parallel mechanisms is studied using topological structure type synthesis method based on the POC set in this paper. Firstly, a general procedure of type synthesis is given and concrete operation of each step is also given by example. Then eleven kinds of 1T1R parallel mechanisms are obtained, and 5 kinds of which are proposed for the first time. Finally, these parallel mechanisms are classified based on the structure characteristics. This type synthesis approach features few calculation rules, simple mathematics operation and definite geometrical and physical meaning, and it is applicable for both non-over-constrained and general over-constrained mechanisms.

Keywords: Position and Orientation Characteristic(POC) set, Parallel Mechanism, Type Synthesis.

1   Introduction

2-DOF parallel mechanisms feature simple structure, easy kinematic and kinetic analysis, which can be applied to automatic devices of food processing, micro-electronics assembly, etc, as well as robot manipulator. The current research of 2-DOF parallel mechanisms are mainly focused on 2T0R parallel mechanisms [1-3] and 0T2R parallel mechanisms [4-7], but fewer on 1T1R parallel mechanism. In the matter of type synthesis, Kong [8] made type synthesis of part-decoupling 2-DOF parallel mechanism with two 1T1R motor patterns based on virtual branched-chain method. Firstly, five kinds of part-decoupling 1T1R parallel mechanisms were synthesized. On this basis, five new mechanisms were synthesized by adding passive kinematic pair and branched-chain, and single motor pattern of each mechanism only needs two active joints, which is superior to the existing mechanisms. Gao[9] synthesized six mechanisms with two branches, which represented by symbolic combining form, and corresponding kinematic sketch of mechanism were given.

Type synthesis of 1T1R parallel mechanisms is studied using topological structure synthesis method based on the POC set in this paper. Eleven kinds of parallel mechanisms are synthesized, and five kinds of which are proposed for the first time. These parallel mechanisms are analyzed and classified based on the structure characteristics.

---

[*] Corresponding author.

X. Zhang et al. (Eds.): ICIRA 2014, Part II, LNAI 8918, pp. 89–99, 2014.

2   General Procedure of Type Synthesis Based on the POC Set

The theories and methods of type synthesis of parallel mechanisms based on the POC set, see references [10]. To save space, here only gives the key steps.

Step1.   Determine POC set M_{Pa} (i.e., motion output type).

Step2.   Synthesize branched chain structure.
POC equation of serial mechanism:

$$M_S = \bigcup_{i=1}^{m} M_{J_i} = \bigcup M_{S_j} . \tag{1}$$

Where M_S is POC set of the end link, M_{J_i} is POC set of the i th kinematic joint, M_{S_j} is POC set of the j th sub-SOC。

POC equation of the parallel mechanism:

$$M_{Pa} = \bigcap_{j=1}^{v+1} M_{b_j} . \tag{2}$$

Where M_{Pa} is POC set of the moving platform of parallel mechanism, M_{b_j} is POC set of the end link of the j th branched chain.

Based on the equation (1)、 (2), topological structures of mechanisms with simple branches(SOC) and complex branches (HSOC) can be designed.

Step3.   Determine combining scheme of branched chains.

Step4.   Determine geometric condition of branch assembly between the moving and base platform.

Based on the POC equation (2) and rationale of branch assembly between the moving and base platform, design the geometric condition of branch assembly between the two platform.

Step5.   Verify DOF of the mechanism
According to DOF formula:

$$\begin{cases} F = \sum_{i=1}^{m} f_i - \sum_{j=1}^{v} \xi_{L_j} \\ \xi_{L_j} = \dim.\{(\bigcap_{i=1}^{j} M_{b_i} \cup M_{b_{(j+1)}})\} \end{cases} . \tag{3}$$

Verify whether DOF of the mechanism satisfies the design requirements.

Step6.   Determine passive kinematic pair of the mechanism
Determine the passive kinematic pair of mechanism based on the judging criteria of passive kinematic pair.

Step7.   Determine active joints of the parallel mechanism.

Determine the active joints of parallel mechanism based on the judging criteria of active joints.

Step8. Determine topological structure of the parallel mechanism.

Including: (1) Topological structure of the branched chains; (2) Topological structure of the moving platform; (3) Topological structure of the base platform; (4) Position of the active joints

Step9. Characteristic analysis of topological structure type of parallel mechanism

Including: (1) Type and coupling-degree of the basic kinematic chain(BKC); (2) DOF type; (3) I-O Decoupling; (4) Other features of topological structure.

3   Type Synthesis for 1T1R Parallel Mechanism

Step1. Determine POC set M_{Pa} of the expected mechanism.

$$M_{pa} = \begin{bmatrix} t^1 \\ r^1 \end{bmatrix}$$

Where t^1 is a finite translation, r^1 is a finite rotation.

Step2. Synthesize branched-chains structure.

(1) POC set of branches

According to equation(2), POC set of parallel mechanism is the intersection of POC sets of all branches. Hence, POC set of branches can be represented as follows.

$$M_{b_i} = \begin{bmatrix} t^1 \\ r^1 \end{bmatrix}, \begin{bmatrix} t^2 \\ r^1 \end{bmatrix}, \begin{bmatrix} t^1 \\ r^2 \end{bmatrix}, \begin{bmatrix} t^3 \\ r^1 \end{bmatrix}, \begin{bmatrix} t^2 \\ r^2 \end{bmatrix}, \begin{bmatrix} t^1 \\ r^3 \end{bmatrix}, \begin{bmatrix} t^3 \\ r^2 \end{bmatrix}, \begin{bmatrix} t^2 \\ r^3 \end{bmatrix}, \begin{bmatrix} t^3 \\ r^3 \end{bmatrix}$$

(2) Structure type of branches

The structure types of SOC and HSOC that satisfy the above POC set are shown in table 1.

Step3. Determine the combining scheme of branches.

Based on the structure types of SOC and HSOC in Table 1, synthesize the combining scheme of branches shown in Table 2.

Taking the combining scheme C-1 shown in Table 2 as example, synthesize a 1T1R parallel mechanism. The combining scheme of branch C-1 is shown as follows.

Branch 1: $SOC\{-P_{11} \perp R_{12} -\}$

Branch 2: $SOC\{-P_{21} \perp R_{22} \parallel R_{23} -\}$

Table 1. Structure type of SOC and HSOC

M_{b_i}		SOC		HSOC
$\begin{bmatrix} t^1 \\ r^1 \end{bmatrix}$	1	$SOC\{-P \perp R-\}$	14	$HSOC\{-P^{(6R)}-R-\}$
$\begin{bmatrix} t^2 \\ r^1 \end{bmatrix}$	2	$SOC\{-R\|R\|R-\}$	15	$HSOC\{-\Diamond\left(P^{(4-4R)},P^{(4-4R)}\right)$ $-R-\}$
	3	$SOC\{-P \perp R\|R-\}$	16	$HSOC\{-\Diamond\left(P^{(3R-2P)},P^{(3R-2P)}\right)$ $-R-\}$
$\begin{bmatrix} t^1 \\ r^2 \end{bmatrix}$	4	$SOC\{-P-\overset{\frown}{RR}-\}$		
$\begin{bmatrix} t^3 \\ r^1 \end{bmatrix}$	5	$SOC\{-P-R\|R\|R-\}$	17	$HSOC\{-R\|R\left(-P^{(4R)}\right)\|R-\}$
$\begin{bmatrix} t^2 \\ r^2 \end{bmatrix}$	6	$SOC\{-P\perp R\|R\perp R-\}$		
$\begin{bmatrix} t^1 \\ r^3 \end{bmatrix}$	7	$SOC\{-P-\overset{\frown}{RRR}-\}$		
	8	$SOC\{-P-S-\}$		
$\begin{bmatrix} t^3 \\ r^2 \end{bmatrix}$	9	$SOC\{-P\|R\|R-R\|R-\}$		
	10	$SOC\{-P\perp R-\overset{\frown}{RRR}-\}$		
$\begin{bmatrix} t^2 \\ r^3 \end{bmatrix}$	11	$SOC\{-R\|R\|R\perp \overset{\frown}{RR}-\}$		
	12	$SOC\{-R(\perp P)\|R-\overset{\frown}{RR}-\}$		
$\begin{bmatrix} t^3 \\ r^3 \end{bmatrix}$	13	$SOC\{-S-P-S-\}$		

Table 2. Combining scheme of branched-chains

A. Structures of all branches are same	
	——
B. Structures of part branches are same	
B-1	$\cdot 1 - SOC\{-R \parallel R-\}$ $\cdot 2 - SOC\{-R \parallel R \parallel R-\}$
B-2	$\cdot 1 - SOC\{-R \perp P-\}$ $\cdot 2 - SOC\{-S-P-S-\}$
...	
C. Structures of all branches are different	
C-1	$\cdot 1 - SOC\{-P \perp R-\}$ $\cdot 1 - SOC\{-P \perp R \parallel R-\}$
C-2	$\cdot 1 - SOC\{-P \perp R-\}$ $\cdot 1 - HSOC\{-P-P^{(4R)}-R-\}$
C-3	$\cdot 1 - SOC\{-P \perp R-\}$ $\cdot 1 - SOC\{-S-P-S-\}$
C-4	$\cdot 1 - SOC\{-P-\overbrace{RRR}-\}$ $\cdot 1 - HSOC\{-P-P^{(4R)}-R-\}$
C-5	$\cdot 1 - SOC\{-R \perp P-\overbrace{RRR}-\}$ $\cdot 1 - SOC\{-P \perp R \parallel R-\}$
C-6	$\cdot 1 - HSOC\{-P^{(6R)}-R-\}$ $\cdot 1 - SOC\{-S-P-S-\}$
C-7	$\cdot 1 - HSOC\{-P^{(6R)}-R-\}$ $\cdot 1 - HSOC\{-R \parallel R(-P^{(4R)}) \parallel R-\}$
C-8	$\cdot HSOC\{-\Diamond(P^{(4-4R)}, P^{(4-4R)})-R-\}$ $\cdot 1 - SOC\{-P \perp R-\}$
C-9	$\cdot HSOC\{-\Diamond(P^{(3R-2P)}, P^{(3R-2P)})-R-\}$ $\cdot 1 - SOC\{-P \perp R-\}$
...	

Step4. Determine the geometric condition of branches assembly between the moving and base platform.

(1) Determine *POC* set of the end link.

$$\text{Branch 1}: \ M_{b1} = \begin{bmatrix} t^1\left(// P_{11}\right) \cup \left\{ t^1\left(\perp R_{12}\right)\right\} \\ r^1\left(// R_{12}\right) \end{bmatrix}$$

$$\text{Branch 2}: \ M_{b1} = \begin{bmatrix} t^2\left(\perp R_{23}\right) \\ r^1\left(// R_{23}\right) \end{bmatrix}$$

(2) Establish POC equation of the parallel mechanism.

$$M_{Pa} = \begin{bmatrix} t^1 \\ r^1 \end{bmatrix} \Leftarrow \begin{bmatrix} t^1\left(\| P_{11}\right) \cup \left\{ t^1\left(\perp R_{12}\right)\right\} \\ r^1\left(\| R_{12}\right) \end{bmatrix} \cap \begin{bmatrix} t^2\left(\perp R_{23}\right) \\ r^1\left(\| R_{23}\right) \end{bmatrix}$$

(3) Determine the geometric condition of branches assembly between the two platforms.

The intersection of POC sets of the two branches should retain one independent rotational element and one independent translational element to achieve the one-dimensional translation and one-dimensional rotation of the moving platform, i.e., the geometric condition of branches assembly is $R_{12} \| R_{22} \| R_{23}$.

(4) Draw up the kinematic sketch of mechanism.

Based on the topological structure feature of the branches and the geometric condition of branches assembly between the two platforms, draw up the kinematic sketch of mechanism, shown in Figure 1.

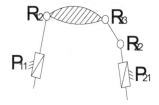

Fig. 1. $SOC\left\{-P_{11} \perp R_{12}-\right\} \oplus SOC\left\{-P_{21} \perp R_{22} \| R_{23}-\right\}$

Step5. Verify DOF of the mechanism.

(1) Determine the number ξ_{L_1} of independent displacement equations of the first independent loop.

$$\xi_{L_1} = \dim.\left\{ M_{b_1} \cup M_{b_2}\right\}$$

$$= \dim.\left\{ \begin{bmatrix} t^1\left(\| P_{11}\right) \cup \left\{ t^1\left(\perp R_{12}\right)\right\} \\ r^1\left(\| R_{12}\right) \end{bmatrix} \cup \begin{bmatrix} t^2\left(\perp R_{23}\right) \\ r^1\left(\| R_{23}\right) \end{bmatrix} \right\}$$

$$= \dim.\left\{ \begin{bmatrix} t^2 \\ r^1 \end{bmatrix} \right\} = 3$$

(2) Determine DOF of the mechanism.

$$F = \sum_{i=1}^{m} f_i - \sum_{j=1}^{v} \xi_{L_j} = 5 - 3 = 2$$

Step6. Determine passive kinematic pair of the mechanism.

Judge whether the joint R_{22} is passive kinematic pair based on the judging criteria.

(1) Suppose the joint R_{22} is fixed, then get a new mechanism. The topological structure type of branch 2 becomes $SOC\{-P_{21} \perp R_{23} -\}$, and the POC set becomes

$$M_{b1} = \begin{bmatrix} t^1 (// P_{21}) \cup \{t^1 (\perp R_{23})\} \\ r^1 (// R_{23}) \end{bmatrix}.$$

(2) Determine DOF of the new mechanism.

①Determine the number ξ_{L_1} of independent displacement equations of the first independent loop.

$$\xi_{L_1} = \dim.\{M_{b_1} \cup M_{b_2}\}$$

$$= \dim. \left\{ \begin{bmatrix} t^1 (// P_{11}) \cup \{t^1 (\perp R_{12})\} \\ r^1 (// R_{12}) \end{bmatrix} \cup \begin{bmatrix} t^1 (// P_{21}) \cup \{t^1 (\perp R_{23})\} \\ r^1 (// R_{23}) \end{bmatrix} \right\}$$

$$= \dim. \left\{ \begin{bmatrix} t^2 \\ r^1 \end{bmatrix} \right\} = 3$$

② Determine DOF of the mechanism

$$F = \sum_{i=1}^{m} f_i - \sum_{j=1}^{v} \xi_{L_j} = 4 - 3 = 1$$

(3) Judge the passive kinematic pair

DOF of the new mechanism is 1, but DOF of the original mechanism is 2, so the joint R_{22} is not passive kinematic pair.

Step7. Determine the active joints of parallel mechanism.

Judge whether joint P_{11} and P_{21} of the mechanism in figure 1 can be active joints simultaneously based on the judging criteria of active joint [10].

(1) Suppose the joint P_{11} and P_{21} are fixed, then get a new mechanism.

The topological structure of branches become $SOC_1\{-R_{12} -\}$, $SOC_2\{-R_{22} // R_{23} -\}$, i.e., planar 3R mechanism, and its DOF is 0.

(2) Judge whether joint P_{11} and P_{21} can be active joints simultaneously.

Because DOF of the new mechanism is 0, joint P_{11} and P_{21} on the same platform of the mechanism in figure 1 can be active joints simultaneously based on the judging criteria of active joint.

Step8. Determine topological structure of the parallel mechanism.

(1) Topological structure of branches

   Branch 1: $1 - SOC\{-P \perp R-\}$

   Branch 2: $1 - SOC\{-P \perp R // R-\}$

(2) Topological structure of the two platforms

   The moving platform: $R_{12} \| R_{23}$

   The fix platform: arbitrary layout

Step9 Characteristic analysis of topological structure type of parallel mechanism.

(1) Type and coupling degree of BKC

① Determine SOC_1

$$SOC\{-P_{11} \perp R_{12} \| R_{23} \| R_{22} \perp P_{21} -\}$$

② Determine constraint degree Δ_1 of SOC_1

$$\Delta_1 = \sum_{i=1}^{m} f_i - I_1 - \xi_{L_1} = 5 - 2 - 3 = 0$$

③ Determine BKC of mechanism and its coupling degree κ

According to the judgment method of BKC, this mechanism includes one BKC, and its coupling degree κ is 0.

$$\kappa = \frac{1}{2} \sum_{j=1}^{v} |\Delta_j| = 0$$

(2) DOF type

This mechanism has complete DOF based on the judging criteria of DOF type.

(3) Input-output decoupling

We can see that the mechanism has partial input-output decoupling based on the decoupling theory of topological structure.

4   Classification of Structure Type

According to the above synthetic approach, select other combining schemes in Table 2, and eleven mechanism types can be synthesized in the same way, shown in table 3. Then, according to branch type, branch complexity, decoupling and coupling degree, the mechanism in Table 3 are classified as follows to provide reference for designer.

Table 3. Topological structure of 1T-1R parallel mechanisms

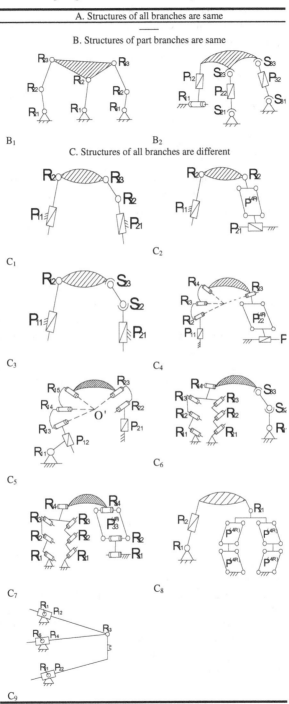

(1) Branch type

① These is no parallel mechanism that structures of all branches are same.

② These are parallel mechanisms that structures of 2 branches are same, i.e., mechanism B_1 and B_2 shown in Table 3.

③ These are parallel mechanisms that structures of all branches are different, i.e., mechanism C_1~C_9 shown in Table 3.

(2) Branch complexity

Mechanism C_2, C_4, C_8 and C_9 in Table 3 all have complex branches.

(3) I-O Decoupling

Mechanism C_1, C_2 and C_3 in Table 3 all have I-O decoupling.

(4) Kinematic (dynamic) complexity

The coupling degree of mechanism B_1 and B_2 are both 1. The coupling degree of mechanism C_1~C_9 are all 0. Obviously, the kinematic and dynamic analysis of mechanism with low coupling degree is easy.

5   Conclusions

(1) Type synthesis of 1T1R parallel mechanism is studied using topological structure synthesis method based on the POC set. Eleven kinds of parallel mechanisms are synthesized, and six kinds of which, i.e., B_1[10], B_2[11], C_1[9], C_2[9], C_3[12] and C_4[9], have been reported in references, and other five kinds are synthesized for the first time based on the above method.

(2) This method features few calculation rules, simple mathematics method, definite geometrical and physical meaning, and it is applicable for both non-over-constrained and general over-constrained mechanisms.

Acknowledgments. This research is sponsored by the *NSFC*(Grant No. 51075045, 51375062, 51405039).

References

1. Clavel, R.: A fast robot with parallel geometry. In: Proceedings of the International Symposium on Industrial Robots, pp. 91–100 (1988)
2. Huang, T., Li, Z., Li, M., Chetwynd, D.G., Gosselin, C.M.: Conceptual Design and Dimensional Synthesis of a Novel 2-DOF Translational Parallel Robot for Pick-and-Place Operations. J. Mech. Design 126(3), 449–455 (2004)
3. Yuan, J., Zhang, X.: Optimal Reduction Ratio Selection for the Driving System of High-speed Parallel Manipulator. Journal of Mechanical Engineering 45(2), 255–261 (2009)
4. Zhang, Y.: Motion Control of the 2-dimensional and 2-DoF Translational Parallel Manipulator. South China University of Technology (2010)
5. Hu, J.: Research on dynamic analysis and vibration control of flexible robot mechanism. South China University of Technology (2010)

6. Yuan, J., Zhang, X.: Dimensional Synthesis of a Novel 2-DOF High-Speed Parallel Manipulator. Journal of South China University of Technology (Natural Science Edition) 35(12), 39–44 (2007)
7. Wang, S., Ehmann, K.F.: Error model and accuracy analysis of a six-DOF Stewart platform. Journal of Manufacturing Science and Engineering 124(2), 286–295 (2002)
8. Huang, T., Whitehouse, D.J., Chetwynd, D.G.: A Unified Error Model for Tolerance Design, Assembly and Error Compensation of 3-DOF Parallel Kinematic Machines with Parallelogram Struts. CIRP Annals - Manufacturing Technology 51(1), 297–301 (2002)
9. Huang, T., Chetwynd, D.G., Mei, J.P., Zhao, X.M.: Tolerance Design of a 2-DOF Overconstrained Translational Parallel Robot. IEEE Transactions on Robotics 22(1), 167–172 (2006)

Error Modeling and Simulation
of a 2- DOF High-Speed Parallel Manipulator

Junyang Wei, Xianmin Zhang, Jiasi Mo, and Yilong Tong

Guangdong Provincial Key Laboratory of Precision Equipment and Manufacturing
Technology, South China University of Technology, 510640, Guangzhou, P.R. China
wei.junyang@mail.scut.edu.com
{zhangxm,mo.jiasi,tong.yl}@scut.edu.com

Abstract. An error model of a 2-DOF high-speed parallel manipulator is presented, which reveals the relationship between input geometric parameter errors and output pose errors. Based on this, the distribution of pose errors in the workspace is demonstrated and analyzed through numerical simulation. The influence coefficient is defined so that the effect separate geometric errors have on the pose errors of the end effector can be better compared. This would benefit guiding the manufacture and the assembly of prototypes. In addition, the structure of an improved manipulator combining the 2-DOF manipulator with a rotating mechanism is developed.

Keywords: Error Modeling, Parallel Manipulator.

1   Introduction

Compared to serial robots, parallel robots have the advantages of high stiffness, high accuracy and good dynamic properties. Specifically, low mobility parallel robots are widely applied in pick-and-place industry operations. Among them, the Delta mechanism is a well-known example capable of spatial translational movements [1]. A 2-DOF translational parallel manipulator named "Diamond" for pick-and-place operations is presented and applied to a device for quality inspection of rechargeable batteries [2].

Another kind of novel 2-DOF translational parallel manipulator is introduced [3]. A serial of work around this manipulator is done as well [4-6]. As shown in Fig. 1, the manipulator is driven by two active limbs. Two passive limbs are used to restrict the orientation of the platform and to improve the stiffness of the parallel manipulator for convenience of the end effector. In article [6], the kinematic model of the mechanism is simplified into a planar five bar mechanism, ignoring the geometric parameters of the platform. Unfortunately, the geometric errors of the passive limbs are unavoidable due to manufacture error, assembly error, etc., which consequently changes the constraints of the platform and becomes the source of output orientation error. Under this condition, driving the mechanism according to the simplified kinematic model would further cause output position error of the end effector. Thus, error modeling is needed.

X. Zhang et al. (Eds.): ICIRA 2014, Part II, LNAI 8918, pp. 100–110, 2014.
© Springer International Publishing Switzerland 2014

In the field of error modeling, much research has been done and applied in practice. For instance, the authors present the error models for a six DOF Stewart platform in article [7]. Researchers develop a unified error model of a three DOF parallel machines in article [8]. Tolerance design of "Diamond" is also presented [9].

The primary purpose of this paper is to develop an error model for the manipulator in Fig. 1 not only to evaluate the positioning accuracy but also to calibrate the prototype.

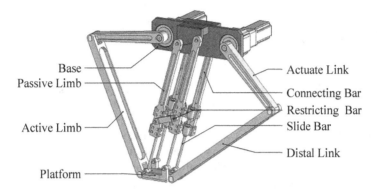

Fig. 1. The CAD model of the novel 2-DOF translational parallel manipulator [3]

2   Inverse Kinematic

The analytical expression of the inverse solution for kinematics equation is listed as Equations (1) and (2), which are needed for calculating the pose error. When the geometry of the manipulator is ideal, it can be regarded as a 5R mechanism, with both active limbs shifting a certain distance [3].

$$\theta_1 = \cos^{-1} \frac{x + e}{\sqrt{(x + e)^2 + y^2}} + \cos^{-1} \frac{L_a{}^2 - L_d{}^2 + (x + e)^2 + y^2}{2L_a\sqrt{(x + e)^2 + y^2}} \tag{1}$$

$$\theta_2 = \pi - \cos^{-1} \frac{-x + e}{\sqrt{(x - e)^2 + y^2}} - \cos^{-1} \frac{L_a{}^2 - L_d{}^2 + (x - e)^2 + y^2}{2L_a\sqrt{(x - e)^2 + y^2}} \tag{2}$$

Where L_a is the length of actuate links, L_d is the length of distal links and e is the equivalent distance: $e = (m - c)/2$.

3   Error Modeling of the Manipulator

The schematic of the kinematic model is shown in Fig. 2 after simplifying the platform as axial force links.

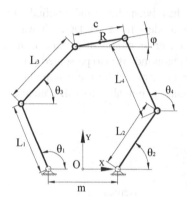
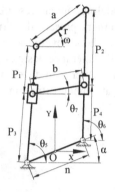
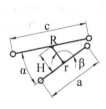

Fig. 2. the schematic of the active limbs   **Fig. 3.** The schematic of the passive limb   **Fig. 4.** The schematic of the platform

Each of the two active limbs consists of an actuate link, a distal link, an input revolution joint at the base, and a revolution joint at the platform. L_1, L_2, L_3, L_4 denote the lengths of each actuate links and distal links respectively. θ_1, θ_2, θ_3, θ_4 represent the orientations of each link relative to the x-axis of the coordinate. The distance between the two input revolute joints is m. This segment is collinear with the x-axis and its middle point coincides with the coordinate origin of the mechanism. The distance between the two revolute joints at the platform is c. The output point of the mechanism $R(x, y)$ i.e. the end effector, coincides with the middle point of c. φ expresses the orientation of the end effector platform relative to the x-axis.

Note that we only discuss the configuration in which the manipulator operates. In this configuration, θ_1 increases while θ_2 decreases as y decreases ($y > 0$ and θ_1, $\theta_2 > 0$).

From the geometry of the mechanism, we will obtain:

$$-\frac{1}{2}m - L_1 \cos(\pi - \theta_1) + L_3 \cos\theta_3 + \frac{1}{2}c \cos\varphi = x \tag{3}$$

$$L_1 \sin(\pi - \theta_1) + L_3 \sin\theta_3 + \frac{1}{2}c \sin\varphi = y \tag{4}$$

$$\frac{1}{2}m + L_1 \cos\theta_2 - L_4 \cos(\pi - \theta_4) - \frac{1}{2}c \cos\varphi = x \tag{5}$$

$$L_1 \sin\theta_2 + L_4 \sin(\pi - \theta_4) - \frac{1}{2}c \sin\varphi = y \tag{6}$$

φ is assumed to be a given quantity and the positions of the two input revolute joints at the base are assumed to be ideal, i.e. the length error of m is not considered. Differentiating the two sides of Equations (3) to (6) yields:

$$A_1 P_1 = B_1 Q_1 \tag{7}$$

Where A_1, B_1 are the error mapping matrixes, P_1 is the given error vector, Q_1 is the unknown error vector.

$$P_1 = [\delta L_1 \quad \delta L_2 \quad \delta L_3 \quad \delta L_4 \quad \delta\theta_1 \quad \delta\theta_2 \quad \delta c \quad \delta\varphi]^T$$
$$Q_1 = [\delta x \quad \delta y \quad \delta\theta_3 \quad \delta\theta_4]^T$$

The passive limbs are next taken into account. The orientation of the platform is actually constrained by the passive limbs, whose principle is applying a four-bar parallelogram linkage. The orientation is no longer kept unchanged when the length errors exist, preventing the passive limbs from being an ideal parallelogram. Therefore, the modeling of the passive limbs should be considered.

As shown in Fig. 3, each passive limb consists of two connecting bars, two slide bars, two revolute joints at the base, two revolution joint at the platform and a restricting bar connecting ends of the two connecting bars. P_1, P_2, P_3, P_4 respectively denote the lengths of each connecting bar and slide bar. b denotes the lengths of the restricting bar. θ_5, θ_6, θ_7, separately represent the orientations of each bar relative to the x-axis.

The distance between the two revolute joints grounded at the base is n. The middle point of this segment coincides the coordinate origin of the mechanism. The orientation of n relative to the x-axis is α. The distance between the two revolute joints at the platform is a, of which the middle point is $r(x_s, y_s)$. ω expresses the orientation of a relative to the x-axis.

The geometrical relationship can be written as:

$$-\frac{1}{2}n\cos\alpha + (P_1 + P_3)\cos\theta_5 + \frac{1}{2}a\cos\omega = x_s \tag{8}$$

$$-\frac{1}{2}n\sin\alpha (P_1 + P_3)\sin\theta_5 + \frac{1}{2}a\sin\omega = y_s \tag{9}$$

$$\frac{1}{2}n\cos\alpha + (P_2 + P_4)\cos\theta_6 - \frac{1}{2}a\cos\omega = x_s \tag{10}$$

$$\frac{1}{2}n\sin\alpha + (P_2 + P_4)\sin\theta_6 - \frac{1}{2}a\sin\omega = y_s \tag{11}$$

$$P_1\cos\theta_5 + a\cos\omega = P_2\cos\theta_6 + b\cos\theta_7 \tag{12}$$

$$P_1\sin\theta_5 + a\sin\omega = P_2\sin\theta_6 + b\sin\theta_7 \tag{13}$$

According to Fig. 4, the distance between R and r is H. β represents its angle relative to a. Then we have:

$$\omega = \alpha + \varphi \tag{14}$$

$$x_s + H\cos(\omega + \beta) = x \tag{15}$$

$$y_s + H\sin(\omega + \beta) = y \tag{16}$$

The parameters of the platform and the position parameters of all the revolute joints at the base are assumed to be ideal, i.e. the errors of the parameters mn$\alpha\beta$H do not exist. Differentiating simultaneous Equations (3) to (6), Equations (8) to (16) leads to an error model of the mechanism as follows

$$A_2 P_2 = B_2 Q_2 \tag{17}$$

Where

$$P_2 = [\delta L_1 \quad \delta L_2 \quad \delta L_3 \quad \delta L_4 \quad \delta\theta_1 \quad \delta\theta_2 \quad \delta c \quad \delta P_1 \quad \delta P_2 \quad \delta a \quad \delta b]^{\mathrm{T}}$$
$$Q_2 = [\delta x \quad \delta y \quad \delta\varphi \quad \delta\theta_3 \quad \delta\theta_4 \quad \delta\theta_5 \quad \delta\theta_6 \quad \delta\theta_7 \quad \delta P_3 \quad \delta P_4]^{\mathrm{T}}$$

4   Simulation

In the simulation, the values of the parameters for the mechanism are selected as shown in Table 1. The workspace is set to a rectangular area symmetric about the y-axis, with a size of 540mm×385mm. The center point of the workspace is 572.5mm from the origin.

Table 1. The values of parameters for the simulation

Name	Value	Name	Value	Name	Value
L_1	390 mm	φ	0°	m	260 mm
L_2	390 mm	P_1	360 mm	n	40 mm
L_3	420 mm	P_2	360 mm	α	45°
L_4	420 mm	a	40 mm	β	90°
c	80 mm	b	40 mm	H	10 mm

Since analytic expressions of the kinematics of the five bar mechanism exist, every output point in the workspace and the corresponding input actuating angle is regarded as accurate.

Using Equations (1) (2) (7), an estimation of separate effect that each error of the parameters has on the end effector is available. As an example, the results of the errors at point P(-150,450) are shown in Fig. 5 to Fig. 7.

4.1   Influence Coefficient

Notice that connecting lines of errors of parameters have good linearity and pass thorough the origin compared with position errors. Take the slopes β as indexes to reflect the effect levels of the errors of parameters.

$$\beta_{ij} = \frac{n\sum X_i Y_j - \sum X_i \sum Y_j}{n\sum X_i^2 - (\sum X_i)^2} \tag{18}$$

In Fig. 6 and Fig. 7, the actual value of the length error is adopted as the unit of the x-axis, which is available for estimating the effect of the assembly errors. Meanwhile in real-world engineering production, the dimensional accuracy of parts is usually reflected by the dimensional tolerance. When parts are produced to the same grade of tolerance, the larger the dimension is, the greater the error allowed. Therefore, it is better to use the Length Error of Part/ Length of Part ratio as the unit of the x-axis to estimate the effect of the length errors. Then, in Equation (18) :

$$X = [\delta L_1/L_1 \quad \delta L_2/L_2 \quad \delta L_3/L_3 \quad \delta L_4/L_4 \quad \delta c/c]$$

$$Y = [\delta x \quad \delta y]$$

The influence coefficient (IFC) is defined as:

$$k_{ij} = \frac{|\beta_{ij}|}{\sum_{i=1}^{5}|\beta_{ij}|} \tag{19}$$

All the IFCs add up to one. The larger the IFC is, the greater the effect that the corresponding length error has on the position error. Fig. 8 and Fig. 9 show the IFC of δL_3 distribution on position errors in the workspace.

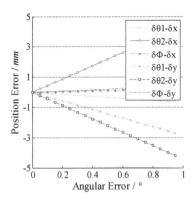

Fig. 5. The input angle errors and the orientation error of the platform versus the position errors

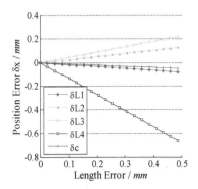

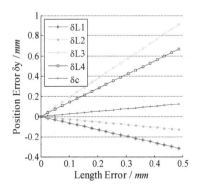

Fig. 6. The length errors of the active limb versus the position error along the x-axis

Fig. 7. The length errors of the active limb versus the position error along the y-axis

Similarly, the IFCs of other length errors are calculated. The results are shown as the 3D surface diagrams in Fig. 10 and Fig. 11.

Because of the symmetry of the mechanism, the IFC distributions of the length errors of the two actuate links $\delta L_1, \delta L_2$ and the length errors of the two distal links $\delta L_3, \delta L_4$ mirror each other along the y-axis. Meanwhile the IFC distribution of δc is also axial symmetric. In most area of the workspace, the position errors are primarily affected by the length errors of the distal links. The one closer to the y-axis has more influence on δx, the one on the opposite side has more effect on δy. The position errors are relatively insensitive to the other length errors.

Using Equation (17), the IFCs of the length errors of the passive limbs is calculated as well. From Fig. 12, it can be seen that in the workspace the effects on the orientation error is dominated by the length errors of the two connecting bars. Also as the value of x increases, the IFC grows at a slow pace. The IFCs of δP_1, δP_2 overlapped in the figure because the signs are eliminated by absolute calculation. δa has the least effect on the orientation error while the effect of δb is greater on the left side of the workspace.

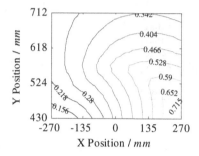

Fig. 8. The IFC (δL_3) distribution on the position error along the x-axis in the workspace

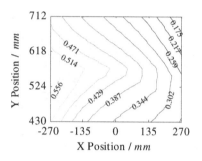

Fig. 9. The IFC (δL_3) distribution on the position error along the y-axis in the workspace

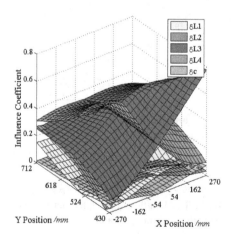

Fig. 10. The IFCs distribution on the position error along the x-axis in the workspace

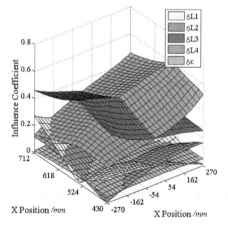

Fig. 11. The IFCs distribution on the position error along the y-axis in the workspace

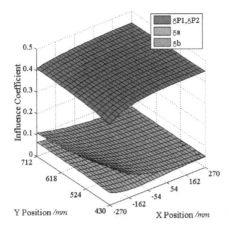

Fig. 12. The IFCs distribution on the orientation error in the workspace

4.2   Pose Error Distribution in the Workspace

According to the error model in Equation (17), the values of error were set as shown in Table 2. We obtained the distribution of the pose error in the workspace, which is shown in Fig. 12 to Fig. 14.

It can be seen that the pose error in the workspace changes smoothly. Except for a small area on the right side of the workspace, the position error δx is within -1mm, δy is within ± 0.5mm and the orientation error is within -1° . The accuracy on the left side of the workspace is higher than the right side.

The orientation error distribution in the workspace is asymmetrical because the passive limb is rotated angle α (0<α <90°). Under these conditions, the orientation errors on the left and right sides of the workspace are larger since the passive limb is closer to the singular configurations (when the angle $\theta_5\theta_6$ is 0° or 90°). Although the length errors of the active limbs contribute to the position errors, the influence of the orientation error should not be neglected. The trends of the position errors follow the trend of the orientation error while the values of the length errors are of the same order of magnitude in the simulation.

In summary, the manipulator still meets the requirement pick-and-place operations when errors as above exist. However, the dimensional accuracy of the passive limbs needs to be paid attention to. The workspace of the manipulator has a favorable operation area dependent on the parameters especially the angle α, which should be considered in motion control.

Table 2. The values of parameters' errors for the simulation

Name	Value	Name	Value	Name	Value
δL_1	0.05 mm	δL_4	-0.10 mm	δP_2	-0.02 mm
δL_2	0.06 mm	δc	0.05 mm	δa	0.03 mm
δL_3	0.08 mm	δP_1	0.05 mm	δb	-0.06 mm

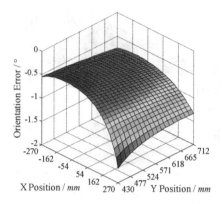

Fig. 13. The distribution of the orientation error in the workspace

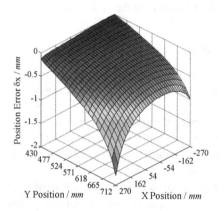

Fig. 14. The distribution of the position error along the x-axis in the workspace

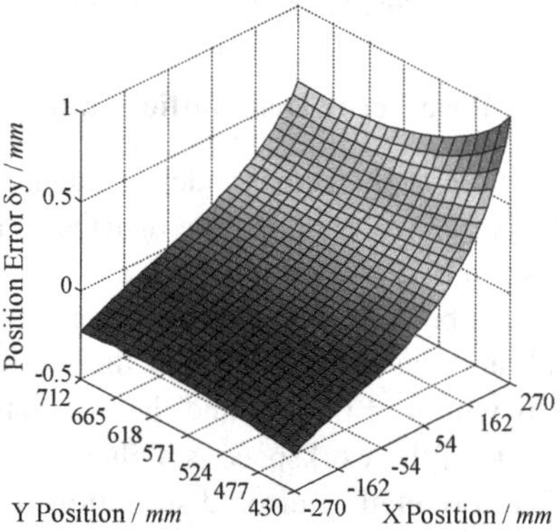

Fig. 15. The distribution of the position error along the y-axis in the workspace

5   Improvement of the Manipulator

As the distance between the two revolute joints at the platform, i.e. c is decreased, the position error of the end effector decreases proportionally. When c is decreased to 0, the position error and orientation error is decoupled. In this case, there are no interferences between the position errors and the orientation error. The former is directly affected by the actuate limbs and the latter by the passive limbs, which would help simplify the kinematic model and benefits the assembly adjustment and the mechanism calibration.

Based on this, a simple hybrid manipulator combining the 2-DOF manipulator with a revolute joint is developed as shown in Fig. 15. Generally speaking, mass of the parts is one of the factors that limit the dynamic property of the manipulator. Overweight components may undermine the acceleration ability of the manipulator.

Among the components, the mass of the actuate motors and reducers usually occupies large proportion. Therefore, the hybrid manipulator is designed to reduce the mass of the rotational base. The actuate motors are fixed (the frame fixing the motors is not illustrated here) along the rotating axle of the revolute joint in order to reduce the rotational inertia of the manipulator. The synchronous belt drive and the bevel gear drive are adopted to actuate the active limbs.

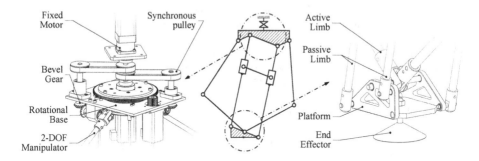

Fig. 16. The geometry of the improved manipulator and two details of CAD prototype

6   Conclusion

In this paper, an error model of a 2-DOF high-speed parallel manipulator is presented and simulated numerically.

The influence coefficient is defined so that the effect separate geometric errors have on the pose errors of the end effector can be better compared. This helps guide the manufacture and assembly of prototypes. Among the parameter errors, the length errors of the distal links and the length errors of the connecting bars have the greatest effect on the pose error.

Based on the error model developed, the distribution of pose errors in the workspace reveals the relationship between input parameter errors and output pose errors and provides theoretical support for calibration of the manipulator.

In addition, the structure of an improved manipulator combining the 2-DOF manipulator with a rotating mechanism is developed. The dynamic character of the manipulator is yet unknown and needs further research to demonstrate the potential application value of the improved manipulator.

Acknowledgments. This research was supported by the National Natural Science Foundation of China (Grant No.91223201), the Natural Science Foundation of Guangdong Province (Grant No.S2013030013355), Project GDUPS (2010) and the Fundamental Research Funds for the Central Universities (2012ZP0004). These supports are greatly acknowledged.

References

1. Clavel, R.: A fast robot with parallel geometry. In: Proceedings of the International Symposium on Industrial Robots, pp. 91–100 (1988)
2. Huang, T., Li, Z., Li, M., Chetwynd, D.G., Gosselin, C.M.: Conceptual Design and Dimensional Synthesis of a Novel 2-DOF Translational Parallel Robot for Pick-and-Place Operations. J. Mech. Design 126(3), 449–455 (2004)
3. Yuan, J., Zhang, X.: Optimal Reduction Ratio Selection for the Driving System of High-speed Parallel Manipulator. Journal of Mechanical Engineering 45(2), 255–261 (2009)
4. Zhang, Y.: Motion Control of the 2-dimensional and 2-DoF Translational Parallel Manipulator. South China University of Technology (2010)
5. Hu, J.: Research on dynamic analysis and vibration control of flexible robot mechanism. South China University of Technology (2010)
6. Yuan, J., Zhang, X.: Dimensional Synthesis of a Novel 2-DOF High-Speed Parallel Manipulator. Journal of South China University of Technology (Natural Science Edition) 35(12), 39–44 (2007)
7. Wang, S., Ehmann, K.F.: Error model and accuracy analysis of a six-DOF Stewart platform. Journal of Manufacturing Science and Engineering 124(2), 286–295 (2002)
8. Huang, T., Whitehouse, D.J., Chetwynd, D.G.: A Unified Error Model for Tolerance Design, Assembly and Error Compensation of 3-DOF Parallel Kinematic Machines with Parallelogram Struts. CIRP Annals - Manufacturing Technology 51(1), 297–301 (2002)
9. Huang, T., Chetwynd, D.G., Mei, J.P., Zhao, X.M.: Tolerance Design of a 2-DOF Overconstrained Translational Parallel Robot. IEEE Transactions on Robotics 22(1), 167–172 (2006)

Fuzzy PD Compliance Control of 6-DOF Robot Using Disturbed Force Sense

Zhang Tie[1], Wang Bo[2], and Lin Junjian[1]

[1] School of Mechanical & Automotive Engineering, South China University of Technology, Guangzhou, Guangdong Province, China
[2] School of Electronic Information Engineering, Tianjin University, Tianjin City, China

Abstract. The article deals with force control when the force sense is disturbed, using a lowpass filter and a fuzzy PD controller. The force sense is installed on the endpoint of the robot having six degrees-of-freedom (6-DOF), with a probe mounted on it. The robot is equipped with position servos and uses velocity command to control the force. From the experimental result, the lowpass filter is proved to be necessary and the performance of the fuzzy PD controller is much better than the traditional PD controller.

Keywords: 6-DOF Robot, Position Servos, Fuzzy PD, Force Control.

1 Introduction

There are great demands on force control applications in the manufacturing such as grinding, polishing and precision assembly. To apply these functions to the industrial robots, it is necessary to use a force sense with well performance. Nowadays, the force sense can reach the precision of 0.5% or less, but it is a very sensitive component, and easily disturbed in the electromagnetism environment. Effective methods to avoid the disturbance have been taken action such as shielding the actuators, but sometimes we may not meet such condition.

In the early years, many researches are focus on robots with torque servos, and actually there are some advantages in using torque to control the servos in force control. First, the desired force can be easily changed to actuators' desired torques through Jacobian equations. And then, it can prevent the servos from overload when collision happened to some extent, as it controls the currents directly. However, in the industrial area, most of the commercialized industrial robot manipulators are equipped with only position servos, it is necessary to do some force control researches on the close-loop of the servos.

Richard Volpe and Pradeep Khosla[1] had theoretical analyzed the force control in the force-base and position-base respectively. I.H.SUH[2] who noted the importance of applying force control in the industrial robots, proposed the fuzzy adaptive control method, and succeeded to let the endpoint be able to adapt the soft, medium, and hard 3 different kinds of surface. And then, many applications using compliance control arithmetic had been proposed [3-5]. But the articles above are established in one robot. Recently, J. Kruger[6] was successful of using dual arm robot to achieve assembly.

X. Zhang et al. (Eds.): ICIRA 2014, Part II, LNAI 8918, pp. 111–118, 2014.
© Springer International Publishing Switzerland 2014

With an excellent capability of the force sense, we can apply nearly all kinds of arithmetic which used in tradition robot control. But there are some reasons that restrict the force control in large-scale applications. The large gain between desired position and desired force is the major problem in the force control. As a small disturb from the force sense may cause a large force change after it goes through the controller. Actually, the force sense would be disturbed by electromagnetic when the powerful servos are running nearby. In this situation, sometimes it is hard to distinguish whether the received signal is disturbed. Hence, it is worth considering of designing a commercial, efficient and stable force controller which can decrease the problems above.

This paper proposes a solution which requires only a fuzzy PD control action on the force error, and a lowpass filter action on the force measurement. It can be easily implemented for real time force control because it only needs to measure the force signals, without knowing any detailed mathematical models when compared with other intelligent arithmetic.

2   System Schematic (Plus Physical Model)

The block diagram of the robot manipulator's control system is shown in Fig.1. The system is comprised of the electrical part and the mechanical part. The electrical part contains a computer and a motion controller. The computer collects the inputs from the motion controller and feedbacks the outputs after calculating through specific arithmetic. The mechanical part includes a 3-dimension force sense, and a 6-DOF robot manipulator. The force sense is fixed on the endpoint of the robot, and a probe is attached to the force sense to serve as an end effecter, with it the applying forces on the environment can be transmitted to the force sense directly.

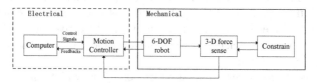

Fig. 1. Schematic diagram of the control system

A block diagram of the control process is shown in Fig.2. Once set the desired force value on the computer, the computer will generate an error force signal through reading the present force by communicating with the motion controller. Note that, because of the noise, the present force is not the present value of the force sense. The computer would filter the noises in the signals using the method below to generate the real present force. Then, according to the force errors, the controller recognizes an output, which means the linear velocity of the endpoint on the force direction in the operating space. With the combination of the linear velocities of the other directions and angular velocities, we can gain the velocities of the operating space. Then, it would be transformed to the velocities in the joint space using the equation below:

$$\dot{q}(n+1) = J^{-1}(q(n))\dot{x}(n+1) \quad . \tag{1}$$

where $\dot{q}(n+1)$ refers the velocities of the joint space, which is a 1×6 matrix, represents the 6 axis's velocity respectively, and $\dot{x}(n+1)$ is velocities of the operating space at the time n+1, which is also a 1×6 matrix, but the upper 1×3 matrix is the linear velocities and the lower 1×3 matrix is the angular velocities, $J^{-1}(q(n))$ is the inverse Jacobian matrix according to the joint angles in the time n, and it is a 6×6 matrix.

The motion controller will serve the actuators on the robot according to $\dot{q}(n+1)$. At the time n+1, we can get a new force value and repeat the process above.

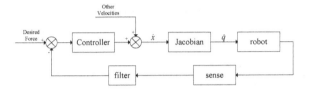

Fig. 2. Schematic diagram of the control process

3   Controller Design

3.1   Character of Lowpass Filter

The force sense is a very sensitive component. It can be accurate if there are no electromagnetic devices such as actuators nearby, but we cannot always avoid this situation. If the sense is disturbed, though it will not be disturbed all the time, the signals will mix many noises, some of them can be distinguished, such as the value exceeds the range of the measurement, or the value generates a peak at a very short time. However, the worst situation which we are unwilling to deal with is that we cannot judge whether the signals are disturbed, and what extent they are disturbed.

To decrease the influence of the noises, we can complement an effective filter according to the characteristic of noise. There are many researches on the filters, and most of them require plenty of calculation. Actually, we need to know about the type of the noise, so that we can select a useful and effective filter.

After observing amount of the signal information, we conclude that most of the noises are high frequency noises, and their frequency is different from the sample frequency obviously. Thus, we propose to choose a lowpass filter in the end, which with an expression shown in eq.(2):

$$L(s) = \frac{a}{s+a} \quad . \tag{2}$$

where a represents the cutoff frequency.

3.2   Using PD Control

There are plenty of research works on force control using fuzzy theory, but mostly of them are focused on the torque servos, the rests discussed on the position servos, which are combined fuzzy with adaptive, neutral network, and so on. However, in the industrial area, PID is still the most wide-spread method. Single PID control may be embarrassed in some complex situations, but it can be ameliorated through integrating other advanced arithmetic. The combination of fuzzy and PID has been proved of its advantages in many succeed progresses. So we propose a fuzzy PID controller on the force control.

As we use Jacobian matrix to transfer the joint space velocity to the operating space velocity above. Assuming the force control model as a spring-damper model, we must integrate the operating space velocity to acquire the force on the end effecter, which will turn the system to a Type I system. One of the advantages is we can achieve the zero steady-state error to a constant force input without using integral control theoretically, moreover, knowing about the type of the system will instruct us to design a proper controller.

To avoid turning the system to Type 0 or Type II, we suggest using P/PD controller instead of other PID combinations. It seems P controller is a better choice, because of the derivation of the PD controller will amplify the noise of the sense, which will reduce the stable of the system. Considering passing through a sudden large noise, though the use of the lowpass filter will smooth the noise. Compared with the same parameters, PD controller will export a larger control signal than P controller. However, no matter what noises pass through, errors must be generated. Though PD controller presents worse in this aspect, that PD controller provides the excellent capability of correct error would compensate its disadvantage, and such ability is more important when the probe is moving on the constrain or the outside force is influencing the end effecter.

3.3   Fuzzy PD Control

According to Section 3.2, PD controller can improve the ability of correcting errors dynamically, but it presents worse on dealing with the noise. To fix such a disadvantage, we consider importing fuzzy rules for the parameters of the PD controller in such a way that the controller can improve the stability when it is close to the zero error, and improve the dynamic performance when the error is increased.

To design such a Fuzzy PD controller, we focus on connecting the controller's parameters to the errors. Assuming that $F_e = F_r - F_a$, and $F_{et} = d(F_r - F_a)/dt$, which represent the present error and the derivative of the error respectively. Treat F_e and F_{et} as the input variables of the fuzzy rules, K_p and K_d as the outputs. All of them are established in five linguistic values: PL (Positive Large), PS (Positive Small), ZE (Zero), NS (Negative Small), NL (Negative Large), where F_e and F_{et} use the Gauss fuzzy membership functions. To make a general use controller, we set their range is -1 to 1, that we can use multipliers k_{F_e} and $k_{F_{et}}$ to adjust the actually range. K_p and K_d use the triangle fuzzy membership functions, the range is 0 to 1. Also, adjusted by k_{F_e} and $k_{F_{et}}$. The rules are described in Table.1 and Table.2.

Table 1. The fuzzy rules for K_p

		F_{et}				
		NL	NS	Z	PS	PL
	NL	VL	VL	L	VL	VL
	NS	L	L	M	L	L
F_e	Z	M	S	VS	S	M
	PS	L	L	M	L	L
	PL	VL	VL	L	VL	VL

Table 2. The fuzzy rules for K_d

		F_{et}				
		NL	NS	Z	PS	PL
	NL	VL	L	M	L	VL
	NS	L	M	S	M	L
F_e	Z	S	VS	VS	VS	S
	PS	L	M	S	M	L
	PL	VL	L	M	L	VL

With the use of the fuzzy PD controller above, we can predict the force control process below: at the beginning, assuming the initial force is zero, as the force error is large, K_p and K_d is large, so the endpoint will gain an excellent dynamic characteristic to close to the reference force. As the force error growing smaller, K_p and K_d will also become smaller. If there is a noise at this time, when it passes through the lowpass filter, with the time-delay character, the filter will prevent the filtered force signal and error growing too fast, neither as K_p and K_d. In this way, the controller will generate a much smaller response than traditional PD controller that ensures the smooth of the endpoint. As the peak noise passed, the signals return to normal, and the error will stop growing, neither as K_p and K_d, the controller will gain a better response to make the control force return to the desired force.

3.4   Evaluate the Controller

In order to evaluate the fuzzy PD controller, define J as the performance function, and it can be described as:

$$J = \frac{1}{n}\sqrt{\sum_{k=0}^{n}\left(F_r\left(k\right)-F_a\left(k\right)\right)^2} \ .$$

where n represents the continuous number of the samples after the rising time, J is called the performance value. Considering the noises of the signals, we can apply this function to the signals which come out of the lowpass filter. This function can clearly express the fluctuation of the signals in the control process. A smaller J means a better effectiveness of the controller. Note that, the function can only reflect the performance parameter among the controllers with the same n in the same control environment.

4   Experimental Results

The effectiveness of the combination of the fuzzy PD controller and the lowpass filter will be tested on a real 6-DOF industrial robot, whose actuators are worked in position mode. The experimental set-up is shown in Fig.3. The force sense is installed on the endpoint of the manipulator, with the measurement range of 30N. The probe which is made of aluminum is installed on force sense. The force sampling rate is 20ms, and the motor served period is 2ms. The experiment is set in a serious electromagnetic environment, with several AC actuators working around without insulation.

Fig. 3. Whole view of the robot use in the system

Fig.4 shows the effect of the traditional PD controller, with the parameters $K_p = 1$ and $K_d = 0.5$, which are decided after several experiments. The red line represents the original samples while the blue line represents the filtered samples. Although the lowpass filter can clean up most of the noise, the peak noise is still the main unstable reason. There must be some oscillations when a peak noise passes through. At this experiment, n is given as 564, and the J is 0.2169.

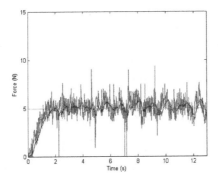

Fig. 4. Force control using tradition PD controller

To adjust the fuzzy PD controller parameters, we set multipliers $k_{F_e} = 0.1$ and $k_{F_{et}} = 0.05$, and with the multipliers, F_e and F_{et} can satisfy the range of the fuzzy inputs. In the other hand, we choose $k_{F_e} = 0.1$ and $k_{F_{et}} = 0.05$, in order to correspond to the traditional PD controller above. Run the robot with this controller, the effectiveness of the fuzzy PD controller is shown in Fig.5, as well as the K_p and K_d's varieties can be referred to Fig.6. Also set $n = 564$, and J is 0.0110.

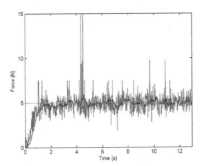

Fig. 5. Force control using fuzzy PD controller

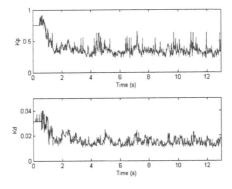

Fig. 6. Change of K_p and K_d in the fuzzy PD control

Compared with the results of the two trials, as analyzed in Section 3.2 and Section 3.3, the traditional PD controller has larger oscillations than fuzzy PD controller, according to the comparison of the performance value and the figures. The former fluctuates between 4.5N and 5.5N in the experiment which has a larger J, while the latter can keep the precision below 0.2N in this experiment, with a much smaller J.

5   Conclusions

PD controller has an excellent dynamic characteristic, but it would become the weakness while the signal is noisy. A fuzzy PD controller is proposed to obtain good force output responses regardless of the noises. The arithmetic is tested by using a 6-DOF commercialized industrial robot equipped with the position servos. From the experimental result, fuzzy PD controller shows better responds than traditional PD control in the disturbed signals.

Acknowledgment. This work was supported by the 863 Key Program (NO.2009AA043901-3), Science and Technology Planning Project of Guangdong Province, China (No. .2012B090600028) and Industry, University and Research Project of Guangdong Province (No. 2012B010900076). Strategic and Emerging Industrial project of Guangdong Province (No. 2011A091101001), Science and Technology Planning Project of Guangzhou City, 2014Y200014.

References

1. Volpe, R., Khosla, P.: A Theoretical and Experimental Investigation of Explicit Force Control Strategies for Manipulators. Automatic Control 38(11), 1634–1650 (1993)
2. Suh, I.H., Eom, K.S., Yeo, H.J., et al.: Fuzzy Adaptive Force Control of Industrial Robot Manipulators with Position Servos. Mechatronics 5(8), 899–918 (1995)
3. Zollo, L., Siciliano, B., Laschi, C., et al.: An experimental study on compliance control for a redundant personal robot arm. Robotics and Autonomous Systems 44, 101–129 (2003)
4. Bigge, B., Harvey, I.R.: Programmable springs: Developing actuators with programmable compliance for autonomous robots. Robotics and Autonomous Systems 55, 728–734 (2007)
5. Kim, H.-S., Park, J.-J., Song, J.-B., et al.: Design of safety mechanism for an industrial manipulator based on passive compliance. Journal of Mechanical Science and Technology 24(11), 2307–2313 (2010)
6. Kruger, J., Schreck, G., Surdilovic, D.: Dual arm robot for flexible and cooperative assembly. CIRP Annals – Manufacturing Technology 60, 5–8 (2011)

Research Scheduling Problem of Job-Shop Robotic Manufacturing Cell with Several Robots

Yujun Yang[*], Chuanze Long, and Yu Tao

Key Laboratory of Computer Integrated Manufacturing System of Guangdong Province
Institute of Mechanical& Electronic Engineering,
Guangdong University of Technology, Guangzhou, China
yjyang@gdut.edu.cn

Abstract. To solve the job-shop scheduling problem of robotic manufacturing cell with multiple robots, population initialization method and neighborhood search mechanism of its solution genetic algorithm are studied. The objective is to determine a schedule of machine and transport operations as well as assignment of robots to transport operation with minimize the maximum completion time. To improve the quality of the initial population, heuristic initialization method based on 3 vectors scheduling combining dispatching rules is proposed. Neighborhood structure is constructed based on the critical path. Experimental results show that the algorithm is effective.

Keywords: Job-shop, Disjunctive graph model, Genetic algorithm, Neighborhood search.

1   Introduction

In order to reduce lead time and labor costs for industry, many advanced material handling techniques have been used in intelligent manufacturing environment. As a part of advanced technology, robot is widely found in logistics, semiconductor manufacturing, human intelligence, and etc. The development of effective and efficient robotic manufacture cell schedules is an important study area. Robotic scheduling problem has been received much attention in the literature in recent years, however most of models belongs to flow-shop scheduling problem [1-2]. In actual manufacturing environments is so complex that research the job-shop scheduling problem with transportation time is more theoretical significance and application value.[1]

2   Problem Formulation

In this section we give a formal definition of scheduling problem in a job-shop environment as follow: a set of n jobs $J = \{J_1,...,J_n\}$ process on a set of m

---

* Corresponding author.

X. Zhang et al. (Eds.): ICIRA 2014, Part II, LNAI 8918, pp. 119–123, 2014.
© Springer International Publishing Switzerland 2014

machines $M = \{M_1,...,M_2\}$, transported by a set of r transport robots $R = \{R_1,...,R_r\}$. And the following additional constraints are taken into account:

Each job J_i is ordered set of n_i operations denoted $o_{i1}, o_{i2}, ..., o_{in}$ for which machine orders are dedicated;

1. each machine can process(without preemption) only one job at one time;
2. each machine can process(without preemption) only one job at one time;
3. Transportation operation $t^{R_r}_{u_i,k^{u_i},k+1}$ must be considered for robot R_r when machine operation o_{ik} is processed on machine u_{ik} and machine operation $o_{i,k+1}$ is processed on machine $u_{i,k+1}$, these transportation times are depend on the jobs(robots) and the machines between which the transport takes place;
4. Each transportation operation is assumed to be processed by only one transport robot which can handle at most one job at one time;
5. To avoid deadlock, we assume that the input/output buffer capacity is unlimited in the system.
6. No additional time is required to transfer job from machine to the unlimited output buffer and the same for transfer job from the input buffer to machine.

The objective is to determine starting times for each operation, in order to minimize the makespan $C_{max} = Min\{ \max_{i=1,n} (C_i) \}$ while satisfying all the constraints and precedence, where C_i denotes the completion time of the last operation $o_{i,n}$ of job J_i.

3   Disjunctive Graph

In this section the well-know disjunctive graph model for the job-shop problem proposed by Phlippe[3] is extended to the job-shop problem with several robots. The disjunctive graph of job shop problem with several robots can be defined as $G = (V_M \cup V_t, C \cup D_m \cup D_R)$. In disjunctive graph of this problem, a set of vertices V_m containing all machine operations and a set of vertices v_t is the set of transport operations obtained by an assignment of robot to each transport operation and two dummy nodes 0 and *.the graph consists of a set of conjunctions c representing procedure constraints. Disjunction for the machine D_m and assignment of robots to transport operation D_R.

1) Selection MS represents the repeat vector of machine operations. In this vector, the first j appeared in MS repersents the first operation of task, and the second appears indicate the task second operation of task, and so on.

(2) Selection TS represents the repeat vector of transportation operation.

(3) Selection RA represents the robot task allocation vector.

Example. We consider an instance of a 4×4 job-shop problem, $m = 4$ machines, $n = 4$ jobs, 16 machine operations and 12 transport operations transport by r robots. MS,TS ,RA are given in figure 3-1.

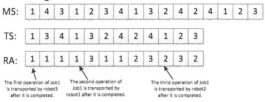

Fig. 1. Instance of MS,TS,Ra

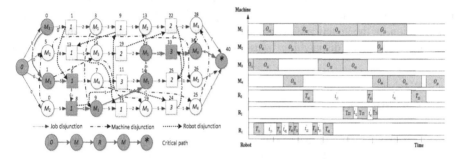

Fig. 2. Disjunctive graph model and Gantt chart

There is a no idle feasible schedule (empty trip is not idle) in fig 3-2, which constitute a longest path from 0 to * in the acyclic graph, such a uninterrupted path is called a critical path and the operations on critical path are called critical operations. The makespan c_{max} is entirely depended on the length of the longest path, so that slight alter of the critical operations may change the value c_{max}. In order to improve the value of c_{max}, a appropriate neighborhood structures need to be built based on critical path. In order to improve the value of c_{max}, a appropriate neighborhood structures need to be built based on critical path. Such an approach had been presented for the single machine problem [6] and adapted to job shop scheduling problems [5]. The same principle is used in this scheduling problem, machine block、 robot block find in critical path and Corresponding neighborhood structures is established.

4   Improved Genetic Algorithm

In order to improve the quality of the initial population, heuristic initialization method based on 3 vectors scheduling combining dispatching rules is proposed. It is possible to use a representation similar to the Bierwirt's [4] representation : a chromosome is a sequence of operations. According to the traditional job shop scheduling problem, the first vector is to scheduling the machine operations.   Then, the second vector considers transportation operations by inserting to the former chromosome. The last is dispatching the transportation to robots based on heuristic rule. As instance of a 4×4 job-shop problem mention above, the first vector regard as *MS*: 1 2 3 4 3 4 2 1 4 3 1 2 2 1 4 3 ;the second vector regard as *TS*:3 4 2 1 4 3 1 2 2 1 4 3;If the number of robot is 3, the third vector regard as *RA*:1 1 1 1 2 2 2 2 3 3 3 3.

4.1   Local Search

As the traditional genetic algorithm to solve scheduling problems is a global search capability, but is easy to fall into premature convergence. However, Local search can avoid an excessive and premature convergence rate by a progress invasion .In order to make full use of each of their characteristic, a improved Genetic algorithm is proposed. Local search approach is good at adjust the chromosome to improve the quality of solution and accelerate the convergence speed which had been crossover and mutation by the genetic algorithm. In this paper, the local search neighborhood is based one block and dispatching transportation operation to each robot on critical path.

5   Computational Results

Numerical experiments are based on instances first introduced by Philippe [3]. In all instances, empty trip and loaded trip duration are differences for the robot and the buffer size of machine is unlimited. In the following, Robot 2 transportation times are 2 times greater than transportation time of robot 1; Robot 3 transportation times are 4 times greater than transportation time of robot 1,   Robot 4 transportation times are 8 times greater than transportation time of robot 1.

Table 1. Comparison with Philippe

	Philippe		improved genetic algorithm	
Robot num	initial	Optimal	initial	Optimal
1	85	54	85	40
2	141	45	75	45
3	183	45	65	40
4	225	45	65	40

As we can seen from table5-1, the initial solution performance of genetic algorithm with local search strategy is better than Philippe' when   the number of robots is 2,3,4. The reason of this result is that the proposed algorithm in this paper mainly used heuristic rules which transportation operations are average assigned to each robot. For optimal solution, the improved genetic algorithm is performance excellent than Philippe' algorithm when the number of robot is 1、3、4.By comparison with Philippe algorithm experimental results on changes in the number of robot and transportation time, the genetic algorithm shows its effectiveness.

6   Concluding Remarks

This paper presented different genetic algorithm with local search approaches for a generalization of job shop scheduling problem which several robot transportation times are taken into account. Corresponding mathematical optimization model and disjunctive graph model was established and heuristic dispatching rules is proposed to solve this scheduling problem. By comparison with Philippe algorithm experimental results on changes in the number of robot and transportation time, the genetic algorithm shows its effectiveness. Some future work will direct to efficient algorithms and search strategy.

Acknowledgements. This project was supported by National Natural Science Foundation of China (Grant NO.51105082), National Key Technologies R&D Program of China (Grant NO.2012BAF12B10) and the key technologies of Guangdong Province strategic emerging industries (Grant NO.2011A091101003).

References

1. Dawande, M.W., Geismar, H.N., Sethi, S.P., et al.: Sequencing and scheduling in robotic cells: Recent developments. Journal of Scheduling 8(5), 387–426 (2005)
2. Dawande, M.W., Geismar, H.N., Sethi, S.P., et al.: Throughput Optimization in Robotic Cells, pp. 1–413. Springer, Germany (2007)
3. Philippe, L., Mohand, L.: A Disjunctive Graph for the job-shop with several robots. In: MISTA Conference, pp. 285–292 (2007)
4. Bierwith, C.: A generalized permutation approach to job-shop scheduling with genetic algorithms. OR Spektrum 17, 87–92 (1995)
5. Hurink, J., Knust, S.: A fast tabu search algorithm for the job shop problem. Management Science 42(6), 797–813 (1996)
6. Caumond, A., Lacomme, P., Moukrim, A., Tchernev, N.: An MILP for scheduling problem in an FMS with one vehicle. European Journal of Operational Research 199(3), 706–772 (2009)

Research on Robotic Trajectory Automatic Generation Method for Complex Surface Grinding and Polishing

Shengqian Li, Xiaopeng Xie[*], and Litian Yin

School of Mechanical & Automobile Engineering,
South China University of Technology, Guangzhou 510640, China
xiexp@scut.edu.cn

Abstract. To solve the problem of high intensity, long time-consuming and low efficiency by on-line programming in robot for free-form surface grinding and polishing, in addition, it isn't programmed by on-line programming in robot for more complex surface. A method of robotic machining trajectory automatic generation for complex surface grinding and polishing is presented, the Non-Uniform Rational B-splines(NURBS) curve fitting and adaptive sampling algorithm are taken as core of the method, which could realize trajectory planning of robot with off-line programming, Finally, the result of the computer simulations show that the method is simple, effective, practical and reliable. Meanwhile, which could not only generate grinding and polishing trajectory for arbitrary complex surface, but also can improve the quality, precision and efficiency of machining.

Keywords: Complex Surface, Robot, Motion Trajectory, Non-Uniform Rational B-Splines (NURBS) Curve, Adaptive Sampling.

1   Introduction

With the more improvement of people's living standards, the more important the visual aesthetic effect of product surface for people. In the during of manufacturing of complex surface, grinding and polishing is a very important process, surface polishing machining well or bad which has a direct influence on the appearance and quality of products. Then the manual working is a main method in tradition grinding and polishing, which is high intensity, long time-consuming and low efficiency, moreover, machining quality is kept up well very hardly. With the CNC and CAD/CAM technology developing constantly[1-3], as well as robots are widely used in recent year,   that make industrial robot be indispensable in the field of automatic production, so researching and application are important directions to use industrial robot[4 5]. Because of robot own character of flexibility, openness and so on, that just as   human wrist, however, robot is very fit to polish surface instead of manual, especially the more complex surface is grinded and polished, the more robotic advantages are shown [5 6]. When robot is finished grinding and polishing for the complex surface, the motion trajectory must be planning, so it is very important to

---

[*] Corresponding author.

X. Zhang et al. (Eds.): ICIRA 2014, Part II, LNAI 8918, pp. 124–135, 2014.

research a method of trajectory generation, which is one of hot topics of the robotic key technology. So that the method of trajectory generation have been an internationally researched in present. To have been adopted the offset section method in the CAD model by Kiswanto G and et al., they used a group of parallel planes to cut the curve surface to get a line section as motion trajectory of robot[7 8]. A adaptive sampling strategy have been adopted for free-form surface is based on CAD model by He Gaiyun et al., while according to curvature character, sampling points are adaptive distributed in curve, that is better approach geometrical characteristic of curve or surface[9].

To be get motion trajectory in the on-line programming for the traditional grinding and polishing robot, this method have many disadvantages, that not only with high intensity, long time-consuming, low efficiency and inflexible, but also can't teaching programs for the complex surface. On the contrary, the motion trajectory is generated in off-line programming for the CAD modeling, that not only have high precision trajectory, low time-consuming and easy to program, but also can teaching programs for the complex surface[10]. Therefore this is obviously improving the grinding and polishing efficiency in trajectory planning process, as well as to reduce work intensity. In order to address all above problems, the factors of complex surface curvature is considered, moreover, advantage and disadvantage of the predecessors are summarized[11], so that a method of robotic machining trajectory automatic generation for grinding and polishing complex surface is presented in this paper, this method is combined CAD technology,   and taken Non-Uniform Rational B-Splines(NURBS) curve fitting method and adaptive sampling algorithm as the core. that is applied to automatic generate the motion trajectory in off-line programming for robot. Finally, the trajectory generation algorithm is completed in programming, and is taken computer simulation experiment in this paper.

2   Researching of the Trajectory Generation Problem of Robot for Complex Surface Grinding and Polishing

The trajectory planning is basic of robotic trajectory tracking and controlling, and it is also basic of performing tasks of robot. Robotic trajectory is displacement, velocity and acceleration of manipulator end in motion process. However, the path is a sequence of robotic position and posture which isn't considered to along with time changing. In order to complete the booking tasks, the trajectory must be planed firstly, however, how to generate motion trajectory? At present generation motion trajectory methods which include parametric method, cross-section method, guiding surface method and so on[12]. However, the robot manipulator end don't along with the ideal trajectory, namely that is only along with tangential path which approximate the ideal trajectory to motion. In other words, the generating trajectory must be discretized to approximate at last.

In the past, robotic polishing trajectory is simple for simple surface machining, which can not give play to the superiority of robot, In order to give play to the advantage of robot, then robot must be polished for more complex surface. But the complex surface is neither expressed to it's mathematical model by the simple nonparametric mathematical, nor is expressed by parameter mathematical, but that are grouped together for these surfaces, Therefore, the mathematical model of complex surfaces is difficult to express on mathematical expression. With CAD/CAM technology advancing, it is providing lots of capabilities for complex surface model designing. According to analysis of trajectory generation problems, A method of robotic machining trajectory automatic generation for complex surface grinding and polishing is presented, this method is combined CAD technology, and taken NURBS curve fitting and adaptive sampling algorithm as the core. The process of this method as follows: First, the complex surface CAD model is designed in the UG software, to be used discrete points to express the line-section which is generated by the cross-section method. According to machining precision, number of discrete points could increase or reduce. Secondly, to be fitted discrete points by the NURBS curve, it make the fitting curve more approach to the ideal curve. Finally, to be got discrete points set by the adaption sampling, it is achieving to automatic generate trajectory for off-line programming. and improving polishing quality and precision for complex   surface.

3   Algorithm of Grinding and Polishing Trajectory Generation for Complex Surface

The offset-section method is grinding tool along line-section which is generated by the surface and offset-section plane in the machining, as shown in figure3.1. This method can quickly generate grinding and polishing trajectory for robot. but it only suit to grind and polish the minor curvature surface. and its advantage is easy to generate the machining trajectory. However, its disadvantage is hard to control the distance in between section planes, and polishing feeding step, with that more cutter location discrete points must be distributed in the minor curvature surface but less in the greater curvature surface, resulting in poor of quality and precision of polishing for surface. Therefore, in order to generate more reasonable grinding and polishing trajectory, so line-sections which are generated by the cross-section method must be dealt again.

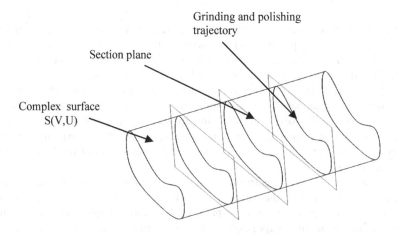

Fig. 1. Offset-section Method Generate Trajectory

3.1   Extracting Data Points from Complex Surface

To be extract data points, that is basic of curve fitting and curvature calculation when complex surface is completed CAD modeling. In this paper, to be modeled complex surface by CAD/CAM technology, then the cross-section method is adopted to get lots of line-sections, and to be taken a sequence of discrete data points to express each of the line-sections, which is getting ready for fitting the NURBS curve as fallow.

3.2   Mathematical Model of Line-Section

According to data points which are extracted from the series of line-sections in previous chapter, to be got NURBS curves of line-sections from discrete data points by the NURBS fitting method.

3.2.1   The NURBS Curve

To be extracted data points from each of the line-section, after that to be got the pitch-point vector U by the cumulative chord length parameterization, and to be got the matrix equation group from data points and pitch-point vector U which were put into the formula of NURBS,then the matrix equation group is solved to get controlling points. Finally, to be got the NURBS curve from the pitch-point vector U , data points, and controlling points.

(1) **Definition of a** p **Times   NURBS   Curve**[13]

$$P(u) = \frac{\sum_{i=0}^{n} N_{i,p}(u)\omega_i d_i}{\sum_{i=0}^{n} N_{i,p}(u)\omega_i} \qquad a \le u \le b \qquad \qquad ...(3.1)$$

Where d_i is the controlling point, which is formed the controlling polygon of the fitting curve. $\omega_i > 0$ is the weight factor of d_i. the $\omega_i > 0$ size can impact the curve shape, but the ω_i is assigned 1 in this paper, namely, $\omega_i = 1$. It show that every controlling points impact the curve shape in the same effect. $N_{i,p}(u)$ is the ith base function of the p times B-Spline in the U . it is defined as fallow:

$$\begin{cases} N_{i,0}(u) = \begin{cases} 1 & u_i \le u \le u_{i+1} \\ 0 & others \end{cases} \\ N_{i,p}(u) = \dfrac{u-u_i}{u_{i+p}-u_i} N_{i,p-1}(u) + \dfrac{u_{i+p+1}-u}{u_{i+p+1}-u_{i+1}} N_{i+1,p-1}(u) \\ rule \dfrac{0}{0} = 0 \end{cases} \qquad (3.2)$$

(2) Pitch-Point Vector Calculation

After a series of discrete points set $\{P_K\}(k=0,1......,n)$ were got from every line-sections, when every discrete points set were fitted the NURBS curve, the every $\{P_K\}(k=0,1......,n)$ was assigned a parameter value which was ordered assembled the U, the U is the pitch-point vector. At present, there are such parameters of the common methods, which are uniform parameterization method, accumulated chord length parameterization method, centripetal parameterization method and so on.

1 The Uniform Parameterization Method

$$u_0 = u_1 = u_2 = u_3 = 0$$

$$u_{k+3} = \frac{k}{n} \quad (k=1,2......,n-1)$$

$$u_{n+3} = u_{n+4} = u_{n+5} = u_{n+6} = 1$$

so that

$$U = \{0,0,0,0,u_4,......u_{n+2},1,1,1,1\}$$

2 The Accumulated Chord Length Parameterization Method

A chord is the ligature which is one point between adjacent one in the discrete point set of every line-section, the chord length L is sum of every chords of every line-sections. and then below.

$$u_0 = u_1 = u_2 = u_3 = 0$$

$$u_{k+3} = u_{k+2} + \frac{|P_k - P_{k-1}|}{L} \quad (k=1,2......,n-1)$$

$$u_{n+3} = u_{n+4} = u_{n+5} = u_{n+6} = 1$$

so that $$U = \{0,0,0,0,u_4,......u_{n+2},1,1,1,1\}$$

3 The Centripetal Parameterization Method

To be let

$$R = \sum_{k=1}^{n} \sqrt{|P_k - P_{k-1}|}$$

$$u_0 = u_1 = u_2 = u_3 = 0$$

$$u_{k+3} = u_{k+2} + \frac{\sqrt{|P_k - P_{k-1}|}}{R} \quad (k=1,2......,n-1)$$

$$u_{n+3} = u_{n+4} = u_{n+5} = u_{n+6} = 1$$

so that $$U = \{0,0,0,0,u_4,......u_{n+2},1,1,1,1\}$$

By the character of NURBS curve fitting, in order to make the curve which is fitted by line-section discrete points more approximate the ideal contour line, with that the more points must be distributed along the line-section with greater curvature, and less points along minor curvature. So the accumulated chord length parameterization method is adopted to calculate the pitch-point vector in this paper.

(3) Definition and Reverse Calculation of Controlling Points

When $\{P_K\}(k = 0,1......,n)$ were got from the line-section, every $\{P_K\}(k = 0,1......,n)$ was assigned a parameter value which was ordered assembled the U. To put the parameter value U into formula3.1, then the matrix equation group is got as fallow.

$$\begin{bmatrix} N_{0,n}(u_0) & N_{1,n}(u_0) & \cdots & N_{n,n}(u_0) \\ N_{0,n}(u_1) & N_{1,n}(u_1) & & N_{n,n}(u_1) \\ \vdots & \vdots & & \vdots \\ N_{0,n}(u_m) & N_{1,n}(u_m) & \cdots & N_{n,n}(u_m) \end{bmatrix} \begin{bmatrix} d_0 \\ d_1 \\ \vdots \\ d_n \end{bmatrix} = \begin{bmatrix} p_0 \\ p_1 \\ \vdots \\ p_k \end{bmatrix} \quad ....(3.3)$$

It is observed that the elements all of the B-spline's basic function, and which are only related to the node value. so all of controlling points d_n of fitting curve were solved from these equations.

In conclusion, to be got a p times NURBS curve by with the controlling points d_n, the fitting curve times p and the node vector U which were put into the formula3.1.

3.3   Self-adaptive Sampling Base on Curvature for The Curve

In order to improve the precision and efficiency of grinding and polishing for the complex surface, the grinding and polishing trajectory which was generated by the fitting curve must be discretized into the abrasive belt polishing contactors at last, then to be got the motion path from connecting every contactors to the others, the motion path is impacted by distribution of contact points, so a method base on the curvature characteristic is presented, this method make that more points must be reasonable distributed along the line-section with greater curvature, and less points along minor curvature. As shown in figure3.2, in order to balance the system, with that mass point must be more closed to the pivot with heavier quality, and mass point must be more farther away from the pivot with lighter quality. To be got inspirations from this, and being taken the curvature function $K(u)$ as the each mass points, in this way, discrete points of adaptive sampling is generated by the theory of leverage balancing .

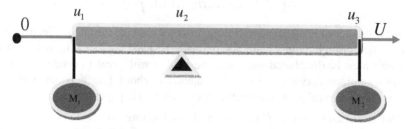

Fig. 2. Schematic diagram of leverage balancing system

(1) When the lever system is balancing, that have the relationship as fallow

$$M_1(u_2 - u_1) + M_2(u_2 - u_3) = 0 \qquad \ldots\ldots\ldots(3.4)$$

It can be introduced:

$$u_2 = \frac{M_1 u_1 + M_2 u_2}{M_1 + M_2}$$

This above formula could be written more simply as fallow:

$$\sum_{i=1}^{J}(u_2 - u_i)M_i = 0 \qquad \ldots\ldots\ldots(3.5)$$

(2) Calculation formula of curvature for NURBS curve

$$K(u) = \frac{|P'(u) \times P''(u)|}{|P'(u)|^3} \qquad \ldots \quad \ldots\ldots(3.6)$$

Then the characteristic function is shown that the curvature degree of the curve:

$$C(u) = \frac{K(u) - \min K(u)}{\max K(u) - \min K(u)} + \alpha \qquad \ldots\ldots\ldots(3.7)$$

Where $\alpha \geq 0$ is expressed the level that adaptive points distribution is impacted by the curvature, the more α, the less adaptive points distribution is impacted by the curvature, and tend to the uniform parameterization distribution with α increasing at last.

(3) Balancing Formula of NURBS Curvature

Because of being taken the curvature function $K(u)$ as the each mass points, so that the balancing formula of NURBS curvature is got from the formula3.5:

$$\sum_{j=1}^{k}(u_i - u_{ij})C(u_{ij}) = 0 \qquad \ldots\ldots\ldots(3.8)$$

Where K is adaptive sampling points number, $C(u_{ij})$ is curvature at position u_{ij}. u_i is balancing pivot, namely, that is sampling points.

(4) Iterative Solution

$$u_i^{(s+1)} = \frac{\sum\limits_{j}^{L} c(u_i^s)u_i^s}{\sum\limits_{j}^{L} c(u_i^s)} \qquad \cdots\cdots\cdots(3.9)$$

In the formula of iterative solution process, whose convergence condition is

$$|u_i^{(s+1)} - u_i^s| < \varepsilon$$

Where ε is the error of the curve fitting, s is the iteration number.

According to the generation process, in NURBS curve's parametric space, a series of adaptive sampling points u_i are got by the convergence condition. Finally, there is one-to-one correspondence between a series of the adaptive sampling points u_i and a series of the adaptive sampling points P_k which were generated by the NURBS curvilinear equation formula3.1.

4   Implementation of Algorithm and Instance Simulation

4.1   Description of Algorithm for Trajectory Automatic Generation

According to above of conclusion demonstrated that robot is used grinding and polishing complex surface, the algorithm flow of motion path automatic generation for robot as fallow:

4.2   Instance Simulation

4.2.1   Being Generated Grinding and Polishing Trajectory of Complex Surface

The master system is developed which was adopted secondary development of UG software to combine Visual C++6.0, then this algorithm is programmed and realized to generated grinding and polishing trajectory in this system. As shown in fig.4.1, which is a motion trajectory was automatic generated for one of any complex surfaces in the system.

4.2.2   Adaptive Sampling Method Compare with Common Others

The powerful function of MATLAB is used in this paper, that can more clearly explain the effect of various common sampling algorithm.

The NURBS curve is robotic motion trajectory, which is got from fitting discrete points of line-section. Then NURBS curve is sampled into cutter location discrete points set by adaptive sampling method, uniform parameterization method, and equal arc-length parameterization method. Meanwhile the effect of this methods sampling are simulated and compared by using the MATLBA.

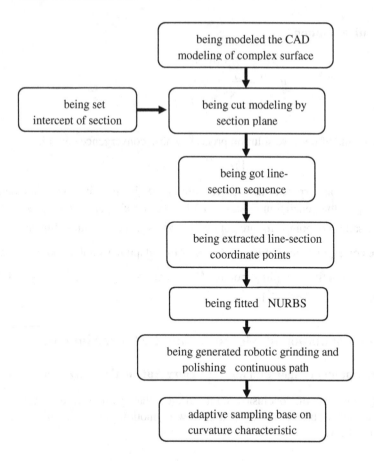

being modeled the CAD modeling of complex surface

being set intercept of section → being cut modeling by section plane

being got line-section sequence

being extracted line-section coordinate points

being fitted   NURBS

being generated robotic grinding and polishing   continuous path

adaptive sampling base on curvature characteristic

Fig. 3.   The Principle Diagram of Trajectory Generation

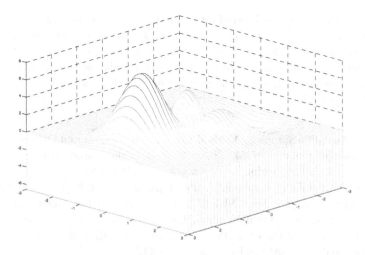

Fig. 4. Robotic grinding and polishing trajectory

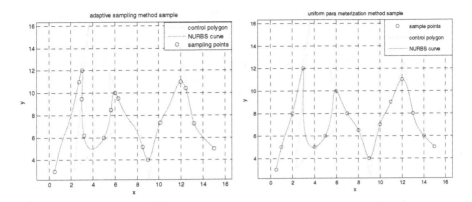

Fig. 5. Adaptive sampling method sample   **Fig. 6.** Uniform parameterization method sample

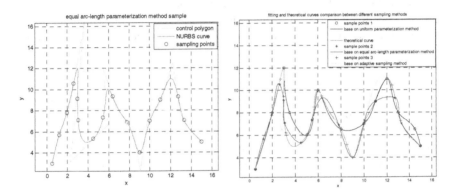

Fig. 7. Equal arc-length parameterization method sample

Fig. 8. Fitting and theoretical curves between different sampling methods

To be analyzed from above diagrams, By means of the self-adaptation sampling method, the distribution of sampling points accords with that the more sampling points are distributed along the surface with greater curvature, and less points along minor curvature. So that these sampling points are realized self-adaptive distribution along with change of curvature, and also the grinding and polishing trajectory is optimized. Therefore the effect of grinding and polishing is improved.

In addition comparing with three sampling method from above diagrams, sampling interval of self-adaptation sampling method is the shortest than other in the greatest curvature, it indicate that these sampling points could more highlight the curve characteristic, which are got by self-adaptive sampling method. Meanwhile it indicate that sampling points are got by self-adaptive sampling method which is better than others methods. Therefore, these sampling points could more highlight the curve characteristic and more reflect the machining trajectory information to improve grinding and polishing effect.

In order to explain which method is more approach to theoretical curve, however, each ten data points which are sampled from theoretical curve by adaptive sampling method, uniform parameterization method and equal arc-length parameterization method, as well as to fit each of them for NURBS curve. That as shown in figure4.5.

We known that fitting curve of adaptive sample method is the most approach to the theoretical curve from figure4.5. it illustrate that precision of adaptive sample method is more better than others, so that it can improve precision of grinding and polishing.

5   Conclusion

It is focused on the problem of high intensity, long time-consuming and low efficiency by on-line programming of robot in free-form surface grinding and polishing in this paper, In which, the complex surface CAD model is designed in the UG software, to be used discrete points to express the line-section which is generated by the cross-section method, according to machining precision, the number of discrete points could increase or reduce. Secondly, to be fitted discrete points by the NURBS curve, it make the fitting curve more approach to the ideal curve. Finally, to be got discrete points set by the adaption sampling, it is achieving to automatic generate the trajectory for off-line programming. and improving the polishing quality and precision for the complex surface. and overcoming many of shortcomings for adopting on-line programming.

Finally, the result of the computer simulations show that self-adaptive sample method is simple, effective, universal, practical and reliable. and which could not only automatic generate grinding and polishing trajectory for arbitrary complex surfaces, as well as can improve the quality, precision and efficiency of machining.

References

1. Wang, Q., Liu, Y.: Integrated Manufacturing Technology of CAD/CAM. Northeastern University Press, Shenyang (January 2001)
2. Wang, X., Chen, S., Chen, L.: The application and development of Technology for Mechanical CAD/CAM. Mechanical Industry Press, Beijing (January 2000)
3. Ning, R., Xu, H. (eds.): The Technology of CAD/CAM In Machine Manufacturing, pp. 10–23. Beijing Institute of Technology Press, Beijing (1991)
4. Ji, Z., Saito, K., Kondo, T., et al.: A new method of automatic polishing on curved aluminium alloy surface at constant pressure. International Journal of Machine Tools and Manufacture 35(12), 1683–1692 (1995)
5. Kuo, R.J.: A robotic die polishing system through fuzzy neural networks. Computers in Industry 32(3), 273–280 (1997)
6. Guvenc, L., Krishnaswamy, S.: Force controller design and evaluation for robot-assisted die and mould polishing. Mechanical Systems and Signal Processing. 9(1), 31–49 (1995)
7. Feng, H., Teng, Z.: Iso-planar piecewise linear NC tool path generation from discrete measured data points. Computer-Aided Design 37(1), 55–64 (2004)

8. Kiswanto, G., Lauwers, B., Kruth, J.: Gouging elimination through tool lifting in tool path generation for five-axis milling based on faceted models. The International Journal of Advanced Manufacturing Technology 32(3), 293–309 (2007)

9. He, G., Jia, H.: Adaptive sampling strategy for free-form surface based on CAD model. Advanced Materials Research 6, 541–544 (2012)

10. Li, L., Li, R.-W., Zou, Y.-B.: Algorithm of Industrial Robot on Path Generation for Complex Surface Machining. Machinery Design & Manufacture 11, 175–178 (2013)

11. Yu, L., Bo, Z., Li, D., et al.: Feedrate system dynamics based interpolation for NURBS curve. Journal of Mechanical Engineering 45(12), 187–191 (2009)

12. Yu, W.: Research of path planning technology of the automatic grinding and polishing die and module with free-form surface. Hunan University, Hunan (2005)

13. Piegl, L., Tiller, W.: The NURBS Book, 2nd edn. Springer, Heidelberg (2012)

The Control System Design of A SCARA Robot

Nianfeng Wang, Jinghui Liu, Shuai Wei, Zhijie Xu, and Xianmin Zhang

School of Mechanical and Automobile Engineering,
South China University of Technology,
Guangzhou, China

Abstract. According to the SCARA robot's control requirements in term of reliability and accuracy, AC servo motors are chose as the end effect and a double-CPU control system is built, which is consist of an industry control computer (IPC) as the host machine and a motion control card as the lower machine. With the Denavit-Hartenberg (D-H) modeling method of the robot, the kinematics and inverse kinematics analysis are implemented based on a Robotics Object Oriented Package in C++, ROBOOP. The interface of the control system is designed applying Microsoft Foundation Classes (MFC) and the hardware communication and control task is realized relying on the dynamic link library (DLL). Based on the double-CPU control system, the algorithm of the motion interpolation is designed. The trajectory optimization of the end effecter is added to the robot control strategy applying the Bezier curve. The robot accuracy is tested using a laser tracker.

Keywords: Industry robot, control system, motion interpolation, trajectory optimization.

1 Introduction

As electromechanical equipment, industry robot plays a key role in the domestic Industrial upgrading process, which not only can help forming a new intelligence industry, but also can be used to reform the backward production line so as to promote the automation level. The data from international federation of robotics (IFR) [1] shows that the installation number of industry robot in china in 2011 has increased by 51% (22577) compared to 2010, which is the highest increasing rate of short term industry robot installation during the past 50 years. IFR believes it's just a matter of time that china became the greatest robot demanded market. The SCARA robot is a kind of selective compliance assembly robot arm [2] and is highly used as assembly robot.

The major performance index of robot is repeat positioning accuracy and dynamic performance. As the first robot generation[3], series manipulator developed quickly and mature structures had been formed already, the joint of which is mostly consist of

X. Zhang et al. (Eds.): ICIRA 2014, Part II, LNAI 8918, pp. 136–145, 2014.
© Springer International Publishing Switzerland 2014

servo motor and reducers with high speed ratio and good accuracy, such as harmonic reducer[4] and RV reducer[5]. However, the control system still plays a very important role in maintaining the accuracy requirement since the mechanical clearance cannot be ignored. In recent years, many researchers are devoted to building a so call open control system [6~8], which is convenient for function expansion compared to the closed control system developed by robot companies like YASKAWA and ABB. Some researchers [9, 10] carry out motion simulation applying Robotics tool box in Matlab [11]. In this case, the kinematics and dynamics code programmed in Matlab can be put in a DLL file, which is ready for calling by other programming language. It complicates the control system design and limits the function expansion.

The Robotics Object Oriented Package in C++, ROBOOP [12], can be applied to avoid the problem of programming. ROBOOP is an open source tool box based in a matrix calculation library in C++, NEWMAT. ROBOOP is perfectly compatible with MFC, which makes it very convenient to design the interface of the control system. In the aspect of hardware, the control system is consist of an industry control computer (IPC) as the host machine and a motion control card as the lower machine, so as to build a double-CPU system. Those functions such as human computer interaction, motion calculation and real time management are realized in the IPC, while the transmission of pulse signal (pulse and direction) and IO management in the lower machine. It makes it an easy task to add future functions to the control system, such as machine vision and multi robot cooperation, etc. A fairly good dynamic performance and a rapid response can be realized because of the use of AC servo motor as the end effecter. To attain the accurate trajectory of the robot, the position control mode is set in the system. The motion interpolation consists of linear interpolation and circle interpolation. A new linear interpolation is developed to meet the demand of constant speed of the robot joint. Bezier curve is a smooth curve with explicit formulas, which can be used to link the different trajectory in Cartesian space so as to improve the robot dynamic performance.

2   The Mechanism and Kinematics of the SCARA Robot

There are three rotational joints and one translational joint to form the serial mechanism. The reduction mechanism in the rotational joints is harmonic reducer except the third axis, which make it achievable to get a high reduction ratio in a small space. The translational joint consists of a ball screw and a ball spline. Synchronous belts are used for the link between AC servo motors and the following mechanism. Figure 1 shows the robot's link schematic.

The key point of robot kinematic is the transformation of position and orientation between joint space and Cartesian space, which is the so-call kinematic and inverse kinematic. To solve the kinematic problem, the structure of the robot is simplified into

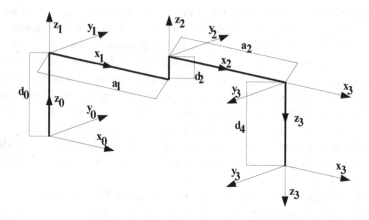

Fig. 1. The links schematic diagram of the robot. The base coordinate system is built on the base of the robot, while other coordinate systems are built on each corresponding joints.

a few linkages which are under specific mathematical description. A robot linkage can be described with four mathematical parameters, which are the link length, the link twist, the link offset and the joint angle. The link length is the length of the common perpendicular between the two axes. The link twist is the intersection angle between the two axes in the plane which is perpendicular to the common perpendicular. The link offset is the distance along the common axis's direction of the two adjacent linkages. The joint angle is the angle which is between the two adjacent linkages around their common axis. The link length and the link offset are used to describe the relation between the two joint axes, while the other parameters between the adjacent linkages. This is the so called Denavit-Hartenberg (D-H) modeling method of the robot. [13]The parameters of the SCARA robot are showed in the table 1.

Table 1. DH parameters of SCARA robot

Axis	α_{i-1}	a_{i-1}	d_i	θ_i
1	0	0	200mm	θ_1
2	0	225mm	41.3mm	θ_2
3	0	125mm	d_3	0
4	0	0	0	θ_4

The model of the robot can be built through the robot D-H parameters, which is the base of robot calculation, such as robot kinematics and the inverse kinematics. The deduction of the related formulas can be found in papper14 [14].

3   The Framework of the Control System

IPC is kind of special computer designed to industrial environment with a good performance in the aspect of anti electromagnetic interference and stability. Hence IPC is chosen as the host machine to complete the kinematics, trajectory planning and upper management, etc. The lower machine is a motion control card based on a DSP chip, which is used to accomplish servo control. Four AC servo motors with absolute encoders are used in the robot body as the control targets of the control system. A teach pendant and cameras can be added into the system to improve the control system. The double-CPU control system integrates the flexible of IPC and the stability of motion control card, which can not only realize the basic functions of the industrial robot, but also make it convenient to expansion of other related functions. The host machine, IPC, mostly manages the non-real-time tasks such as system initialization, parameter setting, off-line programming and interface interaction. The motion control card as the lower position machine output pulse of each axis and receives the IO signal and feedback signal.

Figure 2 shows the basic principle of the robot control system. The host machine is at the top of the control system which is base on the C++ programming language. As a matter of fact, the user commands the control system. IPC help the user to accomplish the commands. The motion commands are transfer into the motion control card through PCI bus. Then the respective pulse signals are sent into each servo drive.

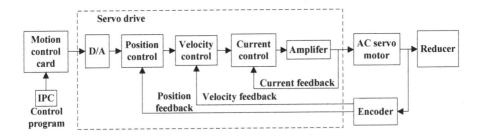

Fig. 2. SCARA robot's control principle

The mode of signal input is pulse and direction input, namely the so-called single pulse input. The basic principle of the single pulse input is to control the motor's direction by verifying the electrical level of the direction signal, while keeping the input pulse. Servo drive is playing the role of a signal amplification tool, which can transfer the pulse signal into the three-phase electrical signal to the servo motor, so as to drive the servo motor rotation. Meanwhile, the encoder of the servo motor returns position information to the control system, making a semi closed loop control.

The software framework of the SCARA robot's control system is showed in Figure 3, which can be divided into two parts: the user level and the system level. The user level includes the operating system and the user interface providing for users. The system level is accomplished inside the control system, transferring the operating command to the motor, completing the control tasks and accepting the feedback.

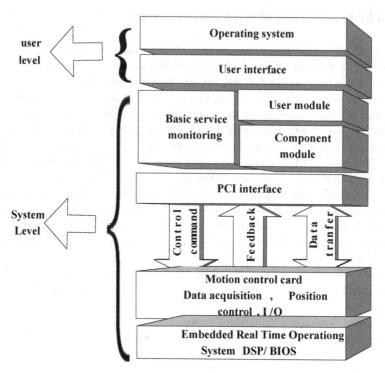

Fig. 3. SCARA robot's software layer architecture

The overall architecture of SCARA robot control system is showed in Figure 4, which can be divided into two parts: software layer and hardware layer. The software layer is implemented under the MFC environment. All functions including the motion parameters, motion operation, off-line programming, simulation and robot vision are developed in MFC. During the actual operation, the MFC and the hardware layer are connected by dynamic link library.

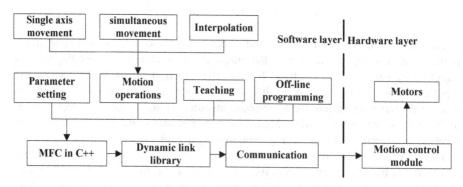

Fig. 4. The map of the control system

4   The Design of Motion Interpolation

The simplest case of robot motion is a trajectory from point to point in Cartesian space. The corresponding joint coordinates of starting point and end point can be calculated through inverse kinematics. The rotation angles of each joint are exactly the coordinates' subtraction of the two points in joint space, which is needed to convert into quantitative pulse to drive the servo motors through motion control card within the same time. The specific way is drawn in the diagram shown in figure 5. The point-to-point motion in Cartesian space is the basis of other complex robot movements.

The motion interpolation of robot's tip is a combination of many point-to-point motions in Cartesian space. The discrete points chosen in the interpolation are exactly located within the required trajectory before interpolation. Substantially, the interpolation is accomplished after continuously point-to-point motions. However, the actual situation is not quite simple. The issues needed to be considered include the distance between the two nearby points in different part of the required trajectory, the link up of velocity between discrete points or different trajectory, the acceleration and deceleration of the different trajectory parts and the appropriate adjustment of the trajectory, etc.

Linear interpolation and circular interpolation is the two common motion of robot. Any trajectory in Cartesian space can be fitted by using these two basic motions. One of the most important indexes is the possible minimum distance between two neighbor points in continuous motion interpolation, namely, the minimum time cell of two discrete points, which is deeply affected the motion interpolation precision. Different velocity in Cartesian space can have great impact on the index, which means an adaptive step size algorithm is the key to the next step work.

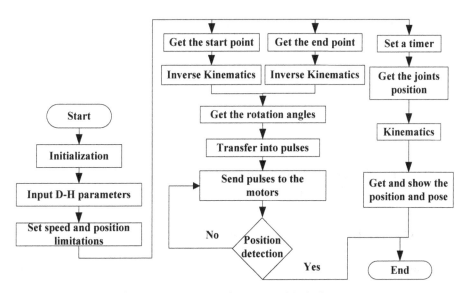

Fig. 5. The Point-To-Point program block diagram

5   The Trajectory Planning Using Bezier Curve

The required trajectory is defined by the user through off-line programming or online programming. However, the preliminary trajectory cannot be used directly to reproduce, which is because that there will be a sharp point between different trajectories leading to velocity mutation and robot vibration. To improve the motion performance, specific control program is designed to modify the preliminary trajectory in the backstage so as to attaining smooth transition between different trajectories.

Bezier curve is a smooth curve controlling by a few specific point, which can be easily obtained with stable numerical method. Whatever a straight line or a curve can be described in mathematics. Formula 1 is the two time formula of Bezier curve, where P_0 is the starting point and P_2 the end point, while P_1 is the control point.

$$B(t)=(1-t)^2 P_0+2t(1-t) P_1+t^2 P_2, t\in[0,1] \tag{1}$$

The sharp point between different trajectories can be eliminated by replacing the nearby part of trajectory around the sharp point with a specific Bezier curve. The four different situations are shown in the Figure 6. The transition parts are replaced by specific Bezier curves in dash form.

The control point (P_1) is the intersection point of the two trajectories and the other two points are chosen along the preliminary trajectories. The exact expression of the intermediate Bezier curve can be obtained, from which we get the new discrete interpolation point.

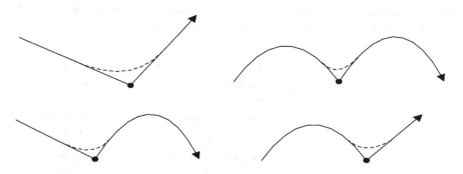

Fig. 6. The transition between two trajectories

The two time formula of Bezier curve can also be used to deal with the spatial transition between spatial line and circle.

6   Experiment Using Laser Tracker

In the experiment, we try to realize a simple motion with the online teaching method. The motion includes four key points taught by user.

1. Jog to point (338.02, 52.77, 110, 0) from the current position.
2. Go straight to (200, 52.90, 110, 0) in linear motion.
3. Move along an arc through the point (291.56, 135.79, 110, 0) and (338.02, 52.77, 110, 0).

The robot and its control cabinet are shown in figure 7. In the experiment, we use Leica AT901B laser tracker to measure the robot tip's motion, which is shown in Figure 8. The motion is in 3 dimensions (XYZ), including two lines, an arc and a Bezier curve transition part.

Fig. 7. The robot and its control cabinet

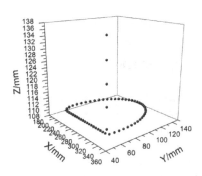

Fig.8. The trajectories of the robot tip

In Figure 9, the robot trajectory is shown in X-Y plane, included two curves: one is command trajectory wanted and the other is the actual position measured by laser tracker. The robot joints' position curve, speed curve and acceleration curve are shown in Figure 10, 11 and 12 respectively.

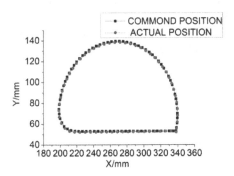

Fig. 9. the instructed and actual trajectories

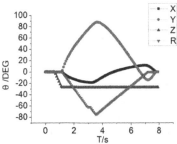

Fig. 10. Joint position curve

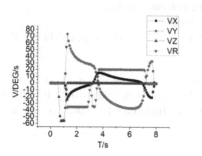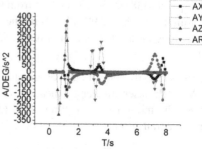

Fig.11. Joint speed curve   **Fig.12.** Joint acceleration curve

7   Conclusions

A double-CPU control system is designed based on the C++. Both the programming efficiency and running speed of the system are enhanced using the ROBOOP toolbox. The multi language compatibility problem is avoided at the same time.

Motion interpolation program in Cartesian space is successfully designed, including linear and circular interpolation. A better dynamic performance of the robot is obtained by the application of Bezier curve in trajectory planning so as to realizing the smooth connection between different trajectory, which can avoid the intersected sharp point and the varying degrees of vibration.

With the help of laser tracker, motion curves are obtained from experiments. The robot is stable and reliable.

Acknowledgement. This work is supported by National Natural Science Foundation of China(Grant Nos. 51205134, 91223201), Science and Technology Program of Guangzhou (Grant No. 2013JA300008), Research Project of State Key Laboratory of Mechanical System and Vibration (MSV201405), Natural Science Foundation of Guangdong Province（S2013030013355）, and Fundamental Research Funds for the Central Universities(Grant No. 2013ZM012).

References

1. Litzenberger, G.: Executive summary of World Robotics (2012)
2. Makino, H., Furuya, N.: SCARA robot and its family. In: Proc. 3rd Int. Conf. on Assembly Automation, pp. 433–444 (May 1982)
3. Brogårdh, T.: Present and future robot control development–An industrial perspective. Ann. Rev. Cont. 31(1), 69–79 (2007)
4. Ueura, K., Slatter, R.: Development of the harmonic drive gear for space applications. European Space Agency-Publications-ESA SP 438, 259–264 (1999)
5. YingHui, Z., WeiDong, H., JunJun, X.: Dynamical model of RV reducer and key influence of stiffness to the nature character. In: 2010 Third International Conference Information and Computing (ICIC), vol. 1, pp. 192–195. IEEE (June 2010)

6. Suoxian, L.: Development of Robot Control Software Based on VC++ and PMAC. Micro. Info. 5, 088 (2008)
7. Farooq, M., Wang, D.B.: Implementation of a new PC based controller for a PUMA robot. J. Zhejiang Uni. SCI. A 8(12), 1962–1970 (2007)
8. Bin, X.U.: Research on the Control System of Robot-arm Based on Motion Control Card. J. Hefei Uni. (Natural Sciences) 3, 022 (2010) (in Chinese)
9. Wang, Z.X., Fan, W.X., Zhang, B.C., Shi, Y.Y.: Kinematical analysis and simulation of industrial robot based on Matlab. Jidian Gongcheng/ Mech. & Elect. Eng. Mag. 29(1), 33–37 (2012)
10. Luo, J.J., Hu, G.Q.: Study on the Simulation of Robot Motion Based on Matlab. J.-Xiamen Uni. Nat. Sci. 44(5), 640 (2005)
11. Corke, P.: A robotics toolbox for MATLAB. IEEE Rob. & Aut. Mag. 3(1), 24–32 (1996)
12. Gordeau, R.: Roboop–a robotics object oriented package in C++ (2005)
13. Denavit, J.: A kinematic notation for lower-pair mechanisms based on matrices. Trans. of the ASME. J. App. Mech. 22, 215–221 (1955)
14. Craig, J.J.: Introduction to robotics: Mechanics and control, pp. 144–146. Pearson/Prentice Hall, New York (2005)

Trajectory Planning with Bezier Curve in Cartesian Space for Industrial Gluing Robot

Zhijie Xu, Shuai Wei, Nianfeng Wang[*], and Xianmin Zhang

School of Mechanical and Automobile Engineering, South China University of Technology, Guangzhou, China
menfwang@scut.edu.cn

Abstract. This paper deals with the trajectory planning problem of a 6-DOF gluing robot to achieve a smooth and controllable tracking performance during the transition path between two basic paths, which mean straight lines and circular arcs, in Cartesian space. First, Bezier curve, which is intuitive and easy to control, is employed in the trajectory planning to generate the interpolation points, which need good accuracy and stability of velocity in Cartesian space. Then, the velocity controlling plan for the end-effector during transition path would be discussed, and it offers great controllability and stability of velocity on the transition paths. Finally, a comparison of the performance both in joint space and Cartesian space between trajectory planning with cubic spline in joint space and that with Bezier curve in Cartesian space is presented to show the advantages of Bezier curve.

Keywords: gluing robot, trajectory planning, Bezier curve, cubic spline, Cartesian space, joint space.

1   Introduction

Many studies have been done on the trajectory planning of 6-DOF industrial robot. However most of them focus on the global optimization of trajectory planning in joint space or in Cartesian space to improve the efficiency, smoothness and accuracy of the robot, having little concern about the accuracy and stability of velocity of the end-effector during transition path in Cartesian space [1]. Moreover most of the applications of the robot, such as wielding and transferring, have no requirement on the robot performance during the transition path between two basic paths [2] which are straight lines and circular arcs, other than the smoothness and continuity of the joint angle positions [3].

While in gluing process, the paths in Cartesian space are required to be smooth and continuous and the velocity of end-effector is required to be stable and controllable in order to make the flow line of the glue continuous and spread evenly during the transition path. Continuity and uniform distribution of the flow line, the two most important factors to evaluate the quality of gluing process, are affected by many

---

[*] Corresponding author.

X. Zhang et al. (Eds.): ICIRA 2014, Part II, LNAI 8918, pp. 146–154, 2014.

factors, such as the thickness of the flow line, the viscosity of the glue and the performance of the robot [4]. Furthermore, concerning the quality of the flow line, the path and velocity of the end-effector in Cartesian space should be intuitive and easy to operate to adjust the flow line with different thickness and viscosity.

In order to fit the requirement mentioned above, Bezier curve is employed in trajectory planning in Cartesian space during the transition path, whose characteristics bring great benefits to control the accuracy and curvature of the transition path, and control the velocity of the end-effector during the transition path by changing the parameters mentioned in chapter 2 of this paper.

In the first part of this paper, a trajectory planning method with Bezier curve and a velocity controlling method for it would be introduced. In the second part, the trajectory planning with cubic spline is presented. A comparison between the performance of trajectory planning with Bezier curve in Cartesian space and that with cubic spline in joint space, transition paths, velocity of the end-effector and joint angle position, is made in the last part of the paper.

2   Trajectory Planning with Bezier Curve in Cartesian Space

In order to improve the controllability and stability of the gluing robot during transition path, trajectory planning with Bezier curve, which is intuitive and easy to observe the path and attitude of the end-effector of the robot, is employed in this case.

2.1   Generating the Path

In this part, Bezier curve is employed to generate the transition path between two basic paths. To improve the smoothness of the transition path, we involve Bezier curve, so that the jerk of the resulting trajectory is continuous [5].

The function expressions of Bezier curve can be obtained as:

$$B(t) = \sum_{i=0}^{n} \binom{n}{i} P_i (1-t)^{n-i} t^i = \binom{n}{0} P_0 (1-t)^n t^0 + \binom{n}{1} P_1 (1-t)^{n-1} t^1 + \ldots$$
$$+ \binom{n}{n-1} P_{n-1} (1-t)^1 t^{n-1} + \binom{n}{n} P_n (1-t)^0 t^n, t \in [0,1].$$

(1)

where:

P_0, P_n : starting point and finishing point of Bezier curve,

$P_1 \ldots P_{n-1}$: controlling points of Bezier curve,

t : the parameter that influence the distribution of the interpolation points [6].

In our case, cubic Bezier curve which has two controlling points to control the curvature would be suitable. And its function expressions would be:

$$B(t) = P_0 (1-t)^3 + 3P_1 t (1-t)^2 + 3P_2 t^2 (1-t) + P_3 t^3, t \in [0,1].$$

(2)

As fig.1 shows, line1 and line2 are the basic paths, and the small circles on transition path indicate the interpolation points. P_1, P_2 and P_3 indicate the starting point, controlling point and finishing point respectively.

One of the advantages of Bezier curve in Cartesian space is the convenience of controlling the transition path in Cartesian space by changing the parameters of Bezier curve function [7].

In fig.2, P_1 and P_2 are the starting point and finishing point of 3 Bezier curves whose controlling points were also indicated in the figure in 3 cases.

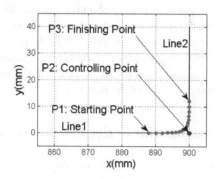

Fig. 1. Transition path generated with Bezier curve in Cartesian space

In fig.3, P_1 is the controlling point of 3 Bezier curves, whose starting points and finishing points were also indicated in the figure in 3 pairs.

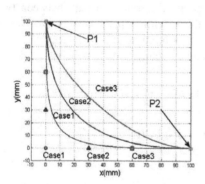
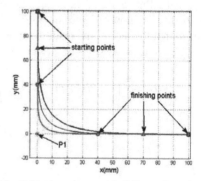

Fig. 2. Bezier curves with different controlling points

Fig. 3. Bezier curves with different length of transition part

According to fig.2, when the starting point and finishing point stay the same, the longer distance between the starting point and controlling point, the higher curvature of the Bezier curve would be. According to fig.3, when the controlling point stay the same, the shorter distance between the starting point and controlling point, the higher curvature of the Bezier curve would be.

By changing the two parameters mentioned above, it is really convenient for the user to choose the accuracy class of the transition path, which plays an important role in gluing procedure[8,9].

2.2   Speed Controlling

Mostly, the end-effector velocity (magnitude and direction) would change during the transition path between two basic paths. Take fig.1 for an example, when the transition angle is around 90°, the end-effector velocity along former basic path would decrease from V_1 (former speed) to 0, and the speed along later basic path would increase from 0 to V_2 (later speed). While in fig.4, it's a different situation, the transition angle is around 180°, and only the speed would change from c to V_2. While in fig.5, the transition angle is around 0°. The speed would slow down from V_1 at P_1 to 0 at the controlling point P_2, and then it would change the direction and increase from 0 at P_2 to V_2 at the finishing point P_3.

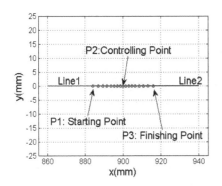 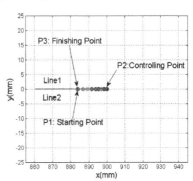

Fig. 4. Transition path when transition angle is 180°

Fig. 5. Transition path when the transition angle is 0°

We can draw a conclusion that, the speed changing during the transition path would be influenced by the highest curvature along the path, which would be influenced by transition angel and the length of the transition part, the former speed (V_1) and the later speed (V_2). The transition velocity can be expressed as follows:

Despite the transition angle or the transition angle is 180°, the speed would be:

$$V_i = V_i + (V_2 - V_1) * t_i \tag{3}$$

$$t_i = i / n \tag{4}$$

where:

n : the number of the interpolation points during transition path,

V_1 : the starting speed of transition,

V_2 : the ending speed of transition.

Concerning the transition angle and the length of transition part, the velocity during transition path would be:

$$V_i' = V_i / slowscale_i \qquad (5)$$

where:

$$V_i = V_1 + (V_2 - V_1) * t_i \qquad (6)$$

$$slowscale_i = \frac{n/2 - |i - n/2|}{n/2} * slowscale_{max} + 1 \qquad (7)$$

$$slowscale_{max} = \frac{180 - \theta}{180} * scale + 1 \qquad (8)$$

θ and n indicate the transition angle and the number of the interpolation points during transition path. And *scale* is determined by the length of the transition part, which will affect the slowest speed of the motion during transition path.

3   Trajectory Planning with Cubic Spline in Joint Space

In order to reduce the soft impact caused by exporting torques, we involve the cubic spline in the interpolations in joint space, which exerts an up to third-order continuous differentiation feature on the joint angle *Position* (θ_i) – *Time* (t_i) curves [10].

Take fig.6 for an example, to finish the trajectory planning of the path in fig.6, the joint angle *Position* (θ_i) – *Time* (t_i) curves would be like the curves in fig.7. Obviously, there are two sudden changes both in Cartesian space and joint space, which need to be avoided in the gluing procedure.

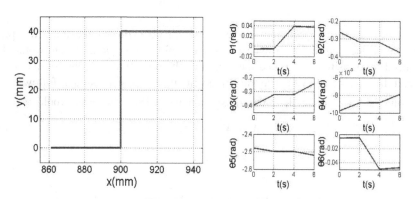

Fig. 6. The paths before trajectory planning   **Fig. 7.** Joint angle Position –Time curvesbefore trajectory planning

Trajectory planning in joint space with cubic spline was employed in this case, in which the path of the end-effector would be like the one in fig.8. And the joint angle *Position* (θ_i) – *Time* (t_i) curves would be like the ones in fig.9.

After trajectory planning with cubic spline in joint space, as fig.9 shows, joint angle positions have good smoothness to reduce the soft impact on the actuators, and the path in Cartesian space would have good smoothness and the interpolating points distribute evenly, which are good features for the interpolation. But the transition part would be difficult to control in detail, because the interpolation would be done in joint space and it needs forward kinematic calculation, which is not intuitive, to transform into Cartesian space[11].

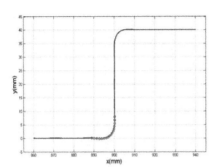 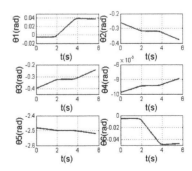

Fig. 8. Transition path after trajectory planning with cubic spline

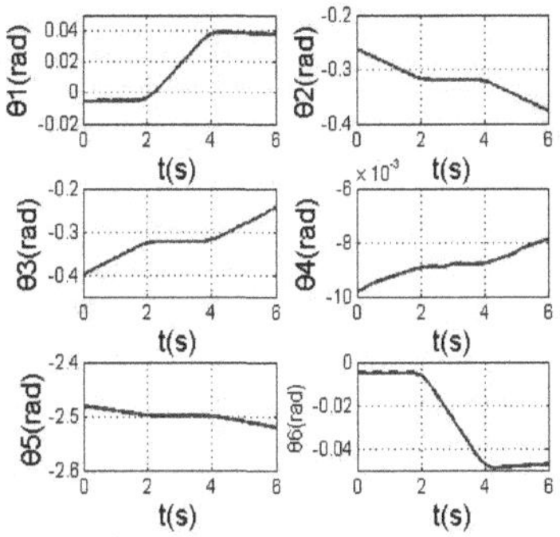

Fig. 9. Joint angle Position – Time curves after trajectory planning with cubic spline

4   Performance of the Trajectory Generated by Bezier Curve and Cubic Spline

Take fig.6 for an example, Bezier curve is employed. The path after trajectory planning would be like fig.10. And the joint angle *Position* (θ_i) – *Time* (t_i) curves are shown in fig.11.

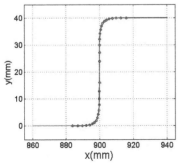

Fig. 10. Transition path after trajectory planning with Bezier curve in Cartesian space

Fig. 11. Joint angle *Position –Time* curves after trajectory planning with Bezier curve

Moreover, in Cartesian space, the *Velocity of the end-effector* (V) – *Time* (t) curves would be like the curves in fig.12. The velocity of the path generated with cubic spline has less stability than the one with Bezier curve. And the velocity of the path generated with Bezier curve could be easily controlled through the *slowscale*$_{max}$ (S) which is a parameter in (8) mentioned in chapter 2 of this paper.

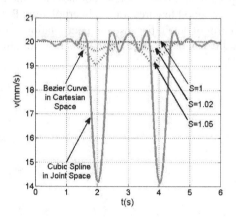

Fig. 12. Velocity of the end-effector after trajectory planning

Concerning the joint angle *Position* (θ_i) – *Time* (t_i) curves in fig.9 and that in fig.11, obviously, the curves have similar shapes which have good smoothness at the transition part to reduce the soft impact on the actuators.

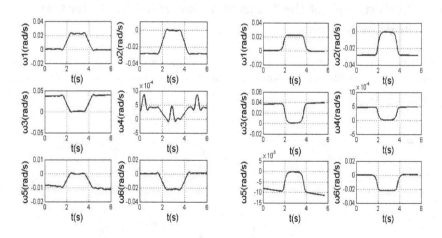

Fig. 13. Angular Velocity – Time curves after trajectory planning with cubic spline

Fig. 14. *Angular Velocity – Time* curves after trajectory planning with Bezier curve

Although the joint angle Position (θ_i) – Time (t_i) curves are similar in shape, the Angular Velocity of the joints (ω_i) – Time (t_i) curves of them are different as it is shown in fig.13 and fig.14. And the curves in fig.13, which is of the path generated with cubic spline in joint space, have better smoothness than that in fig.14, which is of the path generated with Bezier curve in Cartesian space. Although the curve of ω_4 in fig.13 has worse performance than that in fig.14, the changes of ω_4 are really small, which is about $5*10^{-4} \, rad \, / \, s$ and it makes small impact on the result.

We can draw a conclusion that in the joint space trajectory planning with cubic spline has better performance than that with Bezier curve.

5   Conclusion

This paper introduces a method of trajectory planning with Bezier curve in Cartesian space for the gluing industrial robot. In order to make the flow line spread evenly and continuous, this kind of trajectory planning with Bezier curve would be suitable for the gluing work. It is convenient to control the path and the orientation of the end-effector, as the accuracy class and curvature of the path could be easily controlled by changing the positions of controlling points, starting and finishing points of the Bezier curves. Also, the velocity of the end-effector with interpolation of Bezier curve in Cartesian space has better performance than the one with interpolation of cubic spline in joint space in controllability and stability. Moreover, in the joint space, the joint angle position-time curves of these two kinds of trajectory planning are similar in shape, which have good smoothness to reduce the impact. Last but not least, the angular velocity – time curves of trajectory planning with Bezier curve have worst performance than that of trajectory planning with cubic spline, even though the difference is not so significant.

Next step of the research should employ Bezier curve and cubic spline in one trajectory planning task to make the path in Cartesian space smooth and controllable by the use of Bezier curve, and at the same time make the position, angular velocity and angular acceleration of the joints continuous by the use of cubic spline.

Acknowledgement. This work is supported by National Natural Science Foundation of China(Grant Nos. 51205134, 91223201), Science and Technology Program of Guangzhou (Grant No. 2013JA300008), Research Project of State Key Laboratory of Mechanical System and Vibration (MSV201405), Natural Science Foundation of Guangdong Province（S2013030013355）, and Fundamental Research Funds for the Central Universities(Grant No. 2013ZM012).

References

1. Stilman, M.: Global manipulation planning in robot joint space with task constraints. Robotics 26, 576–584 (2010)

2. Liao, X., Wang, W., Lin, Y.: Time-optimal trajectory planning for a 6R jointed welding robot using adaptive genetic algorithms. In: Computer, Mechatronics, Control and Electronic Engineering (CMCE), vol. 2, pp. 600–603 (2010)
3. Xiao, Q.G., Ji, D.W.: Trajectory planning theory and method of industrial robot. In: Computer Research and Development (ICCRD), vol. 2, pp. 340–343 (2011)
4. Ma, K., Zhang, B.: The application of robot in the automated gluing system. Journal of North China Institute of Science and Technology 1, 47–52 (2007)
5. Ozaki, H., Hirano, K., Iwamura, M.: Improvement of trajectory tracking for industrial robot arms by learning control with B-spline. In: Assembly and Task Planning, pp. 264–269 (2003)
6. Ju, M.Z.: HONG, L.W.: Inserting control points for Bezier curve approximation. In: Control Conference (CCC), vol. 31 (2012)
7. Hou, J.H.: Two parameters extension of quartic Bézier curve and its applications. In: Computer Science and Automation Engineering (CSAE), vol. 1, pp. 191–194 (2011)
8. Xiao, G., Xu, X.: Study on Bezier Curve Variable Step-length Algorithm. Physics Procedia 25, 1781–1786 (2012)
9. Faraway, J.J., Reed, M.P., Wang, J.: Modelling three-dimensional trajectories by using Bézier curves with application to hand motion. Journal of the Royal Statistical Society: Series C (Applied Statistics) 56, 571–585 (2007)
10. Kelly, R., Davila, V.S., Perez, J.A.: The Grid: Control of robot manipulators in joint space. Springer (2006)
11. Stilman, M.: Task constrained motion planning in robot joint space. In: Intelligent Robots and Systems, pp. 3074–3081 (2007)

CogRSim: A Robotic Simulator Software

Yao Shi[1], Chang'an Yi[2,*], Yuguang Yan[1], Deqin Wang[1],
Guixin Guo[1], Junsheng Chen[1], and Huaqing Min[1]

[1] School of Software Engineering, South China University of Technology,
510006 Guangzhou, China
[2] School of Computer Science and Engineering, South China University of Technology,
510006 Guangzhou, China
`yi.changan@mail.scut.edu.cn, yichangan.ok@gmail.com`

Abstract. Robot simulator is highly necessary and useful because it could provide flexibility and repeatability in robotic research, and the user is able to freely exert his imagination to design robot environment and algorithms. This paper presents CogRSim, a robot simulator which offers four features that differentiate it from most existing ones: (1) 3-dimensional scene could be viewed synchronously in 2-dimensional panel, which makes the user easy to observe the interaction process between the robots and environment; (2) user-defined experimental result might be displayed in appropriate style similar to Matlab; (3) the user can adjust the simulation speed, pause, restart, slower or faster; (4) some objects can be assembled into a complex one in arbitrary shape and unified physical attribute. CogRSim's architecture and subsystems are described in detail, and then a navigation experiment is carried out to verify its usability.

Keywords: Robot simulator, Open Dynamics Engine, Irrlicht.

1 Introduction

The final aim of robotics research is to make the real robots work correctly and effectively following the algorithms. However, some characteristics of robot hardware, such as low abrasion resistance, high price, difficult to produce and modify, etc, prevent the researchers from freely carrying out experiments. Simulations are easier to setup, less expensive, faster and more convenient to use, as a result, they are often prior to investigations with real robots. A good robot simulator has at least three advantages: decrease the abrade on the real hardware; the researcher could not only change the design of the robot and environment at any time, but also test any new algorithms or even unimaginable ideas; being a good communication media among people from different locations and research fields.

A central principle of cognitive robotics is that effective systems could be designed by eliminating complex internal representations and focusing instead on the direct

---

* Corresponding author.

X. Zhang et al. (Eds.): ICIRA 2014, Part II, LNAI 8918, pp. 155–166, 2014.

relation between environmental stimulus and behavior generation. From this perspective, a good simulation experiment must simultaneously provide three kinds of accurate models: the robot's own geometry, kinematics and sensors; the environment; the physics interaction between the robot and environment. Furthermore, the computational power of computers now has been significantly promoted which makes it possible to run computationally intensive algorithms on personal computers instead of special purpose hardware.

Since 1980s, robot simulator has gained more and more attention and a growing number of commercial and open-source software-simulation tools have been designed both at home and abroad. Existing simulators are either expensive or inflexible in task design, many of them do not contain study material for new users and researchers, and all of these are the motivation to develop CogRSim which is designed to be a free, robust and easy-to-use simulation software for robot research and education.

Besides the traditional functionalities, the main contribution of CogRSim contains four aspects, (i) the two-dimensional (2D) display of the three-dimensional (3D) scene in real time, (ii) flexible display of run-time analysis, (iii) liberal control of simulation speed, such as pause, stop, faster, slower and restart, (iv) assemble some objects into a complex one which has unified physical attribute.

The rest of this paper is organized as follows: section 2 discusses the related work of robot simulators especially the most recent and popular ones, section 3 and 4 present the architecture and subsystems of CogRSim respectively, section 5 gives an example to illustrate the usability of CogRSim, section 6 concludes the whole paper and points out the future trend.

2   Related Work

The robotics community has started to devote increasing interest to simulation tools and their applicability, for example, a workshop at the 2009 IEEE International Conference on Robotics and Automation (ICRA) is about the need for open-source robotic simulation software. Moreover, there are also some papers published talking about simulators [1], [2], [3], [4].

Webots is a commercial development environment used to model, program and simulate mobile robots [5], [6]. With Webots the user can design complex robotic setups with any number of similar or different robots in a shared environment, and the user can choose or set the properties of each object. The robot controllers can be programmed with the built-in IDE (Integrated Development Environment) or with third party development environments. Furthermore, the provided robot libraries enable the user to transfer the control programs to several commercially available real mobile robots. Now, Webots is used by over 1137 universities and research institutes worldwide.

Gazebo is a multi-robot simulator for outdoor environment [7]. With Stage, the user is able to simulate a population of robots, sensors and objects, but does so in 3D world. Because it includes an accurate simulation of rigid-body physics, it could generate both realistic sensor feedback and physically interactions between objects.

Gazebo is developed in cooperation with the Player and Stage projects, and it is compatible with Player. The Player and Stage projects have been in development since 2001, during which time they have experienced wide spread usage in both academic and industry fields. Player provides network-transparent robot control, and Stage is a simulator for large populations of mobile robots in complex 2D domain and is quite capable of simulating the interactions between robots in indoor environments.

Microsoft Robotics Developer Studio (MRDS) is a Windows-based 3D simulator, and it can be used by both professional and non-professional developers as well as hobbyists [8]. MRDS is freely available but not open source, its programming environment is .NET-based and the user could build robotic applications across a variety of hardware equipment. The Visual Simulation Environment, a key part of MRDS, uses Microsoft XNA Framework to render the virtual world and NVIDIA PhysX to approximate interactions between objects within the world. MRDS's current version includes support for both real and simulated Kinect sensor which can be used for navigation and interaction with people.

In fact, some tools that are not dedicated for robot simulation could also provide similar functionalities. For example, Robotics Toolbox encapsulated in Matlab could simulate the kinematics, dynamics and planning for robotic application [9], Robot Operating System (ROS) which is a software framework for robot software development might also offer simulation tools [10]. Furthermore, some simulators are even specific for a particular kind of robot such as Webots for NAO [11] and simulator of iRobot [12], the stages to develop such a robot application include model, program, simulate and finally transfer to a real robot.

3   Architecture of CogRSim

To develop a robot simulator, two basic tools are always elaborately selected. The first is physics engine to simulate rigid body dynamics and features such as joints, collision detection, mass and rotational functions, the other is rendering engine to render the robotic models, objects and environment in 3D scene. In our development tools, these two engines are Open Dynamics Engine (ODE) [13] and Irrlicht [14] respectively which are both free and open source, the operation system is win7, the IDE is Visual Studio 2010, the GUI (Graphical User Interface) is developed by Qt Creator [15] , and Matlab version is 2010b. Fig. 1 shows the relationship between these tools, and the arrows represent control flow.

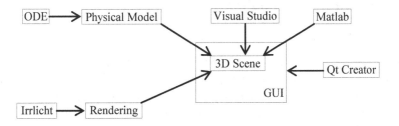

Fig. 1. The relationship between the tools to develop CogRSim

3.1   General Architecture

CogRSim consists of a number of essential modules (subsystems) to make it both easy-to-use and powerful, its general architecture is shown in Fig. 2. Module "World Model" provides the model for robots, objects and other elements in 3D scene, and it is the basis of "Scene Editor" to construct appropriate scene. In 3D scene, an element represents a robot, object, or any other individual thing.

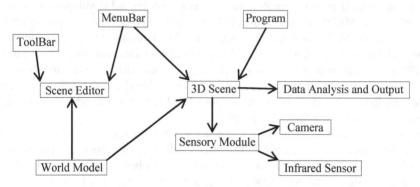

Fig. 2. The general architecture of CogRSim, and the arrows represent data flow

3.2   Module Management

The modules in CogRSim need to communicate with each other, and class "MainManager" is used to manage the communication among them. For each module, there is a pointer inside MainManager to control this module.

Take camera as an example to explain the work flow of MainManager: When method "createCameraAction" sends the signal of creating a camera, MainManager will receive this signal and call method "createCamera" to create a camera, then some parameters will be sent to 3D scene. MainManager has the pointers of 3D scene's display and edit, thus all the work could be finished in MainManager and the consistency could be guaranteed. This work flow can also be described as (1):

$$createCameraAction \rightarrow MainManager \rightarrow createCamera \qquad (1)$$

4   Detailed Implementation

4.1   Main Window

CogRSim's main window is shown in Fig. 3, the user can edit the 3D scene and program to test his idea. Now, the "Code Window" depends on Visual Studio to write code for the user. By dragging the mouse or set the corresponding attributes, the users can also change the viewpoint's position, orientation and zoom as shown in Fig. 4, and this functionality is realized by the main camera in 3D scene. On the other hand, the layout of modules would not be deformed when be dragged, and floating window could be activated, scaled or reopened at any time.

Through ODE, the time flow could be controlled and the simulation speed might be adjusted, for example, if the user needs faster speed instead of higher precision he only needs to decrease the time step. This functionality is illustrated through ControlBar in Fig. 3.

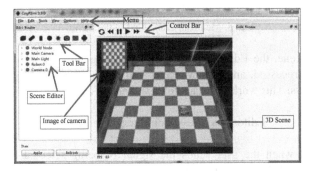

Fig. 3. Main Window and its modules

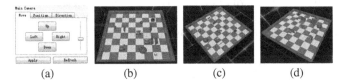

(a)   (b)   (c)   (d)

Fig. 4. Observe the whole scene in different angles. (a) the buttons are used to change visual angles; (b), (c) and (d) are different angles of the main camera.

4.2   Scene Editor

This part is used to edit for and get current information from the 3D scene. In CogRSim, the user could use the classes TreeView, ToolBar, MenuBar to edit 3D scene, and their relationship is shown in Table 1.

Table 1. The core classes related with Scene Editor

Class Name	Introduction of the class
TreeView	A base class for constructing the tree structure in scene editor
ToolBar	A quick way to add certain kinds of elements for 3D scene
MenuBar	Similar to ToolBar, but could define the attribute values in advance

The result of any operation in ToolBar or MenuBar will display in the tree structure, ToolBar is shown in Fig. 3 and MenuBar is shown in Fig. 5.

Fig. 5. MenuBar embedded in the menu

To edit 3D scene, the Editor module only sends signals without caring how the signals will be processed, and MainManager will revoke the corresponding modules for further process. This work flow can be described as (2):

$$Editor\ module \rightarrow MainManager \rightarrow 3D\ scene \tag{2}$$

In TreeView, when the user sends the signal of being selected of a node, the method "refreshNode" will receive this signal and refresh the node information, and the method "itemSelectedSlot" will also receive this signal and then highlight the object or robot in 3D scene. The user can also assemble objects easily, install cameras for objects or robots, as shown in Fig. 6. When deleting the parent node, the children nodes will be deleted together. Furthermore, Light as a shining source could be created in a user-defined position in the 3D scene to send bright.

(a) Assemble three objects in TreeView   (b) The assembled object as a whole

Fig. 6. Assemble some objects into a complex one which has unified physical attribute

4.3   2D Display in Real-Time

When many robots and objects are interacting, the user needs to observe them in a simple form in real time. In this module, all the robots, objects, as well as the trajectory of the robot, are drawn in predefined shape, size, color and name. All the original information are gained from ODE and Irrlicht, and are drawn through Widget of Qt Creator. Class "QPainter" could draw any kind of shape with Qpen or QBrush, and it has various methods such as setBrush(), setPen() and drawRect().

To draw the trajectory of the robot, we use an object called "robot_path" to store its historical coordinates and an integer variable to control the repaint time interval, the initial position of the robot could also be gained and stored in "robot_path". To make the 2D display correctly, we use some "Managers" to manage the robots and objects, as shown in Table 2.

Table 2. Managers to manage the elements in 3D scene

Class name	Function
RobotManager	Gain and store the coordinates of the robot
ObjectManager	Read and store the categories and positions of objects
WallManager	Store the walls in 3D scene

For data obtain, each kind of object has a link list, for example, object list, robot list, obstacle list, etc. From the link list, the system could get the coordinate directly in real time no matter how large or complex the 3D scene is. With data in the link list, the run time information could be re-displayed again and again. Furthermore, the management of all the elements in 3D scene is through the same source, and the information could be gained in real time and in high precision. Fig. 7 shows the 3D scene and its corresponding 2D display, the dark broken line represents the robot's trajectory and the objects are also drawn in predefined style.

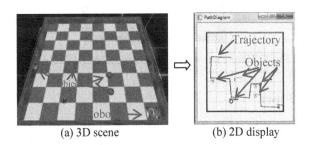

(a) 3D scene   (b) 2D display

Fig. 7. 2D display of 3D scene. The objects in (a) are shown in (b), and the robot's trajectory is also drawn.

4.4   Camera

Camera is an important sensory for a robot to gain outside information, just like the eyes of human beings. In order to view the world in multi-angle, there should be many cameras in different locations and angles. In CogRSim, the user could fix a camera in any position of the robot or object, and the camera will move together with it and could gather dynamic information. The image captured by the camera is displayed in real time, but the user can choose whether to be shown, as well as select and drag any image in the 3D scene.

To realize the above functionalities, three classes, Camera, CameraItem and CameraManager, are used, and they are described in Table 3 in brief.

Table 3. Three classes for camera subsystem

Class name	Introduction of the class
Camera	Attributes and basic methods of a camera, for example, id, position, whether being chosen, whether to be shown, information save.
CameraItem	Store the pointer of each camera, and could get the pointer of each camera object through its id.
CameraManager	Manage all the cameras, the construction and deletion of each camera, the render of the camera information by priority when overlapped.

One example of multi-camera is shown in Fig. 8. There are three cameras in different locations, and the top left picture is from the camera under the robot which could capture the floor and wheels.

Fig. 8. Three cameras in different position and direction

4.5   Infrared Sensor

Infrared sensor is usually set in front of the robot to detect obstacles, and then the robot could decide whether to avoid. Based on irrode which is the combination of Irrlicht and ODE, the infrared sensor in CogRSim is realized to detect the distance of an obstacle in a given direction. There are three classes, RayListener, RayListenerItem and RayListenerManager, to implement that functionality, and they are described in Table 4 in brief.

Table 4. Three classes to realize Infrared Sensor

Class name	Introduction of the class
RayListener	Inherits from IIrrodeEventLsitener of Irrlicht, could handle the event of object detection.
RayListenerItem	Could store the infrared sensor
RayListenerManager	Could manage the infrared sensor

The infrared sensor and its monitor have a one-to-one relationship, as a result, one monitor for a certain infrared sensor will not monitor others. Fig. 9 shows the infrared sensor of a robot, there are two rays emitted from the robot to detect obstacles.

Fig. 9. Robot's infrared sensor and the two rays in red

4.6 Data Analysis and Display

Many kinds of data will be created during experiment, for example, robot speed, object location, images of cameras. The user needs to write code to gather and analyse specific data according to his purpose, and show the result in the right form. In CogRSim, this module provides interfaces to draw different kinds of figures such as curve, box plot and curved surface.

To realize this module is by mixing the programming of C++ and Matlab: first program the script to realize the draw of these figures, then compile the script to DLL (Dynamic Link Library) which could be invoked and further encapsulated by C++. This module is independent and transplantable, and the programmer only needs to invoke these interfaces to draw figures. Fig. 10 shows a typical example of robot speed.

Fig. 10. Curve of robot speed from CogRSim

4.7 World Model

The interaction among robots and objects should meet the principles of kinematics, dynamics and Collision detection, and ODE is chosen to make accurate physical effect. The rigid bodies are generated by composing primitive shapes such as cubes, spheres and cylinders, and each one has been assigned mass, friction, color and texture. In ODE, motor and servo could provide the motivation power, joints could connect bodies together and form complex static ones. There are several types of joints such as universal joints, ball and socket joints, and hinge joints, the users are able to create robots or complex objects via these components.

To build a simple wheeled robot as shown in Fig. 11, it only needs one box as the main body, two cylinders as wheels where each cylinder is associated with the box by a joint, then motor is added to the joint to provide power. However, to build a table which is a static object without any motor ability, it only needs several boxes to glue with joints.

Fig. 11. A two-wheel robot which has power to move and rotate

5   Experimental Validation

In order to illustrate the applicability of CogRSim, a navigation task is carried out on a mobile robot. The task is shown in Fig. 12(a), the thick black lines represent walls and they divide the 8-by-8 maze into two rooms (A, B). Each two neighboring grids are reachable if there is no wall between them. There are four candidate trigger grids (T_1, T_2, T_3, T_4) in room A and four candidate goal grids (G_1, G_2, G_3, G_4) in room B, the start place is a random grid in room A. "trigger" means that when the robot arrives that grid, the two doors will both open immediately as shown in Fig. 12(b). Obstacles will appear dynamically and randomly during the whole task execution, some could be rolled away while the others could not. The robot's task is to first navigate from the start grid to a trigger to make the doors open, then pass a door, and finally to the goal, all following the shortest routine.

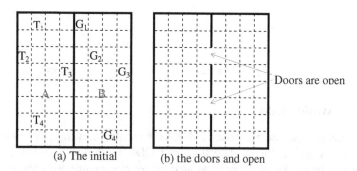

(a) The initial   (b) the doors and open

Fig. 12. The robot's task in a maze environment. When the robot navigates to a trigger, the two doors will both open immediately as shown in (b).

According to the task, we set up experimental environment in Fig. 13. The two-wheel robot has four primitive actions, *North*, *South*, *West* and *East*, and they are always executable.

(a) the doors are closed   (b) the doors are open

Fig. 13. 3D Scene. In (b), the robot has passed the door

The robot's task has been executed for many times, and one test's result is shown in Fig. 14.

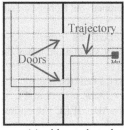

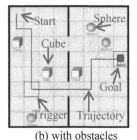

(a) without obstacle   (b) with obstacles

Fig. 14. Robot's trajectory is displayed through the module of "2D display synchronously with 3D scene". (a) the robot finishes its task without obstacle, (b) is the final snapshot of the environment, the spheres could be rolled away by the robot while the cubes should be avoided.

6   Conclusions and Future Work

Simulation is well established in robotics and nearly every researcher benefits from it. This paper presents the overview of CogRSim, a software developed for the simulation of robotic models, sensors, and control in virtual environment. The existing functionalities include world model, scene edit, camera and infrared sensor management, user-defined information analysis, and the control of experiment rhythm. CogRSim has been used to carry out some experiments such as navigation and transportation, and they prove to work well.

In the future, at least three problems will be considered. The first is the cloud storage and sharing, through which the users could store, fetch and run 3D scenes in different hardware, and this is the exploration to study cloud robot in reality [ ]. The second is algorithm library, which could be used and updated by users. The last one is code editor, which integrates the functions of edit and debug.

Acknowledgments. This work is supported by National Natural Science Foundation of China (Grant No. 61372140), and National Training Programs of Innovation and Entrepreneurship for Undergraduates (Grant No. 201310561032). Thanks to open source software Irrlicht and Open Dynamics Engine.

References

1. Staranowicz, A., Gian, L.M.: A survey and comparison of commercial and open-source robotic simulator software. In: Proceedings of the 4th International Conference on PErvasive Technologies Related to Assistive Environments. ACM (2011)
2. Michal, D.S., Etzkorn, L.: A Comparison of Player/Stage/Gazebo and Microsoft Robotics Developer Studio. In: 49th Annual Southeast Regional Conference, pp. 60–66. ACM (2011)
3. Reckhaus, M., Hochgeschwender, N., Paulus, J., et al.: An overview about simulation and emulation in robotics. In: Proceedings of SIMPAR, pp. 365–374 (2010)
4. Carpin, S., Lewis, M., Wang, J., et al.: USARSim: a robot simulator for research and education. In: 2007 IEEE International Conference on Robotics and Automation, pp. 1400–1405 (2007)
5. Webots: robot simulator, http://www.cyberbotics.com/
6. Michel, O.: WebotsTM: Professional mobile robot simulation. arXiv preprint cs/0412052 (2004)
7. Player Project, http://playerstage.sourceforge.net/
8. Microsoft Robotics Developer Studio 4, http://www.microsoft.com/robotics
9. Corke, P.: A robotics toolbox for MATLAB. IEEE Robotics & Automation Magazine 3(1), 24–32 (1996)
10. Powering the world's robots, http://www.ros.org
11. Aldebaran Robotics, http://www.aldebaran.com/
12. iRobot Create Programmable Robot, http://www.irobot.com/create
13. Open Dynamics Engine, http://www.ode.org/
14. Irrlicht Engine, http://irrlicht.sourceforge.net/
15. Qt Project, http://qt-project.org/

A Study of Positioning Error Compensation Using Optical-Sensor and Three-Frame

Yeongsang Jeong, Jungwon Yu, Eun-Kyeong Kim, and Sungshin Kim[*]

Department of Electrical and Computer Engineering,
Pusan National University, Busan, Korea
dalpangi03@pusan.ac.kr,
{garden0312,kimeunkyeong,sskim}@pusan.ac.kr

Abstract. This paper studies about the developed measuring instrument's degree of precision. Through this research, we could convert archer's paradox phenomenon to numerical data and make it use as a performance evaluation element. The instrument uses three frames which composed of line-laser and photodiode sensors for reconstructing full-shape of a flying arrow. Moreover, in order to measure arrows' position precisely which flies about 300km/h, Artificial Neural Network is used for calibration of the measuring instrument. Two grid-plates for calibrating the measuring instrument are installed in first and third frames among three frames. After that, calibration process is performed using converted coordinate data from arrow's position by placing an arrow at holes in the grid-plate. From the suggested arrow position measurement system and the experiment data using the instrument, the automated system for quality control of arrows or performance experiment could be established.

Keywords: Arrow, Measurement, Calibration, Artificial neural network.

1 Introduction

Despite of the research related to analyze performance of arrow is ongoing with complex manufacturing process, both the methodical system for evaluate arrow's performance and the intuitive numerical data are insufficient. For evaluating performance of bows and arrows, there are some currently used methods depending on product reviews and opinions; hunters who have used the specific products for a long time; technicians who produces leisure sports equipment; customers who comments in product reviews. Moreover, the existed research results are only focused on the manufacturing process and the physical properties of arrow [1][2]. Figure 1 shows a flowchart of manufacturing process of arrow. The impact points group of arrow can be obtained from repeated shooting experiments. The experiment steps as follow: First, place shooting paper on the target. Second, change nock angle in

---

[*] Corresponding author.

X. Zhang et al. (Eds.): ICIRA 2014, Part II, LNAI 8918, pp. 167–174, 2014.

specified amount. Third, shoot the arrow. Fourth, go to second step if the stopping criterion is not achieved. However, there are many drawbacks using the shooting paper method, for example, low accuracy; periodically change the shooting paper; hard to represent the relation between impact points in numerical way; and so on. Also, in order to observe archer's paradox while the arrow is flying, a high-speed camera is used for measurement. But there are some problems using the high-speed camera that the device is exceedingly expensive, and only could observe in fixed angle. The research about flying arrow is not enough though there are preceding researches which focused on the escaping moment of the arrow from the bow [3].

This paper studies about the developed measuring instrument's degree of precision which could detect high speed projectile. Through this research, we could solve the problem not only in the impact point group measurement using the shooting paper but also in the archer's paradox phenomenon measurement using the high-speed camera. Furthermore, we could convert archer's paradox phenomenon to numerical data and make it use as a performance evaluation element. In order to convert arrow's position while flying, the square-shaped frame with line-laser [4] and photodiode [5][6] sensors is suggested. Also, for reconstructing and representing the entire shape of flying arrow, the number of frame is decided three. There are errors in initially represented impact points because of the fabrication error of the square-shaped frame, and the effect of line-laser angle. Thus, for more precise measurement of arrow position, the calibration process is performed with Artificial Neural Network. Two grid-plates for calibrating the measuring instrument are installed in first and third frames among three frames. After that, calibration process is performed using converted coordinate data from arrow's position by placing an arrow at holes that have distance between them in one centimeter in the grid-plate which exceedingly and precisely manufactured. The automated system for either quality control of arrow or performance experiments using performance data of arrow is established.

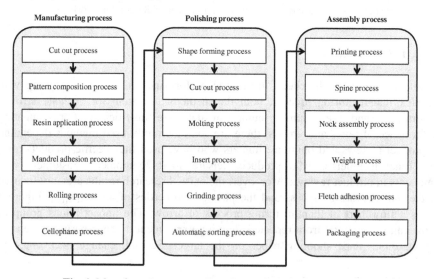

Fig. 1. Manufacturing process flowcharts of arrow manufacturers

2   The Problems of Experimental Environment

The technologies about measuring and analyzing of impact point and trajectory is staying in early stage. And the measurement platform is the shooting paper and the high-speed camera. For analyzing impact point density of arrow, the shooting paper is placed, and the shooting is repeated by changing nock angle. Using the shooting paper, the repeated experiment results, is used for determining performance of arrow. However, there are many drawbacks using the shooting paper method, for example, low accuracy periodically change the shooting paper hard to represent the relation between impact points in numerical way; and so on.

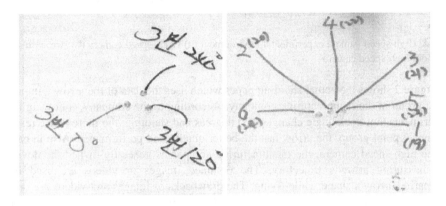

Fig. 2. Analysis of impact point's integration by hand

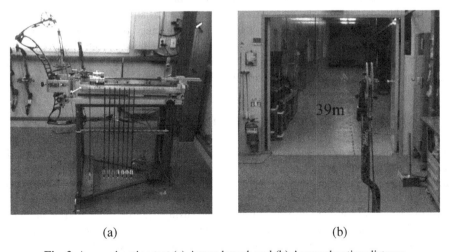

(a)   (b)

Fig. 3. Arrow shooting test (a) Arrow launch pad (b) Arrow shooting distance

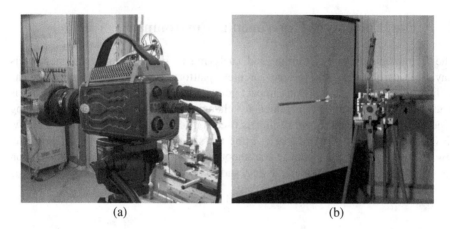

(a)   (b)

Fig. 4. High-speed camera experimental environment (a) High–speed camera (b) Arrow image by shoot high-speed camera

Figure 2 shows the actual shooting paper which uses the data of the arrow's impact point group in the manufacturing company. According to the company, following the experiments that nock angle changes 120 degrees and shooting, the more dense result of impact point group, the arrow has the better quality and performance. Also in case of the high-speed camera, the manufacturing companies generally owned the device for measuring arrow's trajectory. The obtained images or videos are used for comparing arrow's shape while flying. The drawbacks of images and videos are only obtained shape of the object in limited angle. Therefore, it is immensely hard to analyze quality and performance of arrow because there are no data which could convert intuitive and numerical representation, same as the shooting paper method.

Figure 3 (a) shows the launch pad used at the shooting experiment, designed for shooting the arrow using the same strength. Figure 3 (b) shows actual experimental environments from the launch pad to the target: the distance between the launch pad to the parget is about 39 m. Also, Figure 4 (a) represents the high-speed camera, which is cost about one hundred million won. Figure 4 (b) describes the obtained actual image of arrow while flying using high-speed camera.

3   Moving Arrow Position Measurement System

In order to measure the position of flying arrow, they use the shooting paper method for obtaining impact point group and the high-speed camera method for obtaining arrow's shape at video and image form in manufacturing company. However, there are many drawbacks using the shooting paper method, for example, low accuracy and hard to represent in numerical way. In this paper, using the line-laser module and the photodiode sensors, the single frame is implemented for obtaining density of arrow's impact point group. Moreover, the three frame system is implemented for obtaining high speed arrow's shape without using expensive high-speed camera. [7] As shown in Figure 5, the line-laser is placed in L1, L2 position, and the photodiode array

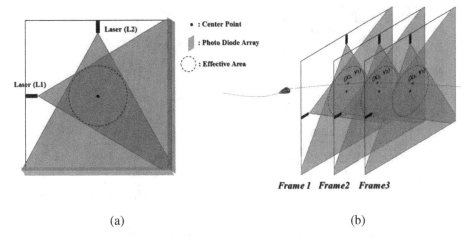

(a)   (b)

Fig. 5. Diagram of flying arrow position measurement system (a) One-frame (b) Three-frame

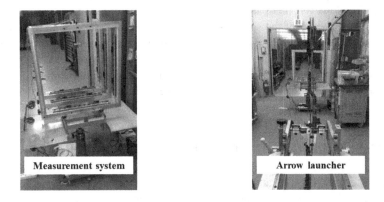

Fig. 6. Arrow position measurement system

sensor is attached at the opposite position of line-laser. The photodiode sensor can convert light energy to electric energy. Thus, the intensity of the light of the line-laser module is converted to voltage level in the photodiode sensor. When the arrow passes through the center of the frame, it can be converted into the coordinates by the position of photodiode sensor and voltage level. But the single frame cannot measure the trajectory of the arrow even if it can derive the impact point group of the arrow. Thus, it is not appropriate to the entire measuring instrument. Therefore, as shown in Figure 6, the problem is solved that add the two more frames in same measuring line. The distance between the frames is changed depending on the length of the measured arrow. And when the arrow reached all three frames' sensing area, the processor which could data and signal processing represents three coordinates. The results could

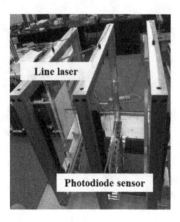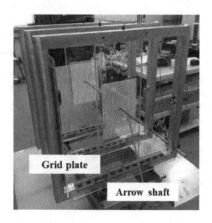

Fig. 7. Data acquisition pictures for calibration using grid plate

make the full shape of arrow as data, and represents as coordinate form: (x1, y1); (x2, y2); (x3, y3). However, there are errors in initially represented impact points because of the fabrication error of the square-shaped frame, and the effect of line-laser angle. Figure 7 shows that the two grid-plates for calibrating the measuring instrument are installed in first and third frames among three frames. In this process, calibration is performed using converted coordinate data from arrow's position by placing an arrow at holes that have distance between them in one centimeter in the grid-plate which exceedingly and precisely manufactured.

4   Calibration Data Configuration and Precision Result

It includes errors that the attached photodiode structure for detecting high speed object, the projection light of line-laser, and the fabrication precision of the frame. Therefore, in order to improve precision of the arrow impact point measurement instrument, a calibration algorithm is applied using the grid-plate.

The grid-plate which has many holes that have distance between them in one centimeter is attached at the frame. Two grid-plates for calibrating the measuring instrument are installed in first and third frames among three frames. After that, calibration process is performed using converted coordinate data from arrow's position by placing an arrow at holes that have distance between them in one centimeter in the grid-plate which exceedingly and precisely manufactured. The calibration algorithm is Artificial Neural Network: the coordinate data is 361 at each frame and the ratio of training data and test data is half and half. Table 1 and Figure 8 show the result that the average error between the coordinate of grid-plate and measured coordinates is improved: 2mm before calibration: 0.4mm after calibration. Figure 9 shows a reconstructed 3D graph of curved arrow shape using three calibrated data.

Table 1. Experimental results before and after calibration

Frame sequence	Before calibration	After calibration	
		Training error	Test error
1st	2.08mm	0.42mm	0.43mm
2nd	2.16mm	0.45mm	0.47mm
3rd	2.24mm	0.41mm	0.44mm

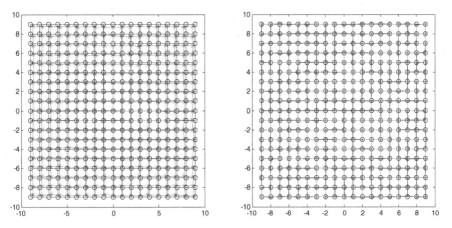

Fig. 8. Representation of before and after calibration data

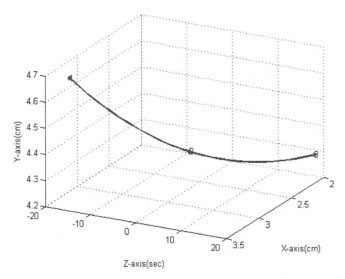

Fig. 9. Represented arrow data obtained from three frame

5   Result

This paper revealed a problem that analyzing a manufactured arrows' performance by lack of systems and technologies which can evaluate the performance of arrow objectively; the analyzing results of arrow depend on an experts' or hunters' subjective opinion currently. Also, it revealed that the technologies that owned by the arrow manufacturing companies are focused on optimizing the manufacturing process using the existed researches which deals with only manufacturing variables and announced via patents. For solving the revealed problems, this paper selected the line-laser module and the photodiode array sensor for exceedingly and precisely measuring impact point and position of more than 300km/h speed object. And using the three frame system, the shape of arrow was reconstructed. When the system could convert the measurement result into the coordinates as output, the properties of the arrow is changed as numerical form, such as archer's paradox which is one of important factor for determining arrow's trajectory. In order to increase reliability of the coordinated form results, the calibration using Artificial Neural Network was applied. The results show that the average error between the coordinate of grid-plate and measured coordinates is improved: 2mm before calibration: 0.4mm after calibration. By the numerical form results, it is possible to compare the quality and the performance of the arrow between other arrows objectively. Also, the automated system could be implemented to arrow manufacturing system which is able to classify exact properties of the produced arrow and purchase recommend system.

Acknowledgement. This work was supported by BK21PLUS, Creative Human Resource Development Program for IT Convergence.

References

1. Lee, I., Baek, G., Kim, S.: The Measurement Automation of Arrow Impact Point Using Machine Vision. In: Korean Society of Manufacturing Technology Engineers KMSTE Spring Conference 2011, pp. 266–267. KSMTE, Korea (2011)
2. Baek, G., Cheon, S.-P., Lee, I., Kim, S.: Parameter Calibration of Laser Scan Camera for Measuring the Impact Point of Arrow. Journal of Manufacturing Engineering & Technology 21(1), 76–84 (2012)
3. Park, J.L.: The behaviour of an arrow shot from a compound archery bow. Proceedings of the Institution of Mechanical Engineers, Part P: Journal of Sports Engineering and Technology 225(1), 8–21 (2011)
4. Lu, S.-T., Chou, C., Lee, M.-C., Wu, Y.-P.: Electro-optical target system for position and speed measurement. Science, Measurement and Technology, IEE Proceedings A 140(4), 252–256 (1993)
5. Liu, J., Yu, L.: Laser-based Apparatus for Measuring Projectile Velocity. In: The Ninth International Conference on Electronic Measurement & Instruments, pp. 2-595–2-598 (2009)
6. Lee, K.-H., Ehsan, R.: Comparison of two 2D laser scanners for sensing object distances, shapes, and surface patterns. Computers and Electronics in Agriculture 60, 250–262 (2008)
7. Jeong, Y., Lee, H., Yu, J., Kim, S.: Measurement and Calibration System of Arrow's Impact Point using High Speed Object Detecting Sensor. In: The Second International Conference on Intelligent Systems and Application, pp. 167–172 (2013)

An Ultrasonic Instrument for Osteoporosis Detecting

Zhou Hongfu[1,*], Zhang Zheng[1], Lin Chengquan[2], Juliana Tam M.Y.[3],
Zhuang Zhenwei[1], and Yao Xinpeng[1]

[1] School of Mechanical and Automotive Engineering, South China University of Technology,
Guangzhou, Guangdong Province, China
[2] Electrical and Mechanical Engineering College, Jingzhou Vocational and
Technical College, Jingzhou, China
[3] State Key Laboratory of Digital Manufacturing Equipment and Technology,
Huazhong University of Science and Technology, China
mehfzhou@scut.edu.cn

Abstract. With the aging population increasing, the Osteoporosis (OP) in aging people is rising, and it has an important significance to take the prevention and treatment of osteoporosis as a way to improve the health quality of the elderly. The research presents an osteoporosis calcaneus bone test device, which designs two ultrasonic probes for test the bone mineral density (BMD), one transmitting probe for sending the ultrasonic signal and the other, received probe, for collecting the ultrasonic attenuates detecting signal pass through bone, also the both probes are designed as a same kind probe for ultrasonic transmitting and receiving selection.

Keywords: Bone mineral density, Osteoporosis Detecting, Ultrasonic Probe, Ultrasonic transmission circuit, Ultrasonic receiving circuit.

1 Introduction

BMD is an important indicator of the strength of human bone. There are more researches showing the accurate rate is about 70-75% for BMD to reflect of bone strength. The measurement of bone density can be reflected the lost bone degrees of osteoporosis patients, it can significantly predict the risk of osteoporotic fractures. For the BMD testing to earlier diagnosis of osteoporosis subjects as an important technology it has been recognized by WHO. By detecting the bone density, it determines whether patients with osteoporosis, which gives the effective guidance to patients for preventing or treatment osteoporosis. Therefore, the BMD measurements to subjects can provide an important reference value in the prevention and treatment of osteoporosis.

There are a variety of detection methods of BMD, including X-ray photographic method, quantitative CT method (QCT) and Quantitative ultrasound (QUS)[1, 2, 3]. QUS method is become more popular in bone density detect comparing to others,

---

* Corresponding author.

X. Zhang et al. (Eds.): ICIRA 2014, Part II, LNAI 8918, pp. 175–182, 2014.

which has more advantages comparing to others for promotion, such as higher accuracy in bone density measurement results, the measurement with repeatable, and non-radioactive.

The research design bone density detecting device with BMD method.

2   Method

In clinical application for avoiding the soft tissue absorbing the ultrasonic signal to effect the BMD test accuracy, in the design it choose calcaneus and tibia bone as test object for they have less soft tissue, it show in Fig 1 [4]. Also some researches choose the bone in forearm for BMD test [5].

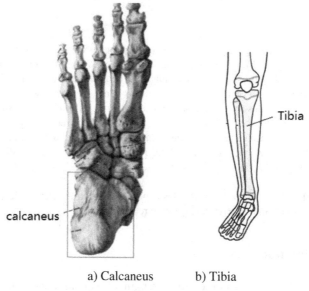

a) Calcaneus   b) Tibia

Fig. 1. Bone for detecting Osteoporosis

With the calcaneus bone as the detection object, the test device has more advantages.

(1) Of the calcaneus in cancellous bone content of 90% and the calcaneus bone losses at a faster speed. In the design choosing the calcaneus it can diagnose earlier osteoporosis, for earlier prevention and treatment of osteoporosis patients. Also when ultrasonic wave passes from the cancellous bone, it attenuates highly, which helps to improve the BMD detection accuracy.

(2) Because of the heterogeneity of bone, it needs to select a big volume bone for bone repetitive scanning, the calcaneus able to meet this condition.

(3) For the calcaneus bone surface smooth, it is easy to position of the ultrasound probe for test. Also for the ultrasonic propagation direction is perpendicular to the bone plane, it is easy to receive the wave by the received probe.

(4) For the soft tissue content of the calcaneus is relatively thin, the noise on the absorption and scattering of ultrasonic waves is relatively small, which can significant decrease the measurement error.

By ultrasonic transmission technology, detecting the calcaneus uses two ultrasound probes, one for emitting an ultrasonic wave, the other for receiving ultrasonic waves. The two probes are place on the both sides of the Calcaneus, which can make the transmitter sent ultrasound signal to the detecting bone and along a straight line pass to the received probe.

For the coupling agent using, in the past, it usually uses a water bath as the coupling agent [6]. And now for the convenience and health, it uses solid coupling agent as the coupling agent. During the BMD test, the transmitting probe emits ultrasonic signal with frequency range between 200 kHz ~ 600kHz. On the other side, the ultrasonic reception signal is received by received probe. And both transmit and receive signals for processing to get the relationship between the frequency and attenuation results. This technique is mainly used in the detection of calcaneal BMD. In Fig 2, it shows the detecting principle. It process the calculatines ultrasonic signal attenuation for detect the bone loss.

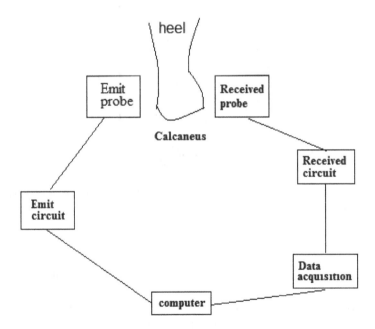

Fig. 2. Calcaneal ultrasonic detecting block

3   The Design of Ultrasonic Probe

With the crystal probe, it tests the calcaneus bone by ultrasonic transmitting and receiving technique, where the two probes placed in the coaxial opposite sides of the calcaneus. In the research, the ultrasonic probe designs into a crystal straight probe which shows in Fig.3, where the probe mechanism designs by Solidworks.

The crystal probe is a piezoelectric ceramic probe which can be used both as transmitted probe and received probe. For transmitting probe, it uses as transfer from electronic voltage signal into mechanic vibration signal. And in verse, for receiving probe, it receives the acoustic sound and transfers it into electronic signal for data analysis.

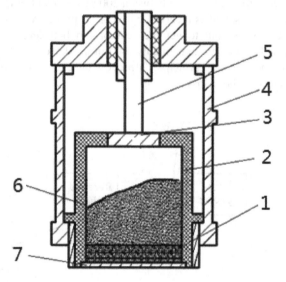

1.   **Piezoelectric ceramic   2. Copper sleeve 3. Cable connect 4. Plastic shell
5. electrical conductivity screw 6. absorbing block   7. Protective film**

Fig. 3. Crystal straight probe

4   Circuit Design of Crystal Straight Probe Detection System

In the ultrasound bone density detection system, the transmitting and receiving circuits are key components, which have directly impact on the device performance, accuracy and detection range of the system. In the research, the maximum energy conversion efficiency and the resistance matching are essentially for the circuits design.

The Calcaneus BMD detecting circuit uses 200-600 kHz ultrasonic sound, and the probe frequency design to 250 kHz for making the ultrasound probe with high sensitivity.

The processing for the calcaneus bone BMD detecting:

(1) Transmitting probe sends ultrasonic pulse wave into the calcaneus bone

(2) The received piezoelectric wafer from the receiving probe detects the attenuation electrical pulse signal

(3) Signal amplifiers, filterers and data acquisition circuit for further data processing.

(4) By data processing, generating the BMD data

4.1   Ultrasonic Transmission Circuit Design

Ultrasonic transmitter requires relatively high performance penetration in bone and high resolution. This is related to the transmission of ultrasonic energy, as well as the ultrasonic pulse width. A transmitting circuit needs to meet these requirements with the following:

1) With a certain amplitude,

2) Steep enough to the signal waveform edge.

The transmitting circuit shows in Fig 4, where the input is internal system ultrasound signal, with 250KHZ frequency, which is generated by system internal ultrasonic generator.

In Fig 4, the amplifier circuit by power MOSFet, Q1, Q2, Q3, and MOSFet Q4.

For the MOSFET Q4, IRFP450, a switching power chip, it triggers by the high voltage pulse from pre-amplifier.

The transmitting circuit is designed by Altium Designer software.

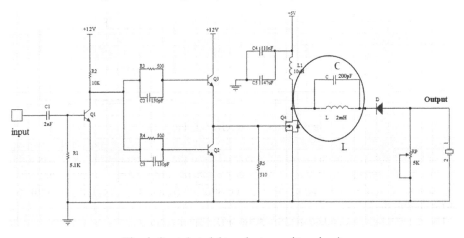

Fig. 4. Crystal straight probe transmitter circuit

The circled part in Fig 4, it is LC oscillator circuit, where the inductor L and capacitor C for frequency selection. According to the oscillation circuit LC formula, the frequency is:

$$f = \frac{1}{2\pi\sqrt{LC}}$$ (1)

For 250kHz ultrasonic signal output, the system select the parameter as:

L = 2mH
C = 200pF
And F=250KHZ

4.2   Ultrasonic Receiving Circuit Design

The ultrasonic waves sending from the ultrasonic transmit probe, it is quickly attenuate after it pass human soft tissue and bone, and the collected voltage signal from received probe is very weak to less than 1 mV volt level. For subsequent analysis, it needs pre-amplifier circuit, a filter circuit and amplifier circuit. In Fig 5, it shows the received circuit architecture.

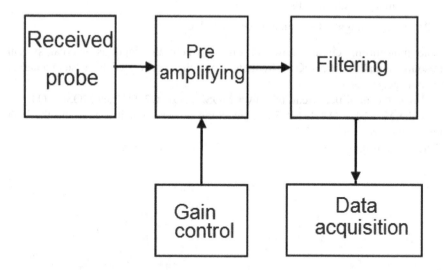

Fig. 5. Ultrasonic receiver circuit block

For signal analysis, acoustic signal received from the ultrasonic probe converts into the electrical signals, after that it process by the pre-amplifying, filtering, and the conversion data for subsequent data processing analysis. The receiving circuit is shown in Fig 6.

The received circuit is designed by Altium Designer software.

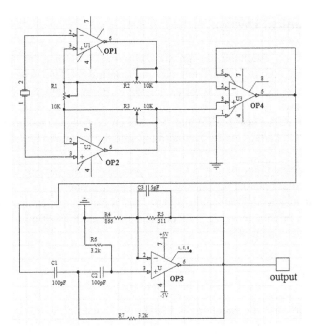

Fig. 6. Straight probe receiving circuit

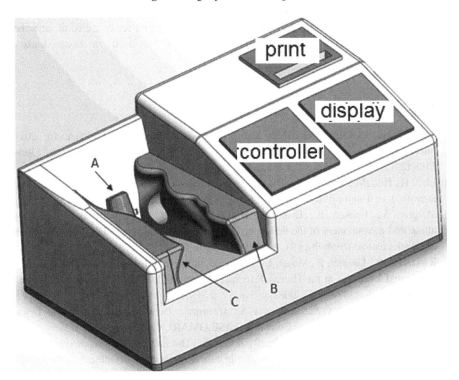

Fig. 7. Mechanic design

5   Mechanical Design

The BMD test rig mechanic design shows in Fig 7, it includes foot positioning module, control module, display module and printing module, where the mechanic design by Solidworks, a CAD software. In Fig 7, the mechanical part A, B, and C descript as followings.

In Fig 7, part A is a heel positioning stop, which is designed to simulate the human heel shape, with a certain elasticity, and moveable for accommodate different sizes of the heel.

Part B is the fixed module for left heel to position on the test rig, and in the ankle-side with a circular opening for placing the ultrasonic probe.

Part C is an elastic and moveable module, and it can be moveable respect to part B and to fit different wideness sizes of feet. Also in the ankle position it opens a hole for placing the ultrasound probe.

6   Conclusion

The research presents an osteoporosis BMD test device, in the device it designs two probes for detecting the calcaneus bone density, one for transmitting the ultrasound signal, the other for receiving the ultrasonic attenuating signal passing from the bone, and the two probes are a same kind probe for ultrasonic transmitting and receiving using. By designing ultrasonic probes, ultrasonic transmit and receive circuit, and the device mechanics parts, it shows this design can be used in bone density measurement.

Reference

[1] Yu, L., Yi, J.: Design of BMD Measurement System based on FPGA. In: 2012 International Conference on Industrial Control and Electronics Engineering, pp. 1385–1388 (2012)

[2] Li, X.H., Hou, M., Yang, H.H.: The Measurement Methods and Influence Factors of Bone Density. J. of Tianjin Institute of Phys. Edu. 20(3), 62–65 (2005)

[3] Mergler, S., Löbker, B., Evenhuis, H.M., Penning, C.: Feasibility of quantitative ultrasound measurement of the heel bone in people with intellectual disabilities. Research in Developmental Disabilities 31, 1283–1290 (2010)

[4] Barkmann, R., Laugier, P., Moser, U.: A Device for in VivoMeasurements of Quantitative Ultrasound Variables at the Human Proximal Femur. IEEE Trans. Ultrason., Ferroelect., Freq. Contr. 55, 1197–1204 (2008)

[5] Kaufman, J.J., Luo, G.M., Siffert, R.S.: Ultrasonic Bone Assessment of the Distal Forearm, Signals, Systems and Computers (ASILOMAR), pp. 1629–1631 (2012)

[6] Vafaeian, B., El-Rich, M., El-Bialy, T., Adeeb, S.: The finite element method for micro-scale modeling of ultrasound, propagation in cancellous bone. Ultrasonic. 54, 1663–1676 (2014)

Avoiding of the Kinematic Singularities
of Contemporary Industrial Robots

Tadeusz Szkodny[*]

Silesian University of Technology, Institute of Automatic Control,
Akademicka 16 St. 44-100 Gliwice, Poland
Tadeusz.Szkodny@polsl.pl

Abstract. The singular configurations of contemporary industrial robot manipulators of such renowned companies as ABB, Fanuc, Mitsubishi, Adept, Kawasaki, COMAU and KUKA, cause undesired stopping. It is the basic defect of these robots. The paper presents simple method of avoiding these singularities. To determine the singular configurations of these manipulators a global form of description of the end-effector kinematics was prepared, relative to the other links. On the base of this description , the formula for the Jacobian was defined, in the end-effector coordinates. Next, a closed form of the determinant of Jacobian was derived. From the formula, singular configurations, where the determinant's value equals zero, were determined. Additionally, geometric interpretations of these configurations were given and they were illustrated. For the exemplary manipulator, small corrections of joint variables preventing the reduction of the Jacobian order were suggested. An analysis of positional errors, caused by these corrections, was presented.

Keywords: kinematics, manipulators, mechanical system, robot kinematics.

1 Introduction

The movement programming in joint space or Cartesian space allow the controllers of such renowned contemporary companies as ABB, Fanuc, Mitsubishi, Adept, Kawasaki, COMAU and KUKA. The following commands PTP, LIN, and CIRC can be applied to programming in Cartesian space. The mentioned commands require start point and end point. For programming in Cartesian space these points must be described in Cartesian coordinates, relative to the base frame, connected to the base of a manipulator. These coordinates can be obtained by the vision system. During the realization of such programmed movement robot happens to stop before reaching the border area, and before reaching the start or the end point. The entrapment takes place when the manipulator reaches the singular configurations. It is without doubt the major problem of modern industrial robots, which makes it impossible for the robots with vision system to properly cooperate with the cameras.

The linear system of ordinary differential equations that describes the kinematics can be applied to programming the robots in Cartesian space. In this system a

---

[*] Author have used the western naming convention, with given names preceding surnames.

X. Zhang et al. (Eds.): ICIRA 2014, Part II, LNAI 8918, pp. 183–194, 2014.
© Springer International Publishing Switzerland 2014

manipulator Jacobian is present. For succesive via points interpolating trajectory programmed in Cartesian space, joint variables of these points can be computed. To these calculations, the standard algorithm for solving the system of linear equations which is one of the elements of a computer software library, can be applied. The lack of protection in software against the reduction of the Jacobian rank in this algorithm, can cause the interruption of the calculations and can cause the robot to stop performing its operations. This rank decreases in singular configuration.

The problem of the inverse kinematics solutions in differential form for singular configuration is presented in [1-6]. In [1-4] for the singular configuration is proposed to use: singular value decomposition techniques SVD of Jacobian, the damped least-squares inverse of the Jacobian DLS or singularity robust inverse of the Jacobian. In the work [6] is presented a method of avoiding singularities by using of dynamical systems of the own motion. The same work presents a solution of the inverse kinematics in singular configurations, obtained using: the method a normal form, Jacobian attached, the zero and the Jacobian space. A common feature of kinematic singularity avoidance methods presented in [1-4,6] is their high computational complexity.

A very simple approach to the problem of determining the kinematic singularities based on the differential description presents the work [5], on the examples manipulators with three degrees of freedom. Simple, because based on closed forms of the determinants of manipulators Jacobian, allowing very simple to determine the joint variables, describing the singular configurations. To solve the inverse kinematics in the field of speed is recommended in [5] simple rules of linear algebra.

Only the methods which do not require a large number of calculations can be of practical use.

This paper presents a very simple method of correction, allowing to avoid the singular configuration of modern industrial robots. In this method, the values of third and fifth joint variables are corrected. These values were determined from the closed form of Jacobian determinant of modern industrial robots, like the work [5]. It was determined the closed form of formulas describing the third and fifth joint variables for the singular configurations of manipulators. On the base of these formulas, one is able to determine whether in the next step of the numerical calculations the freezing of the program, caused by a reduction of the Jacobian rank, will take place. These corrections prevent the reduction of the of Jacobian rank and allow the use of any method of solving the inverse kinematics, including the methods described in the aforementioned works [1,3,4,6].

In the second chapter the kinematic structure and the description of the end-effector kinematics in relation to the other links, including the base of the manipulator was presented. The formulas constituting the differential description of the kinematics were presented in the third chapter. General description of the singular configurations including their illustration and the examples of the calculations of joint variables corrections, preventing the reduction of the Jacobian rank, were presented in the fourth chapter. The fifth chapter summarizes the paper.

2   The Global Description of the Kinematics

Figure 1 illustrates the kinematic scheme of robot manipulators of the majority of modern robots, with numbered links. The link 0 is attached to the ground, other links 1-6 are movable. The link 0 will be called a base link, the last link with number 6 will be called an end-effector. The gripper, or another tool, is attached to this link. A neighboring links are connected by revolute joints. In this scheme were illustrated the co-ordinate systems (frames) associated with links according to a Denavit-Hartenberg notation [2,7,8]. The $x_7y_7z_7$ frame is associated with the gripper. The position and orientation of the links and tool are described by homogenous transform matrices. Matrix $\mathbf{A}_i$ describes the position and orientation of the *i-th* link frame in relation to *i-1-st*. $\mathbf{T}_6$ is a matrix that describes the position and the orientation of the end-effector frame in relation to the base link. Matrix $\mathbf{E}$ describes the gripper frame in relation to the end-effector frame. Matrix $\mathbf{X}$ describes the position and the orientation of the gripper frame in relation to the base link. Parameters Θ_i, λ_i, l_i, α_i are used in the Denavit-Hartenberg notation. For further description of the kinematics joint

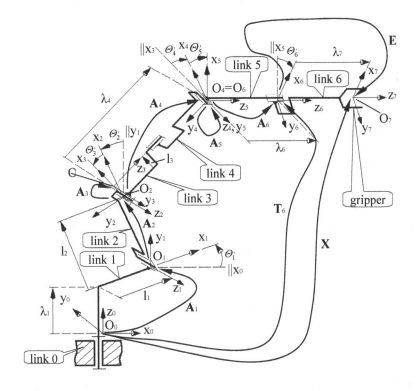

Fig. 1. Kinematic scheme of the manipulator, Denavit-Hartenberg parameters, joint variables and joint frames

variables Θ_i' will be used. Variables $\Theta_i' = \Theta_i$ for $i=1, 3, 4, 5, 6$ and $\Theta_2' = \Theta_2 - 90°$. Matrices A_i are described by equation (1a).

$$A_i = Rot(z, \Theta_i)Trans(0,0,\lambda_i)Trans(l_i,0,0)Rot(x,\alpha_i). \qquad (1a)$$

Matrix E is described by equation (1b).

$$E = Trans(0,0,\lambda_7). \qquad (1b)$$

Matrix $^{i-1}T_6$ describes the end-effector frame kinematics in relation to i-1-st frame. Equation (2a) [9] describes these matrices.

$$^{i-1}T_6 = \prod_{j=i}^{6} A_j, \quad 1 \le i \le 6. \qquad (2a)$$

These matrices are necessary to the differential description of the end-effector kinematics, in relation to the base frame. The description is presented in the next chapter.

3   The Differential Description of the Kinematics

To make the description of the manipulator kinematics independent of the shape of the gripper, the required position and orientation of the gripper frame $x_7y_7z_7$ (described by the matrix X_{req}) are converted to the end-effector frame $x_6y_6z_6$. Correlation $T_{6\,req} = X_{req}E^{-1}$ makes possible the conversion. Therefore, in further considerations, we will focus on the end-effector kinematics.

The movement of the $x_7y_7z_7$ frame in relation to the $x_0y_0z_0$ frame will be described by using displacement differentials $^7dx_7$, $^7dy_7$, $^7dz_7$ of origin O_7, and x-y-z current angle differentials $^7d\varphi_{7x}$, $^7d\varphi_{7y}$, $^7d\varphi_{7z}$ [8,10]. These differentials are described in the $x_7y_7z_7$ frame. Similarly, the movement of the $x_6y_6z_6$ frame in relation to the $x_0y_0z_0$ frame will be described by using the displacement differentials $^6dx_6$, $^6dy_6$, $^6dz_6$ of the origin O_6, and x-y-z current angle differentials $^6d\varphi_{6x}$, $^6d\varphi_{6y}$, $^6d\varphi_{6z}$. These differentials are described in the $x_6y_6z_6$ frame. Further, we will focus on the movement description of the end-effector and therefore one needs to convert the gripper differentials to end-effector differentials. Differential equation $T_6\,^6\Delta_6 E = X\,^7\Delta_7$, which connects end-effector and gripper differentials, results from the work [8]. In this equation $^6\Delta_6$ and $^7\Delta_7$ are differential transformation matrices, respectively, of the end-effector and gripper frame. These matrices have the following forms:

$$
{}^{6}\Delta_6 = \begin{bmatrix} 0 & -{}^{6}d\varphi_{6z} & {}^{6}d\varphi_{6y} & {}^{6}dx_6 \\ {}^{6}d\varphi_{6z} & 0 & -{}^{6}d\varphi_{6x} & {}^{6}dy_6 \\ -{}^{6}d\varphi_{6y} & {}^{6}d\varphi_{6x} & 0 & {}^{6}dz_6 \\ 0 & 0 & 0 & 0 \end{bmatrix}, \quad {}^{7}\Delta_7 = \begin{bmatrix} 0 & -{}^{7}d\varphi_{7z} & {}^{7}d\varphi_{7y} & {}^{7}dx_7 \\ {}^{7}d\varphi_{7z} & 0 & -{}^{7}d\varphi_{7x} & {}^{7}dy_7 \\ -{}^{7}d\varphi_{7y} & {}^{7}d\varphi_{7x} & 0 & {}^{7}dz_7 \\ 0 & 0 & 0 & 0 \end{bmatrix}.
$$

From the above results the equations (3)

$$
{}^{6}\Delta_6 = T_6^{-1}X\,{}^{7}\Delta_7 E^{-1} = E\,{}^{7}\Delta_7 E^{-1}. \tag{3}
$$

From equation (3) and (1b) results equations (4), which makes it possible to compute the end-effector differentials from gripper differentials.

$$
{}^{6}d\varphi_{6x} = {}^{7}d\varphi_{7x}, \ {}^{6}d\varphi_y = {}^{7}d\varphi_{7y}, \ {}^{6}d\varphi_{6z} = {}^{7}d\varphi_{7z},
$$
$$
{}^{6}dx_6 = {}^{7}dx_7 - \lambda_7\,{}^{7}d\varphi_{7y}, \ {}^{6}dy_6 = {}^{7}dy_7 + \lambda_7\,{}^{7}d\varphi_{7x}, \ {}^{6}dz_6 = {}^{7}dx_7. \tag{4}
$$

To describe the end-effector kinematics one will apply Cartesian differential matrix ${}^{6}D_6 = [\,{}^{6}dx_6 \ {}^{6}dy_6 \ {}^{6}dz_6 \ {}^{6}d\varphi_{6x} \ {}^{6}d\varphi_{6y} \ {}^{6}d\varphi_{6z}]^{T}$ and joint differential matrix $dq = [\,d\Theta_1 \ d\Theta_2 \ d\Theta_3 \ d\Theta_4 \ d\Theta_5 \ d\Theta_6]^{T}$ [10]. Equation (5) connecting these matrices is a differential description of the end-effector kinematics.

$$
{}^{6}D_6 = {}^{6}J_6 dq. \tag{5}
$$

In this equation, an end-effector Jacobian ${}^{6}J_6$ is present, described in $x_6y_6z_6$ frame, which has the form (6).

$$
{}^{6}J_6 = \partial\,{}^{6}Y_6 / \partial q, \tag{6}
$$

where

$$
{}^{6}Y_6 = [\,{}^{6}x_6 \ {}^{6}y_6 \ {}^{6}z_6 \ {}^{6}\varphi_{6x} \ {}^{6}\varphi_{6y} \ {}^{6}\varphi_{6z}]^{T}, \quad q = [\Theta_1 \ \Theta_2 \ \Theta_3 \ \Theta_4 \ \Theta_5 \ \Theta_6]^{T}.
$$

The Fig.2 illustrates the differential displacement and the rotation of the end-effector, caused by the differential displacement and the rotation of $x_iy_iz_i$ frame. The vectors ${}^{i-1}d\vec{r}_i$ and ${}^{i-1}d\vec{\varphi}_i$, respectively, describe the displacement and rotation of the x'y'z' frame in relation to $x_{i-1}y_{i-1}z_{i-1}$ frame, caused by differential increase of joint variable $d\Theta_i$. The x'y'z' frame is connected with the i-th link, and coincides with

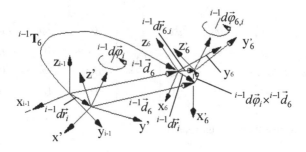

Fig. 2. Differential displacements and rotations of *i-th* and *6-th* frames

$x_{i-1}y_{i-1}z_{i-1}$ frame for $d\Theta_i = 0$. The displacement of $^{i-1}d\vec{r}_{6,i}$ of $x_6y_6z_6$ frame is caused by the displacement of $^{i-1}d\vec{r}_i$ and rotation of vector $^{i-1}d\vec{d}_6$ by an angle $^{i-1}d\vec{\varphi}_i$. Therefore

$$^{i-1}d\vec{r}_{6,i} = {}^{i-1}d\vec{r}_i + {}^{i-1}d\vec{\varphi}_i \times {}^{i-1}d\vec{d}_6 . \tag{7}$$

This movement will be recalculated in relation to the $x_6y_6z_6$ frame, using equations (8).

$$^6dx_{6,i} = {}^{i-1}\vec{a}_6 \cdot {}^{i-1}d\vec{r}_{6,i} , \quad {}^6dy_{6,i} = {}^{i-1}\vec{b}_6 \cdot {}^{i-1}d\vec{r}_{6,i} , \quad {}^6dz_{6,i} = {}^{i-1}\vec{c}_6 \cdot {}^{i-1}d\vec{r}_{6,i} . \tag{8}$$

The vectors $^{i-1}\vec{a}_6$, $^{i-1}\vec{b}_6$, $^{i-1}\vec{c}_6$ are the versors of the $x_6y_6z_6$ frame described in $x_{i-1}y_{i-1}z_{i-1}$ frame. The rotation of the $x_6y_6z_6$ frame is the same as the rotation of $x_iy_iz_i$ frame in relation to $x_{i-1}y_{i-1}z_{i-1}$ frame, which is a vector $^{i-1}d\vec{\varphi}_i$. This vector can be recalculated in relation to the $x_6y_6z_6$ frame by the following equation (9):

$$^6d\varphi_{6x,i} = {}^{i-1}\vec{a}_6 \cdot {}^{i-1}d\vec{\varphi}_i , \quad {}^6d\varphi_{6y,i} = {}^{i-1}\vec{b}_6 \cdot {}^{i-1}d\vec{\varphi}_i , \quad {}^6d\varphi_{6z,i} = {}^{i-1}\vec{c}_6 \cdot {}^{i-1}d\vec{\varphi}_i . \tag{9}$$

In Fig.2 one can see that all links are connected by revolute joints. Therefore, the angles $^{i-1}d\vec{\varphi}_i$ are vectors directed along the z_{i-1} axis, having a length equal to the differential $d\Theta_i$. Therefore

$$^{i-1}d\vec{\varphi}_i = {}^{i-1}\vec{k}_{i-1}d\Theta_i . \tag{10}$$

$^{i-1}\vec{k}_{i-1}$ is the versor of the z_{i-1} axis, described in $x_{i-1}y_{i-1}z_{i-1}$ frame. For such joints, no displacement of the coordinate system x'y'z' in relation to $x_{i-1}y_{i-1}z_{i-1}$ frame takes place, therefore $^{i-1}d\vec{r}_i = 0$. Thus, from equations (7) and (10) one obtains the following:

$$^{i-1}d\overrightarrow{r_{6,i}}=^{i-1}d\overrightarrow{\varphi_i}\times\,^{i-1}\overrightarrow{d_6}=(-^{i-1}\overrightarrow{i_{i-1}}\,\,^{i-1}d_{6y}+^{i-1}\overrightarrow{j_{i-1}}\,\,^{i-1}d_{6x})d\Theta_i\,. \tag{11}$$

The vectors $^{i-1}\overrightarrow{i_{i-1}}$ and $^{i-1}\overrightarrow{j_{i-1}}$ are versors of the x_{i-1} and y_{i-1} axes, respectively, described in $x_{i-1}y_{i-1}z_{i-1}$ frame. $^{i-1}d_{6x}$ and $^{i-1}d_{6y}$ are the x and y - coordinates of the vector $^{i-1}\overrightarrow{d_6}$ in $x_{i-1}y_{i-1}z_{i-1}$ frame. After taking into account equations (11) and (8) one obtains the following:

$$^6dx_{6,i}=(-^{i-1}a_{6x}\,^{i-1}d_{6y}+^{i-1}a_{6y}\,^{i-1}d_{6x})d\Theta_i\,,\quad ^6dy_{6,i}=(-^{i-1}b_{6x}\,^{i-1}d_{6y}+^{i-1}b_{6y}\,^{i-1}d_{6x})d\Theta_i\,,$$
$$^6dz_{6,i}=(-^{i-1}c_{6x}\,^{i-1}d_{6y}+^{i-1}c_{6y}\,^{i-1}d_{6x})d\Theta_i\,. \tag{12}$$

$^{i-1}a_{6x}$, $^{i-1}a_{6y}$ are the x and y - coordinates of the versor $^{i-1}\overrightarrow{a_6}$ in $x_{i-1}y_{i-1}z_{i-1}$ frame. The coordinates of the versors $^{i-1}\overrightarrow{b_6}$ and $^{i-1}\overrightarrow{c_6}$ were marked in a similar way.

From equations (9) and (10) one obtains the following correlations:

$$^6d\varphi_{6x,i}=^{i-1}a_{6z}\cdot d\Theta_i\,,\quad ^6d\varphi_{6y,i}=^{i-1}b_{6z}\cdot d\Theta_i\,,\quad ^6d\varphi_{6z,i}=^{i-1}c_{6z}\cdot d\Theta_i\,. \tag{13}$$

The differential of each coordinate in Cartesian matrix $^6\mathbf{D}_6$ is the sum of the corresponding coordinate differentials caused by differential changes in joint variables $d\Theta_i$. For example $^6dx_6=\sum\limits_{i=1}^{6}\,^6dx_{6,i}$. In the Jacobian matrix $^6\mathbf{J}_6$ the derivatives of Cartesian coordinates of the $x_6y_6z_6$ end-effector frame with respect to the joint variables are present. Derivative $\dfrac{^6\partial x_6}{\partial\Theta_i}=\dfrac{\partial x_{6,i}}{\partial\Theta_i}$ only because the differential $^6dx_{6,i}$ depends on the value of $d\Theta_i$. The situation is similar with other derivatives in this matrix. Therefore, the elements of matrix $^6\mathbf{J}_6$ can be presented in the form of equations (14) and (15), taking into account the relations (12) and (13).

$$\frac{^6\partial x_6}{\partial\Theta_i}=\frac{\partial x_{6,i}}{\partial\Theta_i}=-^{i-1}a_{6x}\,^{i-1}d_{6y}+^{i-1}a_{6y}\,^{i-1}d_{6x}\,,$$
$$\frac{^6\partial y_6}{\partial\Theta_i}=\frac{\partial y_{6,i}}{\partial\Theta_i}=-^{i-1}b_{6x}\,^{i-1}d_{6y}+^{i-1}b_{6y}\,^{i-1}d_{6x}\,,$$
$$\frac{^6\partial z_6}{\partial\Theta_i}=\frac{\partial z_{6,i}}{\partial\Theta_i}=-^{i-1}c_{6x}\,^{i-1}d_{6y}+^{i-1}c_{6y}\,^{i-1}d_{6x}\,. \tag{14}$$

$$\frac{^6\partial\varphi_{6x}}{\partial\Theta_i} = \frac{\partial\varphi_{6x,i}}{\partial\Theta_i} = {}^{i-1}a_{6z}, \quad \frac{^6\partial\varphi_{6y}}{\partial\Theta_i} = \frac{\partial\varphi_{6y,i}}{\partial\Theta_i} = {}^{i-1}b_{6z},$$

$$\frac{^6\partial\varphi_{6z}}{\partial\Theta_i} = \frac{\partial\varphi_{6z,i}}{\partial\Theta_i} = {}^{i-1}c_{6z}.$$

(15)

Matrices $^{i-1}\mathbf{T}_6$, described by equations (2b) can be presented in the form (16).

$$^{i-1}\mathbf{T}_6 = \begin{bmatrix} {}^{i-1}a_{6x} & {}^{i-1}b_{6x} & {}^{i-1}c_{6x} & {}^{i-1}d_{6x} \\ {}^{i-1}a_{6y} & {}^{i-1}b_{6y} & {}^{i-1}c_{6y} & {}^{i-1}d_{6y} \\ {}^{i-1}a_{6z} & {}^{i-1}b_{6z} & {}^{i-1}c_{6z} & {}^{i-1}d_{6z} \\ 0 & 0 & 0 & 1 \end{bmatrix}.$$

(16)

Quantities appearing in formulas (14) and (15) can be replaced by the corresponding elements of matrix $^{i-1}\mathbf{T}_6$, resulting from the form (2b) and corresponding to the form (16). It is easy to notice that index i in equations (14) - (16) is the number of column in matrix $^6\mathbf{J}_6$. After using equations (2b), (14) - (16) and simplifications, we may obtain elements of the end-effector jacobian $^6\mathbf{J}_6$.

4   The Singular Configurations

For the singular configurations of manipulators $^6\mathbf{J}_6$ determinant equals zero. Thus, to determine such configurations one needs to calculate the determinant. The closed form of the determinant was obtained by Symbolic Math Toolbox library of Matlab, it is described by equation (17).

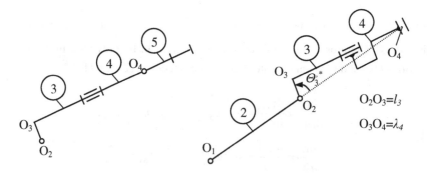

Fig. 3. Configuration at which $\Theta_5' = 0$. The z_3 and z_6 axes are collinear.

Fig. 4. Configuration at which $\lambda_4 C_3 - l_3 S_3 = 0$. Points O_1, O_2 and O_4 are situated on a straight line.

$$det\ {}^6\mathbf{J}_6 = -l_2 S_5 (\lambda_4 C_3 - l_3 S_3)(-l_2 S_2 - l_3 S_{23} + \lambda_4 C_{23} + l_1). \tag{17}$$

In the equation (17) are used following denotations: $S_i = \sin \Theta_i$, $C_i = \cos \Theta_i$, $S_{ij} = \sin \Theta_{ij}$, $C_{ij} = \cos \Theta_{ij}$, $\Theta_{ij} = \Theta_i + \Theta_j$.

The singular configurations appear in zero values of the factors on the right side of equation (17). The joint variables at which it will take place, are described by the equations (18).

$$S_5 = 0 \to \Theta_5^* = 0,\ \lambda_4 C_3 - l_3 S_3 = 0 \to \Theta_3^* = arc\,tg(\lambda_4 / l_3),$$
$$-l_2 S_2 - l_3 S_{23} + \lambda_4 C_{23} + l_1 = 0 \to$$
$$\Theta_{3(1,2)}^{**} = 2 arc\,tg\left(\frac{-l_3 \pm \sqrt{l_3^2 + \lambda_4^2 - (l_2 S_2 - l_1)^2}}{\lambda_4 + l_2 S_2 - l_1}\right) - \Theta_2' \tag{18}$$

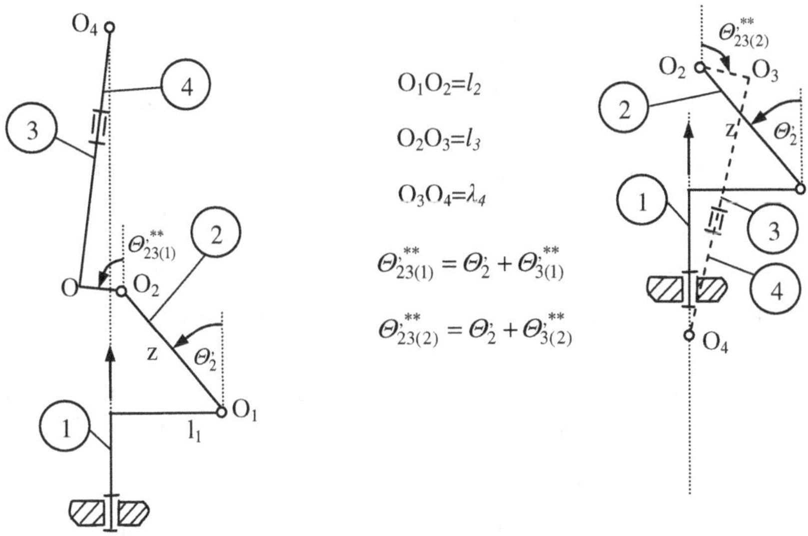

Fig. 5. Configuration at which $-l_2 S_2 - l_3 S_{23} + \lambda_4 C_{23} + l_1 = 0$. Point O_4 is placed on the line passing through the z_0 axis.

Figures 3, 4 and 5 illustrate these configurations.

For example, let's analyze kinematic singularities of the manipulator IRB-1400. The manipulator has the following parameters [11] $l_1 = 150\ mm$, $l_2 = 600\ mm$, $l_3 = 120\ mm$, $\lambda_1 = 475\ mm$, $\lambda_4 = 720\ mm$, $\lambda_6 = 85\ mm$. A ranges of the angle's changes Θ_3' and Θ_5' are the following: $-70° \le \Theta_3' \le 65°$, $-115° \le \Theta_5' \le 115°$. The calculations below were done in Matlab on a PC from an Intel Pentium processor with a frequency of 2 GHz.

Let's assume that $\Theta_1' \div \Theta_4', \Theta_6' = 0°$ for the configuration from Fig. 3. For this configuration $\Theta_5^* = 0°$. In this case the $det^6 J_6 = 0$ and the $rank^6 J_6 = 5$. The rank of the Jacobian $^6 J_6$ is less than 6 and therefore, a standard numerical solution algorithm of the equation (5) will interrupt the calculations. One can avoid zeroing the determinant by increasing or decreasing the angle Θ_5' by a minimum increment $\Delta\Theta_{5\min}'$, resulting from the resolution of the encoder and the gear ratio. Let's assume that a typical resolution encoders is the following $2\pi/4096 = 1.5 \cdot 10^{-3}$ $rad.$, and the value of the gear ratio equals 100. For such data $\Delta\Theta_{5\min}' = 1.5340 \cdot 10^{-5}$ $rad.$ After the corrections $\Theta_5' = \pm\Delta\Theta_{5\min}'$, one obtains $det^6 J_6 = \mp 5.7653 \cdot 10^3$ and $rank^6 J_6 = 6$. Let's estimate the gripper position error caused by this correction. The kinematic scheme in Fig. 1 results in the error not greater than $\Delta\Theta_{5\min}' \cdot (\lambda_6 + \lambda_7)$. Let's assume that a typical value of the parameter gripper is the following $\lambda_7 = 150$ mm. For such the gripper, the $\Delta\Theta_{5\min}'$ causes the change of the gripper position not greater than $3.6 \cdot 10^{-3}$ mm.

For the configuration from Fig. 4 the joint variable $\Theta_3^* = 80.5376778°$. The manipulator IRB-1400 can't reach this variable, because it is outside the range of its changes.

Let's assume that $\Theta_1' = 0°$, $\Theta_2' = 45°$, $\Theta_4' = 0°$, $\Theta_5' = 90°$, $\Theta_6' = 0°$ for a configuration from Fig.5. For these joint variables one obtains $\Theta_{3(1)}^{**} = 13.4676545°$, $\Theta_{3(2)}^{**} = -122.3922989°$. The joint variable $\Theta_{3(2)}^{**}$ is outside the range of its changes. For joint variable $\Theta_{3(1)}^{**}$ $det^6 J_6 = 2.2918 \cdot 10^{-8}$ and $rank^6 J_6 = 5$. The rank of the Jacobian $^6 J_6$ is less than 6 and therefore, a standard numerical solution algorithm of equation (5) will interrupt the calculations. In order to prevent a decrease in the rank of the Jacobian $^6 J_6$ one will increase and reduce the joint variable $\Theta_{3(1)}^{**}$ by a minimum increment $\Delta\Theta_{3\min}'$, equal to $\Delta\Theta_{5\min}'$. After corrections $\pm\Delta\Theta_{3\min}'$ of joint variable $\Theta_{3(1)}^{**}$, one obtains $det^6 J_6 = \pm 4.1854 \cdot 10^3$ and $rank^6 J_6 = 6$. The change in $\Delta\Theta_{3\min}'$ causes an error of the gripper position, which equals max. $14.7 \cdot 10^{-3}$ mm.

5   Summary

Differential description of the manipulator kinematics, presented in this paper, is the base for design of the contemporary software drivers of industrial robots, independent of the software manufacturers robots.

This description can be applied in IRB manipulators series 1000, 2000, 3000, 4000, 6000; in Fanuc manipulators M6, M16, M710, M10, M900; in a KUKA manipulators KR5, KR6, KR15, KR16; in a Mitsubishi manipulators RV-1A, RV-2A, RV-3S, RV-6S, RV12S; in Adept manipulators s300, s650, s850, s1700; in Kawasaki manipulators series M, FS300N, ZHE100U, KF121.

For KUKA KRC3, Adept s300 and Mitsubishi RV-2AJ manipulators the origin of second and third link frame is at the same point $O_2=O_3$ (see Fig. 1). Therefore for these manipulators one has to assume that $l_3 = 0$.

In the controllers of contemporary industrial robots one can write own his/her own applications. These applications can be written using equation (5) derived from this work. In this equation, differentials $d\Theta_i^{\cdot}$ should be replaced by the increments $\Delta\Theta_i^{\cdot}$, and differentials ${}^6dx_6$, ${}^6dy_6$, ${}^6dz_6$, ${}^6d\varphi_{6x}$, ${}^6d\varphi_{6y}$ and ${}^6d\varphi_{6z}$ by corresponding increments ${}^6\Delta x_6$, ${}^6\Delta y_6$, ${}^6\Delta z_6$, ${}^6\Delta\varphi_{6x}$, ${}^6\Delta\varphi_{6y}$ and ${}^6\Delta\varphi_{6z}$. Such discretized equation (5) is the basis for the second stage of the calculation at master level of control [8,10]. During this stage the joint variables of the via-pints are iteratively calculated from the discretized equation (5). At this level of control in the first stage the Cartesian coordinates of the via-points are calculated.

Assume that the joint variables of the previous via-point of trajectory were calculated iteratively and denote them by $\Theta_i^{\cdot(-)}$. For the calculation of the joint variables of the current via-point, also we will apply iterative calculation algorithm. In the next, example the *k-th* iteration of calculations, the initial values of the link variables $\Theta_{i\,start(k)}^{\cdot(-)}$ are equal to the end values $\Theta_{i\,end(k-1)}^{\cdot(-)}$ from the previous iteration. We calculate the Jacobian ${}^6\mathbf{J}_6(\Theta_{i\,start(k)}^{\cdot(-)})$ and increases $\Delta\Theta_{i(k)}^{\cdot(-)}$ from the discretized equation (5). The end values of the joint variables in the *k-th* iteration $\Theta_{i\,end(k)}^{\cdot(-)}$ is the sum $\Theta_{i\,start(k)}^{\cdot(-)}+\Delta\Theta_{i(k)}^{\cdot(-)}$. Subsequent iterative calculations cause the approach of the end values $\Theta_{i\,end(k)}^{\cdot(-)}$ of iteration to a values $\Theta_i^{\cdot}$ representing the solution of the inverse kinematics of the current via-point. If the current via-point is singular, the approaching to its gives rise to the risk of a reduction of the Jacobian ${}^6\mathbf{J}_6(\Theta_{i\,start(k)}^{\cdot(-)})$ rank. These protection's measures are presented in the fourth chapter by using the example of manipulator IRB-1400. They consist in the correction of the angles $\Theta_3^{\cdot}$ and $\Theta_5^{\cdot}$ by the minimum increments $\Delta\Theta_{3\min}$ and $\Delta\Theta_{5\min}$, seen by the servos.

If   during   the   iterative   calculations $\Theta_{3\,start(k)}^{\cdot(-)}$ satisfy   inequality $\left|\Theta_{3\,start(k)}^{\cdot(-)}-\Theta_3^{\cdot*}\right|<\Delta\Theta_{3\min}^{\cdot}$,   we assume: $\Theta_{3\,end(k)}^{\cdot(-)}=\Theta_3^{\cdot*}-sign(\Delta\Theta_{3(k-1)}^{\cdot(-)})\cdot\Delta\Theta_{3\min}^{\cdot}$.   If $\left|\Theta_{3\,start(k)}^{\cdot(-)}-\Theta_{3(1,2)}^{\cdot**}\right|<\Delta\Theta_{3\min}^{\cdot}$,   we   substitute $\Theta_{3\,end(k)}^{\cdot(-)}=\Theta_{3(1,2)}^{\cdot**}-sign(\Delta\Theta_{3(k-1)}^{\cdot(-)})\cdot$ $\Delta\Theta_{3\min}^{\cdot}$.   Also,   for   the   angle $\Theta_{5\,start(k)}^{\cdot(-)}$ satisfying   the   inequality $\left|\Theta_{5\,start(k)}^{\cdot(-)}-\Theta_5^{\cdot*}\right|=\left|\Theta_{5\,start(k)}^{\cdot(-)}\right|<\Delta\Theta_{5\min}^{\cdot}$,   we assume: $\Theta_{5\,end(k)}^{\cdot(-)}=\;-sign(\Delta\Theta_{5\,start(k-1)}^{\cdot(-)})$ $\cdot\Delta\Theta_{5\min}^{\cdot}$. For other joint variables $\Theta_{i\,end(k)}^{\cdot(-)}=\Theta_{i\,start(k)}^{\cdot(-)}$.

Using such protection measures one should check the range of the changes in the gripper position caused by above angles' corrections. For the sample manipulator IRB-1400 suggested corrections can at most result in changes of the position which is the sum of errors, i.e. $3.6 \cdot 10^{-3}$ $mm. + 14.7 \cdot 10^{-3}$ $mm. = 18.3 \cdot 10^{-3}$ $mm.$ The typical positioning accuracy of industrial robots are of the order 10^{-1} $mm.$ Thus, the errors caused by such corrections are acceptable.

If the gripper position errors caused by such corrections were equal to or greater than the required positioning accuracy, inverse kinematics equations in a global form, presented in [12] would need to be used. For these equations, we can apply simple criteria for selecting solutions at the joint variables. Such criteria are much simpler and faster to implement than the criteria proposed in the Jacobian methods [1,3,4,6].

Differential description of the kinematics, including corrections to prevent the loss Jacobian rank, presented in this paper, constitutes the base for the creation of own software, free from basic defect, namely, undesired stopping in singular positions.

References

1. Chiacchio, P., Chiaverini, S., Siciliano, B.: Direct and Inverse Kinematics for Coordinated Motion Tasks of Two – Manipulator System. Journ. of Dynamics Systems, Measurement, and Control. 118(4), 691–697 (1996)
2. Kozłowski, K., Dutkiewicz, P., Wróblewski, W.: Modeling and Control of Robots, ch. 1. PWN, Warsaw (2003) (in Polish)
3. Nakamura, Y., Hanafusa, H.: Inverse Kinematic Solutions With Singularity Robustness for Manipulator Control. Journ. of Dynamics Systems, Measurement, and Control 108(3), 163–171 (2009)
4. Siciliano, B., Sciavicco, L., Villiani, L., Oriolo, G.: Robotics, Modelling, Planning and Control, ch. 3.5.2. Springer, Berlin (2010)
5. Spong, M.W., Vidyasagar, M.: Robot Dynamics and Control, ch. 5.3-5.4. WNT, Warsaw (1997) (in Polish)
6. Tchoń, K., Mazur, A., Dulęba, I, Hossa, R., Muszyński, R.: Mobile Manipulators and Robots, ch. 3.2. Academic Publ. Company PLJ (2000) (in Polish)
7. Jezierski E.: Dynamics and Control of Robots, ch. 2. WNT, Warsaw (2006) (in Polish)
8. Craig, J.J.: Introduction to Robotics, 2nd edn., ch. 4. Addison Wesley Publ. Comp. (1986)
9. Szkodny T.: Kinematics of Industrial Robots, ch. 2,3,4,8.3. Silesian University of Technology Publ. Company, Gliwice (2013) (in Polish)
10. Szkodny, T.: Foundation of Robotics, ch. 2.4. Silesian University of Technology Publ. Company, Gliwice (2012) (in Polish)
11. Szkodny, T.: Basic Component of Computational Intelligence for IRB-1400 Robots. In: Cyran, K.A., Kozielski, S., Peters, J.F., Stańczyk, U., Wakulicz-Deja, A. (eds.) Man-Machine Interactions. AISC, vol. 59, pp. 637–646. Springer, Heidelberg (2009)
12. Szkodny, T.: Inverse Kinematics Problem of IRB, Fanuc, Mitsubishi, Adept and Kuka Series Manipulators. Journ. of Applied Mechanics and Engineering 15(3), 847–854 (2010)

A Review of 3-D Reconstruction Based on Machine Vision

Hong Jin, Fupei Wu*, Chun Yang, Lian Chen, and Shengping Li

Department of Mechatronic Engineering, Shantou University, Shantou,
515063 Guangdong, China
{07hjin,fpwu,12cyang,12lchen4,spli}@stu.edu.cn

Abstract. Reconstructing the three-dimensional (3-D) shape of an object is always a significant study in machine vision. Since it was first proposed by Marr in 1970s, 3-D reconstruction has developed a lot but is still limited in many applications. In this paper, various kinds of theories, algorithms and methods to reconstruct 3-D shape are introduced, classified and analyzed. Then, image acquisition methods, computing methods and their achievements and limits are summarized. In the end, some conclusions and advices for machine vision based 3-D reconstruction are presented.

Keywords: Machine Vision, 3-D Reconstruction, 3-D Shape.

1 Introduction

Machine vision (MV), which was token as a procedure of image process by Marr [1], is an integrated multi-discipline subject, developed from optics, electronics, mechanics, image processing, mold cognition, and computer science etc. MV based 3-D reconstruction is to acknowledge or measure the 3-D shape of the target through 2-D images, in which a lot of research has been done. Thus more devices with vision are applied in different fields, such as intelligent control, industrial inspection, quality control, mobile robot, 3-D animation production etc. Especially in industrial inspection, precise 3-D data is urgently demanded.

A MV system acquires the geometric information of the target, including its shape, gesture, position, and movement, to recognize and understand the obtained information, to reconstruct the target. The general process of how a MV system works is presented. Firstly, illuminated by the specific illuminant or light source. Secondly, acquiring 2-D images of the target using proper image sensors and acquiring methods. Thirdly, computing its 3-D data and reconstructing the target by using advanced tools, technology and algorithms.

According to different methods of image acquisition and processing, the theories about 3-D reconstruction based on MV can be divided into three classes: methods based on stereo vision, methods using shape from shading, and methods of monocular

X. Zhang et al. (Eds.): ICIRA 2014, Part II, LNAI 8918, pp. 195–203, 2014.

vision. This paper provides a review of these methods which are compared and analyzed. Finally, future research trends are presented.

2   Stereo Vision

Creatures with vision can acquire depth or distance information of their environment through their multiple eyes. Stereo vision is a kind of method which imitates them. It uses multiple cameras to acquire images in different directions, then inspects and matches the image points of the same point in the target based on the stereo parallax theory (see Fig.1) and triangulation theory. Using proper advanced tools and algorithms, a 3-D model can be constructed from the depth information of the target's surface [2].

A complete process of stereo vision is generally divided into six stages: image acquisition, camera calibration, feature extraction, stereo matching, 3-D reconstruction and post processing [3] [4].

Image Acquisition. To reconstruct a 3-D model by stereo vision, multiple images are needed by playing cameras in different positions [5] or using a single camera through a concave lens and a convex mirror to produce a stereo pair of images [6]. It can also be realized by rotating or moving the target with a camera and moving the camera while rotating or moving the target [7]. Methods with a single camera decrease the cost, make it easier to lay out cameras and simplify the calculating process compared with the former one.

Camera Calibration. In stereo vision, camera calibration is to build correspondence between image points [8], more than what a camera calibration generally does.

Camera calibration is a classic study and much research has been done. Zhang [9] [10] proposed methods based on known scenes. For changeable scenes, self-calibration methods, based on Kruppa function, calibration by layers [11], and changeable parameters camera calibration [12], improve the accuracy and efficient of camera calibration in a sense, but still not perfect enough.

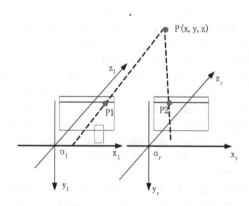

Fig. 1. Stereo parallax theory model [26]

Stereo Matching. Matching the corresponding points in different images and calculating their depth based on parallax theory, are the key technologies [13].

According to matching principles, theories of stereo matching can be divided into three classes [8]: correlation methods, methods based on feature points and global matching of polar lines. They all perform well in specific simulation.

3-D Reconstruction. Referring to the matching points' data, depths of the other points can be calculated using numerical methods. Then a 3-D model can be constructed from depth data using compute techniques.

3 Monocular Vision

The theory that depth information hides in a single 2-D image was first proposed in 1980s and has developed a lot [14] [15] since then. It promotes the development of monocular vision in MV based 3-D reconstruction.

3.1 Camera Parameters

The relationship among the focal length, the position of its focused image point and the corresponding point in the scene is given by the lens formula (1).

$$\frac{1}{f} = \frac{1}{u} + \frac{1}{v} \tag{1}$$

In the equation, f represents the focal length of the camera lens, u is the distance of the target point, and v is the distance of the image point to the lens (see Fig.2). But only points in a certain distance can be shot well enough referring to the image resolution, or the images will too fuzzy to recognize. Based on this, the theories are divided into two parts: depth from focus and depth from defocus.

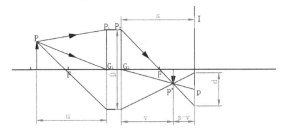

Fig. 2. the camera geometry

P1: first principle plane; P2: second principle plane; P: object point; p: image point;

D: diameter of the lens; d: diameter of blur circle;

G1, G2: first and second principle points of the lens.

3.2   Depth from Focus

When the target is in focus, the algorithms perform well using formula (1). However, the positions of camera and target must be very accurate, and the result precision is always limited by the depth of the field and sensors applied [16].

Many researchers applied optimization algorithms in depth from focus such as Faye Bo Laky method [17], pyramid structure method [18], laplacian method [19] and dynamic search method [20], which all achieved good estimation of the surface with high speed and efficiency in a sense.

This method is simple, but limited and not accurate in applications, many researchers are developing a more universal method which is depth from defocus.

3.3   Depth from Defocus

Depth from defocus decreases errors caused by the offset of the camera or the target, and improves efficiency [3].

In blurred edges, a lot of depth information can be acquired. Pentland [21] [14] roughly solved the problem. Grossman proved it in 1987 [22]. To address the problem easily and efficiently, Subbarao [23] developed Pentland's method into a simpler one assuming that the point spread function of the camera is rotationally symmetric. Sahay [24] proposed a method of harnessing defocus blur to recover high-resolution information. Swain [25] combined fuzzy logic with depth from defocus to improve the accuracy in depth estimation.

In a sense, the defocus methods can obtain accurate data, but it performs badly in texture of complex objects.

4   Shape from Shading

SFS (shape from shading) is an important branch of MV based 3-D reconstruction. Horn [26] [27] first proposed SFS in 1970s. Since then, a great number of papers and approaches have been proposed, but a universal method under realistic assumptions is still lacking. SFS simulates human's eyes mechanisms which use shape information in the image to reconstruct the 3-D shape. According to geometrical optics and physical optics, the intensity of the reflection light reflects the properties of the surface and geometrical structure of the target.

4.1   Methods and Algorithms

The most representative methods and algorithms are classified and compared by Zhang and Jean. Zhang divided them into 4 classes [28]: minimization, propagation, local, and linear based on conceptual differences. The representative methods are displayed and compared in terms of accuracy and time of these methods performing on 4 specific images. Jean classified them into 3 classes based on the mathematical and numerical techniques [29]: methods of resolution of partial differential equations, methods using minimization and methods approximating the image irradiance equation. Their performances are compared and evaluated considering convergence properties, boundary conditions, accuracy, stability and the CPU time of reconstruction. They both drew a conclusion that all SFS algorithms produce generally poor results.

4.2   Reflection Models

Building a proper reflection model is the key part of SFS process, and impacts the image properties directly and difficult to be covered. Many sophisticated models have been proposed, but the basic models are Lambertian Model and Specular Reflection Model.

Lambertian surface reflects light in all directions [28]. The result doesn't have any business with the position of the viewer or the illuminant. It is simple and widely used, but performs badly in terms of rough surface.

Specular Reflection's key property is that the illuminant's incident angle equals to the reflected one. The reflected light in reality is comprised by specular spike and specular lobe, as shown in the Figure 3.

Torrance-Sparrow [30] proposed a refined model, assuming the surface as micro surface where mirror-like small facets are randomly oriented. Nayar [31] developed a reflection model for hybrid reflection which has a diffuse lobe compared with pure specularity, see figure 3. Xiao Dong He [32] proposed a complex physical reflection model which can be used for most of the surface properties and it promotes the development of SFS a lot.

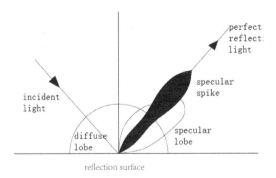

Fig. 3. Reflection Model

5   Conclusions

3-D reconstruction is an important technology in engineering and industry, and the reconstruction result directly affects subsequent design and manufacture [33].

5.1   Performance Comparison of 3-D Methods Based on MV

Table 1 is a brief overview of the advantages and defects of these 3-D reconstruction classes. As shown in Table 1, stereo vision method can be applicable in most of the occasions, but it needs complex devices and algorithms and takes more time to compute. Stereo algorithms suffer from lack of local surface information, but SFS's produce better estimates of local areas. Compared with stereo vision, monocular vision costs less, needs less space and simpler algorithms, but it doesn't perform as good as stereo vision in other aspects. Every method can reconstruct the 3-D shape well in specific situations, but not for all the cases. There is no universal solution. It's better to combine them together. Cryer and his co-workers proposed an algorithm by combining stereo and shape from shading [34]. So some researchers combine monocular and stereo vision to reconstruct the 3-D scenes of the environment [35] and achieve a good result. To obtain accurate data from the images, structured lights are always used to help sharp the details of the target object [36] [37].

5.2   Development Trends

Based on the analysis above, in the near future, the research of 3-D reconstruction based on machine vision can be laid out in the following factors.

1) Refine the algorithms and models. Based on the application in reality, refine every model so that it can perform in applications as well as in simulation.
2) Combine different methods. To make the most use of the information in an image, combine these methods to reach a good result.
3) More sophisticated computing methods. To improve the computing speed, the computer cluster computing, network cloud computing and GPU computing method can be applied to shorten the time of reconstruction.

Table 1. A brief overview of these methods

Methods	Complexity	Automation	Perform ance	Cost
Stereo vision	Higher complexity; Longer time	Full automation Can be achieved	Better	High
Monocular vision	Lower complexity; Short time	A certain degree	Fine	Lower
SFS	Low complexity for the methods with complex algorithms	Full automation Can be achieved	Poor	Low

3-D reconstruction is still in development, there are many problems and difficulties to overcome to a good performance in reality, but the demands for perfect 3-D reconstruction is demanded urgently. With the development and maturation of the techniques, 3-D reconstruction based on machine vision will change the style we produce, inspect products, and the way we live.

Acknowledgement. The project was supported by the National Science Foundation of China (Grantnos.51305247and51175315), Guangdong Natural Science Foundation (Grantno.201301015788), the Fundamental Research Funds for the Central Universities (Grantno.2012ZP0004), and the Foundation for Distinguished Young Talent sin Higher Education of Guang- dong, China (Grantno.012LYM-0062).

References

1. Marr, D: Vision. Freeman and Company (1982)
2. Farquhar, T.H., et al.: An evaluation of exact and approximate 3-D reconstruction algorithms for a high-resolution, small-animal PET scanner. Medical Imaging, 1073–1080 (1998)
3. Liu, S., Zhao, L., Li, J.: The applications and summary of three dimensional reconstruction based on stereo vision. In: International Conference on Industrial Control and Electronic Engineering, pp. 622–623 (2012)
4. Ng, O.-E., Ganapathy, V.: A Novel Modular Framework For Stereo Vision. In: IEEE/ASME International Conference on Advanced Intelligent Mechatronics Suntec Convention and Exhibition Center, Singapore, July 14-17, pp. 857–862 (2009)
5. He, G.-L., Lao, L.: The Multi-vision Method for Localization Using Modified Hough Transform. In: 2009 World Congress on Computer Science and Information Engineering, pp. 34–37
6. Yi, S.: An Omnidirectional Stereo Vision System Using a Single Camera. In: 18th International Conference on Pattern Recognition, ICPR 2006, pp. 861–865. Div. of Electron. & Inf. Eng., Chonbuk Nat. Univ, Jeonju (2006)
7. Di, H., Shang, Q., Wang, S.: A virtual binocular vision range finding method of remote object based on single rotating angle indexing camera. Electronic Measurement & Instruments, 2-846–2-849 (2009)
8. Songde, M., Zhengyou, Z.: Compute Vision. Science Press (1998)
9. Zhang, Z., Deriche, R., Faugeras, O.D., Luong, Q.-T.: A robust technique for matching two uncalibrated Images through the recovery of the unknown epipolar geometry. Artificial Intelligent 78(1-2), 87–119 (1995)
10. Zhang, Z.: Determining the epipolar geometry and its Uncertainty: A review. International Journal of Computer Vision 27(2), 161–195 (1998)
11. Hartley, R., Zisserman, A.: Multiple view Geometry in computer vision. Cambridge University Press (2000)
12. Pollefeys, M., Van Gool, L.: Self-calibration from the absolute comic on the plane at infinity. In: Sommer, G., Daniilidis, K., Pauli, J. (eds.) CAIP 1997. LNCS, vol. 1296, pp. 175–183. Springer, Heidelberg (1997)

13. Scharstein, D., Szeliski, R.: A Taxonomy and evaluation of dense two-frame stereo correspondence algorithms. Int. Computer Vision, 7–42 (2002)
14. Pentland, A.P.: A new sense for depth of field. In: Proceedings of International Joint Conference on Artificial Intelligence, pp. 988–994 (1985)
15. Subbarao, M.: Direct Recovery of Depth-map 11: A New Robust Approach, Technical, Report, Image Analysis and Graphics Laboratory, SUNY at Stony Brook (1987b)
16. Ens, J.: An Investigation of Methods for Determining Depth from focus. IEEE Transaction on Pattern Analysis and Machine Intelligence 15(2), 97–108 (1993)
17. Xiong, Y.-L., Shafer, S.A.: Depth from focusing and defocusing. In: Proc. of IEEE Computer Society Conference on Computer Vision and Pattern Recognition, pp. 68–73 (1993)
18. Darrell, T., Wohn, K.: Depth from focus using a pyramid architecture. Pattern Recognition Letter 11(12), 787–796 (1990)
19. Nayar, S., Nakagawa, Y.: Shape from focus. IEEE Trans. on PAMI 16(8), 824–831 (1994)
20. Yun, J., Choi, T. S.: Fast shape from focus using dynamic programming. In: Proc. of Three-Dimensional Image Capture and Applications, pp. 71–79 (2000)
21. Pentland, A.P.: Depth of scene from depth of Field. In: Proceedings of DAR PA Image Understanding Workshop, Palo Alto (1982)
22. Grossman, P.: Depth from focus. Pattern Recognition Letters 5, 63–69 (1987)
23. Subbarao, M., et al.: Depth Recovery from blurred edge. In: IEEE Computer Society Conference on Computer Vision and Pattern Recognition, pp. 98–503 (1988)
24. Sahay, R.R., Rajagopalan, A.: Harnessing defocus blur to recover high-resolution information in shape-from-focus technique. Computer Vision 2(2), 50–59 (2008)
25. Swain, C., et al.: Depth estimation from image defocus using fuzzy logic. Fuzzy Systems 1, 94–99 (1994)
26. Horn, B.K.P.: Shape from Shading: a Method for obtaining the shape of a smooth Opaque Object from one view, Ph. D thesis, MIT AI lab (November 1970)
27. Horn, B.K.P., Brooks, M.J. (eds.): Shape from Shading. MIT Press, Cambridge (1989)
28. Zhang, R., Tsai, P.S., Cryer, J.E.: Shape from shading: A survey. IEEE Transactions on Pattern Analysis and Machine Intelligence 21(8), 690–706 (1999)
29. Durou, J.-D., Falcone, M., Sagona, M.: Numerical methods for shape-from-shading: A New Survey with bench marks. Compute Vision and Image Understanding 109, 22–43 (2008)
30. Torrance, K.E., Sparrow, E.M.: Theory for Off-Specular Reflection from Roughened Surfaces. J. Optical Soc. Am. 57, 1,105–1,114 (1967)
31. Nayar, S.K., Ikeuchi, K., Kanade, T.: Surface Reflection: Physical and Geometrical Perspectives. IEEE Trans. Pattern Analysis and Machine Intelligence 13(7), 611–634 (1991)
32. He, X.D., Torrance, K.E., Sillion, F.X., Greenberg, D.P.: A comprehensive physical model for light reflection. AUTEX Research Journal (2002)
33. Szeliski, R.: Rapid Octree Construction from Image Sequences, Computer Vision. Computer Vision, Graphics and Image Processing 58(1), 23–32 (1993s)
34. Cryer, J.E., Tsai, P.-S., Shah, M.: Integration of shape from X modules: Combining Stereo and Shading. In: Proceedings of the IEEE Computer Society Conference on Computer Vision and Pattern Recognition, CVPR 1993, pp. 720–721 (1993)

35. Sivaraman, S., Trivedi, M.M.: Combining monocular and stereo-vision for real-time vehicle ranging and tracking on multilane highways. In: 14th International IEEE Conference on Intelligent Transportation Systems (ITSC) (2011)

36. Sun., C., Tao, L., Wang, P., He, L.: A 3D acquisition system combination of structured-light scanning and shape from silhouette. Chinese Optics Letters, 282–284 (May 2006) (in Chinese)

37. Lu, S.L., Zhang, X.M.: Analysis and optimal design of illuminant tin solder joint inspection. Optics and Precision Engineering, 1377–1383 (August 2008)

Research on Surface Mounted IC Devices Inspection Based on Lead's Features*

Wu Hui-hui[1] and Lu Sheng-lin[2]

[1] Dept.of Mechanical and Electronic Engineering, Shunde Polytechnic,
Foshan City of Guangdong Province, 528333, China
[2] Dept. of Mechanical Engineering, Dongguan University of Technology,
Dongguan City of Guangdong Province, 523015, China

Abstract. According to the in-line inspection requirements of the surface mounted IC (integrated circuit) devices on the printed circuit board (PCB), an inspection algorithm based on the lead's features was presented. Firstly, the features of the IC devices with different mounted quality under three colors (red, greed, blue) structure light source were analyzed. Secondly, the lead edge points were extracted with the first color derivative. Then the points were projected to avoid the difficulty of right thresholding. Based on the projections of the edges, the horizontal and vertical borders of the IC leads were obtained by the sliding location window algorithm. After location the leads, the defects such as missing devices, wrong devices, shifts and skews were inspected. Experiments results, from comparing with three existing method, show that our method possesses better performances both on efficiency and speed.

Keywords: automatic optical inspection (AOI), IC devices, leads, color derivative, integral projection.

1   Introduction

Recently, with the developments of surface mounting technology (SMT), the current trends toward miniaturization of devices, denser packing of boards and highly automated assembly equipment make the task of detecting these defects on the PCBs more critical and difficult. Since human inspectors result in low speed and efficiency, as a replacement, several visual inspection systems have been reported with different classification techniques [1-10]. Nevertheless, most attention were paid to the solder joints inspection after reflow [1-3]. However, the later a defect is detected, the more expensive it is to be repaired, thus, early detection (i.e., after surface mounting) is inherently necessary[7].

Teoh et al. [9] ultilized histogram to identify missing devices or misalignment of device. They also proposed employing white pixel count index as criterion to detect device with wrong orientation. The approach was fast, but results in many false alarms since it ignored the features such as color and light. Cripin et al. [12] proposed a template matching and wavelet decomposition technique to detect device absence,

---

* Manuscript August 20,2014;This work was supported in part by the open project of Guangdong provincial   key laboratory of precision equipment and manufacturing technology (50825504);Shunde science and technology planning project.

X. Zhang et al. (Eds.): ICIRA 2014, Part II, LNAI 8918, pp. 204–215, 2014.

which was sensitive to the device rotation. The other decision makers, such as Bayesian frameworks, neural networks with high performances were also applied to detect the devices [4,5]. However, the slow operation made it rarely used in real time system.

In order to develop a fast and reliable inspection strategy, the proposed system and methodology is presented. The defects can be identified properly by the proposed method. The remainder of this paper is organized as follows: In the next subsections the features of devices under three colors (red, greed, blue) structure light are analyzed. In Section 3, a first color derivative process combined with integral projection was presented. Section 4 describes Feature Analyzing and defects detection of IC devices. Experimental results that highlight the potential of the developed algorithm are given in Section 5 and the paper concludes in Section 6

2   Illumination System and Devices Image

2.1   Light Illumination System and Optical Structure

Dedicated illumination system are very crucial in machine vision application. It was reported that the feature distance of the device should be enlarged under the structure light source[11], so a circular tiered illumination system is developed as shown in Fig.1. It consists of three layers of ring-shaped LEDs, a solid-state CCD camera, a color image grabber. The LEDs are coaxially tiered in a sequence of red, green and blue, in ascending distance from the inspection surface.

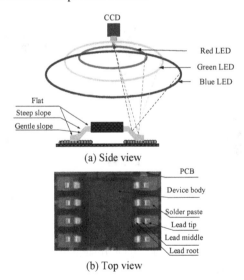

(a) Side view

(b) Top view

Fig. 1. The illumination of vision system

Since the three colored LEDs illuminate the device surface at different incident angles, the specular conditions generated by each ring LEDs are different. Fig.1(a) shows three different specularity conditions according to the configuration of the colored LEDs and the surface inclination. For the red circular LEDs in Fig.1(a), the slope near to the plate meets the specularity condition such that a device surface with this flat appears as a red highlight pattern in a camera image. Other rays emitted from the green and blue LEDs are not admitted into the camera's principal direction, so the camera cannot capture the green or blue rays reflected from the surface. For the green circular LEDs in Fig.1(a), the gentle slope satisfies the specularity condition such that a device surface with this slope appears as a green highlight pattern in a camera image. With the same manner, a device surface with a steep slope appears as blue highlight pattern in a camera image.

Fig.1(b) gives an image of device on the PCB under the illumination system. Specular surfaces, such as leads tip reflect distinct color patterns: red, green and blue, according to their surface slopes. Other rough surfaces, such as the device body and the PCB base plate, diffuse the incident rays randomly in all directions, thus mixing the colors of the incident rays so that the reflections appear as their natural color in white light.

2.2   Device Image Analyzing

There are different types of devices in the SMT. In this paper, we focus on visually inspecting the placement quality of surface mounted IC devices.

In this study, the qualities of mounted device are divided into five classes: *good device* (GD), *missing device* (MD), *device rotation* (DR), *device shift* (DS), *wrong device* (WD). Before inspection, the feature of a good device (reference device) should be learned. Typical synthetic examples of device shapes, and the corresponding highlight color patterns in window images, according to their mounted qualities is depicted in Fig.2.

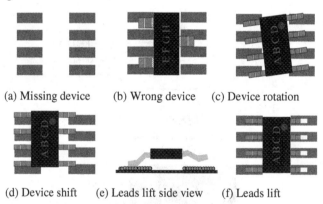

(a) Missing device   (b) Wrong device   (c) Device rotation

(d) Device shift   (e) Leads lift side view   (f) Leads lift

Fig. 2. The color patterns of defect devices

The characteristics of the shape and the corresponding color pattern according to the classes of IC devices are described as follows:

1) The missing device has no leads, and therefore no red pattern can be observed (Fig. 2(a)).

2) For a wrong device, the distance of leads is not match with the requirement of the good device (Fig.2(b)).

3) If the device is rotated, the locations of leads on both sides are asymmetric (Fig. 2(c)).

4) When the device is shift, relatively to the ideal position (i.e., pad), there are displacements in the horizontal or vertical direction (Fig.2(d)).

5) In the case that the device is lift, relatively to the ideal position (i.e., pad), the flat of lead tip is lift and red highlight pattern in the image decreasing, hence the color of lead tip is different (Fig.2(e).

Notice that the visual clues afore-mentioned can be somewhat distinct are the area and position of information of the leads, color features on the device body, therefore, our approach focus on the lead features.

3   Leads Location

In this section, we will discuss how to locate leads. Our first idea might be to use the segmentation algorithm, however, the accuracy of the result we can derive from the segmentation depends on choosing the correct threshold in most cases, this problem is especially grave if the illumination can change. Note that the edges of leads is significantly, therefore, extracted edges will be helpful o locate the leads. It is typically used to determine the position or diameter of an object.

3.1   Location of Left and Right Boundary

We now turn our attention to edges in real images, for simplicity, the first problem we have to address is finding edges in gray image. Let us regard the image for the moment as a 1D function $f(x)$. we know the gray value change significantly if the first derivatives of $f(x)$ differs significantly from 0,i.e., $|f'(x)|>>0$. Considering that the real image is discrete, thus, first derivatives is define: $f_i=f_i-f_{i-1}$. However, this definition is not symmetric. It would compute the derivative at the "half-pixel" positions $f_{i-1/2}$. A symmetric way to compute the first derivative in the 2D image with the size $M*N$ is given by[13] :

$$d(i, j)=[f(i+1, j+1)-f(i-1, j-1)]/2 \tag{1}$$

Where, $f(i+1,j+1)$ is the gray value of point $(i+1,j+1)$, $i=1,2,...M-1$, $j=1,2,...,N-1$.

Notice that the image to process is colored, let F be a $M\times N\times3$ matrix associated with the ROI in the acquired color image, then the *first color derivative* (color derivative for short) is define as :

$$d(i, j)=\frac{1}{3}\sum_{c=1}^{3}[F_c(i+1, j+1)-F_c(i-1, j-1)]/2 \tag{2}$$

Where, F_c represents different color sub-image,i.e,F_1 represents red sub-image,F_2 represents green sub-image and,F_3 represents blue sub-image. F_c $(i+1,j+1)$ is the gray value of point $(i+1,j+1)$ in the sub-image.

This is a very useful in our applications because it is extremely fast. Fig.3(a) displays a line calling horizontal search line along the leads. Its color derivative, computed with Eq.(2), is shown in Fig.3(b). We can see that the edges of leads in the image cause a very large number of local maxima in the absolute value of the color derivative. The salient edges can easily be distinguished by thresholding the absolute value of the color derivative: $d(i,j)|>=t$, where t is the threshold to select the relevant edges. nfortunately, this alone is insufficient to define a unique edge location because there are typically many connected points fulfill this condition. Therefore, to obtain a unique edge position, we must identify the locally maximal value which is called non-maximum suppression. This is called non-maximum suppression.

To extract the edge position, we need to perform the non-maximum suppression. For all color derivative, if $d_{(i,j)}$ satisfied

$$d(i, j) > d(i-1, j) \cup d(i, j) > d(i+1, j) \cup d(i, j) > t \tag{3}$$

$d_{(i,j)}$ is called edge point $e_{(i,j)}$.

In order to fit edge line, we need more edge points, therefore, a lot of search lines are drawing in the image, as shown in Fig.3(c). Along each search line, we can obtain edge points, as shown in Fig.3(d). Then, the integral projection associated with these points are defined as follow:

$$P_i = \sum_{i=1}^{L} \sum_{j}^{S} e(i, j) \tag{4}$$

Where, P_i is a vector with the dimensions $1 \times L$, L is the length of each search line, S is the number of search line. The projection is shown in Fig.3(e).

It is easily found that the peak values of the projection associated with the lead edges, hence, the number of peaks is equivalent to the number of lead boundary, for each peak, its boundary can be located with the max grad of projection. One problem is that because the single peak is usually affected by the noise and the skew of leads, the result is not accuracy. Notice that the width and distance of leads is not changed, based on this, a "*sliding filter window algorithm*" is developed to locate the leads, take the lead shown in the Fig.3(a) for example, the algorithm is detailed as follows:

1) Let element y in the projection P denotes the left boundary of the first lead, supposing the width of lead is W, then its right boundary is $y+W$. supposing the distance of leads is D, then, the left boundary of the second lead is $y+D$, and the right boundary is $y+D+W$, similarly, we can locate the left and right boundary of the 3, 4, $\cdots$, nth lead such as $y+(n-1)*D, y+(n-1)*D+W$. In order to prevent right boundary over the range of the P, the end position of y in the P is restricted to $[1, y_E]$, where, $y_E = M - W - (n-1)*D$.

2) For each element y, calculate the sum of the projecti-on by

$$Sum(y) = \sum_{p=y}^{y+W} P(p) + \sum_{p=y+D}^{y+D+W} P(p) + \cdots \sum_{p=y+(n-1)*D}^{y+(n-1)*D+W} P(p) \tag{5}$$

3) Let the step is 1, sliding y from 1 to y_E, the candidate element having the maximum value of   Eq. (5) is identified as the target element y_m.

$$y_m = \max_{y}(sum(y)), \qquad y \in [1, y_E] \tag{6}$$

y_m is the best left boundary of the first lead. Based on this element, all the left and right boundary can be obtained. The result is shown in Fig. 3(e).

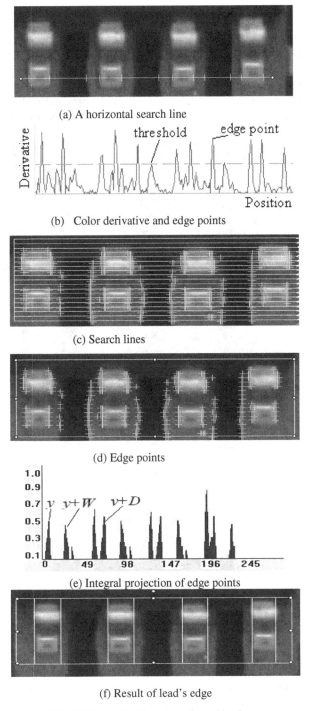

(a) A horizontal search line

(b)   Color derivative and edge points

(c) Search lines

(d) Edge points

(e) Integral projection of edge points

(f) Result of lead's edge

Fig. 3. The vertical location of good leads

3. 2   Location of Top and Bottom Boundary

After location the left and right boundary of leads, we draw some vertical search lines as shown in Fig.4(a). Along each line, the edge points can be obtained and top boundary can be located by the same way discussed in section 3.1. Let the length of lead is L, suppose the top boundary of leads is $Lead_T$, then the bottom boundary is $Lead_T+L$. Comprehensive information with the horizontal and vertical boundary, we finally get the lead position, the location of leads displays in the Fig.4(d).

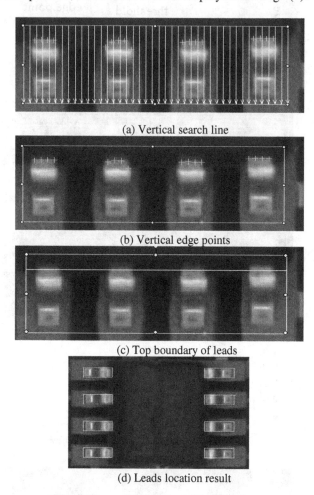

(a) Vertical search line

(b) Vertical edge points

(c) Top boundary of leads

(d) Leads location result

Fig. 4. The vertical location of good leads

4   Feature Analyzing and Inspection

Previously in section 2, the defect device will be detected by analyzing the leads feature. Therefore, during the process of leads location, a set of algorithms are developed and the detail is described as follows.

a) First, in the algorithm for missing device detection. Taking into account the projection of the lead edge points, define:

$$P_m = \frac{1}{M}\sum_{p=1}^{M}P(p) \qquad (7)$$

Where, P_m is the mean of projection, as shown in Fig.5. Comparing it with the mean of projection of reference device, if the difference between the reference device is too large, it can be judged for missing device. Otherwise, the device is acceptable.

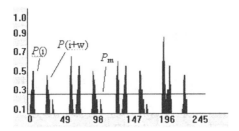

Fig. 5. projection of edge points and mean

b) If the device is present, the number of leads is calculated, since the projection of lead edges is greater than the mean of the projection, so for each point i in the projection P, if $P_i > P_m$, then calculated its next point , the point must satisfied :

$$P_i > P_m \text{ and } P_{i+w} > P_m \qquad (8)$$

Where, W is the width of the lead, as shown in Fig.5. Only if the point i satisfied   the Eq.(8) will be diagnosed as a good one. If not, the solder joint will be diagnosed as the wrong device.

c). In the algorithm for skew device inspection, a symmetry judgment is employed. Let the center of gravity of left leads is G_L, the center of gravity of right leads is G_R, then, the displacement of the center D_G is

$$D_G = |G_L - G_R| \qquad (9)$$

Suppose the ideal displacement is T_G, if $D_G \geq T_G$, the device is rotated. As shown in Fig.6(a), the position of the left and right leads are symmetry, there is no displacement. The inspection of device rotation is shown in Fig.6 (b), it shows that the left and right lead are asymmetry in the horizontal direction since the D_G is large.

(a) Image of skew device　　　(b) Leads location result

Fig. 6. The location of skew device

d). In the algorithm for detecting shift, a center of gravity comparing method is employed. After location the device, the center of gravity of the device can be obtained by

$$M_x = (x_m+L)/2, \qquad M_y = (y_m+W)/2 \qquad (10)$$

Supposed the ideal center is $[C_x,C_y]$ (i.e., the center of the pad), the permission displacement is T_x,T_y, if $|C_x\text{-}M_x| \geq T_x$, the device is horizontal shift; if $|C_y\text{-}M_y| \geq T_y$, the device is vertical shift, otherwise, the device is acceptable.

e). For detecting leads lift, an evaluation function is defined:

$$Fun_A = wR_{A0} + \frac{(1-w)}{n}R_{A1} + \frac{(1-w)}{n}R_{A2} + \cdots + \frac{(1-w)}{n}R_{An} \qquad (11)$$

where, n is the number of leads, R_{A0} represents the ratio of the total area of inspected device and the reference device leads, $R_{A1}, R_{A2}, ..., R_{An}$ represent ratio of area of lead 1,2, ..., n, respectively, w is the weighting factor, the value of their experience:

$$n=2,w=0.5; \quad n=3,w=0.4; \quad n \geq 4,w=0.2 \qquad (12)$$

If Fun_A is less than the threshold T_L, then there is leads lift defect.

Until now, all the common defects have been inspected completed, if all test items are correct, then the devices is judged to be good

5　Experiments Results

In this section, the performance of the proposed algorithm is evaluated by comparing with three existing methods

5.1　Training Procedure

To reduce the time consuming, all the training procedure is processing in the offline system. The training procedure is applied to obtain the feature of the lead in the reference image The precise steps are presented as follows.

Step 1: Selecting the region of interest (ROI) containing the inspection device in the reference image   .

Step 2: Selecting appropriate parameters for different inspection device such as length, width of the leads

Step 3: Our location method is applied to locate the leads in the ROI

Step 4:The feature such as center of gravity, permission shift for the interest device are calculated. these feature are fed into the inspection procedure for classified.

The method used to create the training sets allows us to set the inspection system before implementing the production line. This is very helpful, because it reduces the cost of the inspection and allows us to start the inline inspection in production line and the offline training system simultaneously.

5.2   Performance Evaluation of the Proposed Method

The proposed inspection algorithm has been implemented and tested from the actual placement environment in terms of recognition rate and execution time.

260 actual boards with 2080 different devices are used for testing, after completing enough experiments to adjust some suitable thresholds and train the feature, we finally achieve quite satisfying results are summarized in Table 1. The total success rate is therefore found to be 97.9%. It seems that some classes were misclassified. However, since the cost of misses is more than the false alarm and misclassification for a real inspection system, the misclassification in an appropriate amount of degree will be acceptable.

In order to verify that our algorithm is more efficient the proposed algorithm is compared with methods of neural networks[4], histogram[9], and template matching[12].All the experiments were performed with Micsoft visual studio 2010 on Pentium-4 2.4 G under Windows XP, with 1G of memory.

In terms of time requirements of our approach, Take the device in Fig.3 (the size is 401×281) for example, the average time for each step are summarized in Table 2. The color derivative extraction takes about 16 ms for processing the all leads. The projection and location step takes about 2 ms and the recognition and classifier takes 3 ms. The total consuming time is about 21 ms, which can meet the requirement of real-time inspection.

As shown in Table 3, one finds that the inspection speed of the proposed algorithm is obviously much faster than other latest three methods, which are 85ms, 25ms and 72s. This result shows that the proposed method based on leads feature need much less computing time because most they only need to analyze the leads region. It also indicates that the features extracted and used in the proposed method are simplified and models of the leads types built in the proposed method are more effective to explain the appearance and the shape of the device than the other methods.

As can be seen from Table 3, from the view point of detection device types, one finds that the methods of neural networks, histogram, and template matching can detect only some certain common types of device, such as missing device, wrong device, shift device and acceptable, but the proposed method can detect more other types, including skew and lift. This result means that the proposed method can extract more useful information to explain lead types and its classification is more detailed and useful to the quality control of the manufacturing process.

From these comparisons, it can be concluded that the performances of the proposed algorithm are much better than another latest methods in two aspects, i.e., inspection recognition types and inspection speed. Its more detailed classification and higher recognition rate are more useful to the quality control in manufacturing process, and its faster inspection speed is more helpful to improve the efficiency of the manufacturing process. This indicates that the algorithm is suitable for real-time applications.

Table 1. Classification Results of Testing Data

Type	Number	Results of classification						Correctly classify
		MD	WD	DR	DS	DL	GD	
MD	6	6						100
WD	7	7						100
DR	15	14	1					93.3
DS	11		11					100
DL	17					16	1	94.1
GD	2024					6	2018	99.7
Total	2080	--						97.9

Table 2. The processing time of proposed method

Step of the algorithm	Average time (ms)
Color derivative extraction	16
Projection and location	2
Recognition and classifier	3
Total	21

Table 3. The performance comparison of four different detection methods.

Algorithm	Inspection items	Times(ms)
Neural networks	Missing, wrong, shift, good	85
Histogram	Missing, wrong, shift, good	25
Template matching	Missing, wrong, good	72
Our method	Missing, wrong, shift, lift, good	21

6   Conclusion

In this paper, efficient techniques for surface mounted IC devices inspection have been described. Three colors (red, greed, blue) structure light source is used for efficient extraction device color information. Via the proposed inspection method, The edge points of leads are obtained with the first color derivative and projected to   locate the boundary of leads. After that, a simple Feature Analyzing is shown to be enough for defects detection and classification. Experimental results have shown that the proposed inspection system can identify the defects of the device such as missing, shift, rotation and wrong devices, etc. effectively and achieved a high performance in terms of the recognition rate as well as the execution time. The proposed inspection system is now on the field test.

Acknowledgment. The authors were very grateful to OPT Automatic Technology Corporation and SCI machine vision for providing us a favorable experiment environment.

References

[1] Kim, T.H., Tho, T.H., Moon, Y.S.: Visual inspection system for the classification of solder joints. Pattern Recognition 32(4), 565–575 (1999)

[2] Giaquinto, A., Fornarelli, G., Brunetti, G., Giaquinto, A.: A fuzzy method for globalquality index evaluation of solder joints in surface mount technology. IEEE Trans. Ind. Inf. 7, 115–124 (2011)

[3] Wu, F.P., Zhang, X.M.: An Inspection and Classification Method for Chip Solder Joints Using Color Grads and Boolean Rules. Robotics and Computer-Integrated Manufacturing 30(5), 517–526 (2014)

[4] Acciani, G., Brunetti, G., Chiarantoni, E.: An automatic method to detect missing devices in manufactured products. In: Proceedings of International Joint Conference on Neural Networks, pp. 2324–2329 (2004)

[5] Liu, J.Q., Kuang, H., Ding, S.H.: Research on machine vision system for inspection of SMT chip pins. China Mechanical Engineering 18(16), 1908–1912 (2007)

[6] Wu, H.H., Zhang, X.M., Kuang, Y.C.: Automated Visual Inspection of Surface Mounted Chip Components. In: IEEE International Conference on Mechatronics and Automation, pp. 189–1794 (2010)

[7] Michael, E.Z., Stefanos, K.G., George, A.R.: A Bayesian framework for multilead SMD post-placement quality inspection. IEEE Transactions on Systems, Man, and Cybernetics-Part B: Cybernetics 34(1), 440–453 (2004)

[8] Xian, F.: Testing technology development promoted by HDI packaging technology. Equipment for Electronics Manufacturing 157, 32–35 (2008) (in Chinese)

[9] Teoh, E.K., Mital, D.P., Lee, B.W., Wee, L.K.: Automated visual inspection of surface mount PCBs. In: IECON, 16th Annual Conference of IEEE, pp. 27–30 (1990)

[10] Gallegos, J.M., Villalobos, J.R., Carrillo, G.: Reduced-dimension and wavelet processing of SMD images for real-time inspection. In: Proceeding of the IEEE Southwest Symposium on Image Analysis and Interpretation, pp. 30–35 (1996)

[11] Lu, S.L., Zhang, X.M., Kuang, Y.C.: Optimal illuminator design for automatic optical inspection systems. International Journal of Computer Applications in Technology 37(2), 101–108 (2010)

[12] Cripin, A.J., Rankov, V.: Automated inspection of PCB components using a genetic algorithm template-matching approach. Int. J. Adv. Manuf. Technol. 35, 293–300 (2007)

[13] Steger, C., Urich, M., Wideemamm, C.: Machine vision algorithms and applications, pp. 209–216. Publishing Tsinghua university, Beijing (2011) (in Chinese)

A Fast Coplanarity Inspection System
for Double-Sides IC Leads Using Single Viewpoint

Qiusheng Zhong, Xianmin Zhang, and Zhong Chen

South China University of Technology,
Guangzhou 510640, P.R. China
zhangxm@scut.edu.cn

Abstract. A new single camera vision system for online coplanarity inspection of SOP (Small Outline Packaging) IC leads is presented. First, a novel imaging optical structure with multiple reflection mirrors and a total reflection rectangular prism is developed. This structure characterized in that two side views of IC leads are captured one-time in a small FOV (Field Of View). Second, a fast high-precision algorithm to inspect the coplanarity of IC leads is also proposed. Based on the statistical method, the "3Sigma" criterion is used to judge the quality of IC leads' coplanarity. Finally, the experimental results indicated that the proposed system can inspect IC leads coplanarity in real-time.

Keywords: IC leads, coplanarity inspection, optical structure.

1   Introduction

The coplanarity of IC leads is getting increasing attention throughout the electronics manufacturing industry. If the IC leads have poor coplanarity, the IC can not correctly sit on the pads of PCB during the welding process. The electronic products with this problem may past the terminal detection, but sometimes they could not work normally. Yet, it is impossible to avoid the poor coplanarity altogether in the production process. However, if the IC having poor coplanarity leads continues to assemble on the terminal product, great economic losses will be caused. Hence, some solutions were taken to control the coplanarity of IC leads. They can be classified into four main types: laser based, 3D based, projection based and special layout based methods.

(I) Laser Based Method. Kim [1] proposed an inspection process based on a laser vision system which provide the range of the BGA IC. This system required a Cartesian robot to control of the laser vision sensor moving on the teaching path. Sun[2] presented a line-structured laser scanning measurement technology in SMT-based BGA chip leads' coplanarity. This approach required the high accurate motion control and resulted in slow measurement. Wei[3] proposed a coplanarity inspection of ball grid array (BGA) solder balls based on laser interference. The system required the laser interference equipment to produce laser interference structure.

X. Zhang et al. (Eds.): ICIRA 2014, Part II, LNAI 8918, pp. 216–225, 2014.
© Springer International Publishing Switzerland 2014

(II) 3D Based Method. Johannesson [4] presented a 3D measurement system for coplanarity inspection of the BGA chips based on the sheet of light triangulation principle. Although the system structure was compact, the camera could capture only one line at one time. Yen [5] proposed a 3D measurement system for coplanarity inspection of BGA solder balls. This method could overcome the shortcoming of low inspection speed of physical line-scanning process of the laser scanning technique.

(III) Projection Based Method. Qin[6] proposed a solder ball coplanrity inspection system for BGA package with one CCD camera based on projection method. Since two times imaging with different light, two times fitting for each ball, and motion control unit were all needed, these would make the system slow down. Smeyers [7] presented a special optical system to measure the leads coplanarity of QFP (Quad Flat Package) IC based on dual shadow projection method. Although the system could achieve sub pixel positioning, the four times imaging was time-consuming and calibration procedure had to be implemented.

(IV) Special Layout Based Method. Stanke[8, 9] designed a special layout with four mirrors and a back light source to captured all four chip side views of a QFP IC from one point view. This proposed system has the advantage of high precision and stability, but a static carrier and a special layout are required for IC chip imaging. Hou[10] developed a layout using two mirrors and a ring type LED lighting to measure to double side views of a SOP IC from one point of view. Since a grey scales image template match method was used, the image processing of IC lead position extraction is slow.

Traditionally, for most of the coplanarity measurement system of IC leads, either the structure is complex resulting in high cost, or the imaging quality is poor resulting in complicated algorithm with slow running speed. In this paper, the online coplanarity measurement system for IC leads is presented, which captures double sides of IC leads to be measured using single camera vision system. Hence, a fast algorithm for the coplanarity measurement of IC leads is studied.

This paper is organized as follows. In Section 2, a novel imaging optical structure is presented, which is used for the IC leads coplanarity inspection. In Section 3, the inspection algorithm is proposed. In Section 4, experiments are performed to validate the effectiveness of the proposed system. And the paper is closed in Section 5 by some conclusions.

2   Imaging Optical Structure

The proposed imaging optical structure of coplanarity inspection system for IC leads can be assembled for the simultaneous acquisition of double side views of IC leads in a single frame at a small FOV. As can be seen from Fig. 1 (a), there is an isometric view of imaging optical structure of IC leads coplanarity inspection system illustrating two unattached light sources, two visors, the isolated light plate, the IC sliding rail, the gland, the multiple reflection mirrors (RM1, RM2, RM3, RM5, RM6, RM7, RM8), a prism and a CCD camera arrangement.

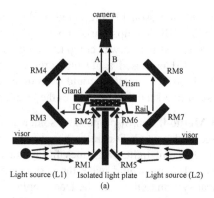

Fig. 1. IC leads image optical structure diagram

In Fig.1(a), as illumination for IC leads, the two light sources L1 and L2, being made of white light LED array, working independently and constructing symmetrically of each other, are provided. As for the left side optical path, based on the backlight illumination, after reflected by RM1 and RM2, L1 provides light that impinges the tip of the left side leads and projects a backlit image onto RM3. After reflected by RM4 again, the projected image can be seen from the left side surface of the prism. Simultaneously, for the imaging principle, the right side optical path is the same as the left side optical path. Thereby, after reflected by RM5, RM6, RM7 and RM8 in turn, L2 provides light that projects the backlit image of the tip of the right side leads on the right side surface of the prism. Consequently, part of the optical signal from L1 is finally incident on the FOV area of the left side of the camera, and the other part of the optical signal from L2 is finally incident on the FOV area of the right side of the camera.

The prism, having two reflective surfaces, is a total reflection rectangular prism. This prism is provided above the IC, and a camera is provided there above, enabling camera to image from two different optical paths. After transmitted by the prism, the image of the left side view of IC leads is incident on the camera via optical path A in Fig. 1, and the image of the right side view of IC leads is incident on the camera via optical path B in Fig. 1.

3   Inspection Algorithm

3.1   IC Lead Tips Searching

The searching algorithm is designed based on the following image features. First, since the gray level difference between the extreme dark leads and very bright background is very large. It is most easy and convenience for the algorithm to identify the target and background. Second, because all IC leads have rising and falling edge feature, it is easy to characterize the start and end of a lead. Finally, the tips of IC leads are much higher than baseline obviously, which can clearly distinguish and effectively restrain the weak jagged edges. So it needs to set a threshold to record and update the possible target points. As a result, by incorporating these features, the search algorithm can executed efficiently.

As for the leads in the lower row, the image is scanned from bottom left to top right column by column. The algorithm of the IC lead tips searching in the lower part is depicted by the flowchart in Fig. 2. As for the leads in the upper row, the searching method is similar to the lower row.

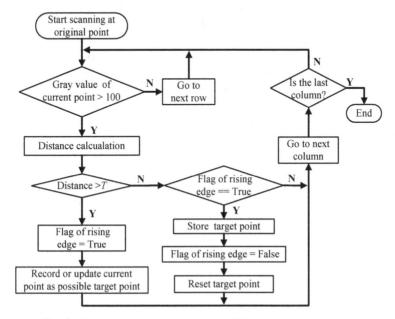

Fig. 2. Flowchart of searching the tips of IC leads in the lower part

3.2   Sub-pixel Positioning

In this research, based on the gray moment method, an innovative four steps means, including estimation, coarse positioning, precision positioning and verification, is utilized to the fast sub-pixel positioning for leads tips. These four steps are as follow.

Step1: estimation procedure. The predicted value is selected as the middle element of the gray vector, and the same element number of both ends is selected. Then, based on the gray moment method, the estimated edge point is calculated. In this procedure, we choice the gray vector with 41 elements.

Step2: coarse positioning procedure. The estimated edge point is selected as the middle element of the gray vector, and enough elements is selected symmetrically for the gray vector construction. During the coarse positioning procedure, the element number of the gray vector is selected as 21.

Step3: precision positioning procedure. The coarse positioning edge point is selected as the middle element of the gray vector, and the less elements is selected symmetrically for the gray vector comparing with the step 2. In the precision positioning procedure, the element number of gray vector is chosen 19.

Step4: verification procedure. The precision positioning edge point is chosen as the center of gray vector, and the fewer element number is selected. In general, during the verification procedure, element number is chosen 17.

3.3   Judgment Criterion

The "3 Sigma" criterion is used to judge the coplanarity is good or not. First, a background image without IC is captured and used to assure the equations of two baselines. After searching the points of baselines, the least squares method is used to fit the equations of baseline, as shown in the formula (1). Second, many images of good coplanarity ICs are selected and the statistical method is used. Then the Sigma of the standoff of IC lead is determined. Third, based on the formulation (2), for the testing image, we calculate the distance from the IC lead tip to corresponding baseline and viewed as the standoff. Finally, according to the "3 Sigma" criterion, datum contained gross error are rejected, shown in formula (3), formula (4) and judgment criteria (5). If the datum of distances can satisfy the requirement of criteria (5), these will be accepted. Otherwise, these will be rejected. After remove the datum having gross error, the number (N) of datum is updated. As a result, based on the formula (3) again, the average of these distances without gross error is calculated, corresponding to the standard in pixels. Then, the coplanarity of IC lead can be inspected.

The two baseline equations can be formulated as

$$\begin{cases} A_1.x + B_1.y + C_1 = 0 \\ A_2.x + B_2.y + C_2 = 0 \end{cases} \tag{1}$$

Then, the equations (2), (3), (4) and (5) can be formulated as

$$\begin{cases} D_i^{upper} = \left| \dfrac{A_1.x_i^{upper} + B_1.y_i^{upper} + C_1}{\sqrt{A_1^2 + B_1^2}} \right| \\ D_i^{lower} = \left| \dfrac{A_2.x_i^{lower} + B_2.y_i^{lower} + C_1}{\sqrt{A_2^2 + B_2^2}} \right| \end{cases} \quad (i = 1,2,3,\ldots,N) \tag{2}$$

$$\begin{cases} \overline{D^{upper}} = \dfrac{1}{n}\sum_{i=1}^{n} D_i^{upper} \\ \overline{D^{lower}} = \dfrac{1}{n}\sum_{i=1}^{n} D_i^{lower} \end{cases} \quad (n=1,2,3,\ldots,N) \tag{3}$$

$$\begin{cases} \sigma^{upper} = \sqrt{\dfrac{1}{n-1}\sum_{i=1}^{n}(D_i^{upper} - \overline{D^{upper}})^2} \\ \sigma^{lower} = \sqrt{\dfrac{1}{n-1}\sum_{i=1}^{n}(D_i^{lower} - \overline{D^{lower}})^2} \end{cases} \quad (n=1,2,3,\ldots,N) \tag{4}$$

$$\begin{cases} \left| D_i^{upper} - \overline{D^{upper}} \right| \le 3.\sigma^{upper} \\ \left| D_i^{lower} - \overline{D^{lower}} \right| \le 3.\sigma^{lower} \end{cases} \quad (5)$$

4   Experiments and Results

The two side views of IC leads image are acquired by the IC leads coplanarity inspection system with a single GIGE camera one-time in 8 bit gray format, having the size of 1600*1200. The lens used in the inspection system is a TEC-M55 from Computar, and the lightings are produced by the two white LED strip light sources. The double-sides 16-pins SOP ICs are used as samples for testing.

4.1   Experiment Platform

As in Fig. 3 (a), there is an inclined angle φ between the imaging optical structure and the horizontal direction. The IC can slide on the inclined rail from top to bottom relying on its own gravity. When the IC is in the FOV, both side views of IC will be captured. The Fig. 3 (b) and Fig. 3 (c) show that the imaging optical structure is compact, and its size is about 5 cm * 5cm * 6 cm (Length * Width * Height).

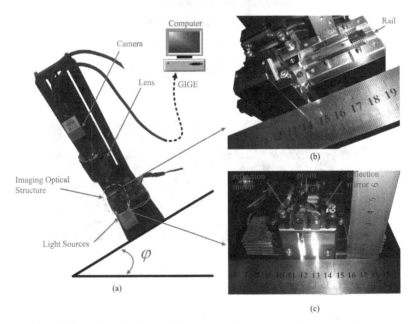

Fig. 3. (a) IC leads coplanarity inspection system. (b) The top view of the imaging optical structure. (c) The side view of the imaging optical structure.

4.2   Statistical Procedure

In the Fig. 4 (a), we need to obtain the two linear equations, respectively. The calculation for the lower baseline is organized as follow. First, the points on the baseline are searched from bottom left to top right. The gray value of point on the baseline is about 100 according to the priori knowledge. Hence, if the gray value of the point is greater than 100, the coordinates of the current point will be recorded and the searching will go to the next column of the search until the search is completed for all columns. Second, based on the center of recorded point, a 41-dimensional gray vector is constituted. Third, based on the gray moment approach, the four-step method is used for the sub-pixel positioning of points on the lower baseline. Finally, by using sub-pixel points on the lower baseline, the least squares method is used to fit the lower baseline. On the principle, the calculation for the upper baseline is the same as the lower baseline. As a result, the equations of the upper baseline and the lower baseline can be obtained as follow.

$$\begin{cases} 0.003459X + Y - 1001.656185 = 0 \\ 0.002225X + Y - 202.642707 = 0 \end{cases} \tag{6}$$

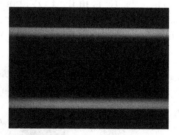

Fig. 4. The background image

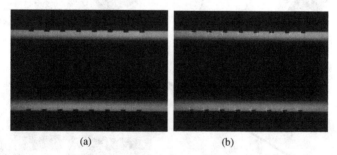

(a)   (b)

Fig. 5. The image with good IC leads.(a) The source IC image. (b) IC image with cross marks for lead tips.

Fig. 5 (a) is one of the good images of IC leads. First of all, the threshold of distance is set as 6. And using the IC lead tips searching method, the IC lead tips are positioning, represented by the green crosses, see Fig. 5 (b). While the two the baseline expressed by the red lines. After that, to positioning result as the center pixel, a 41-dimensional gray vector is constructed. And based on the above four-step sub-pixel positioning algorithm, sub-pixel positioning of the IC lead tips are achieved.

Similarly, the lead tips of 60 IC sample images can be searched and sub-pixel positioned. First, the sub-pixel locations of 480 leads in the upper part, and the other sub-pixel locations of 480 leads in the lower part, can be obtained. Second, on behalf of the standoff value of lead tip, the distance from the IC lead tip to corresponding baseline is calculated. Then two curve graphs of standoff of IC lead tips of the upper part and lower part are shown in Fig. 6. Finally, according to the "3 Sigma" criterion, datum contained careless error are rejected, the average of these distances is calculated and corresponds to the standard standoff value in pixels. As a result, the standard standoffs and the range of the good IC can be seen from the table 1.

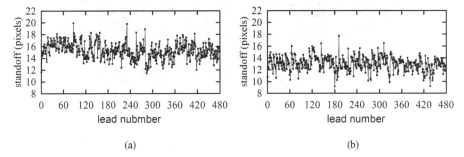

(a)   (b)

Fig. 6. (a) Curve graph of standoff of upper IC leads. (b) Curve graph of standoff of lower IC leads

Table 1. The ranges of standoffs of good IC in pixels

Positions	Standard standoff, H (pixels)	[H-3Sigma, H+3Sigma]
Upper row	15.24	[11.23, 19.25]
Lower row	13.08	[9.66, 16.49]

4.3 Inspection Procedure

The proposed algorithm is implemented on a laptop with a dual-core i5 processor running at 2.27 GHz with 4 GB of RAM and Windows XP SP2. The algorithm is implemented in C/C++, compiled in Visual Studio 2008, and run in the release version. In particular, the function called **Clock()** is used as the evaluation function of the run time. The testing procedure simulates the on-line running environment and calculates the average running time of the continuous loop 1000 times as the evaluation standard.

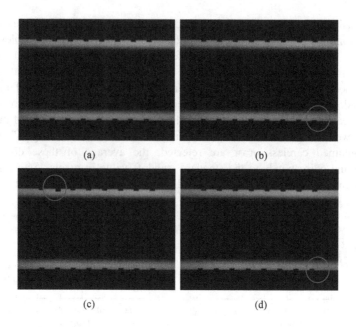

(a)   (b)

(c)   (d)

Fig. 7. Sample images of IC leads. (a) One of the good IC leads image. (b) IC image with lifted lead. (c) IC image with downed lead. (d) IC image with burr lead.

Table 2. Time-consuming of the proposed algorithm with some IC sample images

IC type	Good/Bad	Time-consuming (ms)
IC1	G	2.25
IC2	G	2.46
IC3	G	2.43
IC4	G	2.34
IC5	G	2.41
IC6	G	2.35
IC7	G	2.28
Lifted lead IC8	B	2.26
Downed lead IC9	B	2.38
Burr lead IC10	B	2.39
Average		2.355

Seven good ICs and three faulty ICs are tested. In Fig. 7, some red circles are used to mark the defect pins of these three faulty ICs. The experiment results are list in the table 2. If the coplanarity inspection result is within the range of "3 Sigma", the test IC will be viewed as the good one. Otherwise, it will be viewed as the faulty IC. The lifted lead IC8, the downed lead IC9 and the burr lead IC10 are classified as the bad coplanarity ICs, whereas others are classified as the good coplanarity ICs. In addition, The inspection system takes less than 2.4 ms to get the result for each IC sample. Therefore, our proposed algorithm is reliable for different kinds of IC leads. And the processing speed is fast and stable.

5   Conclusion

The online coplanarity inspection system of IC leads is proposed. First, a novel imaging optical structure enables the simultaneous image acquisition of two side views of IC leads at a small FOV. Second, a fast high accuracy algorithm for IC leads coplanarity inspection is presented. Based on the statistical method, the "3Sigma" criterion is used to judge the quality of IC leads' coplanarity. Finally, the experiment results indicated that the proposed system can fast inspect the leads coplanarity of each IC within 2.4ms. Furthermore, the proposed coplanarity inspection system for IC leads has been applied to the IC production line.

Acknowledgements. This research was supported by the Science and Technology Planning Project of Guangdong Province, China (Nos.2012B011300010 and 2012B091100141), and the Natural Science Foundation of Guangdong Province (S2013030013355).

References

1. Pyunghyun, K., Sehun, R.: Three-dimensional inspection of ball grid array using laser vision system. IEEE Trans. on Electron. Packag. Manufact. 22, 151–155 (1999)
2. Sun, C.K., Shi, H.Y., Qiu, Y., Ye, S.H.: Line-structured laser scanning measurement system for BGA lead coplanarity. In: The IEEE Asia-Pacific Conference on Circuits and Systems (IEEE APCCAS), pp. 715–718 (2000)
3. Wei, Z., Xiao, Z.X., Zhang, X.F., Zhou, H.Y.: Coplanarity inspection of BGA solder balls based on laser interference structure light. In: International Conference on Optical Instruments and Technology (OIT), pp. 82012F–82012F-8 (2011)
4. Johannesson, M., Thorngren, H.: Advances in CMOS Technology Enables Higher Speed True-3D Measurements. In: NEPCON WEST, pp. 181–190 (2000)
5. Yen, H.N., Tsai, D.M.: A fast full-field 3D measurement system for BGA coplanarity inspection. The International Journal of Advanced Manufacturing Technology 24, 132–139 (2004)
6. Qin, J.M., Chen, Y., Wang, F.L.: Solder ball height measurement by projection method. In: 10th International Conference on Electronic Measurement & Instruments (ICEMI), pp. 101–106 (2011)
7. Smeyers, G., Truyens, C.: A lead coplanarity inspection system for surface mounted devices. In: 12th Int. Symposium Electronic Manufacturing Technology (IEMT), pp. 270–277 (1992)
8. Stanke, G., Holm, D., Woehrle, T.: One-point-of-view high-precision and versatile optical measurement of different sides of objects. In: Proc. SPIE, vol. 3824, pp. 313–316 (1999)
9. Stanke, G., Woehrle, T., Schilling, F.: High-precision and versatile optical measurement of different sides of ICs in the confectioning process using only one viewpoint. In: Proceedings of the 24th Annual Conference of the IEEE Industrial Electronics Society, pp. 2425–2427 (1998)
10. Hou, T.H.T.: Automated vision system for IC lead inspection. International Journal of Production Research 39, 3353–3366 (2001)

An Adaptive Enhancement Algorithm
of Materials Bag Image of Industrial Scene

Wei Jia, Ya Wang, Yanbin Liu, Lilue Fan, and Qingqiang Ruan

ZunYi Normal College, Department of Computer and Information Science,
Zun Yi, 563002, China
ncu1013@hotmail.com

Abstract. In this paper, we proposed an adaptive enhancement algorithm based on fuzzy relaxation. Firstly, the OTSU algorithm is used to classify background and objective, and the crossover points for each pixel are defined by the classification results. Then, the concept of fuzzy contrast based on the image normalization is introduced, and the value of fuzzy contrast is defined as a image contrast feature plane. Secondly, at basis of the fuzzy characteristic of the hyperbolic tangent, a novel membership function is proposed, the crossover points and the adaptive function curve can achieve the best by adjusting the control parameters. Finally, the fuzzy contrast feature plane is mapped to gray level plane using the method of linear transformation. The experiment obtains excellent results which is only one time iteration. The linear transformation reduces the lose of the adjacent materials bag image's edge information and improves the operational efficiency. The analysis experimentally demonstrates that proposed algorithm is adaptive and the image details also have been preserved.

Keywords: materials bag, image enhancement, fuzzy set theory, hyperbolic tangent function.

1   Introduction

We have witnessed a rapid development of the vision technology from the lab to practical application [1-3]. In the field of logistics and packaging, and in both regular and automated stereoscopic warehouse (AS/RS), all of the materials bags need to be recognized and picked when it is in or out the warehouse. Obviously, the operation of manual mode is inefficient and error-prone. As automation and intellectualization of the logistics field, robots have been used to handle more complex and changeable process models and various shape material items. In order to overcome the faults that traditional recognition method of sensor system is too complex to recognize materials bag information and status information, the visual technology is used in the field, which can reduce the system complexity, improve efficiency, save the cost, and complete logistics processing jobs under complicated conditions through enhancing the adaptability and reliability of robot in logistics operation.

X. Zhang et al. (Eds.): ICIRA 2014, Part II, LNAI 8918, pp. 226–238, 2014.
© Springer International Publishing Switzerland 2014

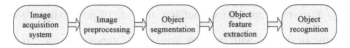

Fig. 1. The flow chart of materials bag detection and recognition

Many factors, such as the climate, light, color, texture, shape, size, location, image semantic variability, fuzzy surface performance and so on, result in ambiguity of the materials bag image and decrease of the edge contrast in the process of materials bag recognition (Figure 1).Thus, it is a key problem how to improve clarity of the scene in machine handling. The application of image enhancement algorithm plays a important role in the materials bag recognition and detection, especially in the edge of the image contrast. Recently, many image enhancement algorithms had been applied to solve the problem, and histogram equalization (HE) is one of the most popular solutions used for image enhancement [4-5], HE performs its operation by remapping the gray level of image based on the probability distribution of the input gray level, HE normally changes the brightness of the image significantly and thus makes the output image, become saturated with very bright or dark intensity values. Lots of factors, including the stronger edge, region, and texture feature of materials bag image, fuzzy characteristics, and the complex and changeable industrial scene, bring about uncertainty for the material bag image processing, understanding and describing. In order to overcome these factors, pal-k et.al [6-8], applied fuzzy set theory to understand and describe the fuzzy feature for the first time. Aiming at the shortcomings of the algorithm, a variety of improved algorithms is proposed. Li [9] proposed a novel fuzzy contrast enhancement algorithm, which does well in extracting the image local contrast, but it make some noise, and lead to image distortion. Wang [10-11] reported single-level and multi-level enhancement algorithm based on the generalized fuzzy theory. Although the algorithm achieves a success in enhancing image contrast of the different gray levels, it does not overcome the problem of materials bag and background region overlap, and the threshold of different image also cannot achieve the ideal effect.

In this paper, we will compute the image contrast, map image contrast feature plane to fuzzy plane by a novel membership function. Then, the adaptive materials bag image enhancement algorithm, will be used to conduct enhancement, meanwhile, the iterations time achieves minimum value.

2   Pal –k Fuzzy Enhancement Algorithm

According to the fuzzy-set theory, the image X of size $m \times n$ has intensity levels L, so it can be converted to the fuzzy matrix of size $m \times n$.

$$X = \bigcup_{i=1}^{m}\bigcup_{j=1}^{m}(u_{ij} / x_{ij}) \qquad u_{ij} \in [0,1](i = 1,2,\cdots,m; j = 1,2,\cdots,n) \tag{1}$$

Where u_{ij}/x_{ij} represents the membership or grade of possessing some appointed brightness by the (i, j) pixel with the gray level x_{ij}. Composed by u_{ij} is called fuzzy feature plane. In the Pal-k's Fuzzy enhancement algorithm, the membership function is described as:

$$u_{ij} = T(x_{ij}) = [((L-1) - x_{ij})/f_d + 1]^{f_e}$$ (2)

Where f_d and f_e are related to u_{ij}, A particular gray level is called the crossover point (u_c) for computing the u_{ij} value. When $f_e=2$, the result can be obtained by the following inverse transformation:

$$x_{ij} = T^{-1}(u_{ij})$$ (4)

$$X' = [x_{ij}]_{m \times n}$$ (5)

Where X is the gray level of the pixel in the result image and $T^{-1}(u_{ij})$ is the inverse transformation.

3   Image Classification Based on OTSU Algorithm

The basic principle of OTSU algorithm [12] is used the variance as a basic judgment, and the minimum image gray level of extra-class variance is computed using the least-squares method, which be regarded as the optimal threshold. The OTSU algorithm has a strong ability to adapt to the environmental change. Moreover, the materials bag volume of industrial scene is large, and the difference between background and foreground image gray level is obvious, so there is no loss of large gray level.

Suppose a image grayness range m, The pixel grayscale I is n, then the image pixel probability is:

$$p_I = n_I / \sum_{I=1}^{m} n_I$$ (6)

Where the foreground image pixel can be described as $r_0 = \{1, 2, \cdots, k\}$, and the background image pixel can be described as $r_1 = \{k+1, k+2, \cdots, m\}$, then the mean value layer is respectively given in the below formal as:

$$w_{r_0} = \sum_{i=1}^{k} p_i$$ (7)

$$w_{r_1} = \sum_{i=k+1}^{m} p_i$$ (8)

$$u_{r_0} = \sum_{i=1}^{k} ip_i / w_{r_0} \tag{9}$$

$$u_{r_1} = \sum_{i=k+1}^{m} ip_i / w_{r_1} \tag{10}$$

Thus the statistical mean can be expressed as:

$$u = w_{r_0} u_0 + w_{r_1} u_1 \tag{11}$$

And extra-class as:

$$\delta(k)^2 = w_{r_0}(u_{r_0} - u)^2 + (u_{r_1} - u)^2 = w_{r_0} w_{r_1}(u_{r_1} - u_{r_0})^2 \tag{12}$$

When the inter-variance of two segmented region is maximum, it is regarded as the optimum separate status, and then determine the threshold value T: T= $\max \delta(k)^2$.

4   Adaptive Image Contrast Enhancement Algorithm

4.1   Image Local Contrast Transformation

Image local contrast reflects on the image local details and the image pixel-pixel features, the difference in value between materials bag image gray level is small. In order to overcome the problem, Dhnawan's method [13] is introduced into our work, the image local contrast transformation can be defined as:

$$f_c = \frac{|u_{ij} - \bar{u}_{ij}|}{|u_{ij} + \bar{u}_{ij}|} \tag{13}$$

Where $\bar{u}_{ij}$ is the average density of center and surround , $|u_{ij} - \bar{u}_{ij}|$ is the local contrast, the matrix of f_c is called contrast plane, the result does not only consider the spatial smoothing of neighborhood average, but also consider the contrast stretch.

4.2   Selection of the Crossover Point

Suppose the materials bag image and background pixels of neighborhood value x_{ij} are defined as x_i and x_j, and the number of pixels are n_1, n_2 ($n_1 + n_2 = m$), then

$$x_{c1} = \frac{1}{n_1} \sum_{i=1}^{n_1} x_i \tag{14}$$

$$x_{c2} = \frac{1}{n_1} \sum_{j=1}^{n_2} x_j \qquad (15)$$

When ($n_1 > 0, n_2 > 0$ (, the crossover point related with x_{ij} can be written as:

$$x_c = \frac{x_{c1} + x_{c2}}{2} \qquad (16)$$

The formula indicates that it makes the crossover point better more adapted to selection, and the edge contrast of the materials bag in the local area can be effectively enhanced.

4.3   A Novel Membership Function

The membership function should meets the following properties: 1) the function should be continuous and derivable at the 'c' point, and a inflection point which is both sides of 'c' point must have concave-convex property, 2) the function curve should exist symmetry point on both sides of the 'c' point.

The hyperbolic tangent function is mainly used in fuzzy control field, and has symmetrical property. The center of symmetry is inflection point. The function can extend image gray level that takes center of symmetry as reference point, then the contrast of most interested part is extended in nonlinear. It make the area of smaller grey value is effectively enhanced, The farther away from the inflection point, the more obviously the grey value changes. The hyperbolic tangent can be defined as follow:

$$g(x) = \frac{e^x - e^{-x}}{e^x + e^{-x}} \text{ and } x \in (-\infty, +\infty) \qquad g(x) \in (-1, 1) \qquad (17)$$

Transform along the y axis

$$g(x) = \frac{1}{2} \left(\frac{e^x - e^{-x}}{e^x + e^{-x}} + 1 \right)$$

$$= \frac{e^x}{e^x + e^{-x}} \text{ and } x \in (-\infty, +\infty) \quad g(x) \in (0, 1) \qquad (18)$$

Then the membership function can be written as:

$$G(x) = \frac{e^{ax-b}}{e^{ax-b} + e^{-ax+b}} \text{ and } x \in (-\infty, +\infty) \qquad G(x) \in (0, 1) \qquad (19)$$

Where 'a' and 'b' is the fuzzy control factor (a>0, b>0), according to the properties of differentiable function, the function G(x) is continuous and derivable in the domain, so it exists the inflection point. It can be computed by

$$G(x)' = \frac{2a}{(e^{ax-b} + e^{-ax+b})^2} \qquad (20)$$

$$G(x)'' = (\frac{2a}{(e^{ax-b} + e^{-ax+b})^2})'$$

$$= \frac{-4a^2(e^{ax-b} + e^{-ax+b})(e^{ax-b} - e^{-ax+b})}{(e^{ax-b} + e^{-ax+b})^4}$$

$$= \frac{4a^2(e^{-ax+b} - e^{ax-b})}{(e^{ax-b} + e^{-ax+b})^3} \tag{21}$$

When $G(x)'' = 0$, Namely

$$x_d = \frac{b}{a} \tag{22}$$

Where the parameter x_d controls the range of function G(x). The hyperbolic tangent function has the same properties with the arctangent function and has concave-convex property. In order to make the function G (x) is closer to range (0, 1), then $b = \pi$. so the crossover point is at the best.

$$x_c = x_d = \frac{\pi}{a} \tag{23}$$

When 'a' value is changeable and $b = \pi$, the function $G_1(x)$ and $G_2(x)$ relationship between the line curve and crossover point is shown in figure 2.

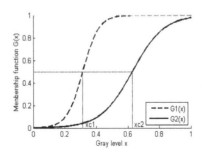

Fig. 2. Relationship between the function curve and the crossover point

Figure 2 shows that crossover point will change when the background and the object pixels in the image change. The smaller value of 'a', the grayscale biased towards the more distribution area, the smaller value of 'a', biased towards the less distribution area. When 'x' is the crossover point, G(x) =0.5, thus it demonstrates that the choice of the crossover point is corrected once again.

4.4   Fuzzy Contrast Image Enhancement

From the above reasoning, when $u_c = G(x_c)$, pal-k's fuzzy function can be defined as:

$$u_{ij} = E_q(E_{q-1}(u_{ij})) \qquad (q = 1, 2, \cdots)$$

And

$$E_1(u_{ij}) = \begin{cases} \dfrac{u_{ij}^2}{u_c} & 0 \le u_{ij} \le u_c \\ 1 - \dfrac{(1-u_{ij})^2}{1-u_c} & u_c < u_{ij} \le 1 \end{cases} \tag{24}$$

Suppose the image contrast after being enhanced is F', and the normalized value of grey value x'_{ij} is u'_{ij}, then, according to the formula (14) getting

$$u'_{ij} = \begin{cases} \dfrac{\bar{u}_{ij}(1-F')}{1+F'} & u_{ij} \le \bar{u}_{ij} \\ 1 - \dfrac{(1-\bar{u}_{ij})(1-F')}{1-F'} & u_{ij} > \bar{u}_{ij} \end{cases} \tag{25}$$

the inverse transformation of u'_{ij} is defined as:

$$x'_{ij} = u'_{ij}(x_{max} - x_{min}) + x_{min} \tag{26}$$

The result reduces the loss of gray grayscale contrast and the amount of calculation.

4.5   The Description of Proposed Algorithm

The proposed algorithm consists of the following operational steps:

Step 1 Make use of the linear formula achieving the image normalization operation.

$$u_{ij} = \dfrac{x_{ij} - x_{min}}{x_{max} - x_{min}} \tag{27}$$

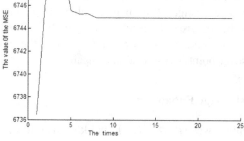

Fig. 3. The relationship between the iterative time and mean square error (MSE)

Step 2 Select image neighborhood of size 3×3 to compute the image contrast F', then the result can be described as an image contrast matrix.

Step 3 Classify and compute the crossover points.

Step 4 Map image contrast F' into the fuzzy plane, then use the formula (24) for image enhancement.

Step 5 Use the formula (25) computes values of u'_{ij}, and use the formula (26) for mapping the fuzzy plane into the gray plane.

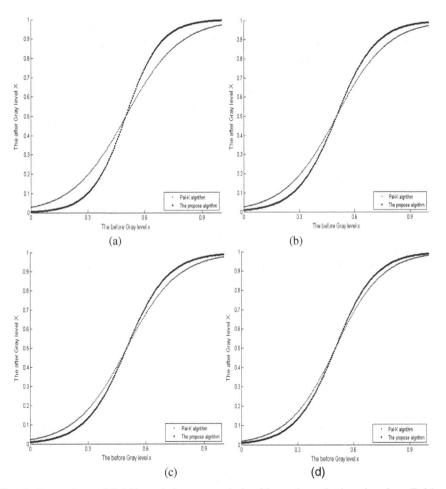

Fig. 4. Comparison of Pal-k's and the proposed algorithm, where the iterative time (Pal-k : figure 4(a) one time, figure 4 (b) two times, figure 4 (c)three times, figure 4(d)four times, Proposed algorithm: figure 4(a b c d) one time)

Figure 3 displays the relationship of proposed algorithm between the iterative time and mean square error (MSE). From the line curve we can see that the proposed algorithm provides a significant improvement when the iterative time is a small number. It can be observed from figure 4, the proposed algorithm is better than the Pal-k algorithm. Actually, when the iterative time is three, the materials image information has seriously lost local information.

5   Experimental Results and Discussions

In this paper, we have applied the proposed algorithm to four kinds of images. The experiments were carried on the image which is from logistics warehouse and factory warehouse. The size of these images is 400*300 pixels. The performance of the proposed method has been tested using well-known image quality measures. Peak-signal-to-noise ratio (PSNR) and mean square error (MSE) are global error measures based on the differences between the original and result image without considering the structural information [14]. Normalized mean square error (NMSE) is complementary to (MSE). The image fidelity (IF) reflects the level of distortion after the image transformation, whereas the structural similarity index measure (SSIM), proposed by Wang et al[15] considers to the visual difference of the image structure, the brightness and the contrast between the original image and the result image, And natural image quality evaluator (NIQE) index is used for measuring mage quality[16-17], it is based on a series of features which is from a simple and successful space domain natural statistic (NSS) model. The range of NIQE value is from 0 to 100 (0 is the best value, but 100 means worst).

Figure 5 shows the simulation results of the gray scale image and histogram of gray scale image. Figure 5(a (1) a (2) a (3) a (4)) reflects the proposed method produce better look and significant enhancement. The brightness is also better than the original image, especially the blurred portions in the image. All of these can be proved from the result shown in the histogram of four kinds of image in figure 5(c(1) c(2) c(3) c(4)).There are the same features that the line curves are smoother than the before histogram shown in figure 5 (b(1) b(2) b(3) b(4)).

The result of figure 5 (a(1) a(2) a(3) a(4)) also reveals good contrast enhancement effect, and there are no distortion in image. The output histogram indicates that the low and high gray levels of the result image achieve smooth distribution.

Figure 6 shows the comparison of average NIQE values and other methods for various images. Further, table 1 gives the quantitative comparison among different methods based on MSE, PSNR, IF, NMSE and SSIM of various images. It can be observed that the proposed method gives superior performance compared to the other observed that the proposed method gives superior performance compared to the other methods. According to table 1, the Wang's method is also very well, even better than the proposed method. But it is obvious that the SSIM value is poorer than proposed method.

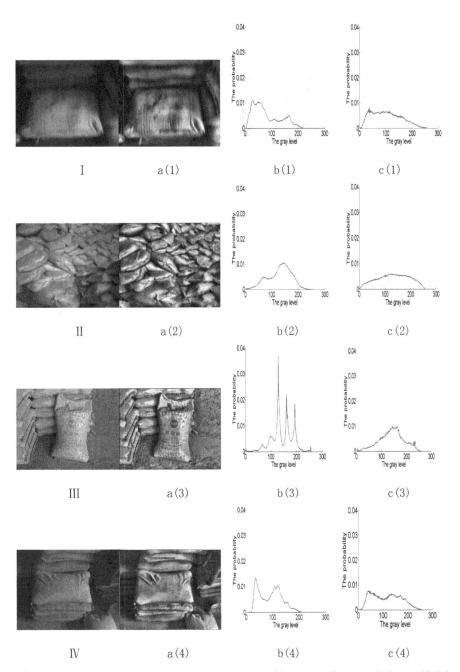

Fig. 5. Simulation results of the gray scale image and histogram of gray scale image.(Original image(I II III IV),Result image(a(1) a(2) a(3) a(4)),Histogram of original image(b(1) b(2) b(3) b(4)),Histogram of experimental image (c(1) c(2) c(3) c(4))).

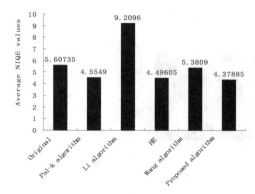

Fig. 6. Comparison of average NIQE values, where the iterative time is equal to one

Table 1. Comparison of enhancement algorithm for different test images, where the iterative Time is equal to one

Image ID	Method	Quality measures				
		MSE	PSNR(db)	IF	NMSE	SSIM
I	Pal-k algorithm	3147.3	11.750	-0.35103	0.64897	0.66345
	Li algorithm	4012.1	10.696	-0.44748	0.55252	0.25744
	HE	2787.5	12.277	-0.31090	0.68910	0.65402
	Wang algorithm	773.30	17.846	-0.08625	0.91375	0.75398
	Proposed algorithm	1098.6	16.320	-0.12254	0.87746	0.78658
II	Pal-k algorithm	1628.4	7.3506	-0.58295	0.41705	0.66254
	Li algorithm	2078.2	14.499	-0.11240	0.88760	0.48683
	HE	1443.8	16.081	-0.07808	0.92192	0.68519
	Wang algorithm	1652.7	15.494	-0.08938	0.91062	0.70368
	Proposed algorithm	1632.4	15.548	-0.08829	0.91171	0.7339⊾
III	Pal-k algorithm	10167	8.0586	-0.47850	0.52150	0.61456
	Li algorithm	2735.6	13.760	-0.12874	0.87126	0.42566
	HE	1122.2	17.630	-0.05281	0.94719	0.63641
	Wang algorithm	1011.9	18.079	-0.04762	0.95238	0.69355
	Proposed algorithm	888.76	18.643	-0.04182	0.95817	0.66411
IV	Pal-k algorithm	2748.2	11.929	-0.27391	0.27391	0.69206
	Li algorithm	2943.7	11.630	-0.29338	0.69225	0.30775
	HE	2230.1	12.836	-0.22226	0.77774	0.66773
	Wang algorithm	1266.1	15.842	-0.13123	0.87377	0.70790
	Proposed algorithm	1256.9	15.326	-0.12527	0.87473	0.78655

6   Conclusions

Image enhancement is the fundamental step of the materials bag of industrial scene detection and recognition. And the proposed approach must have the environmental adaptability. Fuzzy logic has been found many applications in robot vision and

computer vision. In the paper, a fuzzy-based method is presented for material bag image enhancement. The experimental results have demonstrated that it has the robustness for the environment and more adaptive and effective for image integrity. However, since the transformation is from Non-linear to linear (or from linear to Non-linear), this may make the image blur slightly. So our work will keep on in future.

Acknowledgments. This work is supported by the Joint Funds of Gui Zhou Science and Technology Department, Zun Yi Science & Technology Bureau, and Zun Yi Normal College (NO: Qian Zi LKZS[2014]23), the Joint Funds of Gui Zhou Science and Technology Department, Zun Yi Science & Technology Bureau, and Zun Yi Normal College (NO: Qian Zi LKZS[2014]08), the education and research funds of Zun Yi Normal College(NO: 13-08), the education and research funds of Zun Yi Normal College(NO: 13-06).

References

1. Jia, W., Huang, X., et al.: Moving material bag detection method of a fused five frame difference and Gaussian model. Application of Electronic Technique 39(10), 139–142 (2013)
2. Saxena, A., Driemeyer, J., Ng, A.Y.: Robotic grasping of novel objects using vision. The International Journal of Robotics Research 27(2), 157–173 (2008)
3. Elmasry, G., Cubero, S., Molto, E.: In-line sorting of irregular potatoes by using automated computer-based machine vision system. Journal of Food Engineering 122, 60–68 (2012)
4. Su, X., et al.: An image enhancement method using the quantum-behaved particle swarm optimization with an adaptive strategy. Mathematical Problems in Engineering (2013)
5. Yang, Y.-Q., Zhang, J.-S., Huang, X.-F.: Adaptive image enhancement algorithm combining kernel regression and local homogeneity. Mathematical Problems in Engineering 2010 (2011)
6. Pal, S.K., King, R.A.: Image enhancement using fuzzy set. Electronics Letters 16(10), 376–378 (1980)
7. Pal, S.K., King, R.A.: On edge detection of X-ray images using fuzzy sets. IEEE Transactions on Pattern Anaysis and Machine Intelligence (1), 69–77 (1983)
8. Pal, S.K., King, R.A.: Image enhancement using smoothing with fuzzy sets. IEEE Transactions on Systems, Man, and Cybernetics-Part A: Systems and Humans 11(7), 494–501 (1981)
9. Li, J., Sun, W., et al.: Novel fuzzy contrast enhancement algorithm. Journal of Southeast University (Natrual Science Edition) 34(5), 675–677 (2004)
10. Wang, B., Liu, S., et al.: An adaptive multi-level image enhancement algorithm based on fuzzy entropy. Acta Electronica Sinica 33(4), 730–734 (2005)
11. Wang, B., Liu, S., et al.: A novel adaptive image fuzzy enhancement algorithm. Journal of Xidian University (02), 307–313 (2005)
12. Otsu, N.: A threshold selection method from gray level histogram. IEEE Transactions on System, Man and Cybernetics 9(1), 62–66 (1979)
13. Dhnawan, A.P., Buelloni, G., Gordon, R.: Ehancement of mammographic feature by optimal adaptive neighborhood image processing. IEEE Transaction on Med. Imaging 5(1), 8–15 (1986)
14. Hussain, A., Bhatti, S.M., Jaffar, M.A.: Fuzzy based impulse noise reduction method. Multimedia Tools and Applications 60(3), 551–571 (2012)
15. Wang, Z., Bovik, A.C., Sheikh, H.R., et al.: Image quality assessment: from error visibility to structural similarity. IEEE Transactions on Image Processing 13(4), 600–612 (2004)

16. Moorthy, A.K., Bovik, A.C.: Blind image quality assessment: From natural scene statistics to perceptual quality. IEEE Transactions on Image Processing, 20(12), 3350–3364 (2011)
17. Mittal, A., Soundararajan, R., Bovik, A.C.: Making a "completely blind" image quality analyzer. IEEE Signal Processing Letters 20(3), 209–212 (2013)

A Stereo Visual Interface for Manipulating the Grasping Operation of a Fruit-Harvesting Robot

Baixin Liu, Liang Gong*, Qianli Chen, Yuanshen Zhao, and Chengliang Liu

Institute of Mechatronics, Shanghai Jiao Tong University, Shanghai, China
{gxqzez21,gongliang_mi,chenqianli,0120209023,chlliu}@sjtu.edu.cn

Abstract. To facilitate the human-aided recognition and positioning in a human-robot system, we present a design of remote operation interface on the basis of machine vision for fruit-harvesting robot. The interface has two following functions: 1) pointing out an object and commanding the robot's grasping by the point-and-circle graphical user interface (P&C GUI). 2) Online kinematics analysis and robot motion control via establishing the linkage interface between a parameter panel and a virtual robot model. To build up this work, the stereo vision and kinematics emulation are introduced. The stereo vision presents the robotic scenario with deep information and renders it to the operator. The emulation technique provides the pre-operation verification and a direct control scheme. Then implementation and experimentation under laboratory environment are conducted to validate the effectiveness of the proposed paradigm.

Keywords: Stereo visual interface, Fruit-harvesting robot, Machine visual, Emulation technique.

1 Introduction

Industrial robots witness a success by effectively promoting the industrial productivity and capacity. However, this success can't be copied to the agricultural area since the unstructured operation environment compromises a roadblock to robot operations such as targeting, recognizing and positioning. A lot of literatures announce a harvesting robot had been built and intelligent enough to achieve satisfactory result in the laboratory environment, but these robots are less capable of completing complex harvesting tasks in realistic outdoor environment.

Complex environment is the paramount hindrance that the field agriculture robot confronts to. The operating environment for agriculture robot varies in topography and affected by seasonal, natural atmospheric conditions. To tackle this problem, it is promising to combine the human intelligence with the robot diligence by the human-robot cooperation.

A basic controlling interface should include a series of basic and simple functionalities as follows.

---

* Corresponding author.

X. Zhang et al. (Eds.): ICIRA 2014, Part II, LNAI 8918, pp. 239–244, 2014.

- Robot direct control command input: enabling the operators to master every movement of robot in a flexible way.
- Sensor information display: demonstrating the environment information related to the field robot and the state of robot.

Although direct control scheme provides operator with the possible operations of the robot, it is strongly depend on manipulation experience of the operators. Furthermore, it's impossible to control the robot with multiple degrees of freedom. Therefore high command interface will be useful in the task of fruit-harvesting.

Recently researches on high command control, such as voice commands and gesture commands, are active filed in service robot. High command interface aims to make full use of the mature technology as less human participation as possible, realize and extend the system function. On the other hand, it should not increase demands for user skills.

In this paper a new high command controlling paradigm named after 'Point-and-Circle Commands' and a remote operation interface are introduced. The key technology and system architecture supporting this interaction method are given.

2   Key Technologies and System Architecture

The architecture of the whole system is shown in Fig. 1. The communication network between remote client and filed robot employs the TCP/IP protocol to ensure secure delivery of the control command. The remote client is mounted in the control room, mastering the field robot through the wireless network.

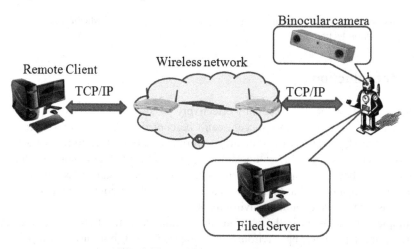

Fig. 1. The architecture of the system

2.1   Communication Network

The communication network is an important part of the remote control system. All the sensing, control and feedback information are transmitted through it. The performance of the robot system is directly affected by network. And it ensures robot can correctly complete jobs in acceptable time.

The 802.11n standard wireless network has the merits of fast networking and simple fault location. Its transmission rate is up to 150 Mbit/s which meets the requirement of image data transmission. And this standard is usually deployed in computer networks which enables terminal equipment access to the control system easily.

2.2   Field Server

As the controller of the robot, field server controls all the motion of robot through Multi-axis motion controller. It also works as a communication server that listening the client connection requirement.

2.3   Binocular Camera

An on-robot binocular camera is used to acquire scenery around it. A stereo image is acquired through processing raw images from the binocular camera. The stereo image is the main sensor information of the system. The vision system applied Space Vision hardware and software solutions which are provided by the Graypoint Inc. from Canada. The vision system has the capacity of 10 microns level accuracy and 25 frames transmission speed. It is eligible for establishing fruit-harvesting system. The hardware is Bumblebee 2 binocular camera and the development toolkits include the "PGR FlyCapture", the "Triclops Stereo Vision SDK" and the Microsoft's MFC.

3   Interaction Channel and Interface

A variety of interactive channel can be seen in Fig. 2: Point-and-Circle Commands, Advanced Text Commands, Direct Control Commands and Configuration Commands. The complexity in controlling the robot is increasing from the top to the

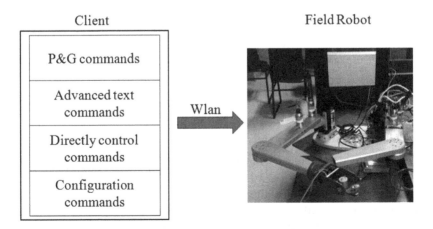

Fig. 2. Four channels of interaction

bottom. High commands, such as P&G GUI and advanced text commands in this system, encapsulate a series of basic predefined operation commands within system that enables novice users to operate the system. The direct control commands and configuration commands are low level commands, which are implemented by virtual robot interface in our system. Fig. 3 is a screenshot of the client main frame containing these four interaction interface.

3.1 Configuration Commands

The interface in the lower left corner of the window supports system and network configuration, advanced text commands and system status monitoring. Modbus group responds to establish the connection between the server and the robot's multi-axis motion controller. And the TCP/IP server group establishes a communication channel for client and server. Advanced text instructions provide a more flexible mode of operation, which allows operator to help the robot out of a software failure. However, the operator needs more skills to utilize the function block.

Fig. 3. A screenshot of the client program's main frame

3.2   Point-and-Circle Command

A stereo image which contains 3D cloud points is displayed on the upper left panel when the image capturing progress is successful executed. The image shows the robot's scene to the operator in real time. The black corroded areas in image are without depth information attached. It is caused by two reasons: 1, failure in stereo matching; 2, depth distance beyond the parameters set of the maximum/minimum distance.

The stereo image panel supports the P&C command. When the operator points out a target fruit on stereo image, a circle encircles it as shown in figure 4. An edge detection operation is applied in the selected and surrounding region to find out the edge of fruit in the image. Then the location and size are affirmed in the image coordinate system. This information is sent to the server and requests a response about the stance of target fruit related to robot.

The interaction method has two advantages: 1) what you see is what you get. It does not involve the operation flow or code in the process of operation. 2) It avoids the pattern recognition in fruit detection which is one of the most challenging jobs for agriculture robot under unstructured environment.

3.3   Direct Control Commands and Movement Simulation

The three-dimensional coordinates of target relative to the robot is converted to joint motion parameters by inverse kinematics in the background. Virtual robot motion simulation is arranged on the upper right corner of main frame. The operator can observe virtual manipulation sequences and decide whether or not to take this action. Also the panel can be used to observe the result of direct control commands which is implemented in the lower right corner of main frame. Dragging the slider or typing parameters in the corresponding test area can move the robot remote.

4   Experimentation

In order to validate the feasibility of the controlling interface, we take an experiment in the laboratory. A laptop as a remote client is connected to the server after affirms that the client and the server are under the same net segment. And then the connection between the server and the robot is established in the Modbus group. Clicke the

Fig. 4. Experiment result

'capture button' or type the text command 'imageg' to get the robot's scene display-ing on the interface as Fig. 3. A fruit in the stereo image is the target in this experi-ment. The operator clicks the target and gets simulation animation showing in the upper right panel. The simulation animation suggests that the task is perfect. The mo-tion is carried out in the real word after clicking the 'Execute button'. Fig. 4 shows the operating result, it demonstrates that the system works well.

5   Conclusion and Future Work

User-friendliness interface plays an important role in a human-robot cooperation sys-tem. In this paper, we proposed a new paradigm of interaction in human-robot interac-tion that utilizes stereo image supported by binocular camera and the integration of virtual robot motion simulation and directly control scheme. By providing this RPG-game-like robot controlling interface, the stereo vision technique is introduced for minimizing the participation of operator.

References

1. Chen, C.Y., Shih, B.Y., Shih, C.H., Wang, L.H.: Human-machine interface for the motion control of humanoid biped robots using a graphical user interface Motion Editor. Journal of Vibration and Control (2012), doi: 1077546312437804
2. Du, G.L., Zhang, P., Yang, L.Y., Su, Y.B.: Robot teleoperation using a vision-based mani-pulation method. In: 2010 International Conference on Audio Language and Image Processing (ICALIP), pp. 945–949. IEEE (November 2010)
3. Nehmzow, U., Recce, M.: Scientific methods in mobile robotics. Robotics and Autonom-ous Systems 24(1-2), 1–3 (1998)
4. Kawamura, K., Pack, R.T., Iskarous, M.: Design philosophy for service robots. In: IEEE International Conference on Systems, Man and Cybernetics, Intelligent Systems for the 21st Century, vol. 4, pp. 3736–3741. IEEE (1995)
5. Kagami, S., Kuffner Jr., J.J., Nishiwaki, K., Sugihara, T., Michikata, T., Aoyama, T., et al.: Design and implementation of remotely operation interface for humanoid robot. In: Pro-ceedings of the IEEE International Conference on Robotics and Automation, ICRA 2001, vol. 1, pp. 401–406. IEEE (2001)
6. Desai, M., Yanco, H.A.: Blending human and robot inputs for sliding scale autonomy. In: IEEE International Workshop on Robot and Human Interactive Communication, ROMAN 2005, pp. 537–542. IEEE (August 2005)
7. Marín, R., Sanz, P.J., Nebot, P., Wirz, R.: A multimodal interface to control a robot arm via the web: A case study on remote programming. IEEE Transactions on Industrial Elec-tronics 52(6), 1506–1520 (2005)
8. Bargen, B., Donnelly, P.: Inside DirectX: in-depth techniques for developing high-performance multimedia applications. Microsoft Press (1998)
9. Yun, S.S., Kim, M., Choi, M.T.: Easy interface and control of tele-education robots. Inter-national Journal of Social Robotics 5(3), 335–343 (2013)
10. Gong, W., Wang, Q., Feng, Q.: The Development of 3D Real-time Visualization Platform Based on DirectX. Machine Tool & Hydraulics 15, 021 (2010)

Effects of Camera's Movement Forms
on Pollutant's Automatic Extraction Algorithm

He Xu[1], Yan Xu[1], Yantao Wang[1], X.Z. Gao[2], and Khalil Alipour[3]

[1] College of Mechanical and Electrical Engineering,
Harbin Engineering University, Harbin, 150001, China
[2] Department of Automation and Systems Technology,
Aalto University School of Electrical Engineering, Aalto 00076, Finland
[3] Department of Mechatronics Engineering, Faculty of New Sciences and
Technologies, University of Tehran, Tehran, Iran
{railway_dragon,wangrilian}@163.com, 453129130@qq.com,
xiao-zhi.gao@aalto.fi, k.alipour@ut.ac.ir

Abstract. The fast and accurate automatic extraction of pollutants on cameras of mobile robots plays a vital role in the follow-up camera's automatic cleaning. Currently, most of the researches focus on the extraction algorithm for pollutants, and there are almost no relevant studies on the effects of camera's movement forms on pollutant's extraction algorithm. Consequently, this paper explores the impact on pollutant's extraction algorithm when the camera is at the same speed, but in different movement forms, which provides some suggestions for the improvement of the pollutant's automatic extraction algorithm.

Keywords: mobile robot, pollutant's extraction algorithm, camera's movement form.

1 Introduction

The robots equipped with vision system can replace humans to do various tasks, such as monitoring, engineering rescue, search and rescue [1], and a variety of work in extreme environments. Fig. 1(a) shows the RHex hexapod robot [2] working in muddy environment, (b) illustrates the Red Snake robot [3] climbing in a pipeline, and (c) is the wheeled robot named Hull BUG developed by United States Navy, which is mainly responsible for hull cleaning [4]. All of these robots work in poor conditions, which makes their vision systems vulnerable to pollution, reducing the quality of acquired images, and even affecting the normal work. Once polluted, it is difficult for humans to go deep into the work environment to clean up the camera. The usual way is terminating the task and commanding the robot to return, however, as the vision system does not work, it also is difficult for the robot to return. Therefore, how to automatically clean the camera in time and ensure the robot steadily acquire high quality image information have become an urgent issue.

X. Zhang et al. (Eds.): ICIRA 2014, Part II, LNAI 8918, pp. 245–255, 2014.
© Springer International Publishing Switzerland 2014

Fig. 1. Robot visual pollution

Currently, cleaning for the robot's cameras mainly relies on labour, and there also are some automatic cleaning systems which rely on timing device. Such systems are unable to determine when to clean, and they would perform the cleaning task once the time node reached. This way will result in serious waste of energy, and the cleaning work itself will affect the normal use of the equipment. Judging from the collected image information by the vision system is the most direct and effective way, thus, this paper focuses on the automatic extraction of pollutants on the mobile robot's camera.

In this paper, pollutants on the camera can be seen as the foreground of the image, so the extraction of pollutants belongs to the foreground extraction. However, it is very difficult to automatically extract the foreground from one image, so some researchers shift their focus to the foreground extraction of video. Compared with image, video can not only provide spatial information, but also time information. Video foreground extraction can be divided into the interactive and the automatic. This paper concentrates on the automatic extraction of the video foreground. Previously, most of the researches concentrated on the extraction algorithm of the video foreground, and mainly can be divided into three classes: optical flow method, background subtraction method and inter-frame difference method. Traditional optical flow method can be divided into sparse optical flow and dense optical flow. Horn-Schunck method [5] is a typical dense optical flow method, which relates each of the pixels to the speed of the image, and calculates the velocity field of the dense optical flow. Lucas-Kanade method [6] is a typical sparse optical flow method. The algorithm only needs the local information around the target, so it can be applied to sparse content. In addition, there are many improved optical flow method, such as pyramid optical flow method [7], differential based optical flow method [8], regional optical flow method [9] and feature optical flow method [10] etc. Common background subtraction method includes multi-frame averaging method [11], statistic based method [12] and Gaussian distribution model based method [13]. Inter-frame difference method is similar to background subtraction method, and there are some improved algorithms, such as three-frame difference method [14] and accumulative-frame difference method [15].

As can be clearly seen, there are quantities of researches about the extraction algorithm, and almost no relevant studies concentrate on the effects of camera's movement forms on pollutant's extraction algorithm. Consequently, this paper explores the impact on pollutant's extraction algorithm when the camera is at the same speed, but in different movement forms.

2   Mobile Robot Platform

Figure 2 presents the mobile robot platform, including personal computer and mobile robot based on embedded system. The robot mainly consists of Arduino controller, camera, router, PTZ, motor drivers, DC motors, also, two driving wheels and one universal wheel, which can make the robot achieve pivot steering. The robot links to PC through wireless, transfers the images collected by camera to PC, and simultaneously receives the instructions from PC to control the robot move and PTZ steer. In the system, the robot is mainly responsible for the acquisition and transmission of video image, while the post image processing is done by PC.

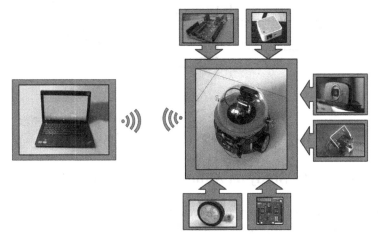

Fig. 2. Mobile robot platform

3   Description of the Automatic Extraction Algorithm for Pollutants

For mobile robot, when there are pollutants on the camera, the pollutants are as foregrounds, and relatively stationary to the camera. The complex environment is as background which is in the global movement state. This case is very different from the traditional fixed camera, and it can be seen as the reverse process of moving target detection, i.e., detecting the stationary target in dynamic scene. This paper takes full advantage of this feature, proposes a pollutant automatic extraction algorithm. Firstly, algorithm deploys three-frame difference method and accumulates the frame difference to remove the background interference. Secondly, the pollutants can be judged by the gray histogram of the accumulative difference images, and once there are pollutants, the adaptive binarization is instantly executed to the difference images, achieving the automatic extraction of the pollutants. The entire algorithm is shown in Fig. 3.

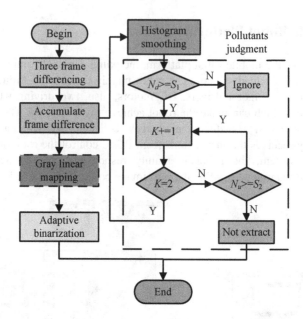

Fig. 3. Process of the automatic extraction algorithm for pollutants

4   Experimental Study on Automatic Extraction of Pollutants

In this part, the effect of different movement forms of the camera on the automatic extraction algorithm of the pollutant is studied at the same speed.

4.1   Movement Description of the Camera

Figure 4 shows different movement forms of the camera, and the movement directions are shown as the arrow. According to the actual movement ability of the mobile robot, this paper will experiment on the pan, horizontal translation and dollying.

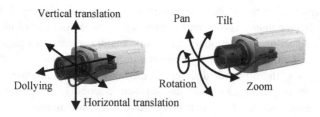

Fig. 4. Movement forms of camera

4.2   Experiment on the Pan

Adjust the angle of the PTZ, to make the camera has a horizontal field of view, Fig. 5 is the experiment process of the pan.

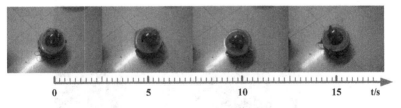

Fig. 5. Experiment process of the pan

Figure 6 denotes the extraction process of the pollutants on the camera in the pan mode.

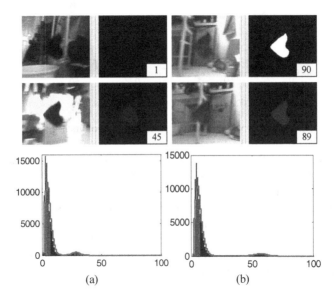

Fig. 6. Extraction process in the pan mode with pollutants on the camera

As shown in Fig. 6, the number on the bottom right corner is the number of the frames. When the images accumulate to 90 frames, the algorithm implements histogram analysis to the accumulative difference images, and if the pollutants are found, binary segmentation is executed. It is can be clearly seen from the figure that the algorithm performs well in the pan mode, the extraction of the pollutant region is complete, and the defocus blur of the pollution dose not affect the algorithm. In the histogram of the 45th and 89th frames(Fig. 6(a,b)), the red ladder line results from the slippage smoothness of the histogram, and through comparison, it is can be found that the histogram changes little in the accumulating process, just stretches to the accumulative frames. The extraction process takes about 15 seconds, which is caused by the wireless video transmission etc.

When there are no pollutants on the camera and the other conditions are the same, experiment is carried out, the entire process is shown in Fig. 7.

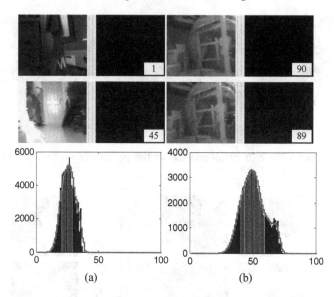

Fig. 7. Extraction process in the pan mode without pollutants on the camera

When there are no pollutants on the camera, the grayscale of the accumulative images are uniform, which means the grayscale of the pixel is relatively concentrated in the histogram. In the 45th frame(Fig. 7(a)), the grayscale concentrates among 10-40, and in the 89th frame(Fig. 7(b)), the grayscale concentrates among 20-80, so the accumulation process is the translation and stretch of the histogram towards the accumulative frames. The original histogram is not smooth, for the histogram of the 45th frame, the effect of the slippage smoothness is remarkable, however, the gray histogram of the 89th frame remains fluctuation among 60-80 after the smoothness, so it is necessary for the algorithm to introduce the filter before judging the trend of the histogram.

For pan, the algorithm has a higher accuracy for judging the pollutants. This is because the gray histogram of the accumulative images forms fast and its shape is stable. So, in the pan mode, in order to shorten the delay time, the number of the frame can be suitably reduced to n=45.

4.3   Experiment on the Horizontal Translation

Rotate the PTZ of the mobile robot to make the robot's moving direction perpendicular to the optical axis of the camera. The conditions of the horizontal translation are same as those of the pan, such the same speed, the same environment, the same size and shape of the pollutant. Fig. 8 is the extraction process of the pollutants in the horizontal translation.

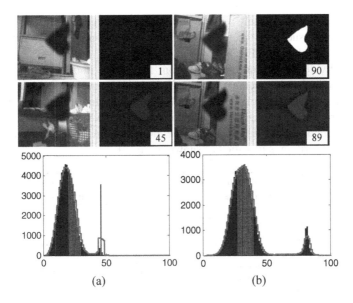

Fig. 8. Extraction process in the horizontal translation mode with pollutants on the camera

As can be clearly seen in Fig. 8, the algorithm also has a good performance in the horizontal translation. In the 45th frame(Fig. 8(a)), the grayscale difference between the pollutants and background is small in the gray histogram. The second rising region of the smoothed histogram starts at about 40, the length is short, which makes it easy to be filtered off, so it will cause misjudgment if the histogram analysis is done at this time. In the 89th frame(Fig. 8(b)), the grayscale difference between the pollutants and background becomes greater in the gray histogram, also, the second rising region of the smoothed histogram is obviously elongated. At this time, the algorithm will have accurate judgment. Also, we did the experiment in the horizontal translation without pollutants on the camera, and we obtained a satisfactory result. So, the original number of the accumulative frame is maintained, and the algorithm is unmodified.

4.4   Experiment on the Dollying

The PTZ is recovered to the initial state, and dollying is executed. In the dollying experiment, the scene in view changes slowly, and the process of accumulative frame

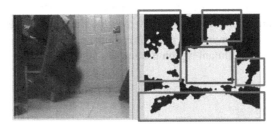

Fig. 9. Extraction result of the original algorithm

difference does not completely remove the object with large and uniform grayscale in the background, such as the wall and floor. The results of the extraction are unsatisfactory, as shown in Fig. 9.

We can divide the region extracted from the image into three sections: 1) the pollutant zone in the blue rectangular box; 2) the background zone with large-area and high grayscale in the red rectangular box; 3) the background zone with low grayscale in the green rectangular box. As the pollutants are close to the camera and basically in a backlight state, so the grayscale of the pollutants zone is generally low. For dollying, the algorithm is modified as shown in Fig. 10:

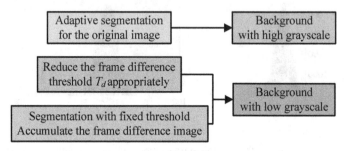

Fig. 10. Modification of the original algorithm

Firstly, before three-frame difference, Otsu adaptive segmentation is deployed to the intermediate frame, to remove the region which is higher than the adaptive threshold. Secondly, the threshold is modified as $T_d = 3$, which can reduce the influence of the low grayscale background. However, the threshold can not be too low, otherwise, it will affect the extraction of pollutants. Thirdly, as there also are some pollutants not separate from the background, which shows that the effect of the Otsu adaptive segmentation is not good, so we consider to use fixed threshold to segment, and the threshold is set as $T_{d2} = \eta \cdot n$, where n is the number of the accumulative frame, and η is scaling factor. If η is too small, the pollutant still cannot be separated, so η should be greater than 0.8 according to many experiments. In order to avoid the extraction of pollutants not complete, one point in the extracted region can be taken as a seed to grow in the original grayscale image [16]. Fig. 11 shows a dollying experiment $T_{d2} = 75$, where (a) is the effect of the adaptive segmentation for the 45th frame, and the floor, the wall and part of the reflective areas do not participate in the algorithm. For the 89th frame (b), the original grayscale histogram fluctuates among 0-10, and the fluctuations are removed after the slippage smoothness, whereby, the pollutants on the camera can be accurately judged. It can be seen that the pollutant region is complete separated by the fixed threshold.

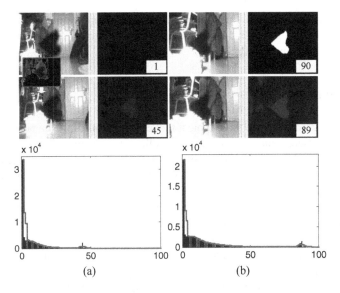

Fig. 11. Extraction process in the dollying mode with pollutants on the camera

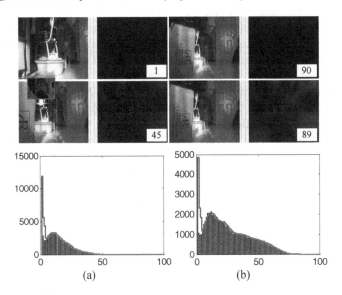

Fig. 12. Extraction process in the dollying mode without pollutants on the camera

Figure 12 shows the extraction process when there are no pollutants on the camera, and the algorithm still uses the modified one. As can be seen from (a), the floor, the door, the box, the chair and other objects have been ignored. Histogram emerges fluctuation among 0-20, and the fluctuation is still there after the slippage smoothness, but it has been filtered out by the filter step S1 and S2 before the histogram analysis, so the algorithm still can judge accurately. In the two dollying experiments, there are large fluctuations in the histogram, which is related to the modification of the algorithm. In order to avoid misjudgment, the filter step can be increased appropriately, taking S1=6.

5   Conclusions

In the paper, according to the three typical movement forms of the camera, we studied the effects of camera's movement forms on pollutant extraction algorithm, and based on the experimental results, some modifications were done to improve the extraction algorithm. In the pan process, the algorithm performed the best, and the number of the accumulative frames can be appropriately reduced to shorten the delay time and improve the algorithm efficiency. For the horizontal translation, the performance of the algorithm was moderate and we remained unchanged. For the dollying, the performance of the algorithm does not ideal, so taking the following measures: 1) deploy the adaptive segmentation to the middle frame image, 2) narrow the threshold value of the three-frame difference, 3) use the fixed threshold to segment the accumulative frame difference image. After improvement, the algorithm can do automatic adjustment according to the movement forms of the camera, to make the time-consuming shrink to a minimum, and extract most accurately.

Further works will concentrate on improve the automatic extraction algorithm according to the rotation and zoom, and find the quantitative relationship between the automatic extraction algorithm and the camera's movement speed.

Acknowledgments. This project is supported by the National Natural Science Foundation of China under Grant 60775060, Specialized Research Fund for the Doctoral Program of Higher Education under Grant 20102304110006, 20122304110014 and Harbin Science and Technology Innovation Talents Special Fund under Grant 2012RFXXG059. Professor X. Z. Gao's research was funded by the Finnish Funding Agency for Technology and Innovation (TEKES).

References

1. Casper, J., Murphy, R.R.: Human-robot interactions during the robot-assisted urban search and rescue response at the World Trade Center. IEEE Transactions on Systems, Man, and Cybernetics, Part B: Cybernetics 33(3), 367–385 (2003)
2. Saranli, U., Buehler, M., Koditschek, D.E.: RHex: A Simple and Highly Mobile Hexapod Robot. The International Journal of Robotics Research 20(7), 616–631 (2001)
3. Wright, C., Buchan, A., Brown, B., et al.: Design and architecture of the unified modular snake robot. In: 2012 IEEE International Conference on Robotics and Automation (ICRA), Saint Paul, Minnesota, USA, pp. 4347–4354 (2012)
4. Souto, D., Faina, A., Lopez-Pena, F., et al.: Lappa: A new type of robot for underwater non-magnetic and complex hull cleaning. In: 2013 IEEE International Conference on Robotics and Automation (ICRA), Karlsruhe, Germany, pp. 3409–3414 (2013)

5. Rother, C., Kolmogorov, V., Blake, A.: "GrabCut": Interactive foreground extraction using iterated graph cuts. In: ACM SIGGRAPH 2004 Papers, pp. 309–314. ACM, Los Angeles (2004)

6. Kumar, M.P., Ton, P.H.S., Zisserman, A.: OBJ CUT. In: IEEE Computer Society Conference on Computer Vision and Pattern Recognition, CVPR 2005, San Diego, California, USA, pp. 18–25 (2005)

7. Jiang, Z., Yi, H.: A feature tracking method based on image pyramid optical flow. Geomatics and Information Science of Wuhan University 32(08), 680–683 (2007)

8. Sánchez, A., Ruiz, J.V., Moreno, A.B., et al.: Differential optical flow applied to automatic facial expression recognition. Neurocomputing 74(8), 1272–1282 (2011)

9. Du, J., Xu, L.: Abnormal behavior detection based on regional optical flow. Journal of Zhejiang University (Engineering Science) 45(07), 1161–1166 (2011)

10. Zhang, Z., Ye, P., Wang, R.: Multi-frame Image Super-resolution Reconstruction Based on SIFT. Journal of Image and Graphics 14(11), 2373–2377 (2009)

11. Min, W., Jiahao, Z., Jinbao, W.: Algorithm for background update in video vehicle detection based on DSP. Electronics Science and Technology 25(05), 129–132 (2012)

12. Peng, Q., Li, H.: Background extraction method based on block histogram analysis for video images. Journal of Southwest Jiaotong University 41(01), 48–53 (2006)

13. Yang, T., Li, J., Pan, Q., et al.: A multiple layer background model for foreground detection. Journal of Image and Graphics 13(07), 1303–1308 (2008)

14. Xu, Y., Zuo, J., Zhang, X., et al.: A video segmentation algorithm based on accumulated frame differences. Opto-Electronic Engineering 31(7), 69–72 (2004)

15. Zhao, S., Zhao, J., Wang, Y., Fu, X.: Moving object detecting using gradient information, three-frame-differencing and connectivity testing. In: Sattar, A., Kang, B.-H. (eds.) AI 2006. LNCS (LNAI), vol. 4304, pp. 510–518. Springer, Heidelberg (2006)

16. Xu, K., Qin, K., Huang, B., et al.: A new method of region based image segmentation based on cloud model. Journal of Image and Graphics 15(05), 757–763 (2010)

Calibration of a Robot Vision System Coupled with Structured Light: Method and Experiments

Xun Deng, Zhong Chen, Jiahui Liang, and Xianmin Zhang

South China University of Technology,
Guangzhou 510640, P.R.China
mezhchen@scut.edu.cn

Abstract. The calibration method of a special robot vision system with a camera and laser light projector mounted on the robot hand is proposed in this paper. The robot kinematic calibration is completed in screw axis identification method using a laser tracker and the robot's geometrical parameters are identified precisely. The robot hand-eye calibration is performed by establishing a chain of transformations of the eye-in-hand system and the homogeneous matrix between the camera frame and robot hand frame is obtained from HALCON. The structured light calibration which determines the pose of the light plane in the camera coordinate frame is completed by analyzing several pattern images in the different height levels with HALCON programming. And the result of the experiment has shown the feasibility of the proposed method.

Keywords: structured light, robot vision system, hand-eye calibration, HALCON.

1 Introduction

The robot vision systems[1] coupled with structured light have been used to perform three-dimensional measurement in scientific and commercial fields such as robotic guidance, arc welding, and virtual environment construction due to their fast speed and noncontact nature[2]. Typically, a robot vision system coupled with structured light consists of a robot, a camera and a projector that projects a pattern on the object, while the camera captures the image of the pattern. By analyzing the captured images of the patterns, dimensional and contour information can be obtained.

Obviously the key to accurate measurement is the proper calibration of each component in the system including robot kinematic calibration[3], hand-eye calibration and light plane calibration. The calibration of the robot[4] is fundamental and usually performed by coordinate measuring devices like CMMs or laser trackers for high precision. The methods proposed in the last few decades for light plane calibration[2] can be classified into two categories: one is two-step calibration, i.e. camera calibration and structured light projector calibration respectively; the other is one-step calibration with some scene points and their correspondence points in images. In this paper, we take the former method. With a calibrated camera, the pose

X. Zhang et al. (Eds.): ICIRA 2014, Part II, LNAI 8918, pp. 256–264, 2014.

of the structured light projector was obtained by fitting the light plane with at least two sets of laser line coordinates. The calibration was performed by HALCON programming which is easier and faster than usual methods.

For eye-in-hand systems[5], several approaches have been proposed. The usual way to describe the hand-eye calibration is by means of 4×4 homogeneous transformation matrix[6]. The well-known hand-eye calibration equation is expressed as

$$AX = XB, \tag{1}$$

where X is the transformation from a camera to a robot hand, A is the camera motion, and B is the robot motion. However, in this paper we don't need to focus on how to solve the equation but to establish a chain of transformations of the eye-in-hand system and the solution will be given automatically by HALCON, which is efficient and reliable.

This paper is organized as follows. In Section 2, describing after the robot kinematic model is derived, how to perform the hand-eye calibration as well as light plane calibration separately with HALCON programming is described detailedly. In Section 3, calibration experiments are performed in order to validate the feasibility of the proposed method. And the paper is closed in Section 4 by some conclusions.

2   Mathematical Model and Method

The calibration of a robot vision system is exactly to determine the pose of the object in the robot base frame $^{base}\mathbf{H}_{obj}$, which is the most important and fundamental task for the system. The calibration method proposed in this paper can be described by the following equation. And the system structure is shown in Fig. 1.

$$^{base}\mathbf{H}_{obj} = {}^{base}\mathbf{H}_{hand} \cdot {}^{hand}\mathbf{H}_{cam} \cdot {}^{cam}\mathbf{H}_{obj}, \tag{2}$$

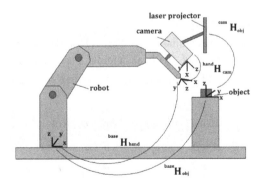

Fig. 1. System structure

where $^{base}\mathbf{H}_{hand}$ means the pose of the robot hand in the robot base frame, $^{hand}\mathbf{H}_{cam}$ represents the relationship between the camera and robot hand, $^{cam}\mathbf{H}_{obj}$ which means the pose of the object in the camera frame. $^{base}\mathbf{H}_{hand}$ can be determined by robot kinematic calibration. $^{hand}\mathbf{H}_{cam}$ can be determined by robot hand-eye calibration. But for an eye-in-hand system, $^{cam}\mathbf{H}_{obj}$ can't be obtained only by camera calibration because the extrinsic parameters of the camera are frequently changed during robot movement. Thus, the laser light projector which can project a thin luminous straight line onto the surface of the object is needed. Then, $^{cam}\mathbf{H}_{obj}$ can be obtained through camera parameters and the pose of the light plane in the camera frame, refers to as the light plane calibration. Therefore, the calibration method of the robot vision system coupled with structured light can be divided into three parts.

2.1   Robot Kinematic Calibration

Due to its simplicity and generalization, the robot uses the D-H model7]. In the D-H model, the four parameters used to fully specify a single joint are: the length of the joint a, the angle of rotation α, the size of the displacement of the joint d and the joint angle θ. And the homogeneous transformation matrix[7] between the adjacent joints is expressed as

$$^{i-1}A_i = \text{Trans}_z(d_i)\,\text{Rot}_z(\theta_i)\,\text{Trans}_x(a_i)\,\text{Rot}_x(\alpha_i)$$
$$= \begin{bmatrix} c\theta_i & -s\theta_i c\alpha_i & s\theta_i s\alpha_i & a_i c\theta_i \\ s\theta_i & c\theta_i c\alpha_i & -c\theta_i s\alpha_i & a_i s\theta_i \\ 0 & s\alpha_i & c\alpha_i & d_i \\ 0 & 0 & 0 & 1 \end{bmatrix}, \tag{3}$$

where c = cosine and s = sine.

However, this model is not suitable for parallel or nearly parallel joints unless replace the distance parameter d with an angular parameter β which represents the angle around y-axis. And the transformation matrix between adjacent joints is defined as

$$^{i-1}A_i = \text{Trans}_z(d_i)\,\text{Rot}_z(\theta_i)\,\text{Trans}_x(a_i)\,\text{Rot}_x(\alpha_i)\,\text{Rot}_y(\beta_i)$$
$$= \begin{bmatrix} c\theta_i c\beta_i - s\theta_i s\alpha_i s\beta_i & -s\theta_i c\alpha_i & s\theta_i s\alpha_i + s\theta_i s\alpha_i c\beta_i & a_i c\theta_i \\ s\theta_i c\beta_i + c\theta_i s\alpha_i s\beta_i & c\theta_i c\alpha_i & s\theta_i s\beta_i - c\theta_i s\alpha_i c\beta_i & a_i s\theta_i \\ -c\alpha_i s\beta_i & s\alpha_i & c\alpha_i c\beta_i & d_i \\ 0 & 0 & 0 & 1 \end{bmatrix}, \tag{4}$$

where $c = \text{cosine}$ and $s = \text{sine}$.

Therefore, for a robotic manipulator the position and the orientation[3] of the robot hand frame with respect to the robot base frame can be represented as

$$T = {}^{0}A_{1}{}^{1}A_{2}{}^{2}A_{3}{}^{3}A_{4}{}^{4}A_{5}{}^{5}A_{6} . \tag{5}$$

The second step in calibrating the robot is to obtain each single axis motion curve[8]. According to the screw axis identification method, each joint is rotated separately while the others are being blocked. For each robot axis, a circle is fitted to these points. From this circle, the robot axis is extracted. Each joint coordinate frame T_{i} can be built according to the robot axis equation based on D-H model or MD-H model. And the homogeneous transformation matrix between adjacent joints can be obtained by the following equation:

$$^{i-1}A_{i} = (T_{i-1})^{-1}T_{i} . \tag{6}$$

Therefore the link parameters can be worked out by equation 3 and 4.

2.2   Robot Hand-Eye Calibration

The calibration proposed in the paper is an eye-in-hand calibration[9] [10] and performed by moving the robot and observing a 2D reference object by the camera at a few different robot poses with HALCON programming. The camera is modeled by the usual pinhole model and applied to the polynomial distortion model in HALCON. There exists a chain of coordinate transformations. In this chain two poses $^{base}\mathbf{H}_{hand}$ and $^{cam}\mathbf{H}_{cal}$ are known. The hand-eye calibration then estimates the other two poses, i.e., $^{cam}H_{hand}$ and $^{base}H_{cal}$ respectively. Therefore the chain results in the following equation:

$$^{cam}\mathbf{H}_{cal} = {}^{cam}\mathbf{H}_{hand} \cdot ({}^{base}\mathbf{H}_{hand})^{-1} \cdot {}^{base}\mathbf{H}_{cal} . \tag{7}$$

For each of the calibration images, we should specify the corresponding pose of the robot hand which is described with position and orientation that includes three angular parameters, referred to as Euler angle. Since the orientation can be described by different sequences of rotations, a parameter is used to recognize the sequence in HALCON. The most commonly used is the Z-Y-X angle sequence and the corresponding rotation transformation matrix is as following

$$\mathbf{R}(\alpha, \beta, \gamma) = \mathbf{R}_{z}(\alpha) \cdot \mathbf{R}_{y}(\beta) \cdot \mathbf{R}_{x}(\gamma)$$
$$= \begin{bmatrix} c\alpha c\beta & c\alpha s\beta s\gamma - s\alpha c\gamma & c\alpha s\beta c\gamma + s\alpha s\gamma \\ s\alpha c\beta & s\alpha s\beta s\gamma + c\alpha c\gamma & s\alpha s\beta c\gamma - c\alpha s\gamma \\ -s\beta & c\beta s\gamma & c\beta c\gamma \end{bmatrix} . \tag{8}$$

The accuracy of the hand-eye calibration can be evaluated by the pose error of the complete chain of transformations in form of a tuple with two elements: the root-mean-square error of the translational part in meter and the root-mean-square error of the rotational part in degree. And the rotation error has a larger impact on the overall precision than the translation error.

2.3 Light Plane Calibration

The points of intersection between the laser line and the camera view depend on the height of the object. Thus, if the object onto which the light line is projected differs in height, the line is not imaged as a straight line but represents a profile of the object. Using this profile, we can obtain the height differences of the object.

To determine the light plane and its pose, we need at least two corresponding straight lines on the plane, in particular one in the plane of the world coordinate frame with 'z=0' and another that differs significantly in z direction. For each position of the calibration plate two images are taken, one showing the calibration plate and one showing the laser line. According to the acquired images, the light plane is fitted and the orientation of the light plane with respect to the camera coordinate frame is obtained. The calculation process above is finished in HALCON programming.

3   Experiments and Results

3.1   Experimental Setup

The robot visual system mainly consists of a camera, a laser line projector, a 6-degree of freedom robot and a computer. The camera and laser projector are mounted on the robot hand by the use of special fixtures. The robot is YR-HP3-A00 of YASAKWA. System device is shown in Fig. 2.

Fig. 2. Experimental Setup

3.2   Robot Kinematic Calibration Results

A laser tracker is utilized in the experiment during the robot kinematic calibration. We have placed the laser tracker as near as possible to the robot for better accuracy. Each circle center coordinate and screw axis orientation vector can be captured through the corresponding software connected to the measuring device. The measured data are listed in Table 1. The identified link parameters are listed in Table 2.

Table 1. Data captured from the laser tracker (mm)

	Joint 1	Joint 2	Joint 3	Joint 4	Joint 5	Joint 6
	3042.838	2983.267	2984.881	2769.598	2804.598	2724.659
C_i	1064.662	983.711	983.110	679.538	679.538	615.894
	-276.597	-649.518	359.660	-272.796	-272.796	-275.761
	0.005128	-0.815725	-0.815866	-0.578155	-0.814883	-0.577854
n_i	-0.001566	0.578416	0.578232	-0.815913	0.579581	-0.815938
	0.999986	0.005329	0.003148	0.004733	0.007246	-0.018159

As shown in Table 1, C_i means circle center coordinates and n_i represents the axis orientation vectors.

Table 2. Link geometric parameters identified

Link No.	α_i /deg	a_i /mm	d_i /mm	θ_i /deg
1	0	0	0	1.570796327
2	-1.571036415	99.86198454	0	-1.570796327
3	0.00022797	289.8619856	-0.086709639	0.000299142
4	-1.570721227	86.23690465	300.4090081	-2.51E-07
5	1.569071616	-0.98001211	0.005521283	-7.08E-06
6	-1.568645424	-1.899993222	-0.09795004	4.92E-05

A laser tracker is utilized to record robot hand position by giving five groups of joint variables. Robot hand coordinates are calculated at the same joint variables by using nominal and identified parameters respectively in the control program. The distances between the true and virtual points are regarded as errors to evaluate the accuracy of the calibration and listed in Table 3.

Table 3. The distances the true and virtual points

Points No.	Before calibration	After calibration
1	1.98261787	0.149415298
2	2.360621007	0.694653861
3	2.89899382	1.00479848
4	3.383122793	1.401059758
5	1.507338242	0.98077312
6	2.263920667	0.501796024

As shown in Table 3, the error is decreased dramatically after calibration which is helpful to the next robot hand-eye calibration. Moreover it proves that screw axis identification method obtain a precise identification of the physical reality of the robot's geometry.

3.3   Hand-Eye Calibration Results

With robot moving to ten different positions, the robot motions and the images of calibration plate were obtained. The poses of the robot hand relative to the base frame are listed in Table 4 and the calibration images are shown in Fig. 3.

Table 4. Poses of the robot hand relative to the base frame

Points No.	X/m	Y/m	Z/m	α /deg	β /deg	γ /deg
1	-0.000184	0.6336288	-0.057165	179.9916	-30.1345	90.0488
2	-0.011471	0.6328875	-0.060842	176.5547	-28.0679	89.2980
3	-0.021115	0.6445409	-0.045429	175.5419	-5.5945	86.8704
4	-0.012766	0.6317892	-0.044914	-165.6042	-24.9480	82.7853
5	0.0089219	0.6200075	-0.045425	166.3300	-22.5697	98.8069
6	0.0265457	0.6132203	-0.041354	165.6205	-25.1105	98.0334
7	-0.000180	0.6337833	-0.056926	179.9823	-32.4425	90.0590
8	-0.012092	0.6066658	-0.074929	164.4112	-51.5740	95.5483
9	0.0030354	0.6179910	-0.049755	145.6041	-30.3056	95.7754
10	-0.025377	0.6280254	-0.064005	144.8454	-26.6897	95.6604

Fig. 3. Calibration images

Input essential parameters in HALCON and perform the program, the intrinsic parameters of the camera and the hand-eye parameters were acquired and shown in Table 5 and Table 6.

Table 5. Camera parameters after calibration

F/m	K1	K2	K3	P1	P2	Error
0.0162894	45.7183	1.2850E+007	-8.7088E+011	0.0007939	0.157394	0.12425

Table 6. Camera and Calibration poses

	X/m	Y/m	Z/m	α /deg	β /deg	γ /deg
Camera Pose	0.0029513	0.0451388	0.169182	359.318	30.1607	266.635
Calibration Pose	0.0042869	0.666662	-0.071888	179.238	359.498	356.391

According to the camera pose relative to the robot hand shown in Table 6, the corresponding homogeneous transformation matrix can be obtained as following:

$$^{hand}\mathbf{H}_{cam} = \begin{bmatrix} -0.0507503 & 0.863129 & -0.502428 & 0.124112 \\ 0.998556 & -0.0527229 & -0.0102908 & 0.00117376 \\ -0.0176072 & -0.502224 & 0.864558 & -0.123546 \\ 0 & 0 & 0 & 1 \end{bmatrix}. \tag{9}$$

Table 7. Hand-eye calibration Error

Quality of the Results	Root Mean Square Error	Maximum Error
Translation Part	0.00484973m	0.00744265m
Rotation Part	1.15652°	2.01205°

As shown in Table 7, the root mean square error is acceptable in calibration and the experimental results demonstrate the effectiveness of the presented methods.

3.4   Light Plane Calibration Results

As shown in Fig. 4, we adapt the lighting in between to get one image with a clearly represented calibration plate and one image that shows a well-defined laser line.

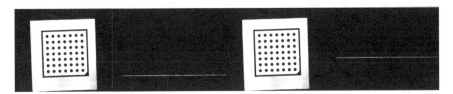

Fig. 4. Calibration plate images and laser line images with different distance in z direction

Table 8. Parameters of light plane pose and calibration error

X/m	Y/m	Z/m	α /deg	β /deg	γ /deg	Mean Residual/m
0.00016539	0.00677468	-0.0118259	240.402	359.918	0.144	0.00011499

The parameters of light plane pose and calibration error are shown in Table 8. The mean residual that evaluate the fit of the light plane show good accuracy of the light plane calibration.

4   Conclusion

In this paper, we proposed a robot vision system calibration method, which computed the robot kinematic parameters with screw axis identification method by using a laser tracker and implemented the hand-eye calibration as well as light plane calibration with HALCON. The robot kinematic calibration showed the validity of algorithm in practice and the results of the robot vision system calibration experiments demonstrated that the method is a feasible and accurate technique.

Acknowledgments. This research was supported by the Science and Technology Planning Project of Guangdong Province, China (Nos.2012B011300010 and 2012B091100141), and the Natural Science Foundation of Guangdong Province (S2013030013355).

References

1. Gao, G., Yang, X.C., Zhang, H.M.: Structured light stereo vision system coupled with laser remanufacturing. In: 2011 International Conference on Control, Automation and Systems Engineering (2011)
2. Liu, Y., et al.: A calibration method for a linear structured light system with three collinear points. International Journal of Computational Intelligence Systems 4(6), 1298–1306 (2011)
3. Wang, W., Li, A., Wu, D.: Robot calibration by observing a virtual fixed point. In: 2009 IEEE International Conference on Robotics and Biomimetics, Guilin, China (2009)
4. Wang, H., Shen, S., Lu, X.: A screw axis identification method for serial robot calibration based on the POE model. Industrial Robot 39(2), 146–153 (2012)
5. Zhuang, H.: Hand/eye calibration for electronic assembly robots. IEEE Transactions on Robotics and Automation 14(4), 612–616 (1998)
6. Zhao, Z., Liu, Y.: Hand-eye calibration based on screw motions. In: 18th International Conference on Pattern Recognition, Hong Kong, China (2006)
7. Santolaria, J., et al.: Rotation error modeling and identification for robot kinematic calibration by circle point method. Metrology and Measurement Systems 21(1), 85–98 (2014)
8. Dai, X.L.: Industrial robot kinematic calibration and research on stiffness identification. South China University of Technology (2013)
9. Zhao, Z.J., Liu, Y.C.: Integrating camera calibration and hand-eye calibration into robot vision. In: 7th World Congress on Intelligent Control and Automation, Chongqing, China (2008)
10. Li, D.M., Zhang, L.J.: CCD camera linear calibration method with a feature of calibration plate. In: 2011 International Conference on Transportation, Mechanical, and Electrical Engineering, Changchun, China (2011)

A Contour Detection Approach for Mobile Robot

Kun Ai[1], Zhiqiang Cao[1], Xilong Liu[1], Chao Zhou[1], and Yuequan Yang[2]

[1] The State Key Laboratory of Management and Control for Complex Systems,
Institute of Automation, Chinese Academy of Sciences,
Beijing 100190, China
{aikun2012,zhiqiang.cao,xilong.liu,chao.zhou}@ia.ac.cn
[2] The College of Information Engineering,
Yangzhou University, Yangzhou, China
yangyq@yzu.edu.cn

Abstract. Safe moving is a basic ability for a mobile robot, and it is beneficial for the robot to avoid the collisions with the environment if it knows the boundaries between the obstacles and free space. In this paper, a contour detection approach is presented. The input image is firstly processed by a Gaussian filter and Sobel edge detector. After it is processed by connectivity-based boundaries extraction, the result is finalized, aided by Canny edge map. The experiments demonstrate the effectiveness of our approach.

1   Introduction

Robot vision system is a crucial part in mobile robot applications. Like human vision system, robot vision system is an important component which is used to percept local environment. Usually, it consists of three parts: image acquisition, image processing and analysis, the output or display. As the contours of the image can describe the shape of the objects in the image, the extraction of the contours is useful in senior visual tasks such as object detection, segmentation and image understanding. There are many methods that can detect contours from an image. Among them, obtaining contours from edge maps of an image is commonly used. One of the problems is that they do not distinguish object contours which are the actual primitives needed in most machine vision applications from edges originating from textured regions [1]. And research on edge detection continues to be a hottest place while a large number of algorithms have been proposed [2]-[10].

One of the capacities of the mobile robot is safe moving in complex and unknown environments [11]. By analyzing the contours which are extracted through the robot vision system, the related information of obstacles is obtained, which can help the robot accomplish the tasks such as obstacle avoidance or object grasping. To achieve this goal, the vision system needs to be effective and fast. However, previous contour detection algorithms based on computer vision are hard to be used in real-time in complex and unknown environments [12]. Grigorescu *et al.* proposed contour

X. Zhang et al. (Eds.): ICIRA 2014, Part II, LNAI 8918, pp. 265–272, 2014.

detection based on non-classical receptive field inhibition [1]. The results show that their method enhances contour detection in cluttered visual scenes. Arbela´ez and Fowlkes proposed a high-performance contour detector, combining local and global image information [13]. However, these algorithms spend much time so that they cannot be used on real robot. We know that most classical edge detectors spend a litter time on extracting edges from images. Therefore, it could be an effective way to extract the contours from the result of an edge detector. For edge detector, there are common criteria [14]. One is low error rate. Moreover, the edge points must be localized well, and the detector has only one response to a single edge. A large number of edge detection algorithms have been proposed. The Roberts [15], Sobel [16], and Prewitt [17] operators detect edges by convolving a gray-scale image with local derivative filters. Marr and Hildreth [18] used zero crossings of the Laplacian of Gaussian operator.

In this paper, we give a contour detection approach from the edge map. Gray level image captured by robot camera is firstly processed by a Gaussian filter and edge operator. After this step, the processed image shall be processed by connectivity-based boundaries extraction to removing the redundant edges. After that, we need to make edges be one pixel wide. Due to the fact that Canny edge map is one pixel wide, we get Canny edge map of the original image. A final result shall be obtained by combining the result obtained by connectivity-based boundaries extraction and Canny edge map.

The remainder of the paper is organized as follow. Section 2 gives the details of our approach. The experimental results are depicted in section 3 and section 4 concludes the paper.

2   The Approach of Contour Detection

In this part we present a contour detection approach for robot, which is based on edge detection. The details of our approach will be described in three parts below.

2.1   Edge Detection

Edge detectors are susceptible to noise in raw unprocessed image data, therefore noise reduction needs to be performed to remove the burrs on a raw image. In our approach, a 5×5 Gaussian filter is used to process the raw image. Now, Prewitt detector and Sobel detector are the most commonly used. From the point of view of suppressing noise, we select the Sobel detector to detect edge. In fact, edge detection is the processing of finding the magnitude and the angle of the gradients of an image. We denote with the output image after preprocessing and edge detection $f_1(x, y)$.

2.2   Connectivity-Based Boundaries Extraction

In this step, morphological operations shall be used. Edges of an image are usually composed by two kinds of edge points. One is generated from boundaries and another is generated from texture regions. For a mobile robot, an image is consisted by obstacles and free space, and it can move safely on free space if the boundaries between obstacles and free space are obtained. The edges generated from texture regions have a negative influence for robot motion, and these edges should be removed from $f_1(x, y)$.

It is obvious that edges from the output of an edge detector are generated by the local information of an image, which means that these two kinds of edge points cannot be distinguished. For object contours, they are continues and mostly are narrow and long, whereas quite a few textures are disorder and discontinued. On this basis, these two kinds of edge points may be distinguished by utilizing these features of all the edges. The detailed procedure to remove the edges generated from texture regions is as follows.

Firstly, the image $f_1(x, y)$ is converted to a binary image $b(x, y)$ with a threshold of T_1. The threshold T_1 is varied according to the image. If it is too small, different edges may be not separate and if it is too big, some edges may be neglected. It is easy to understand that in the image $f_1(x, y)$ edge points have a greater value than other points. Also, object contours are usually very sharp. Therefore, we can find that edge points generated from object contours have greater values than parts of edge points generated from texture regions in an edge image. A suitable selection of the threshold T_1 is required and in this paper, the threshold is selected by analyzing the cumulative histogram of the image $f_1(x, y)$. By this binarization process, some edges generated from light color texture regions can be removed.

An object in an image is represented by a set in mathematical morphology. In a binary image, each element of a set is a 2D-vector whose components are the coordinates of a pixel point. Two neighboring pixel points are said to be connected only when they have the same value. In general, three types of adjacency are considered: 4-adjacency, 8-adjacency and mixed adjacency. In our approach, 8-adjacency is used. Two pixel points, P and Q, are connected if there exists a path of pixel points $(p_0, p_1, ..., p_n)$ such that $p_0 = P$, $p_n = Q$ and $\forall 1 \leq i \leq n$, p_{i-1} and p_i are neighbors.

The image $b(x, y)$ is composed of thousands of connected components which consist of edge points. Next, we need to label and extract these connected components. As our approach shall be used on the robot, a fast algorithm of connected component labeling is required. Stefano *et al.* proposed a simple and efficient connected components labeling algorithm [19]. The algorithm is a two-scan labeling algorithm, and equivalences are processed during the first pass in order to determine the correct state of equivalence classes at each time of the scan. This is obtained by associating a new equivalence class with each new label and by merging the corresponding classes as soon as a new equivalence is found.

The algorithm performs the merging operation by using a simple data structure called Class Array(C). C is a one-dimensional array as large as the maximum label value and containing for each label value its corresponding class identifier.

The labeled result is denoted with $l(x, y)$. The geometric features of each connected component can be acquired by searching the pixels which have the same label.

As some edges originating from texture regions are discontinuous and disorder, we think that each connected component is a separate edge, and the information of each edge is reflected by its connected component. For example, the perimeter and the area of a connected component can be easily obtained in the labeled image $l(x, y)$. It is easy to know that the ratios of the perimeters and the areas of connected components which are generated from object contours are greater than others. In our experiment, we consider the area index of connected components. Assume that there are N connected components in $b(x, y)$ and we denote with the number of the pixel points contained in the ith connected component as $L_i (i=1,2,...,N)$. We need to find a suitable threshold T_2. If $L_i < T_2$, the connected component is considered as the edges originating from texture regions, otherwise, it is considered as an edge originating from object contours. The threshold should be different according to environments. Then, we zero the values of all pixel points contained in the connected components whose numbers of pixel points are less than the threshold T_2. After that, a new binary image $b_1(x, y)$ is obtained.

2.3   Combination with Canny Edge Map

Although the image $b_1(x, y)$ only contains edges originating from object contours, it cannot be the final result of our approach because the edges in $b_1(x, y)$ are very thick. In practice, we would like to have edges with only one pixel wide. Since the angle and the magnitude of the gradient in the image $f(x, y)$ is missed in the process of thresholding, it is impossible to get the accurate edges from the image $b_1(x, y)$. In [14], an approach with non-maximum suppression was proposed. This approach keeps only those pixels on an edge with the highest gradient magnitude. These maximal magnitudes should occur right at the edge boundary, so the edge boundaries are converted to one pixel wide. For this reason we use it in $f_1(x, y)$. To make the edges be continuous and valid, hysteresis thresholding is also adopted. Therefore, the image $f_1(x, y)$ is converted to a binary image $b_2(x, y)$, which shows that all the edges are one pixel wide.

   We aim at obtaining an image which only contains one pixel wide contours. Therefore, the goal can be achieved by taking the intersection with the image $b_1(x, y)$ and $b_2(x, y)$.

   Although the object contours have been generated, the results may be unsatisfied. Some contours may be non-closed, and we need some means to close the contours. Finally, an additional adjustment procedure is applied, and a map reflecting the contours of the input image is outputted. Based on this map, the robot can make a collision-free decision.

3   Experiments

In this paper, some experiments are conducted to testify the contour detection approach.

The experiment I is used to demonstrate the given approach, as shown in Fig. 1. Obviously, edges generated from texture regions are removed by our method. The output just contains the object contours. Experiment II considers more complex environments. The results are shown in the Fig. 2. From these results, we can find that ǒur approach has a good performance. Except removing the texture edges, the light trace is also neglected.

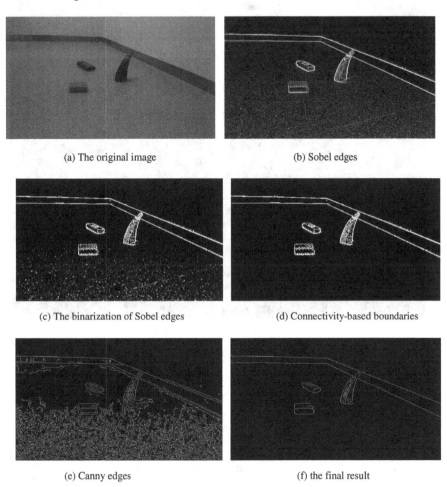

(a) The original image

(b) Sobel edges

(c) The binarization of Sobel edges

(d) Connectivity-based boundaries

(e) Canny edges

(f) the final result

Fig. 1. The demonstration of the given approach

In experiment III, the robot moves by using the given approach to avoid obstacles. There are some randomly chosen objects (potted plant, carton) with different sizes, heights, colors and textures on the floor. From Fig. 3, it is seen that the robot achieves a collision-free motion. Experiment IV is used to test the adaptability of our approach by putting an obstacle when it is moving. Fig. 4 shows the video snapshots of the robot's collision-free motion.

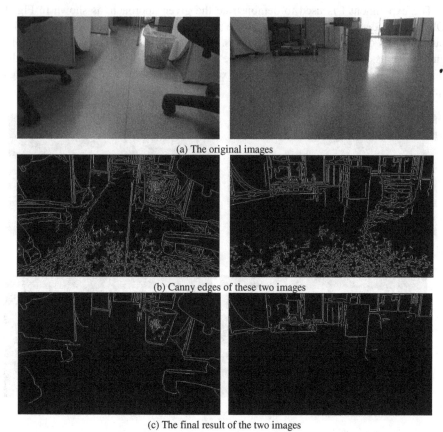

(a) The original images

(b) Canny edges of these two images

(c) The final result of the two images

Fig. 2. The results of experiment II

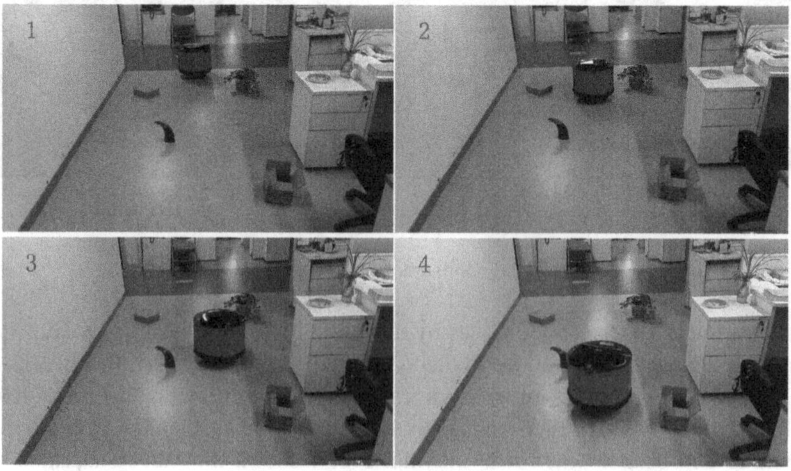

Fig. 3. Video snapshots of experiment III

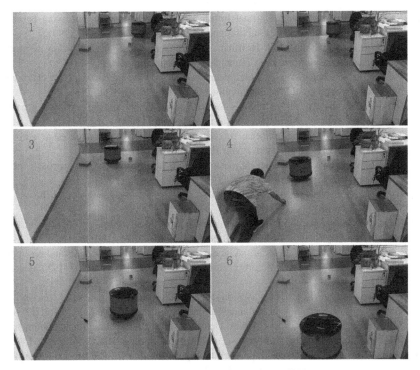

Fig. 4. Video snapshots of experiment IV

4   Conclusion

In this paper, a contours detection approach for robot vision is given. The edges generated by texture regions are removed in the environment, and the edges of the object contours are kept. The validity of our approach is testified by robot experiments. In the future, we shall consider the more complex environments and improve the adaptability.

Acknowledgements. This work was supported in part by the National Natural Science Foundation of China under Grants 61273352, 61175111, 61227804, 60805038.

References

[1] Grigorescu, C., Petkov, N., Westenberg, M.A.: Contour Detection Based on Nonclassical Receptive Field Inhibition. IEEE Transactions on Image Processing 12(7), 729–739 (2003)
[2] Bergholm, F.: Edge focusing. IEEE Transactions on Pattern Analysis and Machine Intelligence PAMI-9, 726–741 (1987)

[3] Sarkar, S., Bowyer, K.L.: Optimal infinite impulse response zero-crossing based edge detection. Computer Vision, Graphics and Image Process 54, 224–243 (1991)

[4] Rothwell, C.A., Mundy, J.L., Hoffman, W., Nguyen, V.D.: Driving vision by topology. In: International Symposium on Computer Vision, pp. 395–400 (1995)

[5] Iverson, L., Zucker, S.W.: Logical/linear operators for image curves. IEEE Transactions on Pattern Analysis and Machine Intelligence 17(10), 982–996 (1995)

[6] Smith, S.M., Brady, J.M.: SUSAN–A new approach to low level image processing. International Journal of Computer Vision 23(1), 45–78 (1997)

[7] Tabb, M., Ahuja, N.: Multiscale image segmentation by integrated edge and region detection. IEEE Transactions on Image Processing 6, 642–655 (1997)

[8] Bezdek, J.C., Chandrasekhar, R., Attikiouzel, Y.: A geometric approach to edge detection. IEEE Transaction on Fuzzy Systems 6(1), 52–75 (1998)

[9] Black, M.J., Sapiro, G., Marimont, D., Heeger, D.: Robust anisotropic diffusion. IEEE Transaction on Image Processing 7, 421–432 (1998)

[10] Meer, P., Georgescu, B.: Edge detection with embedded confidence. IEEE Transactions on Pattern Analysis and Machine Intelligence 23, 1351–1365 (2001)

[11] Liu, X., Cao, Z., Jiao, J., et al.: A general robot environment understanding approach inspired by biological visual cortex. Robotics and Biomimetics, 1953–1957 (December 2013)

[12] De Souza, G.N., Kak, A.C.: Vision for mobile robot navigation: A survey. IEEE Transactions on Pattern Analysis and Machine Intelligence 24(2), 237–267 (2002)

[13] Arbeláez, P., Fowlkes, C.: Contour Detection and Hierarchical Image Segmentation. IEEE Transactions on Pattern Analysis and Machine Intelligence 33(5), 237–267 (2011)

[14] Canny, J.F.: A computational approach to edge detection. IEEE Transactions on Pattern Analysis and Machine Intelligence PAMI-8(6), 679–698 (1986)

[15] Roberts, L.G.: Machine Perception of Three-Dimensional Solids. In: Tippett, J.T., et al. (eds.) Optical and Electro-Optical Information Processing. MIT Press (1965)

[16] Duda, R.O., Hart, P.E.: Pattern Classification and Scene Analysis. Wiley (1973)

[17] Prewitt, J.M.S.: Object Enhancement and Extraction. In: Lipkin, B., Rosenfeld, A. (eds.) Picture Processing and Psychopictorics. Academic Press (1970)

[18] Marr, D.C., Hildreth, E.: Theory of Edge Detection. In: Proc. Royal Soc. of London (1980)

[19] Stefano, L.D.: A simple and efficient connected components labeling algorithm. In: Image Analysis and Processing, pp. 322–327 (1999)

Accuracy of Determining the Coordinates of Points Observed by Camera

Tadeusz Szkodny, Artur Meller, and Krzysztof Palka

Silesian University of Technology, Institute of Automatic Control,
Akademicka 16 St. 44-100 Gliwice, Poland
Tadeusz.Szkodny@polsl.pl

Abstract . In this paper analysis of accuracy of determining the coordinates of points on the plane template in the shape of rectangle is presented. These points lie at the corners of squares with side of 8 *mm*. Images of these points were obtained using Edimax IC-7100P camera from three different points of view and analysed. Position and orientation coordinates of camera relatively to the reference system were calculated. Coordinates of points on the ideal image (without optical distortions) were determined. After reading from the image real coordinates, optical distortion model coefficients of the camera were calculated. After that, errors caused by optical distortion were determined. Next, coordinates read from the image were corrected and coordinates of observed points in the reference system were calculated. Finally, calculated coordinates were compared to them real values and them maximal differences were determined.

Keywords: Computer Vision, Soft Computing, Robot Intelligence.

1   Introduction

One of the basic component of computer intelligence of robots is software, that calculates coordinates of position and orientation of manipulated objects, seen by the camera. Designing of such software must take into account the errors of coordinates read from the camera matrix. These errors cause inaccuracies of calculations of points coordinates in reference system, associated with technical station.

In order to determine accuracy of calculated coordinates of observed points, analysis of errors is needed. These errors are caused by: reading errors of coordinates from the matrix of the camera, optical distortions of the camera, errors of parameters that describes optical system of the camera and errors of calculations. During designing of the vision system, minimization of mentioned errors is needed.

To Edimax IC-7100P camera study, template with points surrounded by yellow circles (Fig.1) is used. These points lie in the corners of squares with side of 8 *mm*. In the figure, x and y axis of reference system are marked in the red colour, whereas X and Y axis of auxiliary coordinate system are marked in the green colour. Auxiliary system is useful to set camera above the template.

X. Zhang et al. (Eds.): ICIRA 2014, Part II, LNAI 8918, pp. 273–284, 2014.
© Springer International Publishing Switzerland 2014

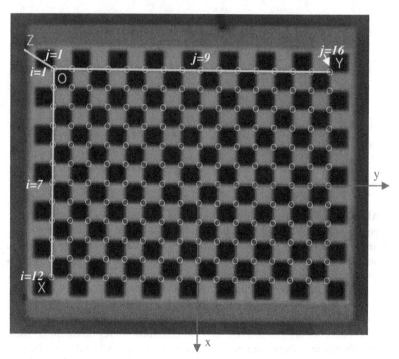

Fig. 1. The template used to camera Edimax IC-7100P study

User can read points coordinates directly from picture in pixels using Microsoft Paint program, but this way is connected with possibility of making mistakes. To decrease risk of making such mistakes, reading of these coordinates in this work has been made automatically using image processing algorithm [1] implemented in C# language, in Microsoft Visual C# 2010 Expres environment.

To compensate optical distortion errors, correction of the read coordinates has been used. Mathematical model of optical distortions used here is presented in works [2,3].

Parameters which describe the coordinate system of camera $x_c y_c z_c$ in reference system xyz are coordinates of position and orientation, focal length and size of the pixel. Calculation of every parameter can be made using iteration methods, which minimize square form of errors [4-5]. Errors of each mentioned parameters occurs in this form. However fundamental disadvantage of these methods is large number of calculations which cause great numerical errors and long time of calculations. In this work, coordinates of position and orientation was calculated using fast and accurate *Camera* algorithm [6]. Precision of calculations of this algorithm amounted to 10^{-6} *mm*. Focal length and size of the pixels were taken from camera datasheet.

In this work, precision of calculation of 192 points from Fig. 1 was analyzed in three different settings of camera above this template. In second chapter, results of calculations of camera position and orientation coordinates are presented. Third chapter contains calculations of mathematical model coefficients of optical distortions. Fourth chapter is about analysis of coordinates calculations errors in reference system. Fifth chapter summarizes all of the studies.

2   Position and Orientation of Camera

Here, calculation of position and orientation coordinates of camera in three different angles with respect to plane of template from Fig.1 is performed. Beginning point O_c of camera coordinates system $x_c y_c z_c$ has been associated with the center of camera matrix. Camera settings are presented in Fig. 2.

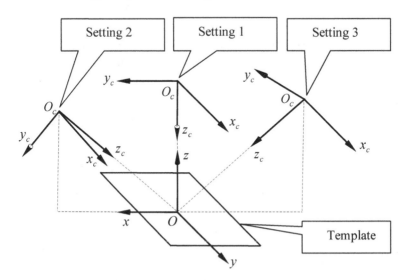

Fig. 2. Camera settings

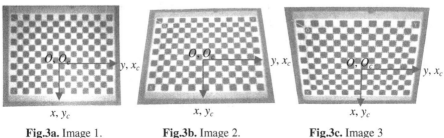

Fig.3a. Image 1.	**Fig.3b.** Image 2.	**Fig.3c.** Image 3

Images of template at setting 1, 2, 3 are presented in the Figures 3a-c. Image 1 was obtained from camera at setting 1, image 2 at setting 2, image 3 at setting 3.

Coordinates of all 192 points from the template are read in pixels with usage of algorithm of image processing [1] implemented in C# language in Microsoft Visual C# 2010 Express environment. From Edimax IC-7100P camera data sheet, can be readed, that size of the pixel is equal to $2.8 \cdot 10^{-3} mm$ x $2.8 \cdot 10^{-3} mm$ and focal length $f_c = 5.01$ mm. After multiplication of coordinates in pixels by the pixel size, we obtain coordinates of points read from the image in mm, in $x_c y_c$ - system.

Coordinates of points in reference system xy are easy to determine. These points are placed in the corners of 8mm by 8mm square.

Position and orientation of camera system $x_c y_c z_c$ with respect to reference system xyz are described by the homogenous matrix $\mathbf{T}_c$ of transformation, like in the equation (1).

$$\mathbf{T}_c = Trans(d_x, d_y, d_z) Rot(z, \gamma) Rot(y, \beta) Rot(x, \alpha) . \tag{1}$$

It is notation of successive transformations with respect to reference system xyz. Mentioned transformations are: rotation about axis x by angle α, rotation about axis y by angle β, rotation about axis z by angle γ, displacement d_z along axis z, displacement d_y along axis y, displacement d_x along axis x [7,8].

These coordinates are determined by *Camera* algorithm [6]. Input parameters of these algorithms are coordinates α, β, γ, d_x, d_y, d_z; coordinates $^c z_A$, $^c z_B$, $^c z_C$ of three points A, B, C from template in the system $x_c y_c z_c$; coordinates $^c x_{Ac}$, $^c y_{Ac}$, $^c x_{Bc}$, $^c y_{Bc}$, $^c x_{Cc}$, $^c y_{Cc}$ of points A, B, C in the system $x_c y_c z_c$ (read from camera); coordinates x_A, y_A, z_A, x_B, y_B, z_B, x_C, y_C, z_C of points A, B, C in the system xyz; focal length f_c; and accuracy of calculations *delta*.

Calculations of coordinates in every camera setting from Fig.2 for 8 sets of points A, B, C was made. Obtained 8 sets of camera coordinates were averaged. Points lying beyond the beginning of coordinate system O_c, but closest to this beginning were chosen. Read coordinates of these points have small errors caused by optical distortions. These points are located on the square with side length of 16 mm which centre approximately coincide with point O_c. Coordinates $z_A = z_B = z_C = 0\,mm$.

The sets of points are as follows:

- set 1: $x_A = 8\,mm$, $y_A = -8\,mm$, $x_B = 8\,mm$, $y_B = 8\,mm$, $x_C = -8\,mm$, $y_C = 8\,mm$;
- set 2: $x_A = 8\,mm$, $y_A = -8\,mm$, $x_B = 8\,mm$, $y_B = 8\,mm$, $x_C = -8\,mm$, $y_C = -8\,mm$;
- set 3: $x_A = 8\,mm$, $y_A = 8\,mm$, $x_B = -8\,mm$, $y_B = 8\,mm$, $x_C = -8\,mm$, $y_C = -8\,mm$;
- set 4: $x_A = 8\,mm$, $y_A = 8\,mm$, $x_B = -8\,mm$, $y_B = 8\,mm$, $x_C = 8\,mm$, $y_C = -8\,mm$;
- set 5: $x_A = -8\,mm$, $y_A = 8\,mm$, $x_B = -8\,mm$, $y_B = -8\,mm$, $x_C = 8\,mm$, $y_C = -8\,mm$;
- set 6: $x_A = -8\,mm$, $y_A = 8\,mm$, $x_B = -8\,mm$, $y_B = -8\,mm$, $x_C = 8\,mm$, $y_C = 8\,mm$;

- set 7: $x_A = -8\ mm$, $y_A = -8\ mm$, $x_B = 8\ mm$, $y_B = -8\ mm$, $x_C = 8\ mm$, $y_C = 8\ mm$;
- set 8: $x_A = -8\ mm$, $y_A = -8\ mm$, $x_B = 8\ mm$, $y_B = -8\ mm$, $x_C = -8\ mm$, $y_C = 8\ mm$.

For initial values of input parameters α, β, γ, d_x, d_y, d_z, ${}^c z_A$, ${}^c z_B$, ${}^c z_C$, x_A, y_A, z_A, x_B, y_B, z_B, x_C, y_C, z_C calculated roughly using geometrical dependencies and ${}^c x_{Ac}$, ${}^c y_{Ac}$, ${}^c x_{Bc}$, ${}^c y_{Bc}$, ${}^c x_{Cc}$, ${}^c y_{Cc}$ read from camera, camera coordinates α, β, γ, d_x, d_y, d_z were calculated with accuracy $delta=10^{-6}\ mm$ [6].

Equation (2a) describes averaged coordinates for camera setting 1 from Fig.2 and equation (2b) describes matrix $\mathbf{T}_c$.

$$\alpha = 179.9774°, \ \beta = -0.0187°, \ \gamma = 89.9424°, \ d_x = -0.0290\ mm, \ d_y = -0.0126\ mm,$$
$$d_z = 236.5727\ mm. \tag{2a}$$

$$\mathbf{T}_c = \begin{bmatrix} 0.0010 & 0.9999 & 0.0004 & -0.0290 \\ 0.9999 & -0.0010 & 0.0003 & -0.0126 \\ 0.0003 & 0.0004 & -0.9999 & 236.5727 \\ 0 & 0 & 0 & 1 \end{bmatrix}. \tag{2b}$$

Equation (3a) describes averaged coordinates for camera setting 2 from Fig.2 and equation (3b) describes matrix $\mathbf{T}_c$.

$$\alpha = 210.1180°, \ \beta = 0.8350°, \ \gamma = 90.2963°, \ d_x = 161.2950\ mm,$$
$$d_y = 5.1845\ mm, \ d_z = 277.9449\ mm. \tag{3a}$$

$$\mathbf{T}_c = \begin{bmatrix} -0.0052 & 0.8650 & -0.5017 & 161.2950 \\ 0.9999 & -0.0028 & -0.0152 & 5.1845 \\ -0.0146 & -0.5017 & -0.8649 & 277.9449 \\ 0 & 0 & 0 & 1 \end{bmatrix}. \tag{3b}$$

Equation (4a) describes averaged coordinates for camera setting 3 from Fig.2 and equation (4b) describes matrix $\mathbf{T}_c$.

$$\alpha = 144.2784°, \ \beta = -0.2332°, \ \gamma = 91.0074°, \ d_x = -201.7442\ mm,$$
$$d_y = -4.9399\ mm, \ d_z = 283.4332\ mm. \tag{4a}$$

$$\mathbf{T}_c = \begin{bmatrix} -0.0176 & 0.8118 & 0.5837 & -201.7442 \\ 0.9998 & 0.0119 & 0.0136 & -4.9399 \\ 0.0041 & 0.5838 & -0.8119 & 283.4332 \\ 0 & 0 & 0 & 1 \end{bmatrix}. \tag{4b}$$

3   Errors of Optical Distortions

For simplification, rows and columns are introduced. Rows are consist of points, lying on lines, which are parallel to axis y on Fig.1. Each row consist of 16 points. Number of rows is equal to 12, according to Fig.1 . Columns are consist of points, lying on lines, which are parallel to axis x on Fig.1. Number of columns is equal to 16. P_{ij} is the point of i-th row and j-th column. Coefficients $^c x_c(i, j)$ and $^c y_c(i, j)$ of image of points P_{ij} of template, read from the camera system $x_c y_c$, have errors $\Delta_* {}^c x_c(i, j)$ and $\Delta_* {}^c y_c(i, j)$, caused by optical distortions. These errors results from mathematical description of distortions by means of coefficients k_1, k_2, k_3, p_1 and p_2 [2,3]. Equations (5a) and (5b) describes these errors. Errors $\Delta^c x_c(i, j)$ and $\Delta^c y_c(i, j)$ can be determine from coordinates read from camera. These are described by equation (5c).

$$\begin{aligned} \Delta_* {}^c x_c(i, j) = {}^c x_c(i, j)[k_1 {}^c r_c(i, j)^2 + k_2 {}^c r_c(i, j)^4 + k_3 {}^c r_c(i, j)^6] + 2p_1 {}^c x_c(i, j) \cdot {}^c y_c(i, j) \\ + p_2[{}^c r_c(i, j)^2 + 2 \cdot {}^c x_c(i, j)^2], \end{aligned} \tag{5a}$$

$$\begin{aligned} \Delta_* {}^c y_c(i, j) = {}^c y_c(i, j)[k_1 {}^c r_c(i, j)^2 + k_2 {}^c r_c(i, j)^4 + k_3 {}^c r_c(i, j)^6] + 2p_2 {}^c x_c(i, j) \cdot {}^c y_c(i, j) \\ + p_1[{}^c r_c(i, j)^2 + 2 \cdot {}^c y_c(i, j)^2], \end{aligned} \tag{5b}$$

$$\Delta^c x_c(i, j) = {}^c x_{ci}(i, j) - {}^c x_c(i, j), \quad \Delta^c y_c(i, j) = {}^c y_{ci}(i, j) - {}^c y_c(i, j),$$
$$^c r_c(i, j)^2 = {}^c x_c(i, j)^2 + {}^c y_c(i, j)^2. \tag{5c}$$

In equations (5c) occurs ideal coordinates of points $^c x_{ci}(i, j)$ and $^c y_{ci}(i, j)$, with no optical distortions. These coefficients can be calculated from homogenous form $\mathbf{r}(i, j)$ of vector that describes point P_{ij} in reference system. We can note that form as follows:

$$\mathbf{r}(i, j) = \begin{bmatrix} x(i, j) \\ y(i, j) \\ z(i, j) \\ 1 \end{bmatrix}. \tag{6}$$

Coordinates of point P_{ij} in reference system, that occurs in equation (6), can be note using indexes as follows: $x(i, j) = (i - 7) \cdot 8 \, mm$, $y(i, j) = (j - 9) \cdot 8 \, mm$. Point P_{79} is the origin O of the reference system. All points are lying on the plane xy, so $z(i, j) = 0$. Since camera matrix $\mathbf{T}_c$ of transformation is known, calculation of homogenous form $^c\mathbf{r}(i, j)$ of vector that describes point P_{ij} in the camera system $x_c y_c z_c$ can be performed.

$$\mathbf{r}(i, j) = \begin{bmatrix} x(i, j) \\ y(i, j) \\ z(i, j) \\ 1 \end{bmatrix} = \mathbf{T}_c{}^c\mathbf{r} \rightarrow {}^c\mathbf{r}(i, j) = \begin{bmatrix} {}^cx(i, j) \\ {}^cy(i, j) \\ {}^cz(i, j) \\ 1 \end{bmatrix} = \mathbf{T}_c^{-1} \begin{bmatrix} x(i, j) \\ y(i, j) \\ z(i, j) \\ 1 \end{bmatrix}. \tag{7}$$

Coordinate $^cx_{ci}(i, j)$ can be obtained from coordinate $^cx(i, j)$ calculated from equation (7). From geometrical dependences, shown in the Fig.4 results dependence (8a) that describes coordinate $^cx_{ci}(i, j)$.

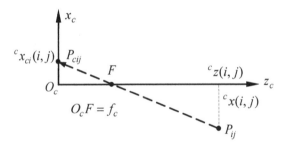

Fig. 4. The coordinates x_c of point P_{ij} and its image P_{cij}.

$$\frac{{}^cx_{ci}(i, j)}{f_c} = \frac{-{}^cx(i, j)}{{}^cz(i, j) - f_c} \rightarrow {}^cx_{ci}(i, j) = -\frac{{}^cx(i, j)}{\dfrac{{}^cz(i, j)}{f_c} - 1}. \tag{8a}$$

Using similiar geometrical dependecies the formula (8b) can be derived.

$$^cy_{ci}(i, j) = -\frac{{}^cy(i, j)}{\dfrac{{}^cz(i, j)}{f_c} - 1}. \tag{8b}$$

Using equations (7), (8a), (8b) coefficients $\Delta^c x_c(i,j)$, $\Delta^c y_c(i,j)$ and $^c r_c(i,j)^2$ occurring in equations (5c) can be calculated. Since these values are known, it allows to apply equations (5a) and (5b) for calculation of coefficients k_1, k_2, k_3, p_1 and p_2. If following sum is created

$$S = \sum_{i=1}^{12}\sum_{j=1}^{16}\{[\Delta^c x_c(i,j) - \Delta_* {}^c x_c(i,j)]^2 + [\Delta^c y_c(i,j) - \Delta_* {}^c y_c(i,j)]^2\},$$

unknown coefficients can be calculated by using minimally square method. Results of these calculations are presented by expressions (9a-11d). Calculations were made for three camera setting as in the Fig.2.

Coefficients for camera setting 1 have following values:

$$k_1 = 0.0120\ mm^{-2}, k_2 + k_3 = 0,\ p_1 + p_2 = 0;\qquad(9a)$$

$$k_1 = 0.0256\ mm^{-2},\ k_2 = -0.0073\ mm^{-4},\ k_3 = 0,\ p_1 + p_2 = 0;\qquad(9b)$$

$$k_1 = 0.0460\ mm^{-2},\ k_2 = -0.0318\ mm^{-4},\ k_3 = 0.0068\ mm^{-6},\ p_1 + p_2 = 0;\qquad(9c)$$

$$k_1 = 0.0443\ mm^{-2},\ k_2 = -0.0302\ mm^{-4},\ k_3 = 0.0064\ mm^{-6},\ p_1 = 0.0014\ mm^{-2},$$
$$p_2 = -0.0012\ mm^{-2}.\qquad(9d)$$

For camera setting 2:

$$k_1 = 0.0043\ mm^{-2}, k_2 + k_3 = 0,\ p_1 + p_2 = 0;\qquad(10a)$$

$$k_1 = -0.0011\ mm^{-2},\ k_2 = 0.0056\ mm^{-4},\ k_3 = 0,\ p_1 + p_2 = 0;\qquad(10b)$$

$$k_1 = 0.0010\ mm^{-2},\ k_2 = 0.0006\ mm^{-4},\ k_3 = 0.0027\ mm^{-6},\ p_1 + p_2 = 0;\qquad(10c)$$

$$k_1 = 0.0007\ mm^{-2},\ k_2 = 0.0021\ mm^{-4},\ k_3 = 0.0022\ mm^{-6},\ p_1 = -0.0010\ mm^{-2},$$
$$p_2 = -0.0003\ mm^{-2}.\qquad(10d)$$

For camera setting 3:

$$k_1 = 0.0026\ mm^{-2}, k_2 + k_3 = 0,\ p_1 + p_2 = 0;\qquad(11a)$$

$$k_1 = -0.0075\ mm^{-2},\ k_2 = 0.0114\ mm^{-4},\ k_3 = 0,\ p_1 + p_2 = 0;\qquad(11b)$$

$$k_1 = -0.0149 \; mm^{-2}, \; k_2 = 0.0279 \; mm^{-4}, \; k_3 = -0.0102 \; mm^{-6}, \; p_1 + p_2 = 0; \quad (11c)$$

$$k_1 = -0.0173 \; mm^{-2}, \; k_2 = 0.0336 \; mm^{-4}, \; k_3 = -0.0172 \; mm^{-6}, \; p_1 = 0.0040 \; mm^{-2},$$
$$p_2 = 0.0002 \; mm^{-2}. \quad (11d)$$

Equations (9a), (10a), (11a) present error description using one coefficient; equations (9b), (10b), (11b) – using two coefficients; equations (9c), (10c), (11c) – using three coefficients; equations (9d), (10d) and (11d) – using five coefficients.

4   Error of Coordinates Calculations

Coefficients described by equations (9a)-(11d) can be applied to calculate errors $\Delta_*{}^c x_c(i, j)$ and $\Delta_*{}^c y_c(i, j)$, described by equations (5a,b). After calculating these errors correction of coordinates of points ${}^c x_c(i, j)$ and ${}^c y_c(i, j)$, read from camera matrix in the coordinate system $x_c y_c$ can be made. By correction it means addition of errors $\Delta_*{}^c x_c(i, j)$ and $\Delta_*{}^c y_c(i, j)$ to the coordinates ${}^c x_c(i, j)$ and ${}^c y_c(i, j)$. Coordinates after this correction are note by ${}^c x_{ccor}(i, j)$ and ${}^c y_{ccor}(i, j)$. Equation (12) describes corrected coordinates.

$$ {}^c x_{ccor}(i, j) = {}^c x_c(i, j) + \Delta_*{}^c x_c(i, j), \quad {}^c y_{ccor}(i, j) = {}^c y_c(i, j) + \Delta_*{}^c y_c(i, j). \quad (12)$$

From corrected coordinates, coordinates of points on the plane of the template in the reference system xy can be calculated. These coordinates can be note by $x_{cor}(i, j)$ and $y_{cor}(i, j)$. Coordinates describes point P_{ijcor} on Fig.5.

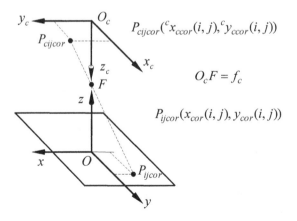

Fig.5. The coordinates x of points P_{ijcor} and P_{cijcor}

In order to calculate coordinates $x_{cor}(i,j)$ and $y_{cor}(i,j)$, coordinates $x_{ccor}(i,j)$, $y_{ccor}(i,j)$, $z_{ccor}(i,j)$ of point P_{cijcor} and x_F, y_F, z_F of focal F in reference system are necessary. Equations (13) and (14) describes that coordinates.

$$\begin{bmatrix} x_{ccor}(i,j) \\ y_{ccor}(i,j) \\ z_{ccor}(i,j) \\ 1 \end{bmatrix} = \mathbf{T}_c \begin{bmatrix} {}^{c}x_{ccor}(i,j) \\ {}^{c}y_{ccor}(i,j) \\ 0 \\ 1 \end{bmatrix}, \tag{13}$$

$$\begin{bmatrix} x_F \\ y_F \\ z_F \\ 1 \end{bmatrix} = \mathbf{T}_c \begin{bmatrix} 0 \\ 0 \\ f_c \\ 1 \end{bmatrix}. \tag{14}$$

Equations (15) describes straight line connecting points P_{cijcor}, F and P_{ijcor}.

$$\frac{x_{cor}(i,j)-x_F}{x_{ccor}(i,j)-x_F} = \frac{y_{cor}(i,j)-y_F}{y_{ccor}(i,j)-y_F} = \frac{z_{cor}(i,j)-z_F}{z_{ccor}(i,j)-z_F}. \tag{15}$$

After substitution $z_{cor}(i,j)=0$, equation (16) is evaluated, that describes searching coordinates.

$$x_{cor}(i,j)=-k_f[x_{ccor}(i,j)-x_F)]+x_F, \quad y_{cor}(i,j)=-k_f[y_{ccor}(i,j)-y_F)]+y_F, \tag{16}$$

$$k_f = \frac{z_F}{z_{ccor}(i,j)-z_F}.$$

Accuracy of determination coordinates of point in the template can be describe by absolute values of differences $\Delta x(i,j)=|x(i,j)-x_{cor}(i,j)|$, $\Delta y(i,j)=|y(i,j)-y_{cor}(i,j)|$ or by the distances $\Delta r(i,j)=\sqrt{\Delta x(i,j)^2+\Delta y(i,j)^2}$. As a reminder: $x(i,j)=(i-7)\cdot 8\,mm$, $y(i,j)=(j-9)\cdot 8\,mm$. Results of calculation of maximal values $\Delta r(i,j)$ for three settings of camera are shown below.

For camera setting 1 from Fig.2 for coefficients (9a) $\max[\Delta r(i,j)]=\Delta r(6,14)=1.5785\,mm$, for coefficients (9b) $\max[\Delta r(i,j)]=\Delta r(6,14)=1.3462\,mm$, for coefficients (9c) $\max[\Delta r(i,j)]=\Delta r(6,14)=1.1716\,mm$ and for coefficients (9d) $\max[\Delta r(i,j)]=\Delta r(6,14)=1.0553\,mm$.

For camera setting 2 from Fig.2 for coefficients (10a) $\max[\Delta r(i, j)] = \Delta r(3,16) = 1.1693\ mm$, for coefficients (10b) $\max[\Delta r(i, j)] = \Delta r(3,16) = 1.1615\ mm$, for coefficients (10c) $\max[\Delta r(i, j)] = \Delta r(3,16) = 1.1579\ mm$ and for coefficients (10d) $\max[\Delta r(i, j)] = \Delta r(3,16) = 1.0909\ mm$.

For camera setting 3 from Fig.2 for coefficients (11a) $\max[\Delta r(i, j)] = \Delta r(12,2) = 1.4181\ mm$, for coefficients (11b) $\max[\Delta r(i, j)] = \Delta r(12,2) = 1.3686\ mm$, for coefficients (11c) $\max[\Delta r(i, j)] = \Delta r(12,2) = 1.3942\ mm$ and for coefficients (11d) $\max[\Delta r(i, j)] = \Delta r(12,2) = 0.8473\ mm$.

It is easy to observe, that coordinates calculation accuracy increases with the amount of optical error model coefficients (5a) and (5b). For each camera setting, the greatest errors appear when only one coefficient k_1 is accounted. On the other hand, the smallest errors are obtained when all five coefficients k_1, k_2, k_3, p_1 and p_2 are applied.

5   Summary

The studies shows, that accuracy of determining the coordinates of points on the plane using camera depends on the optical errors description. The more coefficients from k_1, k_2, k_3, p_1, p_2 is accounted in errors description $\Delta_*{}^c x_c(i, j)$ and $\Delta_*{}^c y_c(i, j)$, described by equations (5a,b), the smaller are coordinates errors $\Delta r(i, j)$ of points determined on the plane. These coefficients depends on camera setting and this is why we account this dependence in the process of determining them.

Studies presented here should be treated as preliminary step of designing vision system with more than one camera. In the system with several cameras, errors model coefficients of every camera in the setting in which it will work should be determined. Afterwards analysis of coordinates determination accuracy on the plane observed by these cameras is necessary to be performed. The greatest error from every camera in the system $\Delta r(i, j)$ determines the accuracy of whole system. Assume that the results presented in this work for three camera settings of one camera are results of three cameras each in one of these settings. In this case, the accuracy of such system of three cameras is determined by maximal from these errors: $1.0553\ mm$ (setting 1), $1.0909\ mm$ (setting 2), $0.8473\ mm$ (setting 3). Therefore the system can determine coordinates with accuracy about $1.1\ mm$.

If accuracy is insufficient, then such system should be treated as vision sensor used to approximate determination of points positions. Errors of position can be reduced by moving camera closer to earlier roughly determined point. It can be made by mounting the camera at the robot manipulator.

Acknowledgement. This publication was supported by the Human Capital Operational Programme and was co-financed by the European Union from the financial resources of the European Social Fund, project no. POKL.04.01.02-00-209/11.

References

1. Zawisza, M.: Vision System for Position the Objects in 3-D Space. M.Sc. Dissertation. Institute of Automatic Control, Faculty of Automatic Control, Electronic and Computer Science. Silesian University of Technology, Gliwice, Poland (2013) (in Polish)
2. Beyer, H.A.: Geometric and Radiometric Analysis of CCD-Camera Based Photogrammetric Closed-Range System. Dissertation, No. 9701. ETH, Zurich (1992)
3. Google: The Calibraction of Camera. etacar.put.poznan.pl/marcin Kiełczewski (2013) (in Polish)
4. Chesi, G., Garulli, A., Vicino, A., Cipolla, R.: On the Estimation of the Fundamental Matrix: A Convex Approach to Constrained Least-Squares. In: Vernon, D. (ed.) ECCV 2000. LNCS, vol. 1842, pp. 236–250. Springer, Heidelberg (2000)
5. Golub, G.H., Van Loan, C.F.: Matrix Computations, 3rd edn. The Johns Hopkins University Press (1996)
6. Szkodny, T.: The Algorithm *Camera* Computing the Object Location. In: Lee, J., Lee, M.C., Liu, H., Ryu, J.-H. (eds.) ICIRA 2013, Part I. LNCS (LNAI), vol. 8102, pp. 637–648. Springer, Heidelberg (2013)
7. Szkodny, T.: Foundation of Robotics. Silesian University of Technology Publ. Company, Gliwice (2012) (in Polish)
8. Craig, J.J.: Introduction to Robotics, 2nd edn. Addison Wesley Publ. Comp. (1986)

Localization Using Vision-Based Robot

Yeol-Min Yun, Ho-Yun Yu, and Jang-Myung Lee

Department of Electrical and Computer Engineering, Pusan National University,
South Korea
{yeolmin7379,hoyun7379,jmlee}@pusan.ac.kr

Abstract. The robot recognizes the position for SLAM (Simultaneous Localization And Map-Building) There are a number of ways, many of the methods recognize the correct position to fuse the various sensors and filter used. There are a number of ways for localization for SLAM of the robot. Some methods use the fused sensor or filter for accurate localization. In this paper, Robot localization using one fixed camera system is proposed. Robot localization using color-based CAMShift algorithm which method of object tracking. The image processing uses the OpenCV library.

Keywords: CAMShift, Vision, Localization, OpenCV.

1 Introduction

Recently Study about SLAM (Simultaneous Localization And Map-Building) has been actively studied. The autonomous driving robot should know the location of current position and the perception of the surrounding environment. The way to find the position of the robot, there are several types of dead-reckoning, RFID, GPS, etc. However, these methods are used in fusion with other sensors because the errors in the calculation process and the surrounding environment occur.

In this paper, the position of the robot was estimated by image processing using one fixed vision system. The present day image processing techniques are used in factory automation, military, medical, remote sensing, surveillance and security in many applications. The image processing was used the OpenCV library. The OpenCV library is widely used in the field of image processing got recognition as processing speed and reliability. The OpenCV library is composed of a function of more than 500, the function of which is usefully employed in the field of various computer vision.

This paper is organized as follows. Chapter 2 is a configuration of the mobile robot system, and CAMShift algorithm. Chapter 3 has verified the validity by experiment. And chapter 4 describes conclusions and future study.

2 Robot Localization Using CAMShift Algorithms

2.1 System Configuration

In this paper, the system of mobile robot was configured for localization of mobile robot. Fig.1 is a whole system block. Control of mobile robot is used the Bluetooth and remote control. Image processing for localization of mobile robot was used the

X. Zhang et al. (Eds.): ICIRA 2014, Part II, LNAI 8918, pp. 285–289, 2014.

OpenCV library. The vision camera was used Logitech C301. MCU of mobile robot was used LM3S8962 (ARM).

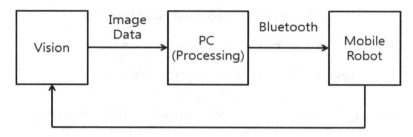

Fig. 1. Whole system

2.2   CAMShift Algorithm

Typical color-based method using image processing techniques has to be MeanShift algorithm and CAMShift algorithm. The MeanShift algorithm is based on the density distribution of the data set to ROI at high speed object tracking. And, the initial search region size and position, repeationg the color data is determined based on the boundary. The object of interest is extracted.   The initial ROI (Region Of Interest) is fixed, the center of ROI moved to the center of data after calculating the center of the ROI and the center of the data. At this time, the resulting vector is MeanShift vector. The new data of interest has been given a ROI interest is given to the movement of the center of the ROI. The central position of the new data of interest and the center position of the new ROI until the same, repeat this process. But, when tracking a target object is smaller than the ROI for the reasons to use a window, and so as to unnecessary operations, the algorithm has the drawback of not suitable for real-time systems. The designed algorithm is CAMShift algorithm. Fig. 2 shows that localization of the robot for CAMShift algorithm.

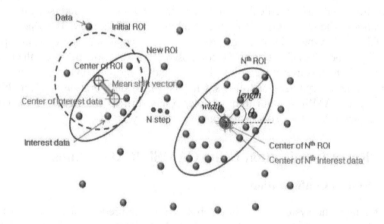

Fig. 2. CAMShift image processing for region of interest

This problem complements the algorithm is designed for real-time systems is CAMShift algorithm. The CAMShift algorithm is the same as a way of tracking MeanShift algorithm CAMShift algorithm is the same as MeanShift algorithm based on the density distribution of the data set that center point of ROI center correspond with point of data of interest through an iterative process. But There is difference considering the size and rotation of the ROI when specifying ROI. This allows the operation of the unnecessary area is removed and the amount of computation is reduced advantages.

2.3 Robot Localization Using CAMShift Algorithms

RGB (Red, Green, Blue) color model is sensitive to light and lighting. Therefore, by using color-based CAMShift algorithm comes to the performance of the robot localization problem may arise. Accordingly, RGB model, instead of the light-resistant HSV models need to be used.   Equation (1), RGB model changes the HSV model.

$$H = \cos^{-1} \frac{\frac{1}{2}((R-G)+(R-B))}{\sqrt{(R-G)^2 + (R-B)(G-B))}}$$

$$S = 1 - \frac{3}{(R+G+B)}\min(R,G,B) \tag{1}$$

$$V = \frac{1}{3}(R+G+B)$$

If the selected ROI, to analyze the histogram of the color data from the camera in ROI, and ROI binarization after the back-projection of the histogram. The white data of binary image apply to CAMShift algorithm, as a result to find the center point of the target. Fig. 3 is result for localization using CAMShift algorithm.

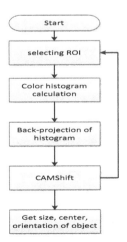

Fig. 3. Recognition of object using CAMShift algorithm

When using CAMShift algorithm has the accuracy. other SLAM method in LRF(Lase Range Finder), ultrasonic sensor and odometry(Encoderm, IMU sensor) require different filters and compensation.

3   Experiment

In this paper, a single sensor was used to find out the exact position of the robot using CAMShift algorithm and vision-camera. Fig. 4 shows the experimental environment. The tracked object has selected the mobile robot has been widely used in the robot.

Fig. 5 shows position data shown by the experimenter. Place the paste on the ceiling vision sensor (camera), and coordinates of the robot is measured by moving the robot after I want to track the ROI setting a desired robot.. An experimental result on average error of 2.76% has occurred. It is possible to recognize the precise location. However, because the vision system invisible situations, it is impossible to know the position of the robot. In addition the precision of data according to the frame of the image differ occurred.

Fig. 4. Experiment environment

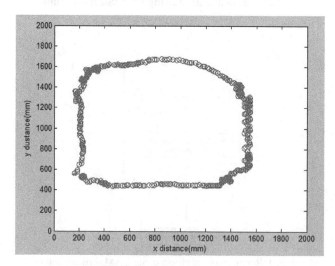

Fig. 5. Experiment Result

4   Conclusion

In this paper, that was localization of the robot using a fixed camera system. Localization of robot was used CAMShift algorithm in an object tracking method. Further, unlike the method of recognition of position, within a constant range, it is possible to know the exact position using a single vision-sensor.

In the future, the disadvantage for missing object recognizing of CAMShift algorithm will improve. And map-building using image processing will study. Also, I will try to apply the other robot besides a mobile robot. Comparison with other position recognition method analysis would proceed.

Acknowledgments. "This research was supported by the MOTIE (Ministry of Trade, Industry & Energy), Korea, under the Industry Convergence Liaison Robotics Creative Graduates Education Program supervised by the KIAT (N0001126)."

"This work was supported by the National Research Foundation of Korea(NRF) Grant funded by the Korean Government(MSIP) (NRF-2013R1A1A2021174) ."

References

1. Hwang, Y.S., Kim, H.W., Lee, D.J., Lee, J.M.: 3D Map Building for slopes based on modeling of Mobile Robot. In: 2014 IEEE International Conference on Indestrial Technology (IEEE ICIT 2014). Haeundae Grand Hotel, Busan (2014)
2. Park, J.H., Lee, J.M.: Beacon selection and calibration for the efficient localization of a mobile robot. Robotica 32(01), 115–131 (2014) ISSN 0263-5747
3. Lee, J.H., Kim, H.S.: A Study of High Precision Position Estimator Using GPS/INS Sensor Fusion. Journal of The Institute of Electronics Engineers of Korea 49(11) (2012)
4. OpenCV, Open source Computer Vision library,
   http://opencv.willowgarage.com/wiki/

Motion Estimation Algorithm for Blurred Image Using Gradient Filter

Qinghua Lu[1], Wei Wei[2], and Weirong Kong[2]

[1] School of Mechanical and Electric Engineering, Foshan University,
Guangdong 528000, China
[2] School of Mechanical and Electrical Engineering,
Jiangxi University of Science and Technology, Jiangxi 341000, China
qhlu@fosu.edu.cn

Abstract. Motion estimation plays an important role in image processing and machine vision. Motion blur has always been considered as degradation on images in existing majority of motion estimation methods. In this paper, a motion estimation technique for blurred image is proposed directly using motion blur images. The motion constraint model for blurred images is constructed based on alpha channel, and the gradient of alpha channel for blurred image is calculated using the filter method to improve the estimator accuracy. Experiment results show the proposed motion estimation approach can offer good estimator performance for blurred image in term of accuracy.

Keywords: Motion estimation, Gradient filter, blurred image.

1   Introduction

Motion image analysis and motion information estimation is becoming more and more popular in image processing and machine vision. The main objective of motion estimation based on images is to estimate accurately the scene or motion according to motion images as possible. In the last two decades, computational vision researchers have proposed a number of algorithms for motion estimation, but majority of motion estimation methods is based on unblurred images [1][2][3][4][5]. In such the methods, to minimize the effect of motion blur on motion estimation, these general techniques employ stroboscopic illumination to "freeze" the measured object or use high-speed imaging device to shorten exposure time. However, the equipment cost or the measurement complexity is increased using these two eliminating motion blur methods.

At present, motion blur has always been considered as degradation on images and little has been done to exploit motion blur as a visual cue for image motion estimation [6], then the image blurred is eliminated by such as filter method and the object motion is further estimated. In reference [7], a space-variant image restoration scheme for interlaced scan images is proposed by estimating motion blur parameters. Dash and Majhi presents an approach of motion parameters estimation for motion blur

X. Zhang et al. (Eds.): ICIRA 2014, Part II, LNAI 8918, pp. 290–298, 2014.

images based on Gabor filter and neural network [8]. Yang et al. presents a blind image deconvolution algorithm for motion deblurring from a single blurred image by using bilateral filtering [9]. It is clearly that the complexity of the motion estimation method is higher, in which motion blur is viewed as degradation on images.

A long time ago, Prof. Burr proposed the motion-from-blur mechanism existing in biological vision [10]. The motion estimation can be performed directly using the blurred images according to the vision mechanism, and there is no need to provide the stroboscopic illumination or high-speed imaging device, thus the cost or complexity for motion estimation is reduced. Recently, several groups measured motion directly using motion blurred images and obtained good results. Xu et al. developed an approach of translational speed measurement based on motion blur information [11]. In reference [6], a method to realize the measurement of rotation parameters from a single blurred image is proposed. Dai and Wu presented motion estimation algorithm for blurred image based on alpha channel model information [12]. However, majority of existing motion estimation methods for blurred image offer poor performance in accuracy.

The paper focuses on the motion estimation directly using motion blurred images. The motion constraint model is constructed using alpha channel information, and the gradient of alpha channel information is computed based on the filter method to improve estimator accuracy.

2   Motion Constraint Model for Blurred Image

Generally, for a motion blurred image I_b, which it is get by the unblurred image I and kernel h, the motion blur constraint can be expressed [12],

$$\nabla I_b|_p \cdot b = I(p + \frac{b}{2}) - I(p - \frac{b}{2}) \ . \tag{1}$$

where p is any position in image, $b = (u, v)^T$ is the displacement vector and T is transpose operator. Here I_b is defined as $I_b = I*h$ and $*$ represents convolution. $\nabla I_b|_p = (\frac{\partial I_b}{\partial x}, \frac{\partial I_b}{\partial y})^T|_p$ represents the partial differential of the blurred image I_b in the x and y directions for any position p.

Equation (1) offers only one given blurred image I_b, and the $I(p + \frac{b}{2})$, $I(p + \frac{b}{2})$ and the displacement b are all unknown components, thus requires further assumptions or less degree of freedom to solve the two unknown components. However, the alpha channel technique applied successfully to image deblurring [13] can be used to reduce the degree of freedom for the motion blur constraint [12].

In such an alpha channel method, the image I can be viewed as a linear combination of foreground image F and background B based on an alpha channel, and the form can be given by,

$$I = \alpha F + (1 - \alpha)B . \tag{2}$$

If the F and B are assumed to be locally smooth, and combine the Equation (1) and Equation (2) according to $I_b = I * h$, then the expression is given by [12],

$$\nabla \alpha_b \cdot b = \alpha(p + \frac{b}{2}) - \alpha(p - \frac{b}{2}) . \tag{3}$$

where $\alpha_b = \alpha * h$ is α channel model of the blurred image I_b, $\nabla \alpha_b = (\frac{\partial \alpha_b}{\partial x}, \frac{\partial \alpha_b}{\partial y})$ represents the partial differential of α_b in the x and y directions. Note that the Equation (3) is equivalent to replacing I in Equation (1) by using the α channel model.

For the blurred image caused by target motion, the alpha values for most pixels are either 0 or 1, and the alpha values is considered to be 0 mostly when $\nabla \alpha$ is also 0. Thus, the Equation (3) can be expressed as follow where $\| \nabla \alpha_b \| \neq 0$ [12],

$$\nabla \alpha_b \cdot b = \pm 1 . \tag{4}$$

The Equation (4) is called the α channel motion constraint model for blurred image, and this model can simplify motion estimation for blurred image.

3   Motion Estimation for Blurred Image Using Gradient Filter

In Equation (4), when the motion parameters are estimated directly using the blurred image, one can see that there is need to obtain alpha channel α_b of the blurred image and calculate the gradient $\frac{\partial \alpha_b}{\partial x}$ and $\frac{\partial \alpha_b}{\partial y}$, here is $\nabla \alpha_b = (\frac{\partial \alpha_b}{\partial x}, \frac{\partial \alpha_b}{\partial y})$. In this paper, the blurred image α_b components are extracted automatically by applying the spectral matting method in reference [14]. Furthermore, the approach to calculating the gradient $\nabla \alpha_b$ has direct effect on the estimator accuracy for motion blurred image.

Researchers originally calculated the image gradients using the first-order difference [12] [15], but this approach provides inferior performance in accuracy for motion estimation. In this paper, the motion estimation is viewed as the signal processing, and the gradient of the alpha channel α_b for blurred image is calculated using the filter method. The gradient $\nabla \alpha_b$ can be estimated in the following way that the row and column of α_b for the blurred image are convolved by applying filters,

$$
\begin{cases}
\dfrac{\partial \alpha_b(x, y)}{\partial x} = \alpha_b(x, y) * g * f^T \\[2ex]
\dfrac{\partial \alpha_b(x, y)}{\partial y} = \alpha_b(x, y) * f * g^T
\end{cases}
\tag{5}
$$

where $\alpha_b(x, y)$ is alpha channel model of the blurred image I_b, g represents derivative filter, f is pre-smoothing filter, T represents transpose and $*$ represents convolution. The way of filter can be expressed as $filter(i) = [x_1, ..., x_i]$, such as it is 5 taps filter when $i = 5$. Such a method can be used to estimate the blurred image motion and no prior knowledge of the target is needed, thus it is available in all applications for the motion estimation for blurred image. One can note that the $\nabla \alpha_b$ values can be calculated as long as the appropriate filters are chosen in Equation (5). Then, the motion parameters can be further obtained by using the EM-like iterative algorithm and the Least Square method [12] according to the Equation (5).

4   Experiment and Analysis

In the experiment, the Central filter [17], the Timoner filter [3] and the first-order difference method [12] are chosen to calculate the gradient of the alpha channel α_b to analyze estimator performance of the proposed motion estimation approach for blurred image. At the same time, the estimator performance comparison of two different blurred images is carried out by using this proposed method. As shown in Fig.1, the first image is IC chip image which [16], and the second image is the classical Lena image [12]. Here, Fig.1 (a), Fig.1 (b) and Fig.1 (c) is unblurred IC image, IC chip blurred image and Lena blurred image, respectively. The IC chip blurred image is blurred by the uniform velocity motion of 10 pixels in the $\theta = -\pi/6$ direction, and the Lena blurred image is blurred with the uniform velocity motion of 15 pixels in the $\theta = \pi/2$ direction.

Furthermore, the threshold is set to 0.1 when the motion parameters are estimated using EM-like iterative algorithm in this experiment [12]. Additionally, $s = \sqrt{|u|^2 + |v|^2}$ and $\theta = \arctan \dfrac{u}{v}$ are used to represent the length and direction of displacement respectively. The tests are repeated 10 times to estimate the mean $\bar{s}$ and the standard deviation δ of each method.

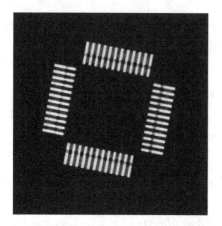

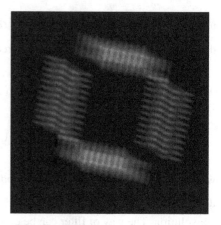

Fig.1(a) IC chip unblurred image   **Fig.1(b)** IC chip blured image

Fig. 1(c) Lena blurred image

Fig. 1. Blurred images

Fig.2 and Fig.3 compares the estimated displacement for the IC chip blurred image and Lena blurred image, respectively, where each different method is repeated 10 times. Here, First Diff , 5-Cen, 5-Tim, 7-Tim and 9-Tim in graph represent the methods that the gradient of the alpha channel α_b is calculated using the first-order difference, 5 taps Central filter, 5 taps Timoner filter, 7 taps Timoner filter and 9 taps Timoner filter, respectively. Additionally, the results of 10 times test are connected with the dash line for the convenience of display. From results shown in Fig.2 and Fig.3, one can see that the estimated displacement using the gradient filter method is more approach to the actual displacement than that of first-order difference method. Fig.2 and Fig.3 also shows the 9-Tim method with 9 taps Timoner filter has better estimator performance than others techniques.

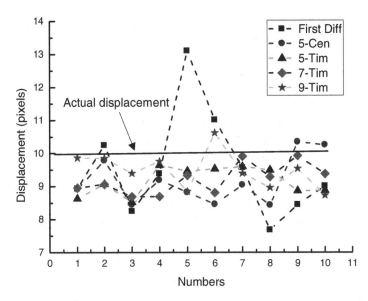

Fig. 2. Comparison of estimated displacement for the IC chip blurred image using the different methods

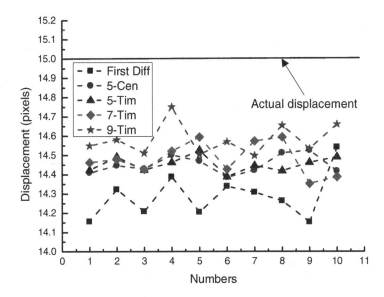

Fig. 3. Comparison of estimated displacement for the Lena blurred image using the different methods

Table 1 and Table 2 compares the mean $\bar{s}$ and the standard deviation δ of the estimated displacement using the methods in Fig.2 for IC chip blurred image and Lena blurred image, respectively, where each method is repeated 10 times. Table 1 and Table 2 show the standard deviation values applied the gradient filter methods are greatly less than that of the first-order difference technique. From standard deviation δ exhibited in Table 1 and Table 2, one can see that while the gradient filter methods can all offer good performance, the 5-Tim approach has smaller standard deviation. Furthermore, for the Lena blurred image, the δ value is less than 0.04 pixels for motions near 15 pixels.

Table 3 and Table 4 exhibit the comparison of the angular estimated using the methods in Fig.2 for IC chip blurred image and Lena blurred image, respectively. Similarly, Table 3 and Table 4 also show the gradient filter methods offer superior performance for angular estimation. At the same time, the angular bias of the 9-Tim method is smaller than that of others methods. In addition, for the Lena blurred image, the angular bias approaches 0.77 degrees in term of the angular motions with 90 degrees.

From the comparison of results shown in Table 1 and Table 2 as well as Table 3 and Table 4, one can see that the motion estimation method based on α channel motion constraint model for blurred image offers a dramatic improvement in estimator performance for the large pixels motions. Especially, the standard deviation of the 5-Tim method is smaller than 0.04 pixels for motions near 15 pixels and the bias of the 9-Tim technique achieves 0.77 degrees for the angular motion estimation with the 90 degrees. This implies that the proposed method can provide good estimator performance for motion estimation for blurred image.

Table 1. Comparison of the mean and the standard deviation of the estimated displacement for IC chip blurred image (pixels)

Type	First Diff	5-Cen	5-Tim	7-Tim	9-Tim
$\bar{s}$	9.5678	8.9232	9.1723	9.2084	9.5014
δ	1.4268	0.6298	0.3837	0.4077	0.5154

Table 2. Comparison of the mean and the standard deviation of the estimated displacement for Lena blurred image (pixels)

Type	First Diff	5-Cen	5-Tim	7-Tim	9-Tim
$\bar{s}$	14.28839	14.45176	14.4531	14.48218	14.58176
δ	0.10588	0.0458	0.0397	0.082	0.0675

Table 3. Comparison of the angular for IC chip blurred image (degrees/°)

Type	First Diff	5-Cen	5-Tim	7-Tim	9-Tim
Angular	-41.5	-36.3	36.3	34.3	34.1
Angular bias	11.5	6.3	4.3	4.1	2.1

Table 4. Comparison of the angular for Lena blurred image (degrees/°)

Type	First Diff	5-Cen	5-Tim	7-Tim	9-Tim
Angular	81.78	88.19	88.45	88.72	89.23
Angular bias	8.12	1.81	1.55	1.28	0.77

5   Conclusions

The motion estimation algorithm for blurred image using gradient filter is developed in this paper. The α channel motion constraint model for blur image is constructed. The motion estimation is considered as the signal processing, and the gradient of α channel for blurred image is calculated by using the filter method. The standard deviation is smaller than 0.04 pixels for motions near 15 pixels and the bias achieves 0.77 degrees for the angular motion estimation with the 90 degrees using the presented method. Experiment results show the proposed motion estimation method based gradient filter can provide good estimator performance for blurred image.

Acknowledgments. This research was supported by the National Natural Science Foundation of China (51105077), the Natural Science Foundation of Guangdong Province under grants S2011010001218 and the Science and Technology Innovation Project of Guangdong Educational Committee (2012KJCX0103).

References

1. Wei, Y.J., Wu, C.D., Dong, Z.L.: Sub-pixel Image Block Matching Based Measurement on Driving Characteristic of a Piezoelectric Actuator. Journal of Mechanical Engineering 46(147), 151–158 (2010) (in Chinese)
2. Jin, C.Y., Jin, S.J., Li, D.C., Wang, J.L.: Measuring In-Plane Micro—Motion of Micro-Structure Using Optical Flow. Trans. Tianjin Univ. 5(1), 9–22 (2009) (in Chinese)
3. Timoner, S.J., Freeman, D.M.: Multi-image gradient-based algorithms for motion estimation. Opt. Eng. 40(9), 2003–2016 (2001)
4. Chen, Z., Hu, X.D., Fu, X., Hu, X.T.: Measurement of MEMS In-plane Motion Based on Fuzzy Set and Wavelet Transformation Technique. Chinese Journal of Scientific Instrument 31(6), 1369–1373 (2010) (in Chinese)
5. Lu, Q.H., Zhang, X.M., Fan, Y.B.: Robust Multiscale Method for in-plane Micro-motion Measurement Based on Computer Microvision. Journal of Mechanical Engineering 31(6), 164–169 (2009) (in Chinese)
6. Li, Q., Wang, S.G.: Analysis on the Rotation from a Rotary Blurred Image Based on Active Visual System. Journal of Shanghai Jiao Tong University 42(2), 227–230 (2008) (in Chinese)
7. Zhao, Z.G., Cheng, S., Wang, G.D., Lu, H.X., Pan, Z.K.: Blind Restoration of Blurred Image Using Motion Estimation. Journal of Optoelectronics Laser 23(10), 2010–2015 (2012) (in Chinese)
8. Dash, R., Majhi, B.: Motion Blur Parameters Estimation for Image Restoration. Optik 125, 1634–1640 (2014)
9. Yang, H.L., Huang, P.H., Lai, S.H.: A Novel Gradient Attenuation Richardson–Lucy Algorithm for Image Motion Deblurring. Signal Processing (2014), http://dx.doi.org/10.1016/j.sigpro.2014.01.023
10. Burr, D.: Mohon Smear. Nature 284, 164–165 (1980)
11. Xu, T.F., Zhao, P.: Object's Translational Speed Measurement Using Motion Blur Information. Measurement 43, 1173–1179 (2010)
12. Dai, S.Y., Wu, Y.: Motion from Blur. In: IEEE Conference on Computer Vision and Pattern Recognition, Anchorage, Alaska, USA, June 24-26 (2008)

13. Jia, J.Y.: Single image motion deblurring using transparency. In: IEEE Conference on Computer Vision and Pattern Recognition, Minneapolis, MN, USA, June 17- 22 (2007)
14. Anat, L., Alex, R.A., Dani, L.: Spectral Matting. IEEE Transactions on Pattern Analysis and Machine Intelligence 30(10), 1–14 (2008)
15. Horn, B.K.P., Schunk, B.G.: Direct methods for recovering motion. International Journal of Computer Vision 2, 51–76 (1988)
16. Wang, S.G., Guan, B.Q.: Measurement of Sinusoidal Vibration from Motion Blurred Images. Pattern Recognition Letters 9(28), 1029–1040 (2007)
17. Barron, J.L., Fleet, D.J., Beauchemin, S.S.: Performance of Optical Flow Techniques. International Journal of Computer Vision 12(1), 43–77 (1994)

Design of a Stereo Handheld Camera Tool
for Single Port Laparoscopy

Kai Xu[1], Zhengchen Dai[1], Bo Feng[2], and Jiangran Zhao[1]

[1] RII Lab (Lab of Robotics Innovation and Intervention), UM-SJTU Joint Institute,
[2] The Affiliated Ruijin Hospital,
Shanghai Jiao Tong University, Shanghai, 200240, China
{k.xu,zhengchen.dai,zjr318}@sjtu.edu.cn, fengbo2022@163.com,

Abstract. SPL (Single Port Laparoscopy) might bring improved surgical outcomes but this new procedure needs effective and functional next-generation surgical tools. Although several robotic systems for SPL have been constructed, only manual SPL tools are currently used in clinical SPL procedures. This paper presents the design, construction and experimental characterizations of a stereo handheld camera tool with integrated illumination for SPL. The camera tool can be folded for the insertion into abdomen through a Ø12mm trocar. Besides providing 3D visualization and illumination of the surgical scene, it can also be unfolded to spare an additional access port for other surgical tools. The actual effectiveness of this camera tool could be further gauged in a surgical setting with other manual SPL tools.

Keywords: Single Port Laparoscopy, handheld tool, stereo camera tool.

1   Introduction

SPL (Single Port Laparoscopy) uses one skin incision (e.g., the umbilicus) for laparoscopic interventions [1]. Compared with traditional multi-port laparoscopy, SPL could bring better surgical outcomes, including lower complication rates, less postoperative pain, shorter hospitalization and better cosmesis under a similar setting [2]. Although the newly introduced NOTES (Natural Orifice Translumenal Endoscopic Surgery) procedures [3] might lead to less invasiveness, NOTES is still quite far away from large scale clinical trials, even assisted by the special tools [4] or the robotic systems [5-11].

Using the newly developed SPL instruments, including the TriPort (Advanced Surgical Concepts), the SILS port (Covidien), the RealHand tools (Novare Surgical Systems), the Cambridge Endo instruments, etc., surgeons have found SPL a viable choice over traditional multi-port laparoscopy [12].

Although robotic systems for SPL [13-17] could be used to ease the difficult hand-eye coordination, these robotic systems are often associated with regulatory hurdles and high costs. Manual SPL tools are cheaper and could prevail more easily even though surgeons need to go through additional trainings.

X. Zhang et al. (Eds.): ICIRA 2014, Part II, LNAI 8918, pp. 299–310, 2014.
© Springer International Publishing Switzerland 2014

This paper hence presents the design of a stereo handheld camera tool with integrated illumination for SPL, as shown in Fig 1. While folded, the tool has a cylindrical laparoscope form with an outer diameter of 12mm. This form facilitates its insertion into abdomen through a Ø12mm trocar. The camera tool can then be unfolded to spare an access port as well as provide 3D visualization and illumination of the surgical scene. Other manual SPL tools could be deployed through this additional access port.

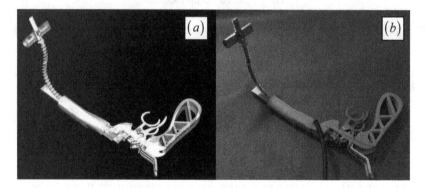

Fig. 1. The stereo handheld camera tool for SPL: (a) CAD model, and (b) prototype

Main contributions of this paper include i) the foldable structure of the camera tool which spares an additional access port for other instruments, ii) the incorporation of a continuum arm for camera head positioning, and iii) the use of mirrors in the camera head which allows the installation of two Ø7.8mm camera chips inside the Ø12mm camera head.

The paper is organized as follows. Section 2 presents the design goals and the overview. Detailed descriptions of the camera tool are presented in Section 3. Section 4 reports the experimental characterization of the camera tool with the conclusions followed in Section 5.

2   Design Goals and Overview

Attempting to address the increasing needs for SPL tools, this paper proposes the design of a foldable stereo handheld camera tool.

When the camera tool is in its folded configuration, the distal part of the tool possesses a Ø12mm cylindrical form so that it can be easily inserted into abdomen through a skin incision (e.g., the umbilicus). After insertion, the camera tool can then be unfolded to spare an additional access port as well as provide 3D visualization and illumination of the surgical scene.

A proposed use of the stereo camera tool could be seen from Fig. 2. The camera tool has unfolded itself and the access port spared inside the camera tool is big enough to pass a standard laparoscopic tool (e.g. the Ø5mm Harmonic scalpel from Ethicon Inc. or other Ø5mm tools). The access port could also be used for water or gas. Two needle-like laparoscopic instruments (e.g. the MiniLap tools from Stryker or the Mini-Laparoscopy tools from Karl Storz) can be inserted into the patient's abdomen for tissue manipulations.

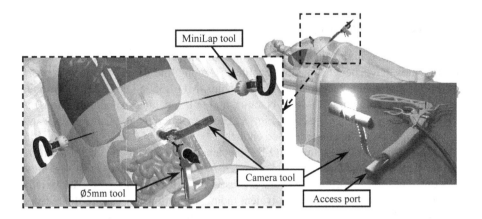

Fig. 2. The proposed use of the handheld camera tool

Considering the intended application of this camera tool, the conceived CAD model and the constructed prototype are shown in Fig. 1. The design's features can be highlighted as follows.

- The distal part of the camera tool only possesses an outer diameter of 12mm in its folded configuration. This is inspired by the Ø12mm endoscopic robot as in [9, 11].
- The use of mirrors allows the installation of two Ø7.8mm camera chips inside the Ø12mm camera head. Relatively big camera chips could bring better image qualities.
- An imaging distance of 100mm to 150mm from the access port to the objects is achieved.
- A 3-DoF continuum arm is incorporated to orient and position the camera head in order to achieve dexterous viewing perspectives of the surgical site.
- Illumination using LEDs are integrated.

Section 3 presents the detailed components descriptions, including the arrangement of the camera chips, the design of a pivoting mechanism, dimension determination of the continuum arm, and the design of the tool handle. Experimental characterizations are reported in Section 4.

3   Components Descriptions

Design of the handheld camera tool include i) the camera head with two camera chips, ii) the pivoting mechanism for the camera head, iii) the continuum arm, iv) the illumination design, and v) the design of the tool handle. These components are described here with sufficient details.

3.1   Design of the Camera Head

It's desired to use high quality camera chips for the camera head. Usually bigger camera chips have better imaging qualities. Among the available products from several major suppliers of miniature camera chips, MO-B3506 (MISUMI Electronics Corp, 640×480 resolution, 50dB S/N ratio, and 0.1 Lux minimal illuminations) is selected, as shown in the inset of Fig. 3. The camera chips at smaller sizes can't provide imaging qualities that are good enough.

In order to fit the two Ø7.8mm MO-B3506 camera chips inside the Ø12mm camera head for stereo visualization, an axial arrangement of the two camera chips were used. As shown in Fig. 3(a), two mirrors are used to reflect the images. The axially arranged camera chips now have equivalent radial orientations because of the mirrors. The distance between the two mirror centers is 11mm and the distance between the two camera chips is 20mm. If γ_1 (the angle between the two mirrors) is equal to 90°, the two equivalent camera chips have parallel axes. By adjusting γ_1, γ_2 (the angle between the two equivalent camera chip axes) can be varied as in Eq. (1).

$$\gamma_2 = 2\gamma_1 - 180° \ . \tag{1}$$

A series of experiments were conducted to adjust the γ_1 angle for better 3D perception for human, as shown in Fig. 4(c). The anaglyph views were shown to several human subjects and they picked one that looked the most natural. The γ_1 angle was then chosen to be 92°.

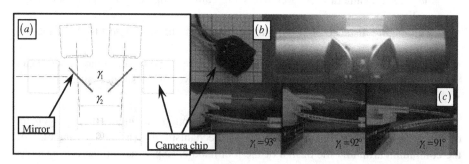

Fig. 3. Camera chip arrangement in the camera head: (a) the schematic, (b) the prototype, and (c) experiments to adjust the camera chip axes

3.2   A Pivoting Mechanism for the Camera Head

The mirrors in the camera head are axially arranged. The camera head has to be rotated for 90°, in order to generate a normal stereo view. A pivoting mechanism is designed as in Fig. 4 to rotate the camera head about the pivoting screw to transform the distal part of the tool from the cylindrical form into the unfolded configuration.

As shown in Fig. 4(a), the actuation block can be pushed and pulled by the actuation rod to move along the pin guide. Outer shape of the actuation block is fabricated to make sure that the camera head is entirely within the Ø12mm cylindrical shape when the camera head is folded. The actuation rod is a Ø0.4mm super-elastic nitinol. When the actuation block is driven, it pushes a driving pin to rotate the camera head between 0° and 90°, as shown in Fig. 4(b).

FlexPCB strips were used to replace the camera chips' original wires so that the wires will not get tangled during the rotating motions of the camera head. As shown in Fig. 4(c), the FlexPCB strips were carefully patterned and arranged inside the camera head so that the rotation of the camera head is not affected by the FlexPCB strips.

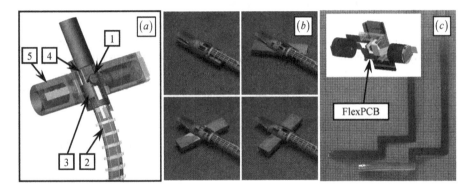

Fig. 4. The pivoting mechanism for the camera head: (a.1) pivoting screw, (a.2) actuation rod, (a.3) actuation block, (a.4) pin guide, (a.5) camera head, (b) the motion sequence, and (c) arrangement of the FlexPCB

3.3   Design of the Continuum Arm

A continuum arm is incorporated so that the camera head can be better positioned and oriented when the handle is constrained by the inserted extra tool though the access port. As shown in Fig. 5, the continuum arm consists of i) actuation rods, ii) a super-elastic nitinol strip, iii) a fixation ring, and iv) several spacers.

The fixation ring is attached at the middle of the nitinol strip. The nitinol strip between the fixation ring and the arm entrance is referred to as segment #1. The nitinol strip between the camera head and the segment #1 is referred to as segment #2.

Two actuation rods are connected to the fixation ring and can slide in the spacers' holes. Pulling of the actuation rod would bend the segment #1 upwards. Another actuation rod is connected to the camera head and can slide in the spacers. Pushing this actuation rod would bend the segment #2 downwards.

The third DoF (Degree of Freedom) of the continuum arm is the translational feed of the segment #1.

Bending shapes of the segments #1 and #2 are assumed to be circular according to the previous studies as in [18, 19]. Using this assumption, some derivations could be carried out to calculate the segments' desired lengths.

The preferred viewing distance of the arm entrance (the access port) ranges from 100mm to 150mm. When the segment #1 is assumed for a 90° bending, the following can be derived.

$$
\begin{cases}
\overline{OC} = \dfrac{2L_1}{\pi} + \dfrac{L_2}{\vartheta}(1 - \cos\vartheta) \\[2mm]
\overline{BC} = \dfrac{2L_1}{\pi} + \dfrac{L_2}{\vartheta}\sin\vartheta \\[2mm]
\overline{CD} = \dfrac{\overline{BC}}{\tan\vartheta} = \dfrac{2L_1}{\pi}\dfrac{\cos\vartheta}{\sin\vartheta} + \dfrac{L_2}{\vartheta}\cos\vartheta
\end{cases}
\tag{2}
$$

$$
\overline{OD} = \overline{OC} + \overline{CD} = \dfrac{2L_1}{\pi}\left(1 + \dfrac{\cos\vartheta}{\sin\vartheta}\right) + \dfrac{L_2}{\vartheta} \ .
\tag{3}
$$

If the segment #2 bends from 0° to 90°, it is preferred for the viewing direction to point towards the middle of the viewing range when $\vartheta = 45°$. Then Eq. (3) gives:

$$
\overline{OD} = \dfrac{4}{\pi}(L_1 + L_2) \rightarrow 125mm \ .
\tag{4}
$$

Many L_1 and L_2 values satisfy Eq. (4). $L_1 = 60mm$ and $L_2 = 40mm$ are used.

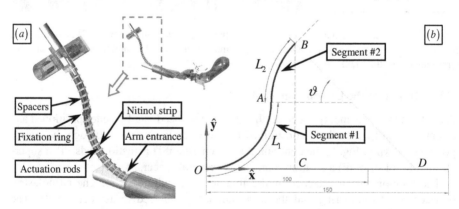

Fig. 5. The continuum arm: (a) structure and (b) dimension calculation

3.4   Onboard LEDs for Illumination

Six LEDs were planned for the illumination of surgical scenes as shown in Fig. 6(a). A FlexiPCB strip was used to mount and power the LEDs as in Fig. 6(d).

Temperature rise could be a concern while using LEDs for illumination. A series of experiments were carried out to investigate the potential heating problem. The LEDs were powered at different voltages in an indoor environment and temperatures of the LEDs were measured.

The LED has a rated voltage of 2.95v. It was found that powering the LEDs at this voltage will always cause heating problems. Then it was decided to power the LEDs at 2.70v, since no heating problems were identified under this voltage.

Fig. 6. Illumination tests with LEDs powered at (a) 2.50v, (b) 2.60v and (c) 2.70v; (d) FlexPCB

3.5   Design of the Handle

The handle is used to hold and manipulate this camera tool. In order to avoid interferences when an additional surgical tool is inserted through the access port, the handle is designed to be horizontally placed as shown in Fig.2 and Fig. 11.

The camera head of the designed handheld tool needs three actuation inputs: i) two inputs for the bending of the segments #1 and #2 of the continuum arm, and ii) one input for the pivoting mechanism.

The three inputs are realized by the triggers #1 ~ #3 as shown in Fig. 7. The trigger #1 actuates two actuation rods of the segment #1; the trigger #2 actuates the actuation rod of the segment #2 by pushing the driving pin which is guided by the actuation slider; and the trigger #3 actuates the actuation rod of the pivoting mechanism. These actuation rods are routed inside the channels in the handle that was fabricated using 3D printing.

All the three triggers have a ratchet-like feature so that the actuation rods could be locked in position. The tooth profiles on the triggers are carefully designed such that the actuation rods will be pushed or pulled approximately 1mm per tooth. Three spring-loaded releases (#1 to #3) could be pressed by a user's thumb to release these triggers and allow the actuation rods to be released.

The use of the handle is explained in Section 4.1.

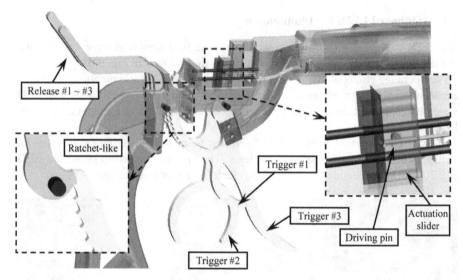

Fig. 7. Design of the tool handle

4   Ex-Vivo Experimental Characterization

After the handheld camera tool is fabricated and assembled, several experiments were conducted to characterize the features of this tool.

4.1   Tool Deployment

The handheld camera tool shall be inserted into abdomen in its folded form as shown in Fig. 8(a). The distal part of the tool has a cylindrical outer shape. The tool could be actuated to unfold itself using the three triggers and then provide illumination and 3D visualization of surgical scenes. This deployment is as follows.

After insertion into abdomen, the handle can further extend the continuum arm through the tube, as shown in Fig. 8(b). Then the trigger #1 can be pulled to bend the segment #1 of the continuum arm upwards. And the trigger #2 can be pushed to bend the segment #2 of the arm downwards. The trigger #3 can be pulled to rotate the camera head to provide a horizontal stereo view of the surgical site. The rotation motion of the camera head could also be seen from Fig. 4(b). The camera tool is then fully deployed as shown in Fig. 8(c).

The LEDs would be powered on to provide illumination. The triggers and the releases can be used together to adjust the continuum arm to a desired pose.

Other surgical tools can be later deployed through the access port spared within this handheld tool. Since the direction of the access port could be constrained by the newly inserted tool, the continuum arm allows dexterous positioning and orienting of the camera head by adjusting the triggers and the releases.

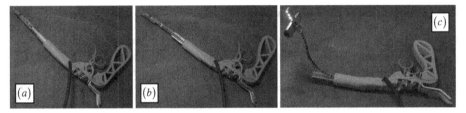

Fig. 8. Tool deployment

4.2   Shape Identification of the Continuum Arm

The dimensions of the continuum arm were calculated in Section 3.3 based on an assumption that both the segment #1 and the segment #2 have circular bending shapes. The experiments reported here were conducted to verify this assumption.

The experimental setup is shown in Fig. 9(a). An optical tracker (Micron Tracker SX60 from Claron Technology Inc.) was used with a pointer. The pointer was pointed at different positions along the continuum arm. The tracker could directly provide the coordinates of the pointer's tip.

The coordinates of the points along the continuum arm were recorded, transformed to $\{\hat{\mathbf{x}}, \hat{\mathbf{y}}\}$ as defined in Fig. 5, and plotted in Fig. 9(b). Two serially connected circular arcs approximate the bent segment #1 and the bent segment #2, whereas a circular arc with a straight line segment approximates the bent segment #1 and the straight segment #2.

The experimental results clearly indicate that the actual shapes of the segments can be well approximated by circular arcs.

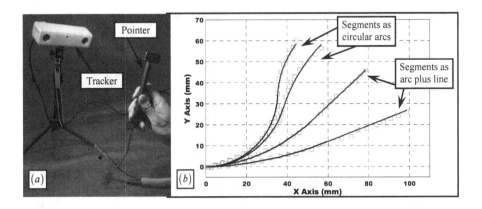

Fig. 9. Shape identification of the tool's continuum arm

4.3   Calibration of the Camera Chips

In order to improve the imaging quality, the camera chips might be calibrated.

The calibration was implemented using the Camera Calibration Toolbox for Matlab. The toolbox uses existing algorithms [20]. As shown in Fig. 10(a), several photos were taken for a calibration board. The corners were repeatedly detected to obtain the distortion correction coefficients. The distortion can be visualized as shown in Fig. 10(b). The calibration was conducted for both camera chips. The parameters were obtained, including the focal lengths, the principal points, the skew coefficient, the distortion coefficients, etc.

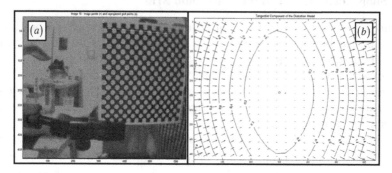

Fig. 10. Camera chip calibration: (a) calibration board, and (b) distortion visualization

4.4   Ex-vivo Trial of the Handheld Camera Tool

The camera tool was also tested in an ex-vivo trial. As in Fig. 11(a), an abdomen under pneumoperitoneum was mimicked by a box. The camera tool was deployed and another tool was inserted. The scene was visualized with illumination provided by the LEDs. The anaglyph view assembled from both camera chips is shown in Fig. 11(b).

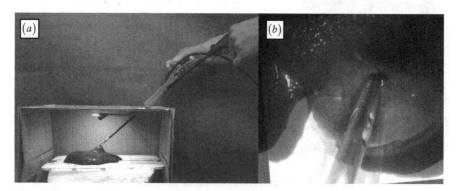

Fig. 11. Ex-vivo trial of the camera tool: (a) setup, (b) anaglyph view

5   Conclusions and Future Work

This paper presents the design, construction, and experimental characterizations of a stereo handheld camera tool with onboard illumination for SPL.

The distal part of the camera tool can be folded into a Ø12mm cylindrical form for the insertion into abdomen. Then it can be actuated to unfold itself using three triggers, providing illumination and 3D visualization of surgical scenes. An additional access port can also be spared inside the tool to insert other surgical tools.

Several sets of experiments were conducted to demonstrate the functionality and characteristics of this handheld camera tool. Particularly, the camera tool successfully visualized a mockup surgical scene using its integrated LEDs. An anaglyph view was generated and another surgical tool was inserted through the spared access port to perform tissue manipulations.

The future work shall focus on improving the reliability and sterilizability of this design so that the tool can be truly gauged in a more realistic setting.

Acknowledgments. This work was supported in part by the National Natural Science Foundation of China (Grant No. 51375295), in part by the Shanghai Rising-Star Program (Grant No. 14QA1402100), and in part by the Shanghai Jiao Tong University Interdisciplinary Research Funds (Grant No. YG2013MS26).

References

1. Navarra, G., Pozza, E., Occhionorelli, S., Carcoforo, P., Donini, I.: One-Wound Laparoscopic Cholecystectomy. British Journal of Surgery 84(5), 695 (1997)
2. Podolsky, E.R., John-Dillon, L., King, S.A., Curcillo II, P.G.: Reduced Port Surgery: An Economical, Ecological, Educational, and Efficient Approach to Development of Single Port Access Surgery. Surgical Technology International (20), 41–46 (2010)
3. Kalloo, A.N., Singh, V.K., Jagannath, S.B., Niiyama, H., Hill, S.L., Vaughn, C.A., Magee, C.A., Kantsevoy, S.V.: Flexible Transgastric Peritoneoscopy: A Novel Approach to Diagnostic and Therapeutic Interventions in the Peritoneal Cavity. Gastrointestinal Endoscopy 60(1), 114–117 (2004)
4. Hu, B., Chung, S., Sun, L.C.L., Kawashima, K., Yamamoto, T., Cotton, P.B., Gostout, C.J., Hawes, R.H., Kalloo, A.N., Kantsevoy, S.V., Pasricha, P.J.: Eagle Claw II: A Novel Endosuture Device that Uses a Curved Needle for Major Arterial Bleeding: A Bench Study. Gastrointestinal Endoscopy 62(2), 266–270 (2005)
5. Abbott, D.J., Becke, C., Rothstein, R.I., Peine, W.J.: Design of an Endoluminal NOTES Robotic System. In: IEEE/RSJ International Conference on Intelligent Robots and Systems (IROS), San Diego, CA, USA, pp. 410–416 (2007)
6. Phee, S.J., Low, S.C., Sun, Z.L., Ho, K.Y., Huang, W.M., Thant, Z.M.: Robotic System for No-Scar Gastrointestinal Surgery. The International Journal of Medical Robotics and Computer Assisted Surgery 4(1), 15–22 (2008)
7. Lehman, A.C., Dumpert, J., Wood, N.A., Redden, L., Visty, A.Q., Farritor, S., Varnell, B., Oleynikov, D.: Natural Orifice Cholecystectomy using a Miniature Robot. Surgical Endoscopy 23(2), 260–266 (2009)

8. Xu, K., Zhao, J., Geiger, J., Shih, A.J., Zheng, M.: Design of an Endoscopic Stitching Device for Surgical Obesity Treatment Using a N.O.T.E.S Approach. In: IEEE/RSJ International Conference on Intelligent Robots and Systems (IROS), San Francisco, CA, USA, pp. 961–966 (2011)

9. Xu, K., Zhao, J., Shih, A.J.: Development of an Endoscopic Continuum Robot to Enable Transgastric Surgical Obesity Treatment. In: Su, C.-Y., Rakheja, S., Liu, H. (eds.) ICIRA 2012, Part I. LNCS (LNAI), vol. 7506, pp. 589–600. Springer, Heidelberg (2012)

10. Tortora, G., Salerno, M., Ranzani, T., Tognarelli, S., Dario, P., Menciassi, A.: A Modular Magnetic Platform for Natural Orifice Transluminal Endoscopic Surgery. In: Annual International Conference of the IEEE Engineering in Medicine and Biology Society (EMBC), Osaka, Japan, pp. 6265–6268 (2013)

11. Zhao, J., Zheng, X., Zheng, M., Shih, A.J., Xu, K.: An Endoscopic Continuum Testbed for Finalizing System Characteristics of a Surgical Robot for NOTES Procedures. In: IEEE/ASME International Conference on Advanced Intelligent Mechatronics (AIM), Wollongong, Australia, pp. 63–70 (2013)

12. Rao, P.P., Rao, P.P., Bhagwat, S.: Single-incision Laparoscopic Surgery - Current Status and Controversies. Journal of Minimal Access Surgery 7(1), 6–16 (2011)

13. Piccigallo, M., Scarfogliero, U., Quaglia, C., Petroni, G., Valdastri, P., Menciassi, A., Dario, P.: Design of a Novel Bimanual Robotic System for Single-Port Laparoscopy. IEEE/ASME Transactions on Mechatronics 15(6), 871–878 (2010)

14. Sekiguchi, Y., Kobayashi, Y., Tomono, Y., Watanabe, H., Toyoda, K., Konishi, K., Tomikawa, M., Ieiri, S., Tanoue, K., Hashizume, M., Fujie, M.G.: Development of a Tool Manipulator Driven by a Flexible Shaft for Single Port Endoscopic Surgery. In: IEEE / RAS-EMBS International Conference on Biomedical Robotics and Biomechatronics (BIOROB), Tokyo, Japan, pp. 120–125 (2010)

15. Lee, H., Choi, Y., Yi, B.-J.: Stackable 4-BAR Manipulators for Single Port Access Surgery. IEEE/ASME Transaction on Mechatronics 17(1), 157–166 (2012)

16. Xu, K., Goldman, R.E., Ding, J., Allen, P.K., Fowler, D.L., Simaan, N.: System Design of an Insertable Robotic Effector Platform for Single Port Access (SPA) Surgery. In: IEEE/RSJ International Conference on Intelligent Robots and Systems (IROS), St. Louis, MO, USA, pp. 5546–5552 (2009)

17. Ding, J., Goldman, R.E., Xu, K., Allen, P.K., Fowler, D.L., Simaan, N.: Design and Coordination Kinematics of an Insertable Robotic Effectors Platform for Single-Port Access Surgery. IEEE/ASME Transactions on Mechatronics 18(5), 1612–1624 (2013)

18. Xu, K., Simaan, N.: Analytic Formulation for the Kinematics, Statics and Shape Restoration of Multibackbone Continuum Robots via Elliptic Integrals. Journal of Mechanisms and Robotics 2(011006), 1–13 (2010)

19. Webster, R.J., Jones, B.A.: Design and Kinematic Modeling of Constant Curvature Continuum Robots: A Review. International Journal of Robotics Research 29(13), 1661–1683 (2010)

20. Zhang, Z.: Flexible Camera Calibration by Viewing a Plane from Unknown Orientations. In: IEEE International Conference on Computer Vision (ICCV), Kerkyra, pp. 666–673 (1999)

A Robust and Precise Center Localization Method for the Strip Target on Microscopic Image

Jindong Yu[1,2] and Xianmin Zhang[1]

[1] South China University of Technology, Guangdong Province Key Laboratory of Precision Equipments and Manufacturing Technology,Wushan Road 381 GuangZhou, China
[2] Guangdong Construction Vocational Technology Institute, Guanghua 2nd Road 638, GuangZhou, China
yujindong@gdcvi.net, zhangxm@scut.edu.cn

Abstract. The center localization of strip is a key to ensure the accuracy of sub-micron line width measurement based on optical microscopic image. The Gaussian function with appropriate scale is used as kernel to convolve with the strip target on the microscopic image, which makes the profile of the strip become a ridge. By computing the moment of order 1st and 2nd, the position and the angle of the ridge line is obtained. The proposed center localization method is robust and precise for positioning the strip object in microscopic images. Experiments show that the novel method can not only avoid interference of noise but also adapt to changing light source. The repeat poisoning accuracy of the proposed method is reach up to 0.003 pixel, and the relative error is 1.5% for measuring the center distance of 10 μ m.

Keywords: Submicron measurement, Center localization, Ridge, Convolution, Moment.

1   Introduction

Sub-micron line width of the photolithography mask is the key size to restrict the quality of IC manufacturing. Measuring sub-micron line width based on computer microscopic image analysis technology belongs to the 2D photogrammetry, which is getting more and more applications [1, 4, 5, 15, 17, 18]. Microscopic image measurement has lots of advantages such as non-contact, convenient, fast, lower cost, etc. Imaging on CCD camera through a optical microscope, the geometry size or relationship between points and lines in a planar are converted into pixels' distance of feature point in the microscopic image. In order to measure the line width it is necessary to accurately survey both the actual distance represented by the pixel and the precise location of the feature point. The PRD (pixel representing distance) is determined by the system magnification factor and can be accurately calibrated by precisely poisoning the center of strip on the optical microscopic resolution power testing board, as shown in Fig.1(a) is the USAF1951 test board. At the same time, the center point along the very section of the submicron grade line of the

X. Zhang et al. (Eds.): ICIRA 2014, Part II, LNAI 8918, pp. 311–320, 2014.

photolithography mask, as shown in Fig.1 (b), is potentially helpful to localizing the edge due to the poor signal to noise ratio (SNR) of microscopic images. Thus, it is important to precisely localize the center point of the strip line in microscopic image.

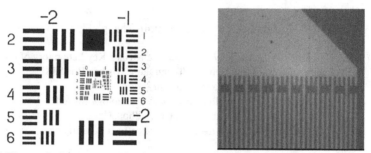

(a) The resolution power testing board (b) The microscopic image of the photo mask

Fig. 1. Strip lines on microscopic images

The strip target possesses large aspect ratio and characteristics of losing stepping edge for the linear object in images applications, such as biological medicine, remote sensing, industrial welding, facial organs recognition, airport runway recognition, construction or pavement crack detection, and so on. Because of several reasons the strip target in microscopic image has unique character of ramp edge with heavy noise. The size of the strip object is comparative with the point spread function of imaging system, under sampling makes the detail characteristics of the object, especially the edges, are aliasing and blurred. In addition, subjected to the space enlarging effect of the optical system the light intensity of objects converging on the CCD element is very weak, and narrow depth of field of high resolution optical objective lens is very easy to make image defocus and blurred. All these factors make most of edge pixels' gray level distribution transitioning slowly into ramp with heavy noise. Articles look on this special kind of edge as roof edge or ramp edge and research new edge detection algorithm [7, 8]. Petrou and Kittler modeled the ramp edge profile with the exponential function which implicitly represents the low-pass effect or bandwidth limitation of the imaging system [10]. ZQ Wang modified the exponential model as a bipolar one and proposed the expansion matching method for ramp edge detection [16]. These methods provide robust edge detector of the unique profile target but the center localization method is absent, especially for the strip in the microscopic image.

The grey value profile of the ideal roof edge is increasing firstly and then decreasing, of which the ridge can be denoted by the zero crossing point of the first directional derivative. Namely, the ridge of the roof edge profile is the center of the strip in special images and reflects the skeleton of targets. To describe the skeleton of objects the known literature proposed perfect algorithms of medial axes transform (MAT) of 2D planar region of any shape [2, 11, 13, 14]. The MAT method is based on the boundary contour which needs to obtain firstly the binary image and the edges. What's more MAT is mainly in order to describe and recognize specific target rather than caring the center poisoning precise. For the microscopic image the SNR is too low to detector edges accurately. Therefore it is necessary to study a novel precise and robust center localization method for the strip in microscopic image. The layout is as follows: The center localization method based on the moment of the ridge transformed from the convolved strip with Gaussian kernel is presented in section 2. Experiments and results show that the proposed method could get robust and precise efficiency in section 3 followed by the conclusion in section 4.

2   Center Locating Algorithm for the Strip Target in Microscopic Image

Because of the above mentioned factors the perfect edge detection algorithm is not work for non step type edge, while the center of the target is not affected and is easy to be localized. Thus, it is possible to measure precisely as long as finding appropriate center localizing method according to pixel gray level distribution. The result of convolved strip with a Gaussian kernel is proved to be a ridge. And the literature indicates the moment computes geometric feature with high precision. If we compute the moment based on the ridge of the object instead of the boundary, it would extract the precise centre of the strip conveniently with no need for binary image or segmentation. In order to construct the computational formula for the localization of the strip it is need to investigate thoroughly the pixel gray profile of the strip target.

2.1   The Strip Target and Its Pixel Gray Profile

Fig.2 presents the optical microscopic image of the resolution power testing board and the pixel gray level profile of the part marked by the red straight line. Due to the noise and the system characteristics the strip target has unique pixel gray profile which is looking like roof with noise. The threshold binary image of the strip has a lot of connected regions instead of only one symmetrical BLOB, and bigger threshold value introduces holes and smaller makes the edge to be out of strip shape.

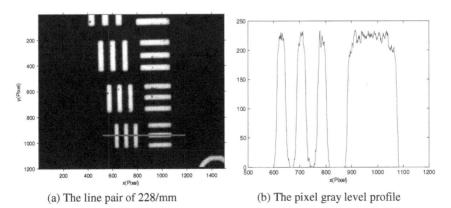

(a) The line pair of 228/mm   (b) The pixel gray level profile

Fig. 2. Pixel gray level profile of the strip in microscopic image

2.2   Transform the Pixel Gray Contour into a Ridge by the Gaussian Convolution Kernel

Through a convolution operation with Gaussian function **G** as kernel, the image **F** will be smoothed. The convolution operation is written in formula as

$$\mathbf{R} = \mathbf{F} * \mathbf{G} \tag{1}$$

Denoted as $R(x_c, y_c)$, the convolved result of any current pixel is calculated as

$$R(x_c, y_c) = \iint\limits_{(x_c - h, y_c - v)\in Neig(x,y)} f(x_c - h, y_c - v)\, g(h,v) dh dv \tag{2}$$

Where (x_c, y_c) is the coordinate of the current pixel. This means that any current pixel gray value $f(x_c, y_c)$ will be replaced by the weighted sum of all pixels in the neighborhood. Surrounding with the current pixel the neighborhood is defined as

$$Neig(x_c, y_c) = \left\{(x,y)\,\middle|\, x\in(x_c - w, x_c + w),\, y\in(y_c - w, y_c + w)\right\} \tag{3}$$

For the noisy pixel it will be averaged by the neighbor pixels, as a result the image will be smoothed. The Gaussian kernel is fit used as the weight function because of the isotropy. The Gaussian kernel takes the form of

$$G(h,v) = \lambda \exp(-(h^2 + v^2)/c^2) \tag{4}$$

Where h and v present the horizontal and vertical coordinates in the 2D domain of the kernel which are corresponding to the pixel in neighborhood; The parameter c denotes the scale factor of the kernel controlling the intension of smoothing effect; The parameter λ plays normalized role making $G(h,v)$ to satisfy equation：：

$$\iint\limits_{-w\le h\le w,\, -w\le v\le w} G(h,v) dh dv = 1 \tag{5}$$

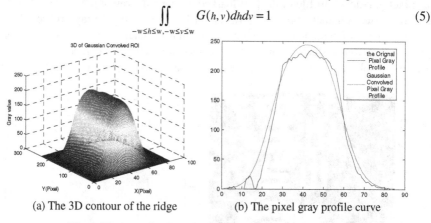

(a) The 3D contour of the ridge   (b) The pixel gray profile curve

Fig. 3. The convolved result of the strip by Gaussian kernel

Actually, except the smoothing effect, the result of the convolved image would be implying a ridge if the width of the neighborhood w can be selected approximately same as the width of the strip and the scale parameter c should be took quarter value of the w .The correlation between the strip and the kernel is biggest when the centers of them coincide. This means that the convolved result is inevitably a ridge and the ridge line denotes the center of the strip. As shown in Fig.3, (a) is the 3D ridge of the convolved strip with the Gaussian kernel scale c equating 4. Fig.3 (b) gives the comparing pixels contour curve orthogonal with the medial axis of the strip.

2.3   Moment Invariant Theory and the Physical and Geometric Meaning

The moment invariants theory was proposed by MK Hu in 1962 and have been frequently used to extract shape feature or recognize object [6]. Based on the method of moment it is easy to compute the geometry feature of the bounded ellipse of some BLOB, such as the center of mass, long axis and short axis, and spindle orientation angle[12]. YZ Niu implemented moment of the BLOB after the threshold segmentation and morphological filter to extract the centroid of the connected regions in microscopic image[9]. CC Chen proposed improved moment invariants based on chain code representation of a shape boundary which needs not all the information of shape boundary associated with the interior region[3]. The literature shows that moment invariants are useful for 2D shape analysis based on the boundary of object. According to MK Hu [6] the moment of order $p+q$ based on the border B of the binary image $b(i, j)$ is defined as

$$m_{pq}=\sum_{j\in B}\sum_{i\in B}i^{p}j^{q}b(i, j) \qquad (6)$$

Where $p+q=0,1,2,\cdots$, the moment of order zero represents the areas of some object, and the moment of order 1st is corresponding with the location of the center of the mass. The coordinates of the the center of the mass can be calculated as

$$\begin{cases} \overline{x}=m_{10}/m_{00} \\ \overline{y}=m_{01}/m_{00} \end{cases} \qquad (7)$$

The central moment of order $p+q$ is defined as

$$\mu_{pq}=\sum_{j\in B}\sum_{i\in B}(i-\overline{x})^{p}(j-\overline{y})^{q}b(i, j) \qquad (8)$$

Based on the central moment of order 2nd the spindle orientation angle of the bounded ellipse of some object can be calculated as

$$\theta=\arctan\left(2\mu_{11}/(\mu_{20}-\mu_{02})\right) \qquad (9)$$

2.4   Center Localization Algorithm Based on the Moment of the Ridge

Being the scale reasonably selected the Gaussian kernel convolves with the strip target which inevitably results in a ridge denoting the center line of the strip. The pixel gray value of the center part is larger than the marginal of the ridge. This will induce the center of the mass is coinciding with the centroid of the strip. According to the physical and geometric meaning, the center location of the ridge can be calculated by the moment of order 1st. Thus plugging the expression (2) into expression (6) and expression (8), the moment of the ridge instead of the boundary of binary image can become true. Defining the moment of order $p+q$ based on the ridge $\mathbf{R}$ as

$$m_{pq_{Ridge}} = \sum_j \sum_i i^p j^q R(i, j) \tag{10}$$

Then the centroid coordinates of the strip is defined as

$$\begin{cases} \bar{x}_{Ridge} = m_{10_{Ridge}} / m_{00_{Ridge}} \\ \bar{y}_{Ridge} = m_{01_{Ridge}} / m_{00_{Ridge}} \end{cases} \tag{11}$$

The central moment of order $p+q$ of the ridge is defined as

$$\mu_{pq_{Ridge}} = \sum \sum (i-\bar{x})^p (j-\bar{y})^q R(i, j) \tag{12}$$

Based on the central moment of second order the spindle orientation angle of the bounded ellipse of the strip can be defined as

$$\theta = \arctan\left(2\mu_{11_{Ridge}} / (\mu_{20_{Ridge}} - \mu_{02_{Ridge}}) \right) \tag{13}$$

Theorfore the novel method proides the center location of the strip as following steps :

(1) Select a zone including the strip target as the region of interesting (ROI).
(2) Implement the convolution operation for the ROI to get the ridge.
(3) Compute the moment of the ridge to get the center location and the orientation angle of the target.

3   Experiment and Result

Center localization algorithm for the strip can work stable and precisely which is the key to guarantee measurement accuracy based on microscopic images. In order to examine the robust and precise performance of the algorithm it is necessary to design several experiences adopting appropriate testing image.

3.1   The Robust Experiment of the Algorithm

The aerial photography of the plane parking on airport is used to test the robust of the center localization method which is taken from the SIPI image database of University of Southern California. Although there contains much of noise the method can locate the center of the target. As shown in Fig.4, (a) shows the original image on which the center localization result is denoted as the red star and the spindle orientation angle is indicated by the blue line. (b) shows the 3D pixel gray level profile of the plane which contains much of noise. (c) shows the 3D pixel gray level profile of the ridge which is the convolution integral of the Gaussian kernel with parameters $w=15$, $c=4$. (d) gives the axial orthogonal pixel gray curves of the original target and the ridge, denoted in blue and red respectively. By comparison the profile of ridge is smooth than the original one and.

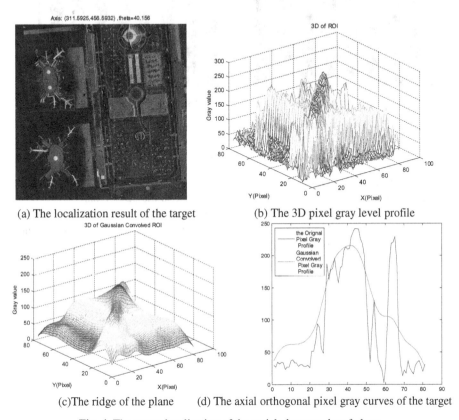

(a) The localization result of the target   (b) The 3D pixel gray level profile

(c)The ridge of the plane   (d) The axial orthogonal pixel gray curves of the target

Fig. 4. The center localization of the aerial photography of plane

For image metrology the changing light intensity is a sensitive factor. Thus two microscopic images shooting on the 100 line pairs per millimeter (LP/mm) under different light intensity are shown in Fig.5 (a) and (b). The magnification of the system is calibrated as 78.26 and the actual PRD is 56.2nm/pixel. Indicated by the red star and blue line, the center point coordinates and the orientation angle of the middle strip under week light is (736.3701,579.6014) (unit: pixel),89.2507, that of stronger light is (736.6544,579.7126), 89.9943. The result shows the biggest difference between two testing times is about 16nm which is satisfying robust to light source change for submicron measurement. Moreover there are also some other disturbing factors such as mechanical vibration and different ROI selected by mouse.

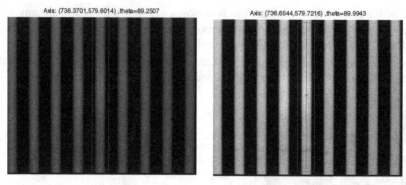

(a) The strip under week light source   (b) The strip under stronger light source

Fig. 5. The center localization of the strip shoot under different light intensity

3.2   The Precise Experiment of the Algorithm

In order to evaluate the precise of the center localization method, two kinds of optical microscopic resolution power test board are used. One is the 228 LP/mm of which two center points are 2/228=87711.93e-6 millimeters apart. Another is the 100 LP/mm of which two center points are 6/100=60000e-6 millimeters. Experimental measurement is executed 20 times and 10 respectively. The testing result is drawn into curves in Fig.6.

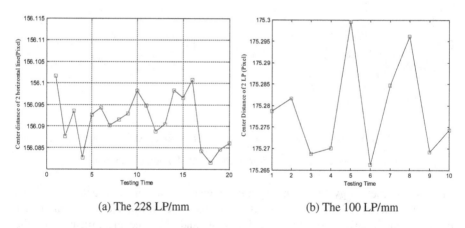

(a) The 228 LP/mm   (b) The 100 LP/mm

Fig. 6. The center distance of two line pairs by the proposed method

The measurement data is processed according to the t distribution. The degree of freedom for 228LP measurement $v = n - 1 = 19$, and that of 100LP is 9. The confidence level $p = 95\%$, the coverage factor $k = 2.093$. As a result the mean of the center distance of two strips on 228LP is 156.092 pixels with the expanded uncertainty $U = kS_{\bar{x}} = 2.093 \times 0.0013 = 0.003$ pixel. Use this result we get the pixel representing distance as:

$$PRD=8.7719/L = (0.0562 \pm 0.0002)\,\mu m/pixel\,.$$

The measured center distance of two strips on 100LP is calculated as $9.85\,\mu m$ while the nominal value is $10\,\mu m$. Thus the full relative error is about 1.5% which indicates that the accuracy of the center localization method can meet requirement.

Table 1.

Statistic items	Center pixel distance of 2 strips on 228 LP	Center pixel distance of 2 strips on 100 LP
Mean (pixel)	157.9476	175.279
Expanded uncertainty (pixel)	0.003	-
PRD (μm/ pixel)	0.0562±0.0002	-
Measured distance (μm)	-	9.85±0.04
Relevant error rate	-	1.5%

4   Conclusion

The center localization of strip in microscopic image is a key to ensure the accuracy of image measurement. Based on convolution integral of the strip target with appropriate scaled Gauss kernel and the compute of moment the robust and precise method is proposed which full the blank of poisoning the center of unique object in microscopic image and others, for example the aerial photography. Experiments show that the novel method is avoided of interferences induced by both noise but also the changing light intensity of source. The center localization expanded uncertainty is as small as 0.003 pixels and the measurement accuracy is 1.5% for the size of $10\,\mu m$.

Acknowledgments. The research was supported by the National Science Foundation of China (Grant No. 91223201, 50825504), the United Fund of Natural Science Foundation of China and Guangdong province (Grant No. U0934004), Project GDUPS (2010), and the Fundamental Research Funds for the Central Universities (2012ZP0004).

References

1. Bacus, J.W., Grace, L.J.: Optical microscope system for standardized cell measurements and analyses. Applied Optics 26, 3280–3293 (1987)
2. Cao, L., Liu, J.: Computation of medial axis and offset curves of curved boundaries in planar domain. Computer-Aided Design 40, 465–475 (2008)
3. Chen, C.C., Tsai, T.I.: Improved moment invariants for shape discrimination, pp. 270–280. International Society for Optics and Photonics, San Diego (1992, 1993)
4. Chen, Z., Liao, H., Zhang, X.: Telecentric stereo micro-vision system: Calibration method and experiments. Optics and Lasers in Engineering 57, 82–92 (2014)
5. Danuser, G.: Stereo light microscope calibration for 3D submicron vision. International Archives of Photogrammetry and Remote Sensing 31, 101–108 (1996)

6. Hu, M.-K.: Visual pattern recognition by moment invariants. IRE Transactions on Information Theory 8, 179–187 (1962)
7. Liang, K.-H., Tjahjadi, T., Yang, Y.-H.: Roof edge detection using regularized cubic B-spline fitting. Pattern Recognition 30, 719–728 (1997)
8. Lindeberg, T.: Edge detection and ridge detection with automatic scale selection. International Journal of Computer Vision 30, 117–156 (1998)
9. Niu, Y., Pan, J., Wang, X.: A method for extracting feature points of micro-calibration sample. In: Sixth International Symposium on Precision Mechanical Measurements, p. 89161F-89161F-89167. International Society for Optics and Photonics (2013)
10. Petrou, M., Kittler, J.: Optimal edge detectors for ramp edges. IEEE Transactions on Pattern Analysis and Machine Intelligence 13, 483–491 (1991)
11. Pinto, F., Freitas, C.: Fast medial axis transform for planar domains with general boundaries. In: 2009 XXII Brazilian Symposium on Computer Graphics and Image Processing (SIBGRAPI), pp. 96–103. IEEE (2009)
12. Prokop, R.J., Reeves, A.P.: A survey of moment-based techniques for unoccluded object representation and recognition. CVGIP: Graphical Models and Image Processing 54, 438–460 (1992)
13. Ramamurthy, R., Farouki, R.T.: Voronoi diagram and medial axis algorithm for planar domains with curved boundaries—II: detailed algorithm description. Journal of Computational and Applied Mathematics 102, 253–277 (1999)
14. Ramanathan, M., Gurumoorthy, B.: Constructing medial axis transform of planar domains with curved boundaries. Computer-Aided Design 35, 619–632 (2003)
15. Wang, B.Z., Liu, W.Y.: Camera calibration method for micro-image measuring system. Opto-Electronic Engineering 2, 030 (2006)
16. Wang, Z., Raghunath Rao, K., Ben-Arie, J.: Optimal ramp edge detection using expansion matching. IEEE Transactions on Pattern Analysis and Machine Intelligence 18, 1092–1097 (1996)
17. Yang, C.H., Han, Y., Liu, B.: Calibration Method for Micro-measurement System Based on Linear CCD. Nuclear Electronics & Detection Technology 12, 025 (2010)
18. Zhao, B., Asundi, A.: Micro-measurement using grating microscopy. Sensors and Actuators A: Physical 80, 256–264 (2000)

The Application of Independent Component Analysis Method on the Mura Defect Inspection of LCD Process

Xin Bi[1], Xiaoping Xu[1], and Jinhua Shen[2]

[1] College of Shanghai Medical Instrumentation,
University of Shanghai for Science and Technology, Shanghai 200093, China
[2] Shanghai Micro Electronics Equipment Co.,LTD., Shanghai 201203, China
bix@smic.edu.cn

Abstract. In the Mura defect inspection for TFT-LCD, the uneven brightness of image have directly influence to the inspection results. In order to adjust the brightness unevenness of LCD image, this paper proposed a new method which combining the homomorphic transform and the independent component analysis method. The homomorphic transform method transformed the multiplicative uneveness into additive one and then the independent component analysis method estimated and separated the mixed source signals and noise signals. The inverse homomorphic transform method estimated signals without noise, and the target image after brightness adjustment was gotten finally. The experiment results show that this method can restrain the brightness unevenness and the moire fringe of image and strengthen the defects.

Keywords: LCD, defect inspection, homomorphic transform, independent component analysis, brightness adjustment.

1   Introduction

In LCD industry, *mura* is visible imperfections appearing as local lightness variation with low contrast and blurry contour on an active LCD panel, which is a Japanese term meaning blemish and is typically lager than single pixels. Once the mura defects are inspected, these panels will be repaired or discarded in LCD process. Now, most of the defect inspections are currently performed by human observer according to the limit samples. But it is difficult to inspect fast and accurate because of the eye perception, environment and subjective factors of observers, and the objective quantification of mura defect is hard to be carried out. Especially, with the continually increasing of the LCD size and resolution, the manual inspection is more and more difficult, and the machine vision inspection has become the focus and nodus.

The mura defects in LCD images for machine vision inspection have irregular shape, blurry contour and low contrast, the images also have textured background and uneven brightness, which make the traditional edge inspection and threshold segmentation method hardly suits to the inspection. Especially, the global uneven brightness and the moire fringe due to frequency aliasing in image acquisition have many influences on the accurate judgments, although they are not defects. So, before

X. Zhang et al. (Eds.): ICIRA 2014, Part II, LNAI 8918, pp. 321–330, 2014.
© Springer International Publishing Switzerland 2014

the segmentation of mura defect, the adjustment of brightness uneveness is very important in order to avoid the error or omission.

In the study of LCD brightness adjustment, the method of histogram equalization [1],[2], background fitting [3],[4] and frequency filtering [5] are applied. The method of histogram equalization is based on the merging of the grey level of similar pixels, and the problems are over-enhancement and details missing. The background fitting method must solve the problem of the large amount of calculation for online inspection. In frequency filtering method, the uneven background and defect are looked as linear relationship, which amount to the linear high pass filter in frequency domain. The shortcoming of this method is the blurring effect in image boundary, and the type and parameters must be selected appropriately, so some prior information about the image features need to be known and it is hard to realize the automatic online inspection. The theoretical study about the eliminating method of moire fringe is rare, and the presented method is the interpolation [6], which also needs the prior information about the image frequency and sample frequency.

The image homomorphic filtering [7] is implemented by homomorphic transform via frequency filtering. This method can suppress the multiplicative noise, but the appropriate choice of filter is key and difficult. The independent component analysis (ICA) method is a newly developing signal processing method in the late 20th century. In this method, the observed signal is looked as the linear combination of some statistical independence signals, and the blind source separation of independent signal is conducted. But, as frequency filtering, the ICA method is based on the hypothesis of linear combination signals [8]. Weidong Jiao etal [9] proposed the signal homomorphic transform and blind source separation method in order to eliminate the multiplicative noise.

In this paper, the homomorphic transform and ICA method is combined aimed to the uneven brightness of LCD image. The homomorphic transform method transformed the multiplicative unevenness into additive one and then the ICA method estimated and separated the mixed source signals and noise signals. Thus, the image brightness adjustment, defect enhancement and moire fringe eliminating method without reference image and prior information for the LCD mura defect online inspection is studied.

2   The Uneven Brightness Analysis of LCD Image

In machine vision inspection, the uneven brightness of LCD images is mainly caused by three factors: uneven lighting, uneven material of LCD and the moire fringe caused by frequency aliasing.

2.1   Lighting Factors

The image acquisition in mura inspection needs the environment of strict lighting. For example, in SEMI standard [10], the environment lighting must be controlled bellow10lux, the precision is $100cd/m^2$, and the error is $\pm0.01cd/m^2$. Especially, for the whole LCD display, the even lighting is needed to avoid the uneven brightness of image.

2.2   Uneven Material of LCD

The LCD itself will be uneven of the whole display because of the uneven backlight or the uneven distribution of liquid crystal materials, which is also a main factor.

2.3   The Moire Fringe Caused by Frequency Aliasing

CCD image acquisition is a process of discrete sampling of continuous image signal. According to Shannon sampling theorem, it is not worth mentioning the aliasing effect of high frequency information on the image, if the brightness change of object is gentle in photography scene. But, if there is fine grid or texture in the scene, then the wavy stripe which is called moire fringe is easily formed in image by aliasing part. There are many pixels matrix in LCD panel which would form repetitive texture background in image and is easy to introduce the moire fringe while image acquisition is executed by CCD.

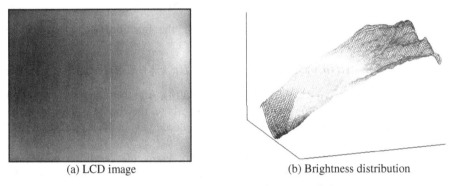

(a) LCD image   (b) Brightness distribution

Fig. 1. An example of LCD image with uneven lightness

Fig.1 (a) shows a LCD image, in which, the repetitive textured background is removed through filtering. And, the brightness distribution of this image is shown in fig.1 (b). From fig.1 (b), it can be seen that the brightness of the entire image nonuniformity is very obvious and it is difficult to distinguish the defect shape. There exits corrugated uneven brightness, which is the moire fringe, on the right side of the image border.

The noise in the image, according to its relationship with the image signal, can be divided into additive and multiplicative noise. The relationship between the additive noise and the image signal is linear, and the noise is unrelated to the image itself. The relationship between the multiplicative noise and image is multiplied, and the noise is associated with the image pixels or its neighborhood pixels. Through analysis of LCD, the uneven lighting factor is unrelated to the original image signal, so it belongs to additive noise. The factors of uneven material of LCD and the moire fringe are multiplicative noise because they are related to the original image. In current, through strict control of lighting online, the uneven lighting can be avoided as far as possible, thus it can not be considered temporary. In this paper, the uneven brightness of LCD image is treated as multiplicative noise considering the other two factors.

3   The Image Brightness Adjustment in Mura Inspection

3.1   Principle of Homomorphic Filtering

It can be assumed that the image $f(x, y)$ is determined by two factors: the illumination field $i(x, y)$ and the reflection coefficient field $r(x, y)$, and the former can be expressed as the product of the later two, which is given by:

$$f(x, y) = i(x, y)r(x, y) \tag{1}$$

In order to separate the illumination and reflectance fields, the logarithm transform is done to formula (1):

$$\ln f(x, y) = \ln i(x, y) + \ln r(x, y) \tag{2}$$

In formula (2), the multiplicative component is transformed to additive one, and the description in frequency domain can be simplified as:

$$G(u, v) = I(u, v) + R(u, v) \tag{3}$$

A transfer function $H(u, v)$ is used to dispose the function $G(u, v)$, it can be described in space domain:

$$S(u, v) = F^{-1}\{H(u, v)G(u, v)\} = F^{-1}\{H(u, v)I(u, v)\} + F^{-1}\{H(u, v)R(u, v)\} \tag{4}$$

It is amount to the frequency filtering in formula (4). And then the index operation is done to get the homomorphic filtering result. In where, the choose of $H(u, v)$ is the key to the filter design and the prior information is needed about the illumination field of image.

For the LCD inspection, the inspector can not know the condition of uneven distribution of LCD material, so the homomorphic filtering method hardly suits to the brightness adjustment in LCD online inspection.

3.2   Principle of Independent Component Analysis

$X=[x_1, x_2, \cdots, x_n]$ is the n random variables and $S=[s_1, s_2, \cdots, s_m]$ is m unknown independent components, $n \geqslant m$, then the relationship of the components between X and S can be described as:

$$X = AS \tag{5}$$

The best approximation of S can be got by solving formula (5). The basic principle of this method is that the solution matrix W is got only through the observed data X, under some statistical hypothesis without knowing the coefficient matrix A and the source matrix S. The goal is to make the components of the output matrix Y as independent as possible, and Y is the best approximation of S:

$$W = YS \tag{6}$$

The requirements of ICA are: (1). Number of the observed signals is greater than or equal to the number of source signals; (2). The source signals are statistically independent; (3). At most one source signal is Gauss.

Therefore, the first demand of ICA method applied in LCD inspection is that the linear statistical independence relationship between the original image and the uneven brightness component. It is not actual.

3.3 The Brightness Adjustment Method Based on Homomorphic Transform and ICA

In this paper, the overall uneven image brightness caused by material uneven of LCD itself and the moire fringe are both treated as multiplicative noise, denoted by L, so the image can be described as:

$$I = J \cdot L \tag{7}$$

In where, I is the acquisition image, J is the source image and L is the brightness error field. Logarithm is done to both sides of formula (7) at the same time, which is the homomorphism transform:

$$\ln I = \ln J + \ln L \tag{8}$$

Formula (8) can also be described as:

$$I' = J' + L' \tag{9}$$

In this way, the known image matrix is expressed as the linear combination of unknown source image and unknown brightness error field. As we known, the image signal is super Gauss distribution, and through homomorphic transform, the source image and the mixed brightness field are statistical independence. So, the ICA method can be used to the blind source separation. In this paper, the classic FastICA algorithm [11] is adopted.

In order to separate the two independent components of source image and brightness field, the number of observations must be greater than or equal to 2. The observed image can be described as follows with ICA:

$$[i_1' i_2', \cdots, i_m'] = A[s_1' s_2', \cdots, s_m'] \quad (m \geqslant 2) \tag{10}$$

In where, A is the unknown mixing matrix and is the independent source for separation. The solution matrix W is obtained using the FastICA algorithm. The estimator of s' is obtained by $\hat{S}' = WI'$ and the estimate independent source can be restored by $\hat{I}' = A\hat{S}'$. Then the exponential transformation $\exp(\hat{I}') = \hat{I}$ is done and the final separation image is gotten. In actual application, mixed multiplicative noise can be more than one.

4   Study on Simulation and Experiment of Brightness Adjustment in Mura Inspection

4.1   The Simulation Study

The simulation method is first used to verify the model proposed in this paper. All the simulation cases in this paper are carried out with the Matlab image processing toolbox.

Figure 2 (a) and (b) are respectively the uneven illumination fields L_1 and L_2 simulated with Matlab, and (c) is the original image: J. The three are multiplied to obtain an image I with uneven illumination, that is $I = J \cdot L_1 \cdot L_2$, which is shown in (d).

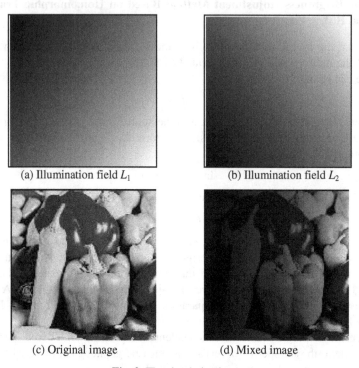

(a) Illumination field L_1 (b) Illumination field L_2

(c) Original image   (d) Mixed image

Fig. 2. The simulation image

In figure 3, the effects of brightness adjustment by the homomorphic transform, the ICA and the method proposed in this paper are compared. In which, (a) is the original image, and (b) ~ (d) are respectively the images adjusted by homomorphic filtering, ICA, and the proposed method. (e) ~ (h) are respectively the 3D brightness distributions of (a) ~ (d). It can be seen from the results that the homomorphic filtering method can reduce the uneven brightness to a certain extent, but the contrast is lowered and the adjustment effect is not ideal. For the ICA method, the overall brightness of the separation image is better adjusted, but there exists serious noise in the image, especially in the border of an object. The proposed method, that combining the homomorphic transform and ICA has an obvious superior effect to the other two. So, because of the multiplicative relationship between the uneven brightness signal and original image signal, a better adjustment effect can be gotten with ICA method after homomorphic transform.

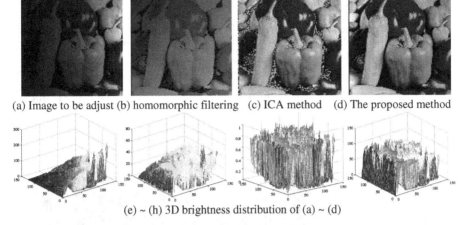

(a) Image to be adjust (b) homomorphic filtering   (c) ICA method   (d) The proposed method

(e) ~ (h) 3D brightness distribution of (a) ~ (d)

Fig. 3. Effects comparison of the proposed method, homomorphic filtering and ICA method

4.2   The Experiment Study

The LCD samples with Mura defects are collected and the experimental platform is set up on the basis of the requirements in SEMI standard. In this paper, the experimental platform is supplied by Philips Mobile Display Systems. For every LCD samples, the bright, 50% gray and dark states of images are captured, which form 3 observation images in ICA. In experiment, for each 2D image matrix, all of its rows are sequentially connected end to end to forming a one-dimensional vector. The three observation images are denoted by I_1, I_2, and I_3, the size are 800×600. Three row vectors which have the dimension of 1×480000 are formed after row scan, denoted by X_1, X_2 and X_3. These three vectors are combined into a 3×4800000 mixing matrix: $Mix=[X_1, X_2, X_3]^T$. The experimental process is shown in figure 4, and the second independent source is assumed as the main feature source.

Figure 5 shows the process of brightness adjustment experiment for a LCD sample, in where, (a) ~ (c) are respectively captured at the states of bright, 50% gray and dark and serving as three observation images in ICA. In (a), the moire fringe can be seen on the right border of image. (d) ~ (f) are the estimated three independent sources after the homomorphic transform and ICA. Through comparison, (e) is thought the most similar to original image and (f) is the moire fringe feature image. So (e) is selected to go through the inverse ICA transformation and exponential transformation, and the output image is obtained. The results are shown in figure 5 (d) ~ (f), which are the restored images of bright, 50% gray and dark state respectively. From figure 5 (g) ~ (i), it can be seen clearly that the moire fringes existing in the right border are eliminated, the uniformities of image background is increased to a certain extent and the defects in the three output images are all been enhanced.

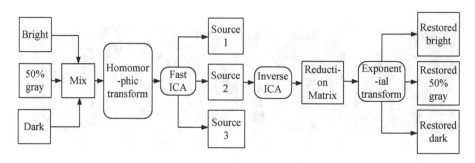

Fig. 4. Flow chart of luminance adjustment experiment for LCD images

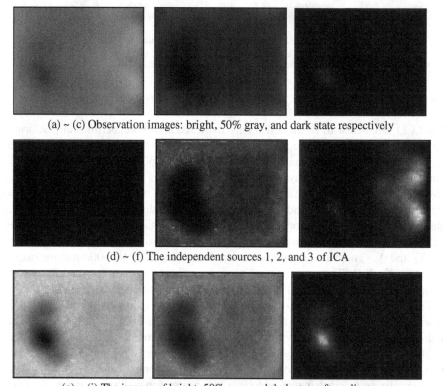

(a) ~ (c) Observation images: bright, 50% gray, and dark state respectively

(d) ~ (f) The independent sources 1, 2, and 3 of ICA

(g) ~ (i) The images of bright, 50% gray and dark state after adjustment

Fig. 5. The example of luminance adjustment results for LCD images

In figure 6, the 3D brightness distribution maps before and after adjustment are shown. In where, (a) ~ (C) are the distribution maps of input observation images at bright, 50% gray and dark state and (d) ~ (f) are the maps of output ones. It can be seen that there exist moire fringes on the border of the three original images before adjustment. Especially in (a), the moire fringe and the overall brightness fluctuation are obvious. Then after adjustment, the moire fringes in all the three images are basically eliminated and the overall uneven brightness of these images is improved visible. So, with the proposed method which combining the homomorphic transform and ICA, the moire fringe can be eliminated successfully and the overall uneven brightness is improved significantly, under the circumstances that the brightness distribution and the frequency aliasing condition are unknown.

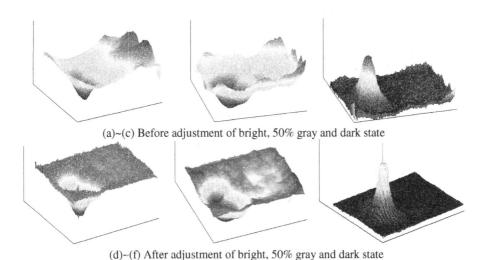

(a)~(c) Before adjustment of bright, 50% gray and dark state

(d)~(f) After adjustment of bright, 50% gray and dark state

Fig. 6. Luminance comparison of images before and after adjustment for the LCD example

5   Conclusions

In this paper, based on the analysis of the brightness uneven reasons of LCD image, the uneven caused by variety of factors are treated as the multiplicative illuminance deviation. By using homomorphic transform, the multiplicative brightness deviation is converted to statistically independent additive image signals. And then, the ICA method is used to the separation of linear mixed signals to remove the noise signals, following, the inverse ICA transform and the inverse homomorphic transform are carried on to the reserved signals, and finally, the target image after brightness adjustment is gotten. Through the simulations and experiments, the proposed method which combining the homomorphic transform and ICA is verified that can adjust the uneven brightness better and lightening the moire fringe caused by varieties of factors. This method dose not requires prior information about the spectrum of the original image and can be applied to the defect online inspection of LCD.

Acknowledgments. This work is supported by the Funding Scheme for Training Young Teachers in Shanghai Colleges under Grant 5114508104.

References

1. Taniguchi, K., Ueta, K., Tatsumi, S.: A mura detection method. Pattern Recognition 39(6), 1044–1052 (2006)
2. Kim, W., Kwak, D., Song, Y., et al.: Detection of Spot-Type Defects on Liquid Crystal Display Modules. Key Engineering Materials 270-273, 808–813 (2004)
3. Lee, J., Yoo, S.: Automatic Detection of Region-Mura Defect in TFT-LCD. IEICE Transactions on Information and Systems 87(10), 2371–2378 (2004)
4. Choi, K., Park, N., Yoo, S.: Image restoration for quantifying TFT-LCD defect levels. IEICE Transactions on Information and Systems 91(2), 322–329 (2008)
5. Chen, S., Chou, S.: TFT-LCD Mura Defect Detection Using Wavelet and Cosine Transforms. Journal of Advanced Mechanical Design, Systems, and Manufacturing 2(3), 441–453 (2008)
6. Wang, H.Z., Chen, J.P., Xu, J., et al.: Research on the disposal of Moire fringe in programmable imaging based on interpolation. Optical Instruments 29(5), 31–33 (2007)
7. Oppenheim, A., Schafer, R.: Homomorphic analysis of speech. IEEE Transactions on Audio and Electroacoustics 16(2), 221–226 (1968)
8. Wang, Z.L., Gao, J., Jian, C.X., et al.: An Independent Component Analysis based Defect Detection for OLED display. Advanced Materials Research 605-607, 724–728 (2013)
9. Jiao, W.D., Yang, S.X., Qian, S.X., et al.: Homomorphic transform based BSS algorithm for multiplicative noise reduction. Journal of Zhejiang University (Engineering Science) 40(4), 581–584 (2006)
10. SEMI D31-1102: Definition of Measurement Index (SEMU) for Luminance Mura in FPD Image Quality Inspection. SEMI 2002 (2002)
11. Hyvarinen, A., Oja, E.: A fast fixed-point algorithm for independent component analysis. Neural Computation 9(7), 1483–1492 (1997)

Modeling of Temperature Control for an Intraperitoneal Hyperthermic Chemotherapy Machine

Teng Wang[1], Xinjun Sheng[1,*], Jianwei Liu[1],
Qiumeng Yang[2], and Xiangyang Zhu[1]

[1] State Key Laboratory of Mechanical System and Vibration,
Shanghai Jiao Tong University 800 Dong Chuan Road, Shanghai 200240, China
[2] Ruijin Hospital, Shanghai Jiao Tong University School of Medicine
{wangteng,xjsheng,mexyzhu}@sjtu.edu.cn,
liujianwei-618@163.com, yangqiumeng@hotmail.com

Abstract. Intraperitoneal hyperthermic chemotherapy (IHCT) combines the efficacies of hyperthermia and chemotherapy on tumor treatment, which achieves satisfactory curative effects in treating abdominal cancer and decreasing the recurrence rate after operation. To design an IHCT machine according to treatment requirements, the temperature control system of chemotherapy liquid is analyzed in detail in this study. A mathematical model on liquid heating process is obtained using the system identification technique. The structure of this model is simple and the fitting accuracy surpasses 92%. The efforts in this study can provide a theoretical basis for liquid temperature control of IHCT.

Keywords: System modeling, System identification, Intraperitoneal hyperthermic chemotherapy, Temperature control.

1 Introduction

Intraperitoneal hyperthermic chemotherapy (IHCT) is a kind of loco regional treatment in which heated liquid with high drug concentration continues cycling after perfusion into patients abdominal cavity. The rationale for IHCT involves direct cytotoxicity of hyperthermia against malignant cells, synergistic effect of hyperthermia and chemotherapy drugs and the washing of peritoneal cavity. As IHCT can destroy free cancer cells in abdominal cavity as well as treat micrometastasis, IHCT has been widely reported in the treatment of gastric cancer, liver cancer, ovarian cancer and pancreas cancer, demonstrating that it has significant clinical effect on these diseases [1,2,3].

The optimal liquid temperature in IHCT ranges from 42.5°C to 45°C, according to a raft of clinical experiments and studies. The curative effect is not obvious with a temperature below 41°C. On the other hand, damage of normal

---

* The work is supported by the Science and Technology Commission of Shanghai Municipality (Grant No. 11441902100).

X. Zhang et al. (Eds.): ICIRA 2014, Part II, LNAI 8918, pp. 331–341, 2014.

tissue occurs with a temperature over 45°C [1]. Therefore, liquid temperature control is crucial in the application of IHCT. In the existing devices reported in [4,5,6], the techniques on liquid cycle, liquid heating and liquid temperature control are not adequate enough and the designs of liquid recycle are not good for practical usage, affecting the clinical effects, and the application of IHCT treatment.

In this paper, an IHCT machine including liquid perfusion and cycle pipe, liquid heating system, liquid temperature measurement system, liquid temperature flow control system and intelligent monitoring system is designed for satisfying the needs of IHCT treatment. A mathematical model of liquid temperature control is deduced by applying the system identification technique. The fitting accuracy of this model surpasses 92%, which is of significant importance for precious control of the liquid temperature.

2   Structure and Working Principle of IHCT Machine

2.1   Design of Pipe System for Liquid Circulation

The liquid circulation system is mainly made up of circulating pipe, pumping pipe, heating bag, collecting bag, three-limb tubes, filter and switches. Through connecting the pumping pipe, heating bag, collecting bag, filter and other parts to circulating pipe via luer taper, the route for liquid circulation is established, see Fig. 1.

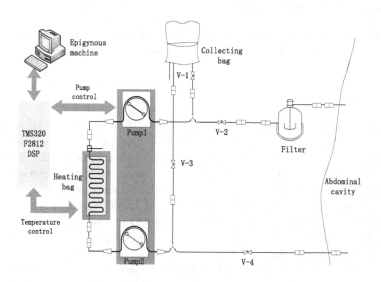

Fig. 1. Structure of IHCT machine system

The circulating pipe is made from medical polymer materials. A disposable aseptic filling bag is used as heating bag. Two peristaltic pumps installed on the pumping pipes generate power for liquid circulation. A well designed collecting bag with S conduit inside is used for liquid heating. In order to monitor liquid temperature in real time, temperature sensors are installed in three-limb tubes. A filter is used for filtering microembolus from the liquid. Four switches V1, V2, V3 and V4 in Fig. 1 are used for accommodating the whole liquid circulation pipe in different treatment phases. Liquid circulates in the pipe and patients abdominal cavity during the whole course of treatment, and pollution can be avoided.

2.2   Temperature Measurement System

Pt resistance temperature detector (RTD) is used for temperature measurement. Once the liquid temperature rises, the resistance of temperature detector Rt decreases linearly. Through connecting Rt in the temperature measurement circuit in Fig.2, the liquid temperature T can be converted to a voltage signal U_{AB} between dot A and B, which can be calculated according to equation(1).

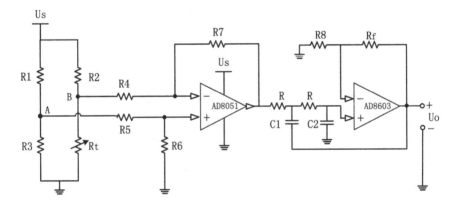

Fig. 2. Temperature measurement circuit

$$U_{AB} = \frac{R2 \cdot R3 - R1 \cdot Rt}{(R1 + R3)(R2 + Rt)} \cdot U_S. \qquad (1)$$

As the amplitude of U_{AB} is very small, an AD8051 chip is used as an amplifier. A second-order active low-pass filter made by AD8603 is used to eliminate high frequency interference. Within the working temperature range of 0-100°C, the range of voltage output U_o can be 2.21-0.5V,which decreases linearly with the temperature.

Voltage output of the temperature measurement circuit is acquired by a digital signal processor (DSP) TMS320F2812. TMS320F2812 consists of 16 channels of analogue to digital conversion, and the input voltage range of each channel is

0-3V, which is capable for sampling voltage from 0.5V to 2.21V. The analogue to digital conversion can be completed in 80ns with a voltage resolution of 12 bit, and the accuracy of temperature measurement can be 0.15°C accommodating with Pt resistance RTD.

2.3   Power Conditioning of Heating Board

A circuit for power conditioning of heating board is designed in Fig.3. The heating board and the bidirectional controlled silicon BTA16-600B are connected in serious to 220V AC power supply. PWM (Pulse Width Modulation) signal generated by TMS320F2812 is applied on the photoelectric coupler MOC3041, which works as an on-off switch in the circuit. By controlling the on-off of BTA16-600B, power of heating board can be adjusted.

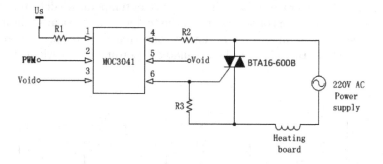

Fig. 3. Heating board power conditioning circuit

2.4   Intelligent Monitoring System on HMI Machine

The Human-machine Interface(HMI) machine with an intelligent monitoring system is connected to TMS320F2812, which works as the slave machine, through serial communication. The HMI machine receives and displays temperature data acquired by the slave machine and sends parameters like treatment phase, objective temperature, heating time to the slave machine. With this interaction, all phases in IHCT treatment can be fully completed.In the meantime, the temperature data is saved, and can be used for the effect evaluation after the IHCT operation.

3   System Modeling and Identification

3.1   Analysis and Modeling on Liquid Temperature Control System

The IHCT machine heats the liquid to an objective treatment temperature by tuning the power of heating board. During a treatment phase, the duty cycle of PWM signal generated by DSP processor changes automatically according to the deviation between the current temperature and objective temperature. By this way, temperature deviation is eliminated and the liquid temperature acheives the objective one. The current of heating board is outputted by MOC3041 chip according to the duty cycle of PWM signal, and the equivalent current I varies proportionally to the duty cycle D, then $I = Imax \cdot D, D \in [0,1]$, Where $Imax$ is a constant value.

The heat exchanging process between heating board and the liquid is illustrated in Fig.4.

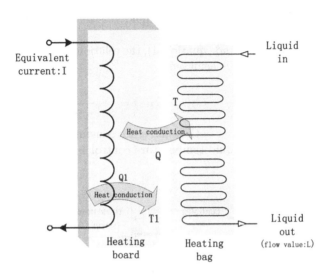

Fig. 4. Heat exchanging process between heating board and the liquid.

Heating board is considered to be a capacitive object. Once it was conducted, heat is generated by resistance wire inside the heating board. Temperature on the shell of heating board rises through heat conduction and heat radiation from the resistance wire. Considering the heat capacity, heating board can be modeled as a first-order inertial element [7,8], transfer function from current I to surface temperature T_1 on the heating board is:

$$G_1(s) = \frac{T_1(s)}{I(s)} = \frac{k_1}{\tau_1 s + 1}. \tag{2}$$

In the heating bag, liquid absorbs heat from the heating board via heat conduction. According to heat conduction functions, temperature variation of the liquid can be calculated in equation(3), without considering the heat conduction from heating board to the air, heating bag to the air and heat radiation loss.

$$\alpha A[T_1(t) - T(t)] = C\rho L \frac{dT(t)}{dt}. \tag{3}$$

In equation(3), T_1 is surface temperature of the heating board, T is the average temperature of liquid in heating bag, α is the coefficient of heat conduction between liquid and the heating board, A is the contact area between heating bag and the heating board, C is heat capacity of the liquid, ρ is the density of the liquid, and L is flow value of the liquid.

After Laplace transform, equation(3) is transformed to equation(4):

$$G_2(s) = \frac{T(s)}{T_1(s)} = \frac{1}{\tau_2 s + 1}. \tag{4}$$

Where $\tau_2 = \frac{C\rho L}{\alpha A}$, as the flow value of liquid L is set to a constant value, τ_2 is a constant value accordingly.

According to equation(2) and equation(4), the relationship between duty cycle D and liquid temperature T is achieved:

$$G(s) = \frac{T(s)}{D(s)} = G_1(s) \cdot G_2(s) = \frac{K_1}{ps^2 + qs + 1}. \tag{5}$$

Where $p = \tau_1 + \tau_2$, $q = \tau_1 \tau_2$, both p and q are constant value.

A block diagram on the liquid temperature control system of IHCT machine is illustrated in Fig.5:

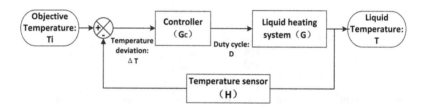

Fig. 5. Liquid temperature control system of IHCT machine

At the beginning of a treatment phase, an objective liquid temperature Ti is set on the HMI machine as an input to the liquid temperature control system. When the temperature deviation ΔT between Ti and the current liquid temperature T is detected by temperature sensor, the controller G_c converts ΔT to duty cycle D proportionally, then $\Delta T = K' \cdot D, K'$ is a constant value. Considering Pt resistance temperature sensor as a pure time delay component, the transfer function of liquid temperature control system is:

$$\frac{T(s)}{Ti(s)} = \frac{G(s) \cdot G_c(s)}{1 + G(s) \cdot G_c(s) \cdot H(s)} = \frac{K}{ps^2 + qs + 1} \cdot e^{-T_d s}. \tag{6}$$

In equation(6), parameters K, p, q, T_d are all undetermined constant values.

3.2   System Identification and Parameter Estimation

As the modeling of liquid heating process deduced above is based on simplification and approximation of the actual heat exchange process, parameters K, p, q, T_d in equation (6) can't be determined by measurement or referring to datasheets. System identification technique is applied in this part in order to estimate model parameters of liquid temperature control system with a high accuracy. System identification is a kind of black box modeling method, the model of an actual system can be achieved by the analysis of system input and output data based on a cost function. System identification can be divided into two categories: structure identification and parameter estimation, depending on whether the system structure is known beforehand or not [10].

Considering the model structure of the liquid temperature control system G_0 has been obtained above in this paper, parameters in this model can be estimated along a specific direction to decrease the cost function between G_0 and G via recursive algorithm, see Fig.6:

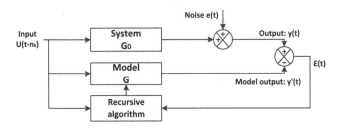

Fig. 6. A principle diagram on system identification

A general identification expression of the liquid temperature control system G_0 is:

$$A(q)y(t) = \frac{B(q)}{F(q)}u(t - n_k) + \frac{C(q)}{D(q)}e(t). \tag{7}$$

Where $y(t)$ is system output at time t, $u(t - n_k)$ is system input, $e(t)$ is white-noise disturbance value, n_k is time delay, q is a translation operator and polynomials $A(q)$, $B(q)$,$C(q)$, $D(q)$, $F(q)$ are defined as:

$$A(q) = 1 + a_1 q^{-1} + \ldots + a_{n_a} q^{-n_a}. \tag{8}$$

$$B(q) = b_1 + b_2 q^{-1} + \ldots + b_{n_b} q^{-n_b+1}. \tag{9}$$

$$C(q) = 1 + c_1 q^{-1} + \ldots + c_{n_c} q^{-n_c}. \tag{10}$$

$$D(q) = 1 + d_1 q^{-1} + \ldots + d_{n_d} q^{-n_d}. \tag{11}$$

$$F(q) = 1 + f_1 q^{-1} + \ldots + f_{n_f} q^{-n_f}. \tag{12}$$

Here n_a , n_b, n_c, n_d , n_f are the orders of the polynomials, a_{n_a}, b_{n_b}, c_{n_c}, d_{n_d}, f_{n_f} are the coefficients of the polynomials.

According to model structure of liquid temperature control system in equation(6), K, p, q, T_d are all time invariant parameters, therefore, polynomials$D(q)$ and $F(q)$ are both taken as 1 in equation(7). As for the model has a single input Ti and a single output T, which can be taken as a SISO(single input single output) system, ARMAX model can be used in this paper:

$$A(q)y(t) = B(q)u(t - n_k) + C(q)e(t). \tag{13}$$

Parameters a_{n_a}, b_{n_b}, c_{n_c} in ARMAX model can be calculated by using least square algorithm (LS-IV) to minimize the cost function, which is defined as follows :

$$V_N = \sum_{t=1}^{N} \epsilon^2(t). \tag{14}$$

Where N is the number of data samples, and the lager N sampled, the more accurate the model becomes.

The cost function minimization process is illustrated in Fig.7:

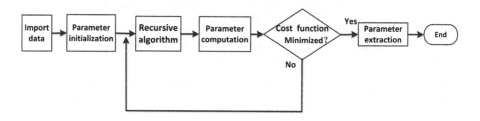

Fig. 7. Process of cost function minimization

Matlab system identification toolbox offers variety kinds of system identification models which can be applied in parameter estimation of the model of temperature control system by sampling system step response data.

3.3   Data Acquisition and Preprocessing

At the beginning of a treatment phase, a constant objective treatment temperature T_i is set as the system input. Liquid temperature acquired by the temperature measurement system is the system output T. Since liquid temperature is a kind of slowly varying signal, sample frequency of the temperature measurement system is set $4Hz$, and temperature data can be recorded via intelligent monitoring system on epigynous machine. As Initial liquid temperature T_0 and temperature rise differ in different experiments, data preprocessing method is used in order to eliminate the influence of different T_i and different temperature rise. Input data and output data are normalized in equation(15):

$$T' = \frac{T - T_0}{T_i - T_0}. \tag{15}$$

In equation(15), T' is the normalized temperature, T is liquid temperature acquired by the temperature measurement system, T_0 is initial liquid temperature and T_i is the objective treatment temperature. The normalized input and output data plotted by matlab are shown in Fig.8:

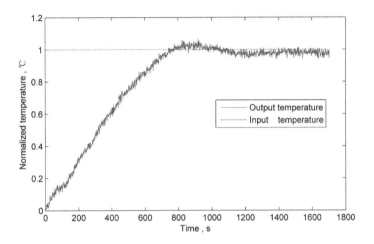

Fig. 8. Normalized input and output temperature

3.4   Parameter Estimation

Parameter estimation can be realised offline in matlab system identification toolbox. As the model structure of liquid temperature control system obtained above belongs to the second-order system with time delay, process models option is used and the model transfer function is defined as G' in equation(16):

$$G' = \frac{Ke^{-Tds}}{1 + (2ZetaTw)s + (Tws)^2}. \tag{16}$$

Where the value of proportional coefficient K is 0.96049, time constant Tw is 236.2703, damped coefficient $Zeta$ is 0.67048, time delay Td is 15. The loss function is 4.33337×10^{-4} and the final prediction error (FPE) is 4.34354×10^{-4}.

By comparing equation(8) with equation(16), parameters in equation(8) can be calculated: $K = 0.96049$, $p = Tw^2 = 5.58237 \times 10^4$, $q = 2ZetaTw = 316.82902$, $Td = 15$.

3.5   Model Verification

In order to verify how the model matches the actual system, step response of the actual system and the model is obtained and illustrated in Fig.9.

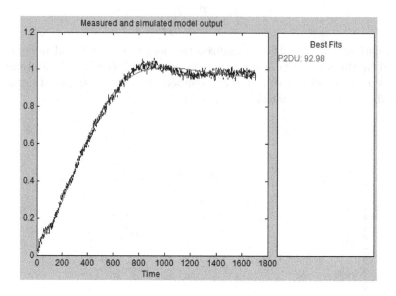

Fig. 9. Measured and simulated model output

Simulated model output plotted in red fits measured output plotted in black well in Fig.9. As the best fit is 92.98%, the model can accurately perform the characteristics of the actual system, and parameters calculated by system identi- fication can be used in the model of liquid temperature control system obtained above.

4   Conclusions

According to the requirement of intraperitoneal hyperthermic chemotherapy treatment, an IHCT machine is designed in this paper. A mathematical model for liquid temperature control is obtained by the analysis on the inner workings of temperature control system. In order to determine model parameters of liquid temperature control, system identification technique is applied and the model can be verified to fit the actual system well. The analysis and modeling on liquid temperature control presented in this paper provide a theoretical basis to further design and optimization of the control strategy in IHCT treatment.

References

1. Ni, B.Q., Zheng, Q.P.: Progress in basic study and clinical applications of intraperitoneal hyperthermic chemotherapy. Huaxia Yixue 20(2), 411–413 (2007)
2. Roviello, F., et al.: Safety and Potential Benefit of Hyperthermic Intraperitoneal Chemotherapy (HIPEC) in Peritoneal Carcinomatosis From Primary or Recurrent Ovarian Cancer. Journal of Surgical Oncology 102, 663–670 (2010)
3. Zhan, G.F., Lei, J.: Progress in preventive and treatment effect of hyperthermic intraperitonral chemotherapy on metastasis of abdominal cavity of gastric carcinoma. China Medical Herald 11(1) (2014)
4. Guan, B.Q., Chen, Y.Z., Pan, G.: Development of an Instrument for Hyperthermal Perfusion Chemotherapy Based on Fuzzy PID Controller. Beijing Biomedical Engineering 25(2), 45–147 (2006)
5. Zhang, Z.G., Li, K.Y., Dong, J.H., Yang, X.D.: Optimization Design of the Machine of Intraperitoneal Thermotherapy and Chemotherapy with Perfusion Based on UC/OS-II. China Instrumentation 26(5), 81–84 (2006)
6. Cui, S.Z., Ba, M.C., Huang, D.W., Tang, Y.Q., Wu, Y.B.: Study and Development of BR-TRG-I Hyperthermic PerfusionIntraperitoneal Treatment System. Journal Press of China Medical Devices 24(9), 7–9 (2009)
7. Lv, G.J.: KMM-Based Temperature Control System for Resistance-Heated Furnace. Techniques of Automation and Applications 21(4), 36–37 (2002)
8. Shao, J.X., Bai, J.Y.: Based on Intelligent Control Resistance Furnace Furnace Temperature Control System. Journal of Electric Power 25(3), 251–254 (2010)
9. Yu, X.F.: Two-wheel Self-balanced Vehicle System Model Identification Based on MATLAB. Mechanical Engineering & Automation 170(1), 67–69 (2012)
10. Hou, M.Y., Wang, M., Wang, L.Q.: System identification and matlab simulation. Science Press, Beijing (2004)

A Diffraction Based Modified Exponential Model for Device-Free Localization with RSS Measurements[*]

Nanyong Jiang[1], Kaide Huang[1], Yao Guo[1],
Guoli Wang[1,2], and Xuemei Guo[1,**]

[1] School of Information Science and Technology
Sun Yat-Sen University, Guangzhou 510006, China
[2] SYSU-CMU Shunde International Joint Research Institute, Foshan 5280000, China
guoxuem@mail.sysu.edu.cn

Abstract. Radio frequency (RF) based Device-free localization and tracking (DFLT) monitors the change in received signal strength (RSS) measurements to locate the targets without carrying any electronic devices in the sensing area covered by a RF sensor network. This paper presents a new modified exponential model to accurately describe the relationship between the RSS measurements and the target state, which can effectively predict the variation of RSS when the target is present on line-of-sight (LOS) path or non-line-of-sight (NLOS) path. Based on the diffraction theory, we first show that the RSS attenuation caused by target mainly depends on two factors: the target-nodes distance and the target-link distance, which can be exploited to depict the change of RSS on LOS and NLOS, respectively. By taking into account these two factors, we then develop our model, and validate it with single link experiments. We finally explore the use of the proposed model with particle filter for DFLT, and demonstrate that our model can improve the DFLT performance by conducting actual experiments.

Keywords: Device-free localization, received signal strength, radio frequency sensor networks, diffraction theory, particle filter.

1 Introduction

Device-free localization and tracking (DFLT) [1] is an emerging technology that enables locating the target without carrying any devices or tags. Lots of sensor technologies, such as ultrasonic sensor, infrared camera and light sensor, can be used for the purposes of DFLT. Recently, there have been a growing interest in DFLT with using the received signal strength (RSS) measurements of radio

---

[*] This work was supported by the National Science Foundation of P. R. China under Grant No. 61375080, and by the SYSU-CMU Shunde International Joint Research Institute Free Application Project under Grant No. 20130201.
[**] Corresponding author.

X. Zhang et al. (Eds.): ICIRA 2014, Part II, LNAI 8918, pp. 342–353, 2014.
© Springer International Publishing Switzerland 2014

frequency (RF) [2]. The basic idea is to monitor the changes in RSS when a target moves across or near the link that connects two sensor nodes. Actually, the presence of a moving target may diffract, reflect, or scatter the RF waves when an intrusion happens [3]. Taking advantage of this feature, the position of the target can be inferred from the RSS measurements. As opposed to the traditional sensor technologies, RF waves can transmit through walls [4], in dark or smoke-filled environments, and protect the privacy at the same time. Thus the DF system has various applications including building security and alarm systems, military and police operations, fire and disaster rescue, and so on.

Several approaches have been proposed for the purpose of RSS based DFLT. One is radio tomographic imaging (RTI) [5], in which the attenuation or motion images are inferred firstly and then the target location is estimated from the images. While such a method can efficiently capture the visual shadowing images of the sensing area, it may introduce additional measurement noise in the two-step process. Another approach is based on RSS models. Exponential model [6], magnitude model [7] and ER model [8] have been presented to describe the target-induced RSS change on radio links and exploit the location information directly. However, the RSS-based models mentioned above only focus on the relationship between the target-link distance and the RSS measurements, while the changes of RSS in case that the target stands at different positions on line-of-sight (LOS) are ignored. Recently, a stochastic model based on diffraction theory [9] is proposed, which takes into account the impact of target cross-section on LOS. However, this model can not provide an effective process for the area outside the LOS. Hence, it is necessary to find a more advanced measurement model to characterize an explicit relationship between the RSS changes and the target locations.

In this paper, we focus on the RSS attenuation both on LOS and non-line-of-sight (NLOS). Based on the diffraction theory, we first model human body and analyze how human body contributes to the attenuation. The theoretical analysis highlights the RSS attenuation caused by target mainly depends on two factors: the target-nodes distance and the target-link distance. These two distance factors can be used to depict the changes of RSS on LOS and NLOS paths, respectively. Inspired by this, we propose a new modified exponential model which considers both of these two factors. A single link experiment is conducted to validate the effectiveness of this theoretic model both on LOS and NLOS paths. We then exploit our model with a particle filter for the purpose of target localization and tracking. Finally, we demonstrate that the proposed modified exponential model outperforms the existing models by means of target tracking experiments in outdoor environments.

The paper is organized as follows. In Section 2, we briefly formulate the model-based DFLT problem. In Section 3, we present the new modified exponential model and conduct a single link experiment to show its validity. The experimental studies in the context of target localization and tracking are conducted to demonstrate the effective of the proposed model in Section 4.

2   Problem Statement

DFLT system is based on a RF sensor network consisting of N sensor nodes and $M = N(N-1)$ unidirectional communication links which spread over the sensing area. In each sampling interval, the N nodes broadcast packets to the neighboring nodes by using a simple token ring protocol. The RSS measurements of these M links will change when a DF person moves into the sensing area since the person can diffract, scatter,and reflect the RF signals. Therefore, the position of DF people can be estimated from the temporal variations of RSS measurements.

Let $\bar{\gamma} \in \mathbb{R}^{M \times 1}$ denotes the average background RSS measurements in the case of missing target. And $\gamma_k \in \mathbb{R}^{M \times 1}$ denotes the instantaneous RSS measurements at time k when the DF target exists in the sensing area. The presence of the target typically causes variation of RSS measurements, which is defined as $\mathbf{z}_k$

$$\mathbf{z}_k = \bar{\gamma} - \gamma_k \tag{1}$$

If we can find a mathematical model which depicts the relationship between the change in RSS and the motion state of the target $\mathbf{x}_k$, the predicted value $\mathbf{y}_k \in \mathbb{R}^{M \times 1}$ can be obtained. Then the error between the measurement value $\mathbf{z}_k$ and the predicted value $\mathbf{y}_k$ is given by

$$\mathbf{e}_k = \mathbf{z}_k - \mathbf{y}_k \tag{2}$$

The more accurate the mathematical model is, the less noise will be introduced. While the predicted value agrees well with the measurement value, $\mathbf{e}_k$ could be assumed to be zero-mean Gaussian, that is, $\mathbf{e}_k \sim N(0, \sigma_\omega^2 \mathbf{I})$.

Most of the RSS-based models (e.g., exponential model [6]) assume people to be a point without considering the impact of human body on LOS. The diffraction model [9] focuses on the attenuation caused by human cross-section on LOS, but the changes of RSS outside the LOS are ignored. It is inevitable to introduce unpredictable noise $\mathbf{e}_k$ for the models mentioned above.

Therefore, our goal is to propose a novel RSS-based model to reduce the noise level. we first model human body in the spatial area (i.e., an ellipsoid area that centered around the LOS) and analyze the impact of human body using the diffraction theory. The theoretical analysis demonstrates that the RSS measurements change when the target stands at different positions on LOS. By combing it with the attenuation outside the LOS which can be depicted as an exponential decay, we propose a new modified exponential model. This new model aims at depicting the changes of RSS both on LOS and NLOS accurately, thus improving DF localization and tracking performance effectively.

3   Modified Exponential Model

3.1   Theoretical Modeling

Consider a single link and a DF person standing on LOS path as shown in Fig. 1. The target is approximated by a conducting cylinder [10], with radius r and height $2h$. The ellipse centered around the LOS path denotes the spatial impact area, which has been widely used [3], [5]. The distance between the transmitting and the receiving nodes is d, and $d_{tx}(\mathbf{x}_k)$, $d_{rx}(\mathbf{x}_k)$ are the distances between the target and the two nodes respectively.

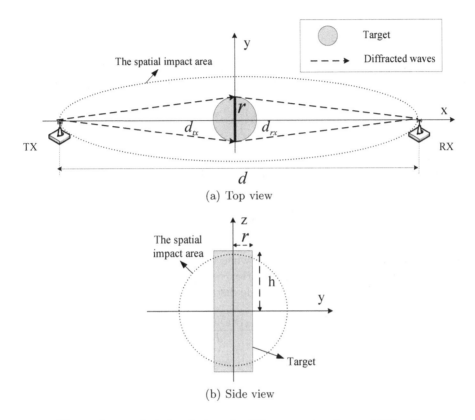

(a) Top view

(b) Side view

Fig. 1. Geometrical description of a DF person standing on LOS path

In this paper, we assume that the RSS attenuation is mainly caused by the diffraction. As shown in Fig. 1, when the target obstructs the LOS path, the radio waves diffract around the target as the RF propagation is obstructed, thus leading to diffraction loss. For simplicity, the human body is assumed to be a knife-edge object. Following the study of diffraction in [11], [12], the RF energy loss can be calculated based on Fresnel-Kirchhoff method.

Let E_f denotes the free space field at the receiver when the target is not present, and $E(\mathbf{x}_k)$ denotes the field when the target stands at $\mathbf{x}_k$ on the LOS path. Then their ratio is the electric field loss, which can be written as:

$$\left|\frac{E(\mathbf{x}_k)}{E_f}\right| = \frac{1}{2}\left|2 - j\int_{-A}^{A}\exp\left\{-j(\frac{\pi}{2}y^2)\right\}dy\int_{-B}^{B}\exp\left\{-j(\frac{\pi}{2}z^2)\right\}dz\right| \quad (3)$$

where A, B are the Fresnel-Kirchhoff diffraction parameters given by

$$A = r\sqrt{\frac{2d}{\lambda_c d_{tx}(\mathbf{x}_k) d_{rx}(\mathbf{x}_k)}} \tag{4}$$

$$B = h\sqrt{\frac{2d}{\lambda_c d_{tx}(\mathbf{x}_k) d_{rx}(\mathbf{x}_k)}} \tag{5}$$

where λ_c denotes the wavelength of carrier frequency. Assuming that $h \gg r$ and $d_{tx}(\mathbf{x}_k)$, $d_{rx}(\mathbf{x}_k) \gg h, r, \lambda_c$, the electric field loss can be approximated as follows

$$\left|\frac{E(\mathbf{x}_k)}{E_f}\right| \approx \frac{\sqrt{2}}{\pi A} \tag{6}$$

The RSS attenuation in decibel is defined as the squared electric field, $|E_f/E(\mathbf{x}_k)|^2$. Thus the variation of RSS on LOS y_{LOS} can be expressed as:

$$y_{\text{LOS}}(\mathbf{x}_k) \approx \frac{\pi^2 r^2 d}{\lambda_c d_{tx}(\mathbf{x}_k) d_{rx}(\mathbf{x}_k)} \tag{7}$$

From (7) we can conclude that the attenuation is determined by the distance between the target and the two nodes when the target obstructs the LOS path. The closer people gets to the transmitter/receiver, the more attenuation happens. Let $\beta(\mathbf{x}_k)$ denotes the target-node distance given by

$$\beta(\mathbf{x}_k) = \frac{d_{tx}(\mathbf{x}_k) d_{rx}(\mathbf{x}_k)}{d} \tag{8}$$

The RSS attenuation y_{LOS} can be simply expressed as:

$$y_{\text{LOS}}(\mathbf{x}_k) \approx \frac{\pi^2 r^2}{\lambda_c \beta(\mathbf{x}_k)} \tag{9}$$

It has been observed that, when the LOS path is vacant from the target but the spatial impact area is partially obstructed, there still exists significant attenuation which could not be ignored. Previous works have shown that the loss can be calculated by a piecewise function [13]. However, it is too complicated for a DFLT system to use this method directly. For simplicity, the exponential model is adopted to depict the attenuation on NLOS path, given by

$$y_{\text{NLOS}}(\mathbf{x}_k) = \psi \exp(-\frac{\lambda(\mathbf{x}_k)}{\sigma}) \tag{10}$$

where ψ is attenuation parameter, and $\lambda(\mathbf{x}_k)$ is the target-link distance function of the certain propagation link, which describes the relationship between target state $\mathbf{x}_k$ and the LOS path. The value $\lambda(\mathbf{x}_k)$ is defined as

$$\lambda(\mathbf{x}_k) = d_{tx}(\mathbf{x}_k) + d_{rx}(\mathbf{x}_k) - d \tag{11}$$

Combining the cases of LOS and NLOS paths, we can conclude that the changes of RSS mainly depend on two factors. One is the target-link distance $\lambda(\mathbf{x}_k)$ and the other is the target-node distance $\beta(\mathbf{x}_k)$.

Therefore, we propose a new model based on diffraction which can be written as:

$$y(\mathbf{x}_k) = \frac{\varphi}{\beta(\mathbf{x}_k)} \exp(-\frac{\lambda(\mathbf{x}_k)}{\sigma}) \tag{12}$$

Where φ and σ are attenuation parameters which can be identified by least square algorithm. The new model depicts the relationship between the changes of RSS and the target state both on LOS and NLOS paths.

3.2   Experimental Validation

We assess the validity of the proposed model using single link measurements. Two radio nodes are set separated by 6m, which use the IEEE.802.15.4 standard for communication. As shown in Fig. 2, the single people stands at different points along the LOS path with intervals of $\triangle x = 0.5m$, and the target stays 10s at each point. Then the person walks through the link along three different paths orthogonal to the LOS path at a low speed. We choose the RSS measurements collected in path 1 to show the attenuation on NLOS path.

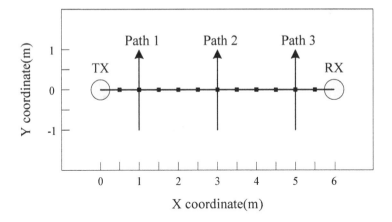

Fig. 2.   The single link experiment setup. Square markers indicate the positions of target standing on LOS path. Arrows indicate the paths of a moving target.

In Fig. 3, the RSS attenuation increases as the target moves close to the transmitter or receiver along the LOS path, and the minimum attenuation is observed in the middle of the link. The new model represented by red solid line fits the measurements well, which shows the attenuation caused by the target

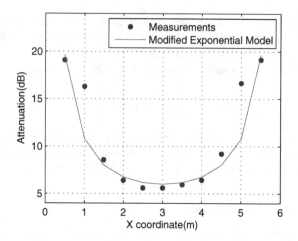

Fig. 3. The RSS attenuation on the LOS path

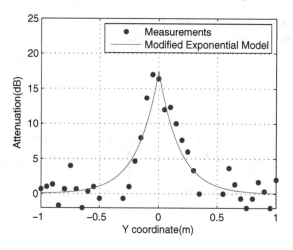

Fig. 4. The RSS attenuation on path1

on LOS path mainly depends on the target-node distance $\beta(\mathbf{x}_k)$. As shown in Fig. 4, the RSS attenuation decreases dramatically when the person moves away from the LOS. It is reasonable to describes this change as an exponential decay in the proposed model. Therefore, our modified exponential model is validated to be effective both on LOS and NLOS path.

4   DFLT with Our Model

4.1   Localization and Tracking Algorithm

Our task is to localize and track DF people from RSS measurements. Sequential Important Re-sampling (SIR) particle filter [14] is adopted to provide the position estimation. The kernel of this method is utilizing some random sample to express the posterior probability density of system variable, and the optimal approximate numerical solution of the system model can be obtained. This property is fit for our system. The target motion can be regarded as a first-order Markov process $p(\mathbf{x}_k|\mathbf{x}_{k-1}, \mathbf{x}_{k-2}, ..., \mathbf{x}_0) = p(\mathbf{x}_k|\mathbf{x}_{k-1})$. It indicates that the current motion state is determined by the last motion state. Here we simply use the autoregressive gaussian (ARG) model to describe the process of target movement. That is

$$\mathbf{x}_{k+1} = \mathbf{x}_k + \sigma_v \mathbf{v} \tag{13}$$

where σ_v is a constant and $\mathbf{v} \sim \mathcal{N}(\mathbf{0}, \mathbf{I})$.

The state $\mathbf{x}_k$ is hidden into the RSS measurements $\mathbf{z}_k$, we use particle filter to track the posterior distribution $P(\mathbf{x}_k|\mathbf{z}_k)$. The pseudo-code of SIR particle filter is specified as Algorithm 1.

Algorithm 1. SIR particle filter

Initialization: $\{x_0^i\}_{i=1}^{N_p}$ uniformly distributed, $\{\omega_0^i\}_{i=1}^{N_p}$
for k=1 to T do
　for i=1 to N_p do
　　predict step: $x_k^i \sim f(x_k^i|x_{k-1}^i)$
　　weight update: $\omega_k^i \propto p(\mathbf{e}_k|x_k^i), \mathbf{e}_k = \mathbf{z}_k - \mathbf{y}_k$
　　normalization: $\bar{\omega}_k^i = \dfrac{\omega_k^i}{\sum_{i=1}^{N_p} \omega_k^i}, i = 1, 2, \ldots, N;$
　　estimation step: $\hat{x}_k \approx \sum_{i=1}^{N_p} \omega_{k-1}^i x_k^i$
　　resample step: $\{\omega_k^i, x_k^i\} \rightarrow \{\dfrac{1}{N_p}, x_k^i\}$
　end for
end for

4.2 Experimental Setup

As shown in Fig. 5, we build a RF sensor network with 24 radio nodes along the perimeter of a 6m×6m square area to calibrate and validate the parametric model. There is 1m interval between two adjacent nodes, and the height of each node is 1m.

The radio nodes are MIB520CB made by Crossbow, which use the 2.4GHz IEEE.802.15.4 standard for communication. A base station is utilized to transmit data or command between the RF sensor and computer. In addition, a simple toking ring protocol is used to collect the RSS data of all possible radio links.

RSS data of all links are gathered when a single person stands at a certain point or walks along a predefined path. These samples are then used to identify the parameters and validate the the new model using least square algorithm. In this paper, we set the parameter $\varphi = 13.82$ and $\sigma = 0.023$ for the experimental scene. We perform the single target localization and tracking experiments using the proposed model, exponential model and diffraction model.

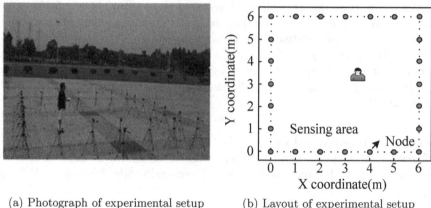

(a) Photograph of experimental setup   (b) Layout of experimental setup

Fig. 5. Experimental platform is built in an open environment at Sun Yat-Sen University. 24 RF sensor nodes are deployed on the boundaries of a 6m×6m sensing area.

Table 1. The Localization error using different models

Model	Mean error (m)	Max error (m)
Diffraction model	0.29m	0.85m
Exponential model	0.21m	0.78m
Modified Exponential model	0.12m	0.25m

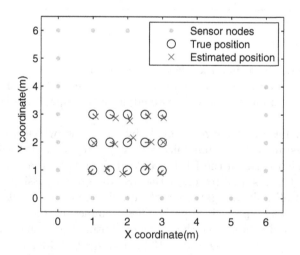

Fig. 6. The 15 groups of localization results using the proposed modified model

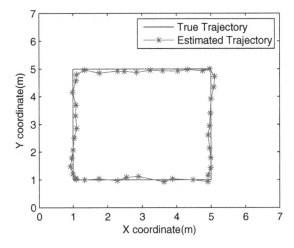

Fig. 7. Tracking result using the proposed modified exponential model

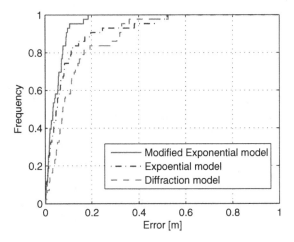

Fig. 8. The cumulative error distribution for tracking using different models

4.3   Localization and Tracking Results

To evaluate the effectiveness of the proposed model, 36 groups of single target localization are tested using different models. Here, the particle filter is used for localization, and each estimated position was taken after 50 iterations with 100 particles, and the average result of three model are listed in Table 1. It can be seen that the new model enhances the accuracy of estimation compared with the other two models. We choose 15 groups of localization results to demonstrate the localization performance using the new modified exponential model, as shown in Fig. 6.

For the tracking issue, we use the ARG model to predict the motion state with the parameter $\sigma_v = 0.5$. Fig. 7 shows the tracking performance using the proposed model. The cumulative error distribution for tracking using different models is shown in Fig. 8, which illustrates the proposed modified model outperforms the diffraction model and the exponential model in tracking performance.

5   Conclusions

We present a new modified exponential model to describe the RSS attenuation caused by people in outdoor environments. The diffraction theory is introduced to analyze how human body contributes to the RSS attenuation in this paper. The theoretical analysis highlights that the RSS attenuation mainly depends on the target-node distance and target-link distance. The proposed model, which combines these two factors, can perform well in handling the changes of RSS both on LOS and NLOS paths. Experiment results show that the modified exponential model achieves high estimation accuracy using SIR particle filter. In future, we will develop the model for multiple targets and apply it in cluttered indoor environments.

References

1. Youssef, M., Mah, M., Agrawala, A.: Challenges: Device-free passive localization for wireless environments. In: Proceedings of the 13th Annual ACM International Conference on Mobile Computing and Networking, pp. 222–229. ACM (2007)
2. Zhang, D., Ma, J., Chen, Q., Ni, L.: An RF-based system for tracking transceiver-free objects. In: IEEE International Conference on Pervasive Computing and Communications, pp. 135–144. IEEE (2007)
3. Patwari, N., Wilson, J.: RF sensor networks for device-free localization: Measurements, models, and algorithms. Proceedings of the IEEE 98(11), 1961–1973 (2010)
4. Wilson, J., Patwari, N.: See-through walls: Motion tracking using variance-based radio tomography networks. IEEE Transactions on Mobile Computing 10(5), 612–621 (2011)
5. Wilson, J., Patwari, N.: Radio tomographic imaging with wireless networks. IEEE Transactions on Mobile Computing 9(5), 621–632 (2010)
6. Chen, X., Edelstein, A., Li, Y., Coates, M., Rabbat, M., Men, A.: Sequential Monte Carlo for simultaneous passive device-free tracking and sensor localization using received signal strength measurements. In: 2011 10th International Conference on Information Processing in Sensor Networks (IPSN), pp. 342–353. IEEE (2011)
7. Nannuru, S., Li, Y., Zeng, Y., Coates, M., Yang, B.: Radio-frequency tomography for passive indoor multitarget tracking. IEEE Transactions on Mobile Computing 12(12), 2322–2333 (2013)
8. Guo, Y., Huang, K., Jiang, N., Guo, X., Li, Y., Wang, G.: An Exponential-Rayleigh model for RSS-based device-free localization and tracking. IEEE Transactions on Mobile Computing PP(99), 1 (2014)
9. Savazzi, S., Nicoli, M., Carminati, F., Riva, M.: A Bayesian Approach to Device-free Localization: Modeling and Experimental Assessment. IEEE Journal of Selected Topics in Signal Processing PP(99), 1 (2013)

10. Ghaddar, M., Talbi, L., Denidni, T., Sebak, A.: A conducting cylinder for modeling human body presence in indoor propagation channel. IEEE Transactions on Antennas and Propagation 55(11), 3099–3103 (2007)
11. Mokhtari, H., Lazaridis, P.: Comparative study of lateral profile knife-edge diffraction and ray tracing technique using gtd in urban environment. IEEE Transactions on Vehicular Technology 48(1), 255–261 (1999)
12. Nyuli, A., Szekeres, B.: An improved method for calculating the diffraction loss of natural and man made obstacles. In: IEEE International Symposium on Personal, Indoor and Mobile Radio Communications, pp. 426–430 (1992)
13. Athanasiadou, G.E.: Incorporating the Fresnel Zone Theory in Ray Tracing for Propagation Modelling of Fixed Wireless Access Channels. In: IEEE 18th International Symposium on Personal, Indoor and Mobile Radio Communications, pp. 1–5 (2007)
14. Arulampalam, M., Maskell, S., Gordon, N., Clapp, T.: A tutorial on particle filters for online nonlinear/non-gaussian bayesian tracking. IEEE Transactions on Signal Processing 50(2), 174–188 (2002)

A Novel Mathematical Piezoelectric Hysteresis Model Based on Polynomial

Jinqiang Gan and Xianmin Zhang[*]

School of Mechanical and Automotive Engineering,
South China University of Technology, Guangzhou, Guangdong 510640, China
1024768638@qq.com, zhangxm@scut.edu.cn

Abstract. In this paper, a novel mathematical piezoelectric hysteresis model based on polynomial is proposed. This model uses quadratic polynomial and linear equation to describe the relationship between input voltage and output displacement precisely. To describe the hysteresis characteristics, experiments are performed with three kinds of input triangular wave signals. The simulation results are compared with the experiment data to demonstrate the validity of this proposed model. The results show that the mathematical model can completely match the hysteresis of the piezoelectric actuator at lower errors.

Keywords: Piezoelectric actuator, hysteresis model, quadratic polynomial, linear equation.

Introduction

Piezoelectric actuators are widely used in high-speed and high-precision positioning applications because of their fast response time, large bandwidth and high output force [1-5]. However, the piezoelectric actuators exhibit hysteresis behavior, creep and nonlinear multi-valued hysteresis phenomenon, which can cause inaccuracy in the system response. To compensate for the piezoelectric hysteresis effects, many models have been developed. These models can be mainly classified into the physics-based models and the phenomenological models. Physics-based models are generally derived on the basis of a physical measure, such as energy, displacement, or stress-strain relationship e.g., the classical Jiles-atherton model [6]. Alternatively, the phenomenological models contain the Preisach model [7-9], the Bouc-Wen model [10, 11], the Prandtl-Ishlinskii model [12-15], the Duhem model [16], the Backlash-like model [17] and the Polynomial model [18]. However, the drawbacks exist such that most of these models are relatively complex. Complicated expressions and intricate identification with a large amount of data are often needed in these models. Another disadvantage is that it is not easy to get the precise inverse model directly.

In order to remedy the above disadvantages, a new mathematical model with quadratic polynomial is proposed to describe the hysteresis behavior in piezoelectric

---

[*] Corresponding author.

X. Zhang et al. (Eds.): ICIRA 2014, Part II, LNAI 8918, pp. 354–365, 2014.
© Springer International Publishing Switzerland 2014

actuators. To characterize the hysteresis effect precisely, simple identification of a set of parameters with the designed signals is needed. In this new model, the linear fitting of hysteresis data in small input voltage is adopted. The validity of this new model is demonstrated by both simulation and experiments.

1   The New Hysteresis Model

This section describes the modeling of hysteresis using quadratic polynomial and linear equation instead of cubic polynomial. In Sun [18], the curve which passes through the initial point is called the reference rising curve $P_n(u(t))$. When the input voltage changes direction from ascending to descending or from descending to ascending, the input voltage value is called the turning voltage U_t.

1.1   Rising Hysteresis Curves

The relationship between the referenced rising curve and the other rising curves $P_n(u(t))$, $P_{i+1}(u(t))$, $P_i(u(t))$ and $P_{i-1}(u(t))$ are shown in Figure 1

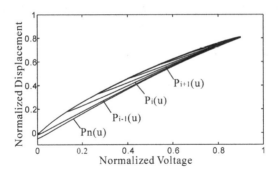

Fig. 1. Rising hysteresis curves

In Figure 1, it is obvious that all the raising curves converge to one point. In this paper, quadratic polynomial is proposed to describe the hysteresis curves in order to simplify polynomial model according to equation (1)

$$y_{(t)} = P_i(u(t)) = a_{i1}u^2(t) + a_{i2}u(t) + a_{i3} \tag{1}$$

Where $y(t)$ and $P_i(u(t))$ are the output displacement of the piezoelectric actuator; $u(t)$ is the input voltage; a_{i1}, a_{i2} and a_{i3} are the polynomial coefficients. But there are some big errors in the initial small input voltage according to the above model directly. In order to reduce these errors, the rising hysteresis curves are segmented according to equation (2)

$$y(t) = P_i\left(u(t)\right) = \begin{cases} a_{i1}u^2(t) + a_{i2}u(t) + a_{i3} & u(t) \geq U_c \\ b_{i1}u(t) + b_{i2} & u(t) \leq U_c \end{cases} \qquad (2)$$

Where U_C is a dividing voltage whose value is fixed. U_C is set as about 8 percent of the maximum input voltage. b_{n1}, b_{n2} are coefficients. Figure 2 shows the segmentation process of the hysteresis curves.

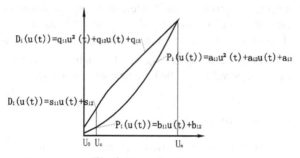

$D_i(u(t)) = q_{i1}u^2(t) + q_{i2}u(t) + q_{i3}$

$P_i(u(t)) = a_{i1}u^2(t) + a_{i2}u(t) + a_{i3}$

$D_i(u(t)) = s_{i1}u(t) + s_{i2}$

$P_i(u(t)) = b_{i1}u(t) + b_{i2}$

$U_0 \ U_c$ U_n

Fig. 2. Piecewise curves

The rising reference curve can be obtained by using quadratic polynomial in full range according to equation (3)

$$p_n(u(t)) = a_{n1}u^2(t) + a_{n2}u(t) + a_{n3} \qquad 0 \leq u(t) \leq U_n \qquad (3)$$

When the input voltage is smaller than U_C, this part of curve can be acquired by using the linear equation according to equation (4)

$$p_c(u(t)) = b_{n1}u(t) + b_{n2} \qquad 0 \leq u(t) \leq U_c \qquad (4)$$

All the rising curves converge to the peak $(U_n, P_n(U_n))$. The input voltage $u(t)$ is larger than or equal to the turning voltage U_i in the rising curves. Every rising curve has this following relationship with the reference curve:

$$a_{i2} = k \cdot a_{n2} = \frac{P_n(U_n) - P_i(U_i)}{P_n(U_n) - P_n(U_0)} \cdot a_{n2} \qquad (5)$$

$$b_{i1} = b_{n1} \qquad (6)$$

According to the value of the turning voltage U_i, the rising curve is divided into two cases.

1) The turning voltage U_i is smaller than or equal to the fixed voltage U_C.

When $U_i \leq u(t) \leq U_C$, this part of curve is described by the linear equation according to equation (7)

$$P_i(u(t)) = b_{n1}u(t) + P_i(U_i) - b_{n1} \cdot U_i \tag{7}$$

When $U_c \leq u(t) \leq U_n$, this part of curve is described by the quadratic polynomial. It passes two points $(U_c, \ P_i(u_c))$ and $(U_n, \ P_n(u_n))$. The point $(U_c, \ P_i(u_c))$ can be obtained according to equation (7). a_{i2} is obtained according to equation (5). Under these conditions, the relationship is obtained:

$$
\begin{aligned}
P_i(u(t)) &= a_{i1}u^2(t) + a_{i2}u(t) + a_{i3} \\
&= \frac{P_n(U_n) - P_i(U_c) - ka_{n2}(U_n - U_c)}{(U_n^2 - U_c^2)}u^2(t) + ka_{n2}u(t) \\
&\quad + P_n(U_n) - ka_{n2}U_n - \frac{P_n(U_n) - P_i(U_c) - ka_{n2}(U_n - U_c)}{(U_n^2 - U_c^2)} \cdot U_n^2
\end{aligned}
\tag{8}
$$

2) The turning voltage U_i is larger than the fixed voltage U_C.

This part of curve is also described by the quadratic polynomial. It passes two points $(U_i, \ P_i(U_i))$ and $(U_n, \ P_n(U_n))$. The point $(U_i, \ P_i(U_i))$ can be acquired by the i-1th descending curve. a_{i2} can be obtained according to equation (5). Under these conditions, the relationship is obtained:

$$
\begin{aligned}
P_i(u(t)) &= a_{i1}u^2(t) + a_{i2}u(t) + a_{i3} \\
&= \frac{P_n(U_n) - P_i(U_i) - ka_{n2}(U_n - U_i)}{(U_n^2 - U_i^2)}u^2(t) + ka_{n2}u(t) \\
&\quad + P_n(U_n) - ka_{n2}U_n - \frac{P_n(U_n) - P_i(U_i) - ka_{n2}(U_n - U_i)}{(U_n^2 - U_i^2)} \cdot U_n^2
\end{aligned}
\tag{9}
$$

1.2   Descending Hysteresis Curves

Figure 3 shows that the reference descending curve $D_n(u)$ and descending curves $D_i(u)$. The descending curve which passes the maximum voltage and the minimum voltage at the first time is named the reference descending curve $D_n(u)$. In Sun Lining's paper [18], all the descending curves converge to one point (U_0, $D_n(U_0)$), whereas curves in the real experiment converge to different point, $D_i(U_0)$. Figure 4 illustrates this characteristic.

The descending hysteresis curves are segmented according to equation (10)

$$y(t) = D_i(u(t)) = \begin{cases} q_{i1}u^2(t) + q_{i2}u(t) + q_{i3} & u(t) \geq U_c \\ s_{i1}u(t) + s_{i2} & u(t) \leq U_c \end{cases} \tag{10}$$

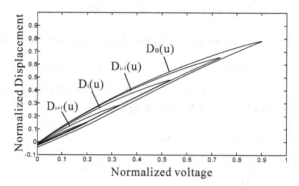

Fig. 3. Descending hysteresis curves

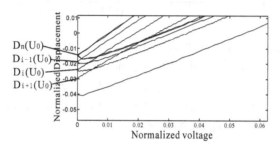

Fig. 4. Hysteresis curves

Where U_C is the same fixed voltage as the rising curves; q_{i1}, q_{i2}, q_{i3}, s_{i1} and s_{i2} are the polynomial coefficients. The reference descending curve can be obtained in the same way like the reference rising curve. The descending curve function is given by

$$D_n(u(t)) = q_{n1}u^2(t) + q_{n2}u(t) + q_{n3} \qquad 0 \leq u(t) \leq u_n \tag{11}$$

$$D_c(u(t)) = s_{n1}u(t) + s_{n2} \qquad 0 \leq u(t) \leq U_c \tag{12}$$

The input voltage $u(t)$ is smaller than or equal to the turning voltage U_i in the descending curves. All the descending curves have these following relationships with the reference descending curve:

$$q_{i2} = q_{n2}, s_{i1} = s_{n1}, s_{i2} = D_i(U_0) \tag{13}$$

$$
\begin{aligned}
D_i(U_0) &= K\left(D_n(U_0) - P_n(U_0)\right) + P_n(U_0) \\
&= \frac{D_i(U_i) - D_n(U_0)}{D_n(U_n) - D_n(U_0)}\left(D_n(U_0) - P_n(U_0)\right) + P_n(U_0)
\end{aligned} \tag{14}
$$

According to the value of the turning voltage U_i, the descending curve is also divided into two cases.

1) The turning voltage U_i is smaller than or equal to the fixed voltage U_C.

This part of curve which passes the point $(U_i, D_i(U_i))$ can be described by the linear equation according to equation (15)

$$D_i\big(u(t)\big) = s_{i1}u(t) + s_{i2} = s_{n1}u(t) + D_i(U_i) - s_{n1}U_i \tag{15}$$

2) The turning voltage U_i is larger than the fixed voltage U_C.

When $U_0 \le u(t) \le U_C$, the descending curve function is given by

$$D_i\big(u(t)\big) = s_{i1}u(t) + s_{i2} = s_{n1}u(t) + D_i(U_0) \tag{16}$$

When $U_C \le u(t) \le U_i$, this part of curve which passes two points $(U_c, D_i(U_c))$ and $(U_i, D_i(U_i))$ can be described by quadratic polynomial. The point $(U_c, D_i(U_c))$ can be acquired according to equation (16). The coefficient q_{i2} can be obtained by equation (13). So, the function of this part of curve is given by

$$
\begin{aligned}
D_i(u(t)) &= q_{i1}u^2(t) + q_{i2}u(t) + q_{i3} \\
&= \frac{D_i(U_i) - D_i(U_c) - q_{n2}(U_i - U_c)}{\left(U_i^2 - U_c^2\right)} u^2(t) + q_{n2}u(t) \\
&\quad + D_i(U_i) - q_{n2}U_i - \frac{D_i(U_i) - D_i(U_c) - q_{n2}(U_i - U_c)}{\left(U_i^2 - U_c^2\right)} \cdot U_i^2
\end{aligned}
\tag{17}
$$

2   Experimental Description

A number of experiments have been conducted on a 1-DOF piezo-driven compliant mechanism (see Figure 5.a). A stack piezoelectric translator PST 150/7/60VS12, which is a preloaded piezoelectric translator (PZT), was used to describe hysteresis characteristic. This actuator which includes an integrated high-resolution strain gauge position sensor (SGS) provides maximum 60μm displacement. The control signal was amplified by an XE-500D controller from Harbin Core Tomorrow Science & Technology Company with an amplification factor of fifteen. The output displacement was measured by strain gauge position sensor (SGS). The dSPACE-DS1104 rapid prototyping controller board equipped with 16-bit ADC and 16-bit DAC was adopted to generate and obtain experiment data. The experiment set up is shown in Figure 6(b). The sampling frequency of the system was set as 10 kHz. All the parameters, which include a_{n2}, b_{n1}, $P_n(U_n)$, $P_n(U_0)$, q_{n2}, s_{n1} and $D_n(U_0)$, can be identified from the experiment data by the direct least square fitting method through the MATLAB offline. In this paper, U_c was chose as 10 V. Figure 7 shows the structure of the experimental system.

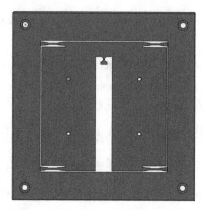

Fig. 5. (a) 1-DOF compliant mechanism

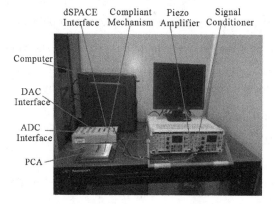

Fig. 5. (b) Experiment set up

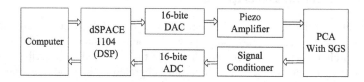

Fig. 6. Structure of experiment control system

3   The Test of the New Model

In order to protect the stack piezoelectric actuator, the maximum input voltage in experiments is 135 V instead of 150 V. The real output displacement is less than 60 μm. In the experiment, both the input voltage and the output displacement were normalized with the values (150 V and 48 μm respectively) so as to process the data easily. Three kinds of input triangular wave signals (signal A, signal B and signal C) were adopted in the experiments for the purpose of demonstrating the validity of the new hysteresis model. Signal A is shown in Figure 7(a) whose peaks are 135 V. Figure 8(a) shows Signal B whose troughs are 0 V. Signal C in Figure 9(a) is irregular. The output displacement was obtained by the strain gauge position sensor (SGS). The simulated displacements were acquired by using the new model. The final results and errors between measured displacements and simulation are shown in Figure 7(b) and (c), Figure 8(b) and (c), Figure 9(b) and (c).

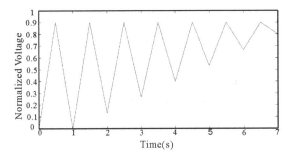

Fig. 7. (a) Triangular wave

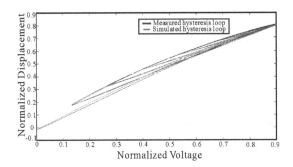

Fig. 7. (b) Contrast between the measured and simulated

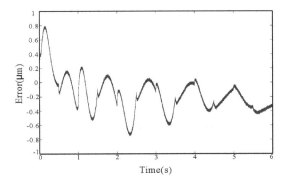

Fig. 7. (c) The errors between the measured and simulated

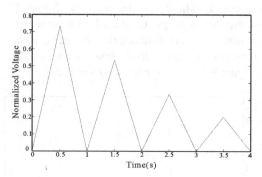

Fig. 8. (a) Triangular wave

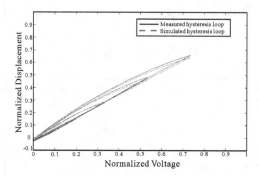

Fig. 8. (b) Contrast between the measured and simulated

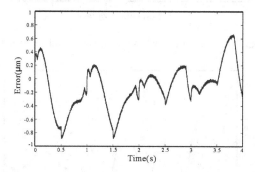

Fig. 8. (c) The errors between the measured and simulated

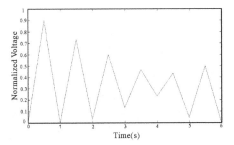

Fig. 9. (a) Triangular wave

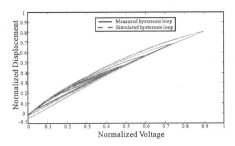

Fig. 9. (b) Contrast between the measured and simulated

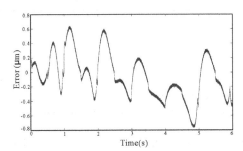

Fig. 9. (c) The errors between the measured and simulated

In order to quantify modeling errors, the maximum absolute modeling error $\Delta(x_i)_{max}$ and the maximum relative modeling error δ_{max} in [10] were introduced into this paper. The average absolute modeling error $\overline{\Delta(x(i))}$ was also introduced. These are defined as follows

$$\Delta(x(i))_{max} = \underset{1 \leq i \leq n}{MAX} \left| x(i) - x_p(i) \right| \tag{18}$$

$$\delta_{max} = \frac{\Delta(x(i))_{max}}{x(i)_{max}} \times 100\%_{100} \tag{19}$$

$$\overline{\Delta(x(i))} = \frac{\sum_{i=1}^{n} \Delta(x(i))}{n} \tag{20}$$

Where i=1, 2, . . . n is the total number of sampling; x(i) and $x_p(i)$ are the measured and predicted output displacements. According to equations (18), (19) and (20), the errors of the three kinds of signals of the input triangular voltages are listed in Table 1. It is observed in Table 1 that the average absolute modeling error is about 0.28μm and the maximum absolute modeling error is less than 0.88μm.

Table 1. The modeling errors

Input signals	Error		
	$\Delta(x(i))_{max}$	δ_{max}	$\overline{\Delta(x(i))}$
Signal A	0.7934μm	2.06%	0.2496μm
Signal B	0.8851μm	2.36%	0.2736μm
Signal C	0.6403μm	1.65%	0.2784μm

4   Conclusion

In the paper presented above, a novel mathematical hysteresis model is established and its performance has been investigated. Adopting the quadratic polynomial and liner equation to describe the hysteresis makes this hysteresis model simple and convenient to use, which also brings great convenience in getting the inverse model. The simulation results agree well with the measured data and demonstrate the validity of the proposed model. In the future, a model-based feedforward will be implemmented with the help of this model to cancel the hystresis.

Acknowledgments. This research was supported by the National Science Foundation of China (Grant No. 91223201, 50825504), the United Fund of Natural Science Foundation of China and Guangdong province (Grant No. U0934004), Project GDUPS (2010), and the Fundamental Research Funds for the Central Universities (2012ZP0004). This support is greatly acknowledged.

References

1. Zhang, D., Chetwynd, D.G., Liu, X., Tian, Y.: Investigation of a 3-DOF micro-positioning table for surface grinding. International Journal of Mechanical Sciences 48(12), 1401–1408 (2006)

2. Tian, Y., Shirinzadeh, B., Zhang, D.: A flexure-based mechanism and control methodology for ultra-precision turning operation. Precision Engineering-Journal of the International Societies for Precision Engineering and Nanotechnology 33(2), 160–166 (2009)

3. Kenton, B.J., Leang, K.K.: Design and Control of a Three-Axis Serial-Kinematic High-Bandwidth Nanopositioner. IEEE/ASME Transactions on Mechatronics 17(2), 356–369 (2012)

4. Wang, D.H., Yang, Q., Dong, H.M.: A Monolithic Compliant Piezoelectric-Driven Microgripper: Design, Modeling, and Testing. IEEE/ASME Transactions on Mechatronics 18(1), 138–147 (2013)

5. Grossard, M., Boukallel, M., Chaillet, N., Rotinat-Libersa, C.: Modeling and Robust Control Strategy for a Control-Optimized Piezoelectric Microgripper. IEEE/ASME Transactions on Mechatronics 16(4), 674–683 (2011)

6. Jiles, D.C., Atherton, D.L.: Theory of ferromagnetic hysteresis (invited). Journal of Applied Physics 55(6), 2115–2120 (1984)

7. Ge, P., Jouaneh, M.: Generalized preisach model for hysteresis nonlinearity of piezoceramic actuators. Precision Engineering-Journal of the American Society for Precision Engineering 20(2), 99–111 (1997)

8. Geng, J., Liu, X.D., Liao, X.Z., Lai, Z.L.: Neural Networks Preisach Model and Inverse Compensation for Hysteresis of Piezoceramic Actuator. In: 2010 8th World Congress on Intelligent Control and Automation (WCICA), pp. 5746–5752 (2010)

9. Song, G., Jinqiang, Z., Xiaoqin, Z., De Abreu-Garcia, J.A.: Tracking control of a piezoceramic actuator with hysteresis compensation using inverse Preisach model. IEEE/ASME Transactions on Mechatronics 10(2), 198–209 (2005)

10. Zhu, W., Wang, D.H.: Non-symmetrical Bouc-Wen model for piezoelectric ceramic actuators. Sensor Actuat A-Phys 181, 51–60 (2012)

11. Xiao, S., Li, Y.: Dynamic compensation and H∞ control for piezoelectric actuators based on the inverse Bouc–Wen model. Robotics and Computer-Integrated Manufacturer 30(1), 47–54 (2014)

12. Al Janaideh, M., Rakheja, S., Su, C.Y.: An Analytical Generalized Prandtl-Ishlinskii Model Inversion for Hysteresis Compensation in Micropositioning Control. IEEE-ASME Transactions on Mechatronics 16(4), 734–744 (2011)

13. Liu, S.N., Su, C.Y.: A Modified Generalized Prandtl-Ishlinskii Model and Its Inverse for Hysteresis Compensation. In: Proceedings of the American Control Conference, pp. 4759–4764. IEEE, New York (2013)

14. Tan, U.X., Win, T.L., Ang, W.T.: Modeling Piezoelectric Actuator Hysteresis with Singularity Free Prandtl-Ishlinskii Model. In: IEEE International Conference on Robotics and Biomimetics, ROBIO 2006, Kunming, pp. 251–256 (2006)

15. Tan, U.X., Latt, W.T., Widjaja, F., Shee, C.Y., Riviere, C.N., Ang, W.T.: Tracking control of hysteretic piezoelectric actuator using adaptive rate-dependent controller. Sensor Actuat A-Phys 150(1), 116–123 (2009)

16. Stepanenko, Y., Su, C.Y.: Intelligent control of piezoelectric actuators. In: IEEE Conference on Decision and Control-Proceeding, pp. 4234–4239. IEEE, New York (1998)

17. Chun-Yi, S., Stepanenko, Y., Svoboda, J., Leung, T.P.: Robust adaptive control of a class of nonlinear systems with unknown backlash-like hysteresis. IEEE Transactions on Automatic Control 45(12), 2427–2432 (2000)

18. Sun, L.N., Ru, C.H., Rong, W.B., Chen, L.G., Kong, M.X.: Tracking control of piezoelectric actuator based on a new mathematical model. Journal of Micromechanics and Microengineering 14(11), 1439–1444 (2004)

Characterization of Presliding
with Different Friction Models

Yunzhi Zhang, Xianmin Zhang, and Junyang Wei

School of Mechanical and Automotive Engineering,
South China University of Technology, Guangzhou, China
{z.yz,xmzhang,wei.jy01}@scut.edu.cn

Abstract. Presliding may be the most difficult regime to characterize in friction phenomenon. Two different friction models which are widely used in practice are discussed in this paper, also there performance on presliding. The disadvantages and advantages of these friction models are compared in this paper, at last, the phenomenon of presliding is detected in one experiment, and the future work on presliding research is proposed.

Keywords: presliding, friction model, stick-slip, nano position.

1   Introduction

Friction is a complicate nonlinear phenomenon exist in nearly every mechanical system, which can cause many problems for mechanical control, even for precise positioning machine. Stick-slip actuated machine is a mechanism actuated by friction, which can achieve nanoscale positioning (Fig. 1). Since the existence of friction, when a small amplitude vibration is applied to basement, the slider will move together. If the input vibration signal is suitable, slider can move towards one direction in nanoscale.

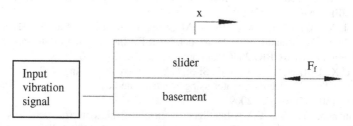

Fig. 1. Schematic diagram of stick-slip actuated machine

In general, the basement is fixed to and actuated by piezoelectric material, because of its high stiffness and high response frequency. If a suitable voltage signal is input to the piezoelectric material, etc. sawtooth like signal, the slider will move towards one direction, and have a nanoscale net displacement. The movement can be described as follows (Fig. 2):

X. Zhang et al. (Eds.): ICIRA 2014, Part II, LNAI 8918, pp. 366–376, 2014.

Phase 1, basement and slider are in the initial position, no voltage signal is applied on the piezoelectric material, and there is no relative moving between the basement and slider;

Phase 2, input a slowly rising voltage signal to the piezoelectric material, which will be elongated, and the basement is pushed forward slowly. Since the friction and the low velocity of basement, the slider will stick to basement and move together.

Phase 3, the input voltage decrease to zero suddenly, since the high response frequency of piezoelectric material, the basement will be pulled back to the initial position simultaneously, the slider will slip forward because of its inertia. So that, when a whole cycle of input voltage finished, slider will move forward in a short distance. If we keep input the sawtooth like voltage to piezoelectric material, the slider will keep moving forward.

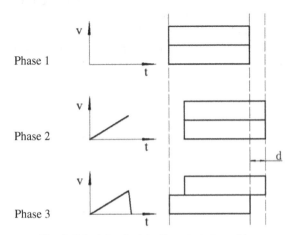

Fig. 2. Principle of stick-slip actuated machine

So as the description above, during the period of stick-slip actuation, the movement state of slider is switches between stick and slip, so as the primary driving force of the mechanism, friction also changes from static friction to dynamic friction again and again, which will cause many obstacles to the smooth moving of slider, such as vibration, hysteresis, and backlash, etc.

In order to control the stick-slip mechanism moves smoothly and precisely, we need to describe and explain the friction phenomenon in a reasonable and simple way, so a sensible friction model is requested.

In the past few decades, many classical friction models had been proposed, such as LuGre model (LGM) and Elasto-Plastic model (EPM). We will introduce these two models in the following sections.

Presliding is a kind of contact state when slider moves from stick to slip, and also maybe the most difficult regime to be modeled by friction model. Many friction model will have some problems when modelling presliding, this will be discussed in the following sections too.

This paper is organized as follows. First, the presliding phenomenon is introduced and some problems in presliding research is discussed in section 2. Next, two classical friction model which are widely used in friction research are introduced in section 3, and also there advantages and disadvantages. Then the performances of these two friction models in presliding modeling are compared and one experiment to detect presliding phenomenon is proposed in section 4. Finally, conclusion and future work about presliding research are presented.

2   Introduction of Presliding

According to the research so far, friction is a result of extremely complex interactions between the surface and the near surface regions of the two interacting materials, and other substances present such as lubricants [1].

According to one widely accepted conclusion, the surface of object cannot be absolutely smooth, there are countless tiny asperities with stiffness, approximately uniformly distributed on the surface. So the contact of two surfaces is indeed contact by these asperities in micro scale (Fig. 3) [2].

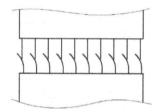

Fig. 3. Contact of two surfaces in micro scale

Asperity model is a wonderful description of contact, and proposes a new direction of friction research. Since the asperity has stiffness, when a tangential external force is applied to one object, the asperities will deform. If the external force is small enough, asperities will have some elastic deformation. If the external force is big enough, when reach the asperity's yield limit, some asperities may have plastic deformation. If the force keep rising, and when most of the asperities deform plastically, object begin to slip.

So in the presliding regime, since the deformation of asperity, friction force appears to be a function of displacement rather than velocity [3] [4].

There are many problems in presliding regime have not been clearly researched, they will be introduced as follows.

1) In Fig.3, in order to simplify the research, LGM and EPM consider that the asperities on one side is rigid body. So that only one side asperities can deform. In practice fact, the two contact material always have the different stiffness, so that asperities on both the two contact surfaces will deform, despite the different degree of deformation. Does this simplified way affect the accuracy of these friction model? Can it be more accuracy if we consider two side deformation?

2) A widely accepted view is that, the shape of asperity may be circular cone like [5] [6]. Since the height and diameter of asperity are not the same, the stiffness of asperities cannot be the same, so even on the same side, not all the asperities can be contacted or deform at the same time, so, when some asperities deform elastically, other asperities may deform plastically or even not deform. How we calculate these deform, can an average value of deform be enough? In practice, the height of asperities is based on the distribution of probability density function [5]. This may be considered in some new friction model building.

3) As we all know, in the static friction regime, the value of friction between the two contact surfaces is equal to the value of external force. But how to confirm the value of the maximum static friction force? The most popular view is that, the value of the maximum static friction equals to the product of static friction coefficient and the normal stress (1).

$$f_s = \mu_s \cdot F_n \tag{1}$$

In practice, we always test the maximum static friction first, and then calculate the value of static friction coefficient. One of the most popular view is that, the static friction coefficient is independent of the size of the contact area [5]. We suppose that asperities distribute on the surface which contact uniformly in the micro scale, so that, the bigger contact surface is, the more asperities contact each other, the larger force is needed to push slider to move. But on the other hand, with the same weight, the larger contact area is, the less pressure exist between the two contacted surfaces, so the less asperities stagger each other, the smaller force is needed to pull slider to move. So whether the friction force is related to the size of contact area, also need more research.

4) In the presliding regime, can the deformation of asperity fully recover? As we know, when most of the asperities deform plastically, slider will slip. So if an external force is applied on the slider, which value is close to the maximum static friction force, when we remove this force, can slider move back to the initial location? The LGM and EPM propose two different description, they will be introduced in the following section, and the real presliding phenomenon will be proposed too.

In the next section, two classical friction models will be introduced and their performance in presliding description will be compared.

3   Introduction of Friction Model

In the past few decades, many friction models had been developed. A suitable friction model can help us to understand the friction theory correctly and predict friction behavior, and also can help us to design, analysis, control, compensate the mechanical system contain friction [3] [4]. So far, there are some very classical friction models, which are widely accepted and applied in research of control system design, two of them are LuGre model [7] [8] and Elasto-Plastic model. They will be introduced respectively in the following part.

3.1   LuGre Model

LuGre model is developed based on Daul model, compared with Daul model, LuGre model can describe the static friction, Stribeck effect, viscous friction, and hysteresis friction. Its equations are as follows [9].

Where σ_0 is the asperity stiffness, and σ_1 is the damping coefficient of asperity, and σ_2 is the damping coefficient in macro scale. v is the relative velocity between two contact surfaces, v_s is the Stribeck velocity, f_s is the maximum static friction force, f_c is the dynamic friction force, and z is the average deformation of asperities.

$$ f = \sigma_0 \cdot z + \sigma_1 \cdot \dot{z} + \sigma_2 \cdot v \tag{2} $$

$$ \dot{z} = v - \frac{\sigma_0}{g(v)} \cdot z \cdot |v| \tag{3} $$

$$ g(v) = f_c + (f_s - f_c)e^{-(\frac{v}{v_s})^2} \tag{4} $$

Different with the Daul friction model use the random amount to describe asperities' deformation, LuGre model describe the deformation of asperity with the average deformation, which make the model more simple and easy to use.

3.2   Elasto-Plastic Model

Elasto-Plastic Model is developed based on the LuGre model. One of its breakthroughs is that, it classified the deformation of asperity into elastic and plastic deformation, and give the way to calculate them. Its equations are as follows [10] [11].

$$ f = \sigma_0 \cdot z + \sigma_1 \cdot \dot{z} + \sigma_2 \cdot v \tag{5} $$

$$ \dot{z} = v(1 - \alpha \cdot \frac{z}{z_{ss}}) \tag{6} $$

$$ z_{ss} = \frac{f_c + (f_s - f_c)e^{-(\frac{v}{v_s})^2}}{\sigma_0} \tag{7} $$

$$ \alpha = \begin{cases} 0 & |z| \le z_{ba} \\ \alpha_m & z_{ba} < |z| < z_{ss} \quad sgn(v) = sgn(z) \\ 1 & |z| \ge z_{ss} \\ 0 & \qquad\qquad\qquad sgn(v) \ne sgn(z) \end{cases} \tag{8} $$

$$\alpha_m = 0.5 \times (\sin(\frac{z - 0.5 \times (z_{ss} + z_{ba})}{z_{ss} - z_{ba}} \cdot \pi) + 1) \qquad (9)$$

Where z_{ss} is the stable asperity deformation when slider is sliding, z_{ba} is the deformation when asperities breakaway. α is a coefficient, which can distinguish the slider is in static regime, or presliding regime, or sliding regime. Other coefficients, such as σ_0, σ_1, σ_2, f_c, f_s, v_s, have the same meaning with the LuGre Model.

According to this model, when z is smaller than z_{ba}, the deformation of asperity will be pure elastic deformation, when the external force is removed, the deformation can recover, and slider can back to the initial position. When z is larger than z_{ba}, and smaller than z_{ss}, the deformation will be elastic mixed with plastic, so when the external force removed, the deformation may not recover all.

This model will do a better performance if $z_{ba}=0.7169z_{ss}$, so that nearly 71.69% of the friction state's motion, presliding displacement is purely elastic [10] [11].

4   Presliding Simulation

4.1   Introduction of Stick-Slip Stage

A mechanism as follows is proposed to process stick-slip actuation.

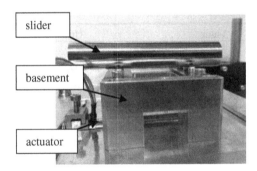

Fig. 4. Stick-Slip mechanism

This mechanism consists of three parts: slider, basement, and actuator. Slider is a chrome plated steel cylinder, which is supported by four ruby balls. The ruby balls are fixed on the basement, which is actuated by actuator (piezoelectric ceramics). Input suitable voltage signal to the piezoelectric ceramics, the actuator will actuate basement to move back and forth in a high frequency, so the slider will move towards one direction.

Since the slider is actuated by friction force supplied by ruby balls, so whether the slider will move a displacement, when the actuated force is smaller than the maximum static friction force? A simulation and one experiment have been done to investigate the performance of these friction models.

4.2   Simulations with Different Friction Model

We use MATLAB/Simulink and Adams to do some co-simulations. The diagram of simulation is shown in Fig. 5.

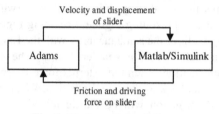

Fig. 5. Diagram of simulation

The external force applied on slider is calculated by Simulink and input to Adams, and the velocity and displacement of slider are output from Adams to Simulink to calculate friction.

In simulation, we propose different external force to the slider, compare the displacement of the slider calculated by different friction models.

The parameters of these simulations are shown in table 1, some of them are according to [9].

Table 1. the parameters of simulations

Parameters	Value	Unit
σ0	10^5	N/m
σ1	$\sqrt{10^5}$	Ns/m
σ2	0.4	Ns/m
m	1.76	kg
μ_c	0.1	-
μ_s	0.15	-
fc	1.72	N
fs	2.58	N
vs	0.001	m/s

As shown in table 1, the mass of slider is 1.76 kg, the dynamic friction force is 1.72 N, the maximum static friction force is 2.58 N.

Two simulations have been done in this paper. In the first simulation, an external force is applied on the slider, which rise slowly to 1.6 N (less than the dynamic friction force), keep for a while and decrease to zero slowly, keep for a while and rise slowly again; and in the second simulation, the same type external force with the same frequency is applied, but the different amplitude, which is 2.4 N (larger than the dynamic friction force f_c, but smaller than the maximum static friction froce f_s). The results of these two simulations are shown as below.

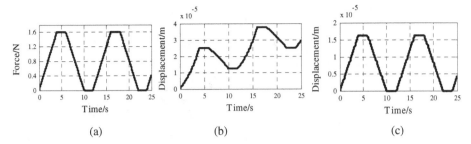

Fig. 6. The external force (a) and the responces of LGM (b) and EPM (c) in first experiment

As shown in Fig. 6, when the external force is smaller than the dynamic friction force, the displacement of slider simulated by LGM is drifting as the external force is applying, and the displacement of slider simulated by EPM has no drift happens. This is as same as the conclusion made in [10] [11].

When the external force is larger than f_c and smaller than f_s, just like Fig. 7 (a) shows, both the results simulated by LGM and EPM are drifting as the external force is applying. But in EPM, the slider has a smaller displacement than LGM.

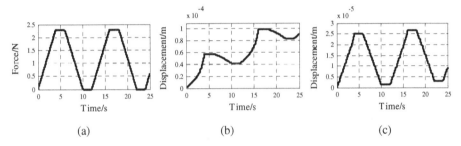

Fig. 7. The external force (a) and the responces of LGM (b) and EPM (c) second experiment

As we all know, if we push a heavy object, when the force is smaller than the maximum static friction force, we could not move this object at all. But an different result is made in these two simulation above, espacially in the second simulation, both the two friction models have the drifting phenomenon.

One experiment will be done in the next section, in order to check whether these two friction model can describe the real presliding phenomenon.

4.3   Experiment

We choose one chrome plated steel cylinder, which length is 80mm and diameter is 30mm, as the slider. A digital force gauge is fixed on one linear stage, and the front of the digital force gauge touch the cylinder, as the Fig. 8 shows. When the linear stage moves forward in a fixed velocity, we can detect the dynamic friction force of the stick-slip stage. The dynamic friction force of this stage is detected and shown as Fig. 9. The mean value of this friction is 0.4751 N, which is calculated by matlab.

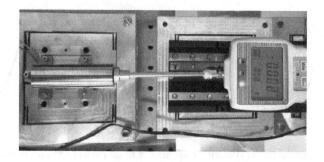

Fig. 8. Detect the dynamic friction of Stick-Slip Stage

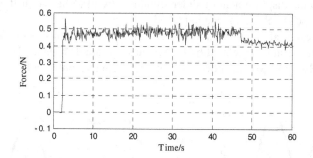

Fig. 9. The dynamice friction force of Stick-Slip Stage

In the next part, we will pull the slider with two forces in different values, one is less than the dynamic friction force, and the other one is between the dynamic and the max static friction force. Laser Interferometer is used to dectect the displacement of slider in different external force. The result is shown as follows.

When the external force is smaller than the dynamic friction force f_c, the deformation of asperities will be pure elastic deformation, so when remove the external force, the displacement of slider will drop to zero and slider back to the initial position, just like the Fig. 10 shows.

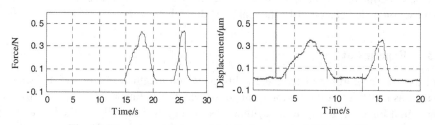

Fig. 10. The external force and the displacement of slider ($F<f_c$)

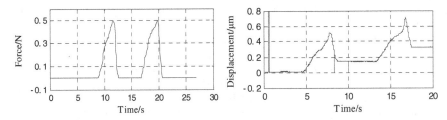

Fig. 11. The external force and the displacement of slider ($f_c<F<f_s$)

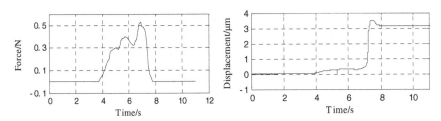

Fig. 12. The external force and the displacement of slider ($F>f_s$)

In Fig. 11, if the external force is larger than the dynamic friction force and smaller than the maximum static friction force, the deformation of asperities will be mixed by elastic and plastic deformation, so when the external force is removed, the elstic deformation recover, and the plastic deformation cannot be revoverd,so the slider will have a small displacment in nanoscale.

At last, in order to be compared with the two conditions above, when the external rises up to bigger than the maximum static friction force, most of the asperities' deformation will be plastic, so the slider slides.

5   Conclusion and Future Work

In this paper, some problems about presliding have been discussed. There are still many problems in this regime, which have not been clearly solved, need more research.

Two simulations and one experiment have been done, and found that, the LuGre Model can not reflect the presliding phenomenon in evry condition, so it is not fit for the research of presliding. And the elasto-plastic model can describe the presliding phenomenon corectly.

In the future, more research should be done on the deformation of asperity, and one new simple friction model should be proposed to describe this complicated friction phenomenon.

References

1. Lampaert, V., Al-Bender, F., Swevers, J.: A generalized maxwell-slip friction model appropriate for control purposes. In: PhysCon, St. Petersburg, pp. 1170–1177 (2003)

2. Haessig, D., Friedland, B.: On the modeling and simulation of friction. In: Proceeding of 1990 American Control Conference, pp. 1256–1261 (1990)
3. Liu, L., Liu, H., Wu, Z., Wang, Z.: Advances in mechanical systems friction model research. Advances in Mechanics 2, 201–213 (2008)
4. Yuan, D., Liu, L., Liu, H., Wu, Z., Wang, Z.: Research Status of sliding friction model. Journal of System Simulation, 1142–1147 (2009)
5. Ping, H., Tianmao, L.: Friction Model Based on Real Contact Area. Journal of South China University of Technology (Natural Science Edition) 10, 109–114 (2012)
6. Greenwood, J.A., Williamson, J.B.P.: Contact of nominally flat surfaces. In: Proceedings of the Royal Society of London. Series A, Mathematical and Physical Sciences, pp. 300–318 (1966)
7. Zhong, B., Sun, L., Chen, L., Wang, Z.: The dynamics study of the stick-slip driving system based on lugre dynamic friction model. In: Proceedings of the 2011 IEEE International Conference on Mechatronics and Automation, Beijing, pp. 584–589 (2011)
8. Sun, L., Wang, S., Song, Y., Li, M.: Model of a 3-dof spherical micromanipulator based on stick-slip. In: Proceedings of the IEEE International Conference on Mechatronics and Automation, pp. 1581–1585 (2005)
9. de Canudas, C., Wit, H., Olsson, K.J.: Astrom. A new model for control of system with friction. IEEE Transactions on Automatic Control, 419–425 (1995)
10. Dupont, P., Hayward, V., Armstrong, B., Altpeter, F.: Single state elastoplastic friction models. IEEE transactions on Automatic Control, 787–792 (2002)
11. Dupont, P., Armstrong, B., Hanyward, V.: Elasto-Plastic friction model: contact compliance and stiction. In: American Control Conference, vol. 2, pp. 1072–1077 (2000)
12. Miao, Y.: Study of slip-stick precision positioner for nanomanipulation in SEM. Harbin Institute of Technology, Harbin (2011)
13. Wang, S.: Research and model of friction drive theory based on stick-slip effect. Harbin Institute of Technology, Harbin (2006)
14. Ishikawa, J., Tei, S., Hoshino, D., Izutsu, M., Kamamichi, N.: Friction compensation based on the lugre friction model. In: SICE Annual Conference, Taipei, pp. 9–12 (2010)
15. Freidovich, L., Robertsson, A., Shiriaev, A., Johansson, R.: Lugre-model-based friction compensation. IEEE Transactions on Control Systems Technology 18, 194–200 (2010)

Fieldbus Network System Using Dynamic Precedence Queue (DPQ) Algorithm in CAN Network

Hwi-Myung Ha, Wang ZhiTao, and Jang-Myung Lee

Department of Electrical and Computer Engineering, Pusan National University, South Korea
{hwimyung7379,zhitao7379,jmlee}@pusan.ac.kr

Abstract. This paper proposes a fieldbus network system for oil tank by using a DPQ (Distributed Precedence Queue) algorithm that collects the state of oil tank, temperature and pressure of tank through the CAN (Controller Area Network) network. The CAN is developed by Bosch corp. In the early 1980's for automobile network. The data from various sensors in the flowing-water tank are converted to digital by the analog to digital converter and formatted to fit the CAN protocol at the CAN module. Nine CAN modules are connected to another CAN Modules through a CAN BUS for the efficient communications and processing. This design reduces the cost for wiring and improves the data transmission reliability by recognizing the sensor errors and data transmission errors. Also each CAN Module processes sensor's error data, transmission errors data and it can be served as a master-slave system as well. Via the real experiments, effectiveness of the Fieldbus Network System using DPQ algorithm in CAN Network is verified.

Keywords: CAN, Fieldbus network, oil tank, DPQ algorithm, Monitoring system, Receiving rate.

1 Introduction

In the oil tank, is important to check the status of the each part and then check the status of observation or information about oil tank. In practically, engineer can check the internal state of the oil tank that means directly exposing with the negligent accident. For the reason of diagnosing the state of the water level, temperature, pressure, water leak rate and so on, it is important to check real-time oil tank monitoring system. Nowadays, the oil tank and variety of applications including fieldbus network are researched actively underway. Accordingly, in the industrial environment of the network, it is needed to have a higher reliability and stability of the system. Therefore CAN (Controller Area Network) communication via the fieldbus network-based real-time monitoring is a dominant interest.

CAN is half-duplex communication and two twist pair wires is used to send short message, which is suitable for the system of high-speed applications. CAN communication is determined by priority. Therefore, problem of CAN is not guaranteed low priority of the node. In this paper, to solve the problem a node of the low Priority receiving rate guarantee and proposed each individual treatment system

X. Zhang et al. (Eds.): ICIRA 2014, Part II, LNAI 8918, pp. 377–385, 2014.
© Springer International Publishing Switzerland 2014

for applying Using DPQ algorithm CAN fieldbus system. Also CAN module can be served as a master-slave, depending on user's role, it can be master or slave, so it will be more efficient implementation of the fieldbus network.

In the second chapter, the basic theory of CAN is presented briefly while in the third chapter, it is described about DPQ algorithm. System organization and system result analysis via experiment will be shown in fourth chapter and fifth chapter, respectively.

This system allows to compare the general CAN Receiving rate by using DPQ algorithm Receiving rate base on the pros and efficiency of DPQ algorithm. As a result base on fieldbus network system to the real-time monitoring system of oil tank using DPQ algorithm, it can be applied to the processing industry.

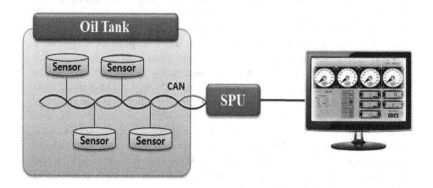

Fig. 1. Oil tank using CAN Fieldbus Network

2 CAN (Controller Area Network) Analysis

2.1 CAN Feature

In 1986, CAN protocol was officially released by Bosch.corp for automobile of controller, sensor, actuator, ABS, audio system, engine control system and so on. It is a serial communication protocol, based on the Human-Machine Interface, it manages the communication of rational way. CAN provided high data rates, stability and reliability according to distance. For example, with the high rates up to 1Mbps, automatic retransmission of collided messages and 15bit Cycle Redundancy Check (CRC) using error detection. So it interconnected variable electric control system and effectively support to distributed control network of real-time control.

The CAN bus which is commonly used in Embedded Systems (or Microcontroller) is formed the communication network between the Microcontrollers. The half-duplex communication connected between two twist pair wires is used to send short message, which is suitable for the system of high-speed applications. In addition to the external factors (such as noise), communication system can minimize the error rate because of the toughness, which shows the high reliability of the system. In theory, 2032 different devices (Embedded Controller) connected to the single communication network is performed, but due to the limitations of CAN transceiver 110 different devices can be connected to a node.

2.2   CAN Frames Structure

Fundamentally, as according to standard division CAN protocol (serial, asynchronous, time sharing multiple clock/data connected) is classified CAN 'A', 'B' of the ISO / OSI (International Standardization Organization/Open Systems Interconnect) model. CAN 2.0A and 2.0B version is also compatible with any previous protocol. CAN 2.0A is composed 11 bits while extended CAN 2.0B is composed 29 bits. Main properties of CAN BUS protocol structure are pecking order, latencies guarantee, flexible configuration, and time synchronization based on multi-source can accommodate. Additionally, Multi-master capability, error detect, transmission, retransmission of error message, time error, division of non-functional permanent and automatic disconnection of the wrong node have attributes as well.

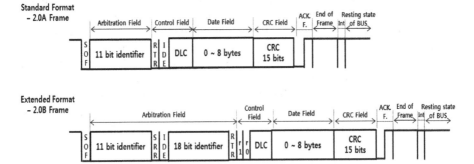

Fig. 2. Configuration of CAN data frames

Table 1. Type of CAN frame

Type	Role
Data Frame	Transports data from the sender to the recipient.
Remote Frame	Sends a remote frame for requesting transmission of a data frame node that is enabled on the bus and has the same value as the identifier of the remote frame.
Error Frame	Any node on the bus will send an error frame when error is detected on the bus.
Overload Frame	It is used in data frames transmitted previously or a data frame to be transmitted subsequently, in order to request additional time difference between the remote frame
Inter Frame	Inter frame is used to separate the remote frame or the data frame that was already transmitted using the inter-frame space.

3   DPQ (Dynamics Precedence Queue) Algorithm

CAN strictly assigns each object exchanged in the network a priority that corresponds to the identifier of the object itself. Even though this algorithm enforces a deterministic arbitration which is able to resolve any conflict which occurs when several nodes start transmitting at the same time, it is clearly unfair. If many nodes are connected on the network, nodes which are low priority rank can continuously lose a transmission opportunity. Namely, If high priority object transmit continuously, finally low priority object can lead to dangerous situations due to missing important message which is relatively not important than high priority object. In accordance with this, it is necessary to have the algorithm which uses a relative priority according to considering low priority nodes although CAN implicitly assigns a priority. Fair behavior, which, for example, enforces a round-robin policy among the different stations, has to be guaranteed to all the objects which exchanged at a given priority level.

In this paper, it is shown that this kind of behavior can be obtained by slightly modifying the frame acceptance filtering function of the LLC sub-layer. In particular, only the significance of the identifier field in the transmitted frame has to be modified in some way. The resulting arbitration mechanism is able to enforce a round-robin policy among the stations which want to transmit a message on the bus, and provides two levels of a priority for the frame transmission services. At this time, both ID of the standard CAN protocol and ID of the DPQ protocol is used to determine the priority of message transmission. Little or nothing has to be changed at the MAC level and in this way it is possible to reuse the same electronics components

3.1   DPQ Algorithm Definition

The basic idea of this CAN fairness control algorithm which is introduced at this paper is to insert into a global queue all the nodes which want to transmit over the shared medium. A node D of which transmission is continuously delayed like Figure 3 creates queue to transmit node standard D and the other nodes which transmit with D, so several queue can be made partially. In this research, two queues will be considered.

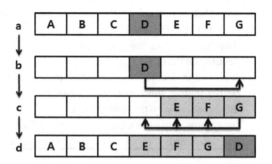

Fig. 3. Generation of a precedence queue in DPQ algorithm

This dynamic precedence queue protocol provides opportunity to create precedence queue to all nodes on the network. Additionally, in case of existing several precedence queues, each precedence queue assigns a priority lank. So they can implement independently. Field at the Figure 5 indicates precedence queue which orders itself of each node. Whenever a node carries out a transmission, it moves to the end of the queue, thus lowering its precedence to the minimum. All the nodes are following the transmitting node in the queue advance by one position, filling up the space which has just been created. Using this round-robin policy, it is implemented to manage collision among the messages. The queue is not physically stored in some specific location. Instead it is distributed among all the nodes on the network. Each node is responsible for storing and updating. Namely, if it comes permission maximum delay time of itself, it creates precedence queue, then it has to change dynamically priority to transmit preferentially with relatively nodes. And a precedence queue has to be disjointed when finished an urgent situation. Like Figure.3, we suppose a network which is composed from A to G. If the node D builds up queue, ID which is entered data frame queue can transmit and designate to 7 by lower 7 byte. At this time, it will be designated precedence priority to higher byte. Then, each node is filtering to enter itself to the queue, and it assigns its queue, After D transmits message which is desired, it will go the last position in the queue. And the other nodes will go the upper queue by order. And the reminder nodes which are transmitted are designated using the upper 1byte in the Figure 4; they will be disjointed or maintained using the upper 1 byte in the Figure 4 after finished all transmission

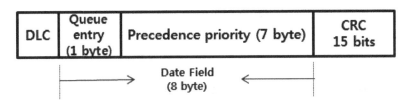

Fig. 4. Structure of a data field for DPQ

At this point, it is necessary to point out the differences between the priority and the precedence. At first, the priority is assigned to an object (message) when it is sent on the network, it is designated transmission order to precedence in the queue when the queue is occurred. Namely, the priority is assigned by importance and emergency, the precedence express transmission order in the queue. Both are related to the deterministic resolution of collisions on the bus, the priority is static value, but the precedence that can't be verified by user is dynamic value. The priority and the precedence of the message are assigned 18 bits ID and 11 bits ID in the each transmission frame. This way is efficient method to resolve when a collision occurs.

3.2 DPQ Algorithm Realization Method

The DPQ algorithm can be implemented without changing any modifications to the basic format of CAN frames. It is used identified field to designate the priority queue.

Since the length of the conventional identifier field defined in the CAN standard is too small, the CAN extended format is adopted. DPQ uses the first 11 bits of the identifier field for its control information, while the remaining lower order 18bits (ID ext.) are used to store newly and dynamically the effective identifier if the exchanged object (EID). The first two bits (t0, t1) must be set at the logical value zero like Figure 5. Then, the protocol is divided by a standard CAN communication and DPQ frames. Therefore, DPQ has always a higher priority than CAN frames, and they can exist at same space. The priority bit P specifies whether the frame has to be transmitted as a high priority frame (P=0) or as a low priority frame (p=1). When T1 and P are used virtually, the priority can be assigned maximize 4 queues. The next 8 bits represent the precedence level of the frame. Namely, these 8 bits show transmission queue order. DPQ, which is used at this research use t0, t1 to distinguish the standard CAN frames, sets each queue Using P, and concludes the precedence in the queue using 8 bits.

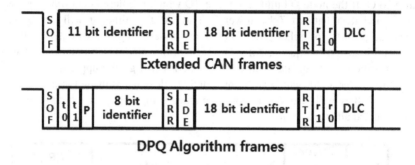

Fig. 5. Format of the header of Extended CAN frames and DPQ frames

4   System Organization and Experiments

4.1   System Organization

The algorithm in this paper is verifying the effectiveness of one SPU (Signal Processing Unit) module and 9 transceiver modules which is connected by CAN bus line. It is constituted nine transceiver module are received to the SPU module and then sent to the PC using RS232 serial communication. Using the custom-built MFC program derives the result. And proposes to the real-time monitoring system of oil tank using DPQ algorithm. Figure 6 shows Oil tank Monitoring system structure

4.2   System Experiments

Figure 7 shows the overall system of Oil tank Monitoring system structure. It is hardware structure of the CAN module and MFC application. Each object has ARM-Cortex M3 (STM32F107) which uses built-in CAN module. Each CAN module can be playing a role as SPU through the CAN BUS. SPU module is available to transmitted integrated processing and individual processing data into the system.

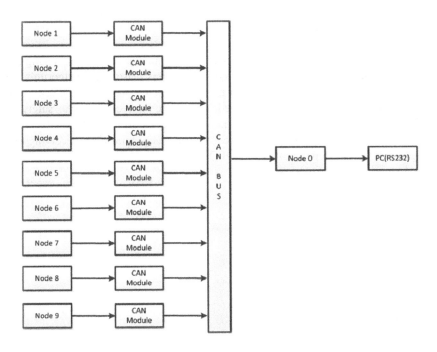

Fig. 6. Oil tank Monitoring system structure

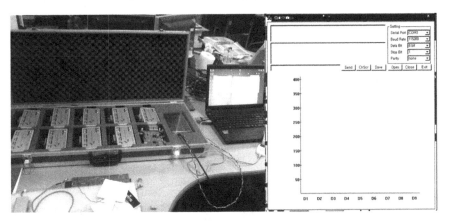

Fig. 7. The organization of system.

CAN module can also running as a master-slave system. Experimental systems are using same hardware and communication and then check the MFC application to real-time monitoring system. Each Device (1~9) uses MFC application and stores in the 1000 buffer with the graphed results

5   Result and Analysis

In case of faster speed CAN communication or large number of devices, having low priority node data is a problem to poorly receive. For this reason, DPQ algorithm can be applied. Relatively low priority node is guaranteed for the receiving rate. Also keeping track of each object is controlled by setting for receiving rate safety. In experiment, the general CAN Receiving rate and DPQ algorithm in CAN Receiving rate through MFC application are compared. Figure 8 is general CAN receiving rate, and figure 9 is CAN receiving rate using DPQ algorithm.

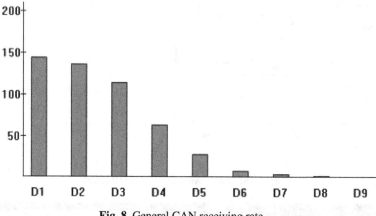

Fig. 8. General CAN receiving rate

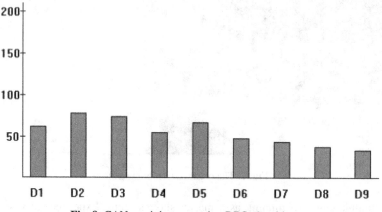

Fig. 9. CAN receiving rate using DPQ algorithm

At the Figure 8, it is CAN mode (2.0B)'s results in experiment, high priority node data is received but relatively low priority node data is received poorly. For this graph, CAN Receiving rate are device 1 is highest and Device is 9 lowest. Because it is priority rule of CAN system.

Figure 9 is the result on experiment of using DPQ algorithm in CAN receiving rate. Unlike the existing general CAN Receiving rate, relatively low priority node data and constant type of CAN receiving rate are guaranteed. For this graph, using DPQ algorithm on CAN Receiving rate is confirmed stable.

6   Conclusion

This paper is using general CAN protocol suitable real-time or distributed processing and it adjusts to the network collision. For this method, to improve the inefficiency of the priority, DPQ algorithm is used following by analyzing the result through real-time oil tank monitoring system. Using DPQ algorithm is a low priority to the node of Receiving rate was found to be guaranteed. Also CAN module can be served as a master-slave, depending on user's role, it can be master or slave. Therefore, this system can be used for the advantage in the industrial applications.

In the future research, how this developed algorithm can be applied conveniently for the other CAN application compatibility will be considered. In term of pros and cons of DPQ algorithm, it is need to in-depth study. It will be need to developed algorithm with efficient management of each object DPQ algorithm.

Acknowledgments. "This research was financially supported by the Ministry Of Trade, Industry & En-ergy(MOTIE), Korea Institute for Advancement of Technology(KIAT) and DongNam Institute For Regional Program Evaluation(IRPE) through the Leading Industry Development for Economic Region."

References

1. Bosch, CAN Specification Version 2.0., Robert Bosch GmbH. Stuttgard (1991)
2. Tindell, K., Burns, A.: Guaranteeing Message Latencies on Control Area Network (CAN). In: Proc. 1st International CAN Conference, Mainz, Germany (September 1994)
3. Lee, H.S., Lee, J.M.: A Dynamic Precedence Queue Mechanism to Assign Efficient Bandwidth in CAN Networks. In: Conference of Advanced Intelligent Mechatronics Monterey, USA (July 2005)
4. Dominique paret, Réseaux multiplexés pour systémes embarqués: CAN, LIN, flexray, saafe-by-wire.., acornpub (2005)
5. Kim, M.H., Lee, J.G., Lee, S., Lee, K.C.: A study on distributed message allocation method of CAN system with dual communication channels. Journal of Institute of Control, Robotics and System, 1018–1023 (October 2010)
6. Jung, J.W., Kim, D.S.: Real-time synchronization algorithm for industrial hybrid networks: CAN and sensor network. Journal of Institute of Control, Robotics and System 16(2) (February 2010)
7. Lee, H., Lee, J.S., Lee, J.M.: Marine Engine State Monitoring System using DPQ in CAN Network. Journal of Institute of Control, Robotics and System, 13–20 (March 2012)
8. Lee, H., Lee, H.W., Lee, J.M.: Real-time Marine Engine State Monitoring Using fuzzy-based Dynamic CAN Priority. Journal of Advanced Science Letters, 370–374 (July 2012)

Research and Implement of 3D Interactive Surface Cutting Based on the Visualization Toolkit

Guoxian Feng, Zhanpeng Huang[*], and Tao Chen

College of Medical Information Engineering, GuangDong Pharmaceutical University,
Guangzhou 510006, China
tozhanpeng@sohu.com

Abstract. Surgical simulation increasingly appears to be an essential aspect of tomorrow's surgery. In this paper we present a preliminary work on virtual reality applied to liver surgery. The area of the liver, intra-hepatic arteries, veins, the portal vein and tumor have been extracted based on the seeded region growing algorithm. Then a realistic three-dimensional image was created, including the liver and the four internal vessels. After that, a user graphical interface was designed to interactively adjust the surgical resection planes. The surgical simulation platform was developed by providing a 3D visualization and virtual cutting system of liver based on personal computer. Our preliminary result shows that the visualization of liver and internal vessels and interactive modifications of the resection plane could present with a vivid profile.

Keywords: Visualization Toolkit, surface intersection, plane cutting, sphere cutting.

1 Introduction

Liver carcinoma belongs to the most wide-spread malignant diseases world-wide. Primary liver cancer (cancer that starts in the liver) affects approximately 1,000,000 people each year; in 2002, it caused 618,000 deaths, while 45 percent of the deaths has happened in China [1]. Often, the only cure for primary liver cancer is liver resection, which is the surgical removal of various types of liver tumors that are located in the resectable portion of the liver. To elaborate a surgical plan for each patient, doctors must retrieve the required indices during preoperative planning based on the planar slices of CT and MR images [2]. However, without knowledge of the major blood vessels and other important structures related to the lesion, surgery cannot be performed curatively and safely. Reconstruction 3D images have been very useful and widely adopted in neurosurgery [3].

Research work in computer-aided liver surgical planning is primarily found in Europe at the French National Institute for Research in Computer Science and Control (INRIA) [4]. Developed by INRIA, the Epidaure system features a physical liver tissue model to simulate cutting within the liver structure using a force-feedback device.

---

[*] Corresponding author.

X. Zhang et al. (Eds.): ICIRA 2014, Part II, LNAI 8918, pp. 386–391, 2014.
© Springer International Publishing Switzerland 2014

Currently, a mimic resection system on liver has been developed at Southern University (SMU), which construct the liver's surface and its internal structures whit MIMICS and performs optional resection with the hepatic device (PHANTOM) [5]. However, both the systems developed by INRIA and SMU rely on the expensive hepatic device, which may be difficult to promote in the small and medium-sized hospitals.

A 3D visualization and virtual cutting system of liver has been developed based on personal computer without additional devices. The system has been designed by using the Visualization Toolkits (VTK) [6] and realized by VC++ 6.0. Moreover, a user graphical interface is provided to observe the spatial relationship among liver and internal vessels by adjusting the transparency of the object.

2   Principles and Methods of 3D Resection

In order to simulate the operation of the liver resection, the cutting plane should be interactively adjusted by the users. And the system needs to demonstrate the spatial relationship of the liver and the internal vessels after the cutting operations, which is useful for surgeons to choose the proper surgery program.

2.1   The Principle of Cutting Operation by the Plane and Sphere

The liver and the internal vessels are visualized by surface rendering method. In order to simulate the resection of lesion, we have to consider the intersection between the cutting plane and the liver. The surfaces of liver and the internal vessels are composed of small triangles, and we need to consider the space relationship between cutting plane or sphere and the triangles of all the surfaces.

The cutting plane is defined as:

$$ax + by + cz + d = 0 \qquad (1)$$

There is a vertex P (x, y, z) in the 3D data space, and the space relationship between the point and the cutting plane could be calculated by:

$$F = ax + by + cz + d \qquad (2)$$

s

The point P is on the cutting plane when F equals to 0. If F is bigger then 0, the point P is in front of the cutting plane. Otherwise, the point P is at the back of the cutting plane. We can retrieve all the vertices of surfaces and determine the remaining vertices based on the relationship between vertices and cutting plane.

For each triangle, there are three vertices P_1, P_2 and P_3. The relationship between the vertices and cutting plane can be calculated by Eq. (2), and the results are defined as F_1, F_2 and F_3. The triangle will be preserved if all the values of three vertices are no smaller than 0, which means that all the three vertices of the triangle are in front of the cutting plane. However, the triangle should be deleted if all the values are no more than 0. For the other situations, we need to delete some vertices and reconstruct the triangle [7].

IF there is only one vertex of the triangle in front of the cutting plane, which is labelled as P_1. And the other two vertices P_2 and P_3 are at the back of cutting plane. Those two vertices P_2 and P_3 will be deleted after the cutting operation. The Intersection points between the cutting plane and the line P_1P_2 and line P_1P_3 are defined as P_4 and P_5 separately, which need to be calculated. As shown in Fig 1 (a), the new triangle consists of vertexes P_1, P_4 and P_5 after the cutting operation. For the other situation, there is only one vertex of the triangle at the back of the cutting plane, and there are two vertexes in front of the cutting plane. As the same, the vertex at the back of the cutting plane will be deleted, and the new triangle consists of $P_1P_4P_5$ after the cutting operation [7]. The schematic diagram is shown in Fig.1 (b).

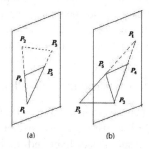

Fig. 1. Intersection among the cutting plane and triangular vertices

In three-dimensional data space, there is a sphere with centre in O(a,b,c) and radius R, which is defined as following:

$$(x-a)^2 + (y-b)^2 + (z-c)^2 = R^2 \qquad (3)$$

where (x,y,z) are the values of three-dimensional coordinate of the points on the surface of the sphere. We define P(x,y,z) as a vertex of the triangle, and the space relationship between the vertex and the sphere could be calculated by:

$$G = (x-a)^2 + (y-b)^2 + (z-c)^2 - R^2 \qquad (4)$$

The vertex is inside the sphere when G is no more than 0. Otherwise, the vertex is outside the sphere. As the same with cutting operation of plane, we can get the cutting results of sphere after the cutting operation.

2.2   Cutting Operation Based on the VTK

The plane and sphere are used to simulate the cutting operation, which has been implemented based on VC++ 6.0 and the VTK. The flow charts of the cutting operation based on the plane are shown in Fig. 2. For the cutting operation of the plane, the object vtkBYUReader is used to read the data files. In order to appoint the data source and the values of cutting function, our implementation is based on two key methods: SetInput() and SetClipFunction() of the vtkClipPolyData class separately. In our application, the values of cutting function are controlled by the vtkPlane, which is an attribution of the vtkImplicitPlaneWidget class.

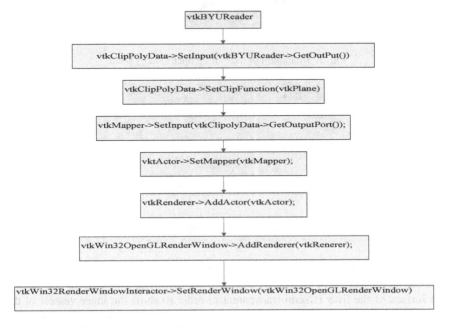

Fig. 2. Flow chart of the cutting operation based on the plane

The cutting operation based on the sphere is the same as the cutting operation by the plane. However, the vtkSphere class is used to specify the SetClipFunction() method of the vtkClipPolyData class, which is controlled by the vtkSphereWidget class.

3   Result

A visualization and simulation system has been developed based on VC++ 6.0 and the VTK. The region growing method has been used to extract the liver, intra-hepatic arteries, veins and the portal vein [8]. Then the contour tiling method has been used for surface reconstruction of the liver and internal vessels. And a user graphical interface was provided to observe the spatial relationship between the liver and blood vessels by adjusting the transparency of the object.

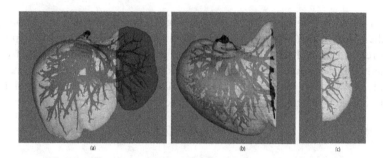

Fig. 3. Simulation of the cutting operation by plane

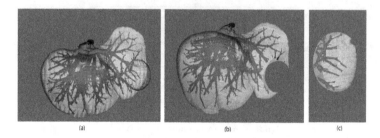

Fig. 4. Simulation of the cutting operation by sphere

Two basic interaction approaches for virtual resections were developed. Users can simulate the operation of cutting by interactively adjusting the plane or the sphere. The surface of the liver is semi-transparent in order to show the inner vessels of the liver. The colour of intra-hepatic arteries is red, and the orange colour and the blue colour are used to indicate the veins and the portal vein separately, as shown in Fig.3 (a). The darker area dynamically shows the cutting area. The results of cutting operation by the selected plane are shown in Fig. 3(b) and Fig. 3(c). As the same, the cutting operations by the sphere are interpreted in Fig. 4. By the visualization of liver, intra-hepatic arteries, veins and the portal vein, the space relationship is clearly demonstrated, which is useful for surgery planning. At the same time, the cutting operations roughly simulate the liver resection, and help the surgeons to comprehend the primary result of the surgery of liver resection.

4   Conclusion

The rapid development of computer technology offers many possibilities in diagnosis and surgery planning. With MDCT, thin slices are obtained with excellent temporal resolution, and precise 3D models can be created. We have successfully developed a PC-based system for liver surgery planning and simulation in 3D environment.

Acknowledgments. This work is supported by the Innovation and Entrepreneurship Training Project for Undergraduate of Guangdong Province (1057313035). And the authors would like to thank Zhu-Jiang hospital for providing the datasets used in this paper.

References

1. World Health Organization. The World Health Report (2002), http://www.who.int/whr/2002/en
2. Numminen, K., Sipila, O., Makisalo, H.: Preoperative hepatic 3D models: virtual liver resection using three-dimensional imaging technique. Eur. J. Radiol. 56(2), 179–184 (2005)
3. Kockro, R.A., Serra, L., Tseng-Tsai, Y., et al.: Planning and simulation of neurosurgery in a virtual reality environment. Neurosurgery 46, 118–137 (2000)
4. Ayache, N.: Epidaure: A Research Project in Medical Image Analysis,Simulation and Robotics at INRI. IEEE Trans. Medical Imaging 22(10), 1185–1201 (2003)
5. Fang, C.-H., Yang, J., Fan, Y.-F., et al.: The Research of vertual hepatectomy. China J. Surg. 45(11), 753–755 (2007)
6. Kitware, Inc. The VTK user's guide. USA, http://www.vtk.org.com
7. Huang, Z., Yi, F., Bao, S.: Three-dimensional visualization and simulation of liver resection. Beijing Biomedical Engineering 29(4), 336–339 (2010)
8. Peng, F., Bao, S.: Segmentation of liver and its vessel based on CT sequence image. Computer Engineering and Applicatiions 45(20), 205–207 (2009)

A Model Integration Architecture and Language in Computer Numerical Control System Development

Xin Huang[1], Fang Li[2,*], Zhiheng Wu[1], Qiyu Chen[1], Erbin Liu[3],
Xianyun Duan[1], Ping Li[1], and Zhiwen Deng[1]

[1] Research Institute of Mechanical and Electrical Engineering,
Guangdong General Research Institute of Industrial Technology
Guangzhou, China
[2] Computer Science and Engineering,
South China University of Technology,
Guangzhou, China
[3] Guangzhou Ruisong Technology Company Limited,
Guangzhou, China
{halps,1f8195,wuzhiheng23,lipingcsu}@163.com,
chen.qiyu.cn@ieee.org, liuerbin@gz-ruisong.com,
jensen_dxy@126.com, 44239953@qq.com

Abstract. As the brain of CNC machine, the CNC system faces a great challenge in meeting various kinds of the demands. In the paper, model integration architecture for CNC system design and development is presented. Then, a CNC modeling language, naming CNCMoL is developed, in which the VFC methodology is used to deal with the modeling complexity, and get performance parameters and acquire reusability. The strategies for CNC meta-modeling are described in detail. The approach is attempts to create an infrastructure to support the CNC system design in a swift way, at the same time acquire some advanced capability of the system such as modularity, flexibility, reusability, etc.

Keywords: Model integration architecture and language, CNC.

1 Introduction

CNC system determines the functions as well as the performances of the CNC machine to a degree. With the trend of increasing function and increasing speed and accuracy, the CNC system development faces a great challenge in meeting various kinds of the demands.

From the industrial point of view, open architecture is the fundamental demand in achieving the modern CNC system. At present, most CNC system incorporate proprietary control technologies that have associated problems: non-common interfaces, higher-integration costs, and specialized training. The function change or adding can be rather difficult in the proprietary control system. For example, in a tuning machine, the tool breakage detection function needs to be added to improve the product quality. On the other hand, different CNC machines call for different control

X. Zhang et al. (Eds.): ICIRA 2014, Part II, LNAI 8918, pp. 392–401, 2014.

system functions. For example, the milling machine will have different function unit in the control system with the laser machine etc. The flexibility of the machine type calls for the flexibility of the function union to form the CNC system for high efficiency control system development process to shorten the time-to-market.

From academic point of view, CNC system is a typical embedded control and computing system with strong multidisciplinary characteristics as well as its rigorous performance requirements, for example, real-time performance and safety criticality. The flexibility function demands call for modularity, reusability and reconfigurability during the system design and implementation.

What's more, due to the performance and function variety of the application, it calls for the different computing platform to realize the demands. For example, the general proposed CPU with windows XP plus a RTOS extension are used. While a lot of DSP combine with FPGA modules are also used. At the moment the development of the CNC system is more like an art than a science, to make the system possess more function as well as performance properties.

Due to increasing complexity of CNC systems and the requirement for shortening the development cycle and increasing variety of design choices, the need arises for innovative approaches and greater tool support for development of such systems. So it would be beneficial if the design and development of the CNC system can follow a systematic way and use proper tool set to fulfill the flexibility, modularity and at the same time guarantee the performance such as speed as well as the accuracy of the machine from the early development stage.

The study shown in this paper attempts to create an infrastructure, which supports the CNC design and implementation in a swift way. The remaining of this paper is organized as follow: A short overview of related work is concluded in section 2. Section 3 gives the model integration architecture for the CNC system design. Then, the strategies to develop the CNC domain modeling language described in detail. Finally, in the last part conclusions of this paper is drawn.

2   Related Work

A lot of effective works in developing open CNC system are implemented since 1990s, such as Open System Architecture for Control within Automation (OSACA) in Europe [1], Open Modular Architecture Controller (OMAC) in the USA [2], and Open System Environment for Controller (OSEC) in Japan [3]. In China, the national standard Open Numerical Control system (ONC) has been implemented since 2003. Recently, Researches on open and component based CNC system are becoming a hotspot in CNC system development. Shige Wang etc. from the University of Michigan present the functional and non-functional classification of CNC system to design component-based CNC system [4]. Wang Heng etc. from Tsinghua University put forward the research on software-based open architecture controller platform, and realize the system function component of lathe CNC system [5]. Liu Tao etc.

from Harbin Institute of Technology develop an open architecture soft-CNC system named HITCNC [6]. Whereas, these researches focus on function modularization and interface standardization of CNC system and pay little attention to the performance maintenance of CNC system, then is not so effective to guarantee the reliability of CNC system.

At the same time, model-based design and component-based design for embedded computer control system is currently exploited in many engineering activities, using several domain-specific tools and frameworks. The OMG (Object Management Group) proposed the MDA (Model Driven Architecture) for software development, which uses UML (Unified Modeling Language) as meta-modeling language, takes the model as the center role in the development process, to realize requirement analysis, system design as well as code generation [7]. Obviously, MDA is a good method to make the development process automatic, then to increase the software development efficiency. Whereas, as a unified language, UML doesn't always adapts to all domain. What's more, the performance description in UML is not so convenient.

DSM is the domain-oriented develop approach to deal with the deficiency of UML modeling. It proposed the Domain specific modeling language (DSML) development to precisely describe the domain, which makes the model transformation and code generation much easier for developer. MetaCase from Finland developed a DMSL on the consumption electron producer and the integration environment MetaEdit+[8]. ISIS researched EMSL[9] that specifically on aviation domain. This method is efficiency in system development.

What's more, how to make the embedded system with better quality and more advanced characteristics is the problem that a lot of people try to solve. Thomas A. Henzinger etc. at the University of California at Berkeley develop model of computation (MOC) theory [10] which allows different domains to be described in the same language and allows the user to combine partial models of different kind into a common heterogeneous model. Job van Amerongen etc. from the University of Twente[11] use two core models of computation to describe the total embedded control system and develop a series of toolsets for system modeling, simulation and verification. Christo Angelov etc. at University of Southern Denmark [12] develop Component-based Design of Software for Embedded Systems (COMDES) framework for distributed embedded system design to obtain the reusability and reconfigurability of the system. These methods are effective in their way to develop the corresponding systems.

3   The Model Integration Architecture for CNC system Design

The design and development infrastructure for CNC system is shown in Figure 1. From the development sequence of the CNC system, it can be classified into 2 stages, domain development stage and application development stage, which is divided by dash line in the figure 1.

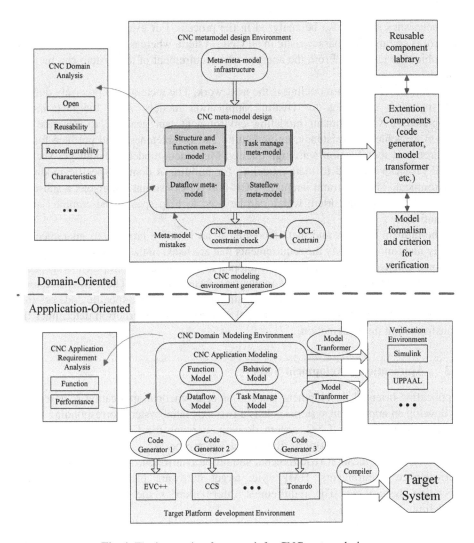

Fig. 1. The integration framework for CNC system design

3.1   Domain Development Stage

Domain development stage is the basis for the application development stage. The meta-model obtained in this stage is the pre-condition for constituting domain modeling environment, which is then used for CNC system modeling. This stage includes domain analysis, CNC meta-model design and some extensive tools development or integration for simulation, verification and code generations.

Domain analysis is the first assignment in the domain development stage. It is a process to analysis the requirements of the CNC domain, for example, the system properties such as open, reusable, reconfigurable etc. What's more, the characteristics of all CNC systems, including the alterable characteristics and immobility characteristics should also be analyzed in the process. For example, the axis motion control is community characteristic in all CNC system, whereas, the axis number is an alterable characteristic. From the analysis, the requirement of the system can be fully found.

Then, the CNC meta-modeling is the next work. The meta-model precisely defines the structures and rules for creating application models. After the meta-model completed, a CNC domain modeling environment can generate to describe the system, to achieve consistency, un-ambiguity and completeness, and produce formal requirements specification with high quality. Meta-meta-modeling languages such as GME [13] can be used to meta-modeling for building a domain specific modeling environment. The constraint language OCL (Object Constraint Language, a subset of UML 2.0) is also supported, which can be used to help specify the complex static semantics.

The domain analysis and meta-modeling is an iterative process, to guarantee the finally meta-model satisfy the requirements that are listed in the domain analysis.

Some extensive tools can also be developed in the domain development stage, such as the code generation interpreters, model transformation tools. These extensive tools realize the transformation from CNC domain model to platform-specific code or other model type. The CNC domain meta-model and target model criterion determines the transformation implementation.

3.2   Application Development Stage

Application-oriented development stage aims at a practical application requirement, to develop an application system such as a two-axis lathe system corresponding to the modeling language for CNC system domain. It owns the CNC domain modeling element, semantic and constrains. It can be a 3-step process.

Firstly, the application requirements should be confirmed, including function and performance requirements, HMI style, target hardware platform and embedded operation system etc. The requirement information is the basis for application modeling.

Secondly, the system application model should be checked and verified to ensure its correctness and reliability. The model will be transformed to some integrated tools such as Simulink or UPPAAL for function simulation and performance verification. These transformations will be performed by different model transformers which have been developed in the domain-oriented stage. The results of the model verification can be used to judge the quality of the model for further modification and optimization.

Thirdly, the verified model will be generated to platform specific source code automatically by the code generator. Different code generators will be used aiming at various target platforms, such as DSP/BIOS platform, WinCE platform, VxWork platform, etc. The generated code can be imported to integration development environment for compiling and linking via no or little modification, which will then be generated to target executive application system.

4   CNCMoL Development

The meta-model precisely defines the structures and rules with the domain specific syntax and semantics. After the meta-model completed, a domain specific modeling language naming CNCMoL can generate to explicitly describe the CNC system. Meta-meta-modeling languages such as MoF[14], Dome[15] and GME[13], can used to meta-modeling for building a domain specific modeling environment. The constraint language OCL (Object Constraint Language, a subset of UML 2.0) is also supported, which can be used to help specify the complex static semantics.

4.1   VFC Methodology Definition

To achieve consistency, un-ambiguity and completeness, and produce formal requirements specification with high quality, a VFC methodology is used for CNC system meta-modeling. VFC methodology means using a multi-view fashion to capture the feature of a domain, and using multi-view components to composite the domain system, then mapping the feature to the components. It can be represented by a set *VFC= <View, Feature, Component>*.

View means to describe and analyze a domain space from different standpoint using different model formalism. It can be represented by a set *View =$\{view_1$, $view_2$,...$view_n\}$*.The multi-view strategy makes the system well performed from the specific perspective, at the same time makes the system more explicitly organized.

Feature is the inherent attributes of system and components captured by multi-view, reflecting the performance of a system. It can be represented by a set *Feature=$\{feature_1$,$feature_2$,...$feature_n\}$*. The feature can be the property such as real time, reliability, safety etc. Each feature contains a set of parameters, which will then be attached to components.

Component is the reusable element that can constitute the system from composition. It is prefabricated function module or subsystem that encapsulate executable industrial algorithms in a form that can be readily understood. Complete applications, can be built from networks of components, formed by interconnecting their inputs and outputs. A *Component* is defined as a port-based object:

Component =$<A_C, E_C, I, O, B_C, T_C>$,where,

A_C is a set of computations, called actions, which implement the components functionality;

E_C is a set of events, to activate the behavior of component;

I is a set of input ports through which the component receives its inputs;

O is a set of output ports through which a component exposes its computation results;

$B_C \subseteq E_C \times I \rightarrow A_C$ specifies a set of behavior of the component;

T_C is a set of attributes of the component. It comes from the parameters of Feature.

4.2   The CNC Meta-Model Design Based on VFC

Utilizing the VFC methodology, the CNC system meta-model can be developed by following steps:

CNC domain view select: The views in CNC system design include behavioral, structural, and system, which is depicted in figure 2. A structural view specifies the hierarchical architecture of a CNC system from the prefabricated component composition to explicitly describe the structure of the system. A behavioral view specifies control of the CNC system using state-oriented and data-oriented model to describe the interaction and activities in the system (Fig4 is the FSM meta-model in behavioral view). A system view specifies the runtime architecture and execution environment for CNC system, including the platform information, communication mode, operation system, storage resource etc.

CNC domain feature appointment: The main feature of CNC system is real time property. A set real time *feature=<Period, WCET, Priority, Deadline>* is used to represent the real time performance of the system. Period is the cycle of a task; WCET is the worst case execution time; Priority is the sequence to execute a task; Deadline is the time that the task should completed.

CNC system composition: It is a process to rationally partition the system to a set of components. To handle complexity, the design must progress in a top-down and evolutionary manner. The components can be described in different abstract level, with a different granularity of detail. With the abstractions, the model design will be a stepwise refinement process. The CNC system is compartmentalized to HMI, PLC, MC in the first level, then the MC component can be divided into code interpretation component, task manage component, interpolation component, etc, which is depicted in fig 3.

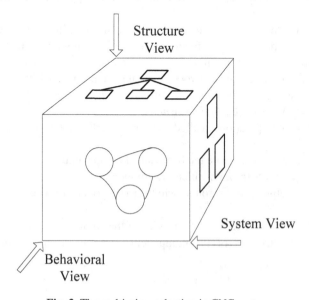

Fig. 2. The multi-view selection in CNC system

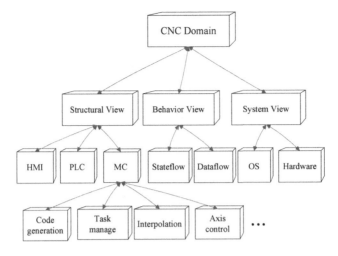

Fig. 3. VFC tree for CNC system development

Using the proposed strategy, a multi-view hierarchical meta-mode is built in GME. A domain-specific modeling language, i.e. CNCML, is generated from the meta-model. Figure 4 is the hierarchical FSM meta-model in behavioral view of a CNC system. It can be instantiated to the model for state transition in CNC HMI model, which is depicted in Figure 5.

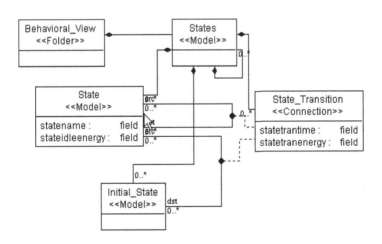

Fig. 4. The HFSM meta-model of CNC system

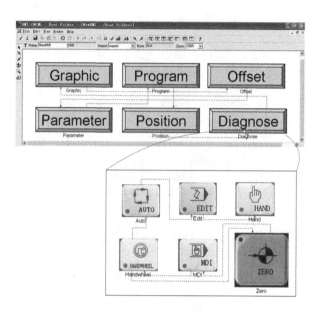

Fig. 5. The FSM model in CNC model environment

5   Conclusion

Model integration architecture for Computer Numerical Control system Design and Development is presented in the paper, which can be used to develop the CNC system design in a swift way, at the same time acquire some advanced capability of the system such as modularity, flexibility, reusability. The CNC modeling language naming CNCMoL are developed, in which the VFC methodology is used to deal with the modeling complexity, and get performance parameters and acquire reusability.

The CNC development environment is then generated by the CNCMoL, in which models are developed. The architecture can use to the fast reconfiguration for the systems, especially for those products that belong to the same product family. Development experience shows that development lifetime can be shortened greatly and development cost can be saved compared to a traditional approach.

Acknowledgment. This work is supported in part by Innovation Fund of Guangdong General Research Institute of Industrial Technology under grant no. 2013A007, and Science and Technology Program of Guangzhou under grant no 2013Y-00100.

References

1. OSACA Work Group. OSACA Handbook PartII: How to Develop OSACA Applications (EB/OL) (1996), http://www.osaca.org/
2. OMAC API Work Group. The OMAC API Open Architecture Methodology (EB/OL), http://www.isd.mel.nist.gov/projects/omacapi/

3. OSEC Consortium, OSEC-II Project Technical Report (October 1998), http://www.sml.co.jp/OSEC
4. Object Management Group. Model Driven Architecture - A Technical Perspective (EB/OL) (July 9, 2001), http://www.omg.org/mda/
5. MetaCase Website (EB/OL), http://www.metacase.com
6. ISIS. ESML (EB/OL), http://www.isis.vanderbilt.edu/Projects/mobies/downloads.asp
7. Wang, S.G., Shin, K.G.: Constructing Reconfigurable Software for Machine Control Systems. IEEE Transactions on Robotics and Automation 18(4), 475–486 (2002)
8. Wang, H., Chen, K., Liu, S.: Research on open architecture controller platform based-on software pattern. Computer Integrated Manufacturing Systems 12(3), 446–450 (2006) (in chinese)
9. Liu, T., Wang, Y.-Z., Fu, H.-Y.: The Open Architecture CNC System HITCNC Based on STEP-NC. In: Proceedings of the 6th World Congress on Intelligent Control and Automation, pp. 7983–7987. IEEE, Dalian (2006)
10. University of California, Berkeley. Ptolemy II Project (EB/OL) (February 02, 2005), http://ptolemy.eecs.berkeley.edu/ptolemyII/
11. Breedvel, P.C.: Port-Based modeling of mechatronic systems. Mathematics and Computers in Simulation 66, 99–127 (2004)
12. Angelov, C., Xu, K., Sierszecki, K.: A Component-Based Framework for Distributed Control Systems. In: EUROMICRO-SEAA 2006, pp. 1120–1127. IEEE, Cavtat (2006)
13. ISIS. GME User Manual, Version 5.0 (EB/OL) (October 10, 2006), http://www.isis.vanderbilt.edu/
14. MOF, Meta Object Facility(MOF) specification V1.4 (Apirl 2002)
15. DOME. Dome Guide Version 5.2.2 (EB/OL) (April 10, 1999), http://www.htc.honeywell.com/dome/inrec.htm/
16. Shu, Z., Li, D., Hu, Y., Ye, F., Xiao, S.: From Models to Code: Automatic Development Process for Embedded Control System. In: 2008 IEEE International Conference on Network, Sensor and Control, Shanya, China (2008)

Design and Implementation of High-Speed Real-Time Communication Architecture for PC-Based Motion Control System

Jian Hu, Hui Wang, Chao Liu, Jianhua Wu, and Zhenhua Xiong

State Key Laboratory of Mechanical System and Vibration, School of Mechanical
Engineering, Shanghai Jiao Tong University, Shanghai 200240, China
{hu-j,351582221,aalon,wujh,mexiong}@sjtu.edu.cn

Abstract. To enhance the efficiency and stability of real-time data acquisition and transmission in motion control process, a high-speed real-time communication architecture based on FPGA and DSP is developed in this paper. The direct-memory-access (DMA) control module initiates DMA transfers over the peripheral-component-interconnect (PCI) bus. After that, two first-in-first-out (FIFO) buffers work as caches and convert data widths between the PCI bus and the local bus. Consequently, the data caches can be accessed by DSP over the local bus under DMA mode. The size of data transferred each time is able to be adjusted according to the application requirements. Experiments are carried out to demonstrate that the proposed communication architecture is capable of achieving high speed and stability in data communication with low consumption of compute resources.

Keywords: DMA transfer, PCI bus interface, dual-clock FIFO, data communication, motion controller.

1 Introduction

With the rapid development of computer and microelectronics techniques, the realization of embedded motion control systems with combination of DSP and FPGA becomes the mainstream for the design of open architecture motion controller based on computer platform [7]. The standard peripheral interfaces for computer equipments, which exchange data with controllers, can be divided into two categories. Serial interfaces like USB and RJ-45 have a simplified way of connection, which seems to be cheap and flexible. But the speed and stability of transmission between devices are not guaranteed by considering the complex validation rules and high traffic on the bus. In contrast, parallel ones like PCI bus have a higher transfer speed with lower error rate and are capable of supplying power to controllers. The efficiency and stability of real-time data acquisition and transmission in motion control process play an important role in pursuing the system characteristics, such as high speed, high precision, good real-time capability and so on. Therefore, improving the communication performance on the PCI bus becomes one issue to be solved in the paper.

X. Zhang et al. (Eds.): ICIRA 2014, Part II, LNAI 8918, pp. 402–413, 2014.

The input/output (I/O) access between the PCI bus and the local bus is usually adopted in data communication of embedded motion control system, which results in low transmission efficiency, coarse time granularity and poor reliability [6]. The software real-time extension (RTX) was installed in some cases to improve the performance of data transmission [5]. However, this method restricts latencies to the microsecond range and leads to over reliance on computer ability [4]. On the other hand, it is only suitable for motion control systems directly controlled by computers due to the presence of clock frequency offset. On the contrary, bulk data transfers under DMA mode from one memory to another without the intervention of processors have a higher average speed and the data cache of FPGA can be accessed by DSP, which can reduce the dependence on the performance of processors clocked by a plurality of low precision crystal oscillators [3]. A steady rate of transmission and the flexibility of communication must be assured to meet the reliability and maneuverability request of control. Therefore, the FPGA implementation of the PCI protocol module is adopted in the communication architecture design to realize DMA transfer, and the size of data transferred should be adjustable.

Besides the approach for enhancing the transmission performance on PCI bus, the other one is to improve that on local bus between DSP and FPGA. In real applications, compared with the WINDOWS operation system environment, DSP has a more powerful real-time processing capability while FPGA is more excellent in precise timing control [1]. Data acquisition and data storage are generally done by FPGA. And then, DSP fetches data and enables essential calculations immediately, whereas the processor of computer is only responsible for the command interaction and data collection. In [2], the core processor of the four-axis motion controller is the DSP chip TMS320F2812, which interacts with external devices via limited number of buses under I/O mode. The floating-point DSP chip TMS320F28335 can adjust the data size transferred each time and achieve better communication performance via new added six-channel DMA bus, which has a base of four cycles/word. In order to improve the real-time performance in data processing and transmission, TMS320F28335 could be chosen as the kernel processor in the communication architecture implementation.

In this paper, a PC-based motion control system with DMA mode realized by a PCI protocol module in FPGA and XINTF of DSP is presented. The size of data transferred each time in the PCI bus and that in the local bus are both able to be adjusted according to the application requirements. The dispatch work can be mostly executed by DSP rather than the computer. The communication architecture is implemented in the self-designed embedded motion control system. Experiments are conducted on two different platforms and the results verify that the scheme of bulk data exchange over a PCI bus under DMA mode enhances the speed and stability performance of real-time data acquisition and transmission in motion control system.

The remainder of this paper is organized as follows: in Section 2, the DMA control module is firstly introduced and the computer application procedure to realize DMA communication between the computer and FPGA is discussed consequently.

Section 3 detailed the proposed cache modules. In Section 4, the hardware connection between FPGA and DSP is described and then the detailed implementation of data transfer is given. In Section 5, the experimental results are further presented to validate the effectiveness of the communication architecture. Section 6 gives the conclusion.

2   Transmission Architecture between FPGA and PC

2.1   Implementation of DMA Control Module

The hardware architecture of motion control system is shown in Fig. 1, which takes a host PC with WINDOWS operation system as the platform. The DSP chip TMS320F28335 and the FPGA chip EP2C8Q208I8N are chosen as main components of multi-axis motion controller to implement control operations. DSP is the core microprocessor responsible for the calculation of data and generation of control signal, while FPGA generates logical modules with parallel structures such as data collectors and buffers. Data like position, analog voltage, external I/O signal status and other parameters is continuously collected and transmitted to the DSP during the control process, and finally sent to PC after calculation.

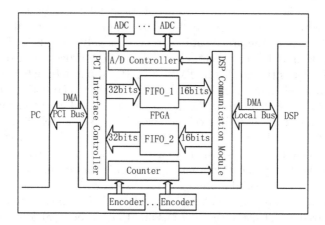

Fig. 1. Hardware architecture of motion control system

To control the DMA transfer on the PCI bus, the PCI bus interface controller is integrated in FPGA. This DMA control module contains a master module, a target module and a parity generation-and-check module. The master module works mainly as a PCI bus manager and is responsible for master DMA-mode data transfer. The PCI interface controller is designed by using languages of VHDL, Verilog and schematics according to PCI Local Bus Specification 2.3. DMA transmission in 32-bit PCI bus is initiated by the master module in FPGA after determining the data storage status of caches. During DMA transfers, the

PCI protocol module initiates data transmission process assisted by the master module. Once the high-speed data transmission is triggered, an internal counter begins to judge whether the transmission is completed. After eliminating miscellaneous work of I/O, the more efficient use of interrupts can solve the bottleneck of the original large volume data exchange process. In this way, the transmission efficiency on PCI bus is significantly enhanced between the computer and motion controller.

2.2   Design of Application Procedure

The system driver development environment comprises of Visual C++6.0, Windows XP DDK and Driver Studio 3.2. To realize the data transmission under DMA mode, a common buffer is added in upper computer and data will be read or write after being moved into or out of the dedicated memory space. The common buffer is used as the kernel mode stack, and its size can be adjusted if necessary.

Once a DMA-writing operation is conducted, data is firstly shifted to the common buffer. Specific transmission parameters are informed to the PCI protocol module integrated in FPGA by the configuration of source address register and byte count register. The computer issues a DMA-writing request to the module. After that, the DMA master controller within the module requests control of the PCI bus, and then initiates a transaction with the memory-read over the PCI bus after permission. Data is finally written into the selected memory location according to the address map of device that is built at system power-up. If the amount of data sent is bigger than the depth of FIFO, transmission will be terminated after the receive FIFO is full.

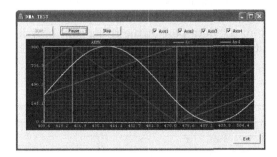

Fig. 2. Real-time data display interface

When a DMA-reading operation is conducted, the application makes a read call to a user-mode dynamic link library (DLL). The component in DLL, which implements the call, starts a kernel-mode thread to wait until the data has been prepared. The thread uses the WaitForSingleObject function. After the desired interrupt signal is captured, the driver configures destination address register and byte count register, and dismisses the interrupt. A DMA-reading request to the module is then issued, notifying the DMA master controller to initiate a transaction with the memory-write. Data will be copied from the kernel mode stack to its user mode stack in computer at the end of the data transfer.

Microsoft foundation class (MFC) is used to program the graphical user interface (GUI) of motion control system. In practice, to ensure that the monitored process is under control, data should be obtained sustainedly from the PCI bus to the user mode stack. However, the thread for capturing the notification signal may be closed unexpectedly. A simple worker thread of data receiving can prevent the background process from being stopped accidentally. Fig. 2 shows the real-time data display interface.

3   Cache Module Architecture for Import and Export

3.1   Description of Cache Module

The communication between the PCI interface controller and DSP is based on the 16-bit local bus at 150MHz operating frequency, while the 32-bit PCI bus at 33MHz operating frequency is used to exchange data between host PC and the PCI interface controller. To solve the problem caused by different width of data and unrelated clocks, dual-clock FIFO buffers instantiated through Quartus II IP core are chosen to interface the PCI bus with the local bus [8]. The Altera parameterizable dcfifo megafunction is used to provide the high-speed data buffer with the width and depth, which is required here for the high-data-rate and asynchronous-clock-domain application. The full and empty flags of the FIFO, caught by the PCI interface controller and DSP, are used to schedule bulk data transfers with start and end times while performing reading and writing operations. Fig. 3 is the typical timing diagram for dual-clock FIFO captured by the system-level debugging tool named SignalTap II Embedded Logic Analyzer. The data is imported from the data port at the rising edge of write clock, and then exported at the output port on the rising edge of read clock when wrreq and rdreq are high.

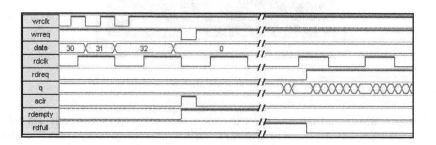

Fig. 3. Timing diagram for FIFO in FPGA

3.2   Implementation of Data Transfer Functions

A set of high-speed dual-clock FIFO modules created with Altera Quartus II development platform is used to provide interconnections between the computer and the DSP chip on motion controller, as is shown in Fig. 4.

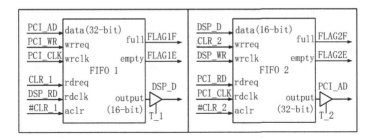

Fig. 4. Structure for dual-clock FIFOs

There are two stages in data transmission process from PC to DSP. In the first stage, data is transferred from the computer to FPGA via PCI bus. Once the DMA transfer is triggered, user-specified data is transferred from computer memory to the 32-bit-wide data port of FIFO 1. Then it is stored and read by DSP from the 16-bit-wide output port. Based on the work principle and time sequence requirement of FIFO 1, the connection of port linked with DSP communication module is different from that with PCI interface controller. The read signal of DSP operates as the clock signal, and the clear signal generated by DSP operates as not only the asynchronous clear signal (aclr) after negation but also the normal read request signal (rdreq). Tri-state gate works as the switch of output port to avoid any interference on local bus. Reading operations are conducted by DSP according to the flag signals of FIFO 1, and data is moved to the internal RAM on DSP for further processing. Each time before DSP enters into the normal work state, FIFO 1 needs initialization to enable the input and output lines and reset all memories. The signal here is triggered by the clear signal of DSP. In the second stage, data is transferred from FPGA to DSP via local bus. After all the data in FIFO 1 is ready to receive, DSP initiates DMA transfers by reference to the address mapping table. During the read cycle, data is sampled on the trailing edge of read signal and latched inside the DSP, which means only one rising edge is generated within the cycle. After the transmission has done, FIFO 1 releases the local bus by deasserting the enable signal of the tri-state gate.

Similarly, the other module FIFO 2 is used in the data transmission process from DSP to PC. DSP controls the acquisition and calculation, and fills the FIFO with the feedback data. An interrupt is generated by the PCI interface controller to the driver afterwards. The driver program, which responds to the interruption, informs the application immediately. The PCI interface controller triggers the DMA transfer after accepting the read request of computer. And then feedback data is transferred from the output port of FIFO 2 to computer memory. The initialization and emptying of the FIFO 2 and the state of tri-state gate are still controlled by signals of DSP. One thing should be noticed that there exists FIFO flag latencies during simultaneous read and write transactions. With the proposed architecture for data import and export, few resources are taken and the high-speed buffers are easy to implement. The depth of high-speed buffers are also adjustable. As a result, commands and feedback information can be exchanged efficiently in real-time data communication.

4   Transmission Architecture between DSP and FPGA

4.1   Hardware Connection between FPGA and DSP

As shown in Fig. 5, the external interface (XINTF) is a non-multiplexed asynchronous bus, which is used to realize seamless connection to the FIFOs instantiated in configurable logic of the FPGA. Furthermore, not only the address bus and the 16-bit data bus but also the read and write control signals are directly connected to the I/O ports of FPGA.

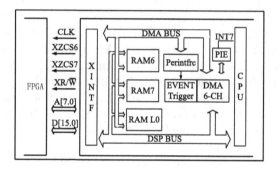

Fig. 5. Connection between FPGA and DSP

The address decoding module in FPGA is realized to help DSP complete the FIFO reading and writing operations under the control of the synchronous clock provided by DSP. The traditional I/O operations in processing other non-real-time data are also assisted by the same module. The occurrence of data overflows in the reading and writing process can then be prevented in this way, ensuring a uniform transmission rate during the DMA transfer.

4.2   Design of DMA Transfer Process

The digital signal control modules of TMS320F28335 contain a six-channel DMA controller, which is capable of performing data transfers not involved by the DSP processor. In this paper, channel 1 and 2 are configured as the receive channel and the transmit channel. The space of FIFOs is mapped into two fixed zones named XINTF Zone 6 and Zone 7, which are programmed with specified wait states, setup and hold timings. Data from FIFO1, which is mapped into Zone 6, is transferred to L6 SARAM integrated in DSP through the DMA channel 1. Data from L7 SARAM is transferred to FIFO 2, which is mapped as Zone 7, through the DMA channel 2. The amount of transmission data and trigger conditions are also configured during the initialization process of motion control system. The DMA capability of DSP allows its processor not to be intecy-clone2007devicerrupted frequently when a set of data is transferred, providing an efficient and convenient data manipulation for real-time communication in motion control process.

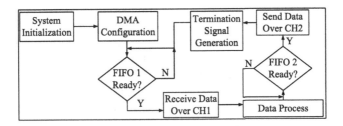

Fig. 6. Flow chart of DMA transfer controlled by DSP

Fig. 6 shows the integral process of the master DMA transfer initiated by DSP. Firstly, the configuration of control registers of XINTF is performed automatically from a power-up or a reset. The size of memory space used and that of data transmitted are also configured at the beginning of each data transfer. Then, the data from FIFO 1 is transferred to L6 SARAM after capturing the read ready signal generated by FPGA. The DSP processor executes algorithms and generates control signals. The processing results and feedback data are stored in L7 SARAM afterwards and written back to FIFO 2 when accessible. Finally, the required data is received by the computer through FIFO 2 before the advent of the next set of data.

5   Experimental Studies

5.1   Experimental Setup

The proposed communication architecture is implemented on the five-axis motion control system. The FPGA and DSP are core modules of the self-designed motion controller. FPGA accomplishes the PCI protocol and dual-clock FIFOs, while DSP implements the operating instructions and the complex control algorithms. The data transfer system with dual DMA channels on the local bus is designed to transfer data between them. This motion controller is installed on the PCI slot for exchanging data with the computer through the PCI bus under DMA mode.

Fig. 7. Industrial PC platform   **Fig. 8.** Mini PC platform

Two different experimental platforms are established to evaluate the overall performance of the communication architecture, in which the DMA mode transmission is compared with the traditional I/O mode one. The first platform contains an industrial PC with Pentium 2.80 GHz CPU, 2 GB DDRII RAM and Windows XP Operating System, as is shown in Fig. 7. The other one contains a mini PC with Atom 1.86 GHz CPU, 2 GB DDRIII RAM and Windows XP Embedded (XPE) Operating System, as is shown in Fig. 8.

5.2   DMA Configuration of Motion Control System

The DMA configuration of the communication on the PCI bus is analyzed firstly in this subsection. The size of data read or written by the computer each time is adjustable and its configuration is discussed below. The performance test of DMA transfers over the PCI bus is carried out to verify the reliability and efficiency of different size data transmission. Data is read by the computer from the FIFO in FPGA in this test. DMA mode is suitable for transmission of a large volume of data. One hundred tests of data transfer over the PCI bus are conducted for each case. Fig. 9 shows the averages and standard deviations of time consumed in data transmissions of different size over PCI bus on industrial PC platform. Thus, to take full advantages of DMA-mode data transfer, 2 K, 4 K and 8 K bytes of data are selected to be transmitted through PCI bus in the test for communication within the overall motion control system.

The DMA configuration of the communication on the local bus within motion controller is then analyzed. The DMA controller of DSP is confirmed to be capable of performing one 16-bit data reading operation every 5 clock cycles and one 16-bit data writing operation every 9 clock cycles in the operating frequency of 150 M, which achieves the highest transmission rate. And in the test, it is set to perform one reading operation every 9 clock cycles and one writing operation every 13 clock cycles. For instance, to exchange 8 K bytes data between DSP and FPGA through 16-bit DMA channel, the consumption of time at each reading operation is 245.76 us and that at each writing operation is 354.99 us.

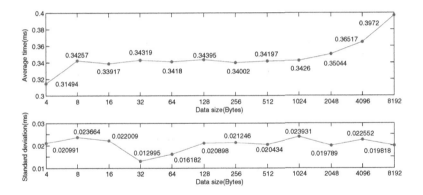

Fig. 9. Time consuming of DMA-mode data transfer on the PCI bus

5.3   Performance Test of Communication Architecture

During the test for communication within motion control system, data is first sent to FPGA by the computer through the PCI bus. And then, the data buffered in one FIFO is read into the internal RAM of DSP through the local bus. After executing simple handling tasks, DSP copies the feedback data to another internal RAM and transfers it back to another FIFO in FPGA. Finally, the computer receives the data and determines whether the feedback data is correct. The value of data sent by the computer is generated repeatedly in an ascending order and a descending order.

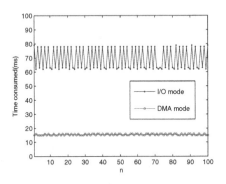
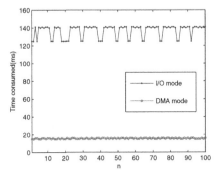

Fig. 10. Time consuming of 2 K Bytes   **Fig. 11.** Time consuming of 4 K Bytes

The performance difference between I/O mode transmission and DMA mode transmission on the same platform is first tested in this subsection. To evaluate the overall system communication performance on industrial PC platform, one hundred tests are conducted for each case. The time required to complete the transmission of 2 K and 4 K bytes of data under two data transfer modes is

Table 1. Time consuming in the data transmission of 8 K bytes

Industrial PC platform (ms)		Mini PC platform (ms)	
I/O mode	DMA mode	I/O mode	DMA mode
265	15	640	16
282	16	641	15
265	16	641	16
281	15	625	15
265	16	640	16
282	15	625	16
281	16	641	15
281	16	641	16
266	15	640	16
281	16	641	15
266	16	625	16
281	15	640	15
281	16	641	16
266	15	625	16
282	16	641	15

shown in Fig. 10 and Fig. 11. The time consumed in data transfer under DMA mode is usually between 15 ms and 16 ms, while that under I/O mode varies a lot among data transmissions of different sizes.

The performance difference of two mode transmissions on different platforms is then tested, and the amount of data tested here is 8 K bytes. With two different host PCs, the time required to complete the transmission of 8 K bytes of data under two data transfer modes is shown in Table 1. The processor of industrial PC here is stronger than that of mini PC, but the operation system Windows XPE on mini PC platform is more stable and reliable. In fifteen tests for each case, the mean value of the time consumed in the traditional I/O data transfer on industrial PC platform is 275.0 ms, while that on mini PC platform is 636.5 ms. The standard deviation values on two PC platforms are 8.0 ms and 7.2 ms respectively. The DMA capability allows the processors of the computer and DSP not to be interrupted frequently when a set of data is transferred, providing an efficient and convenient data manipulation for real-time communication. Thus, the mean values of the time consumed in the DMA data transfer on two PC platforms are both reduced to 15.6 ms and the standard deviation values on two PC platforms fall into 0.5 ms.

Therefore, compared with the traditional way of transmission, the proposed communication architecture can guarantee the efficiency and the stability of data exchange with little reliance on the change of data size and the performance of the host PC, which will contribute much to the achievement of the high-speed and high-accuracy requirements in the motion control process.

6 Conclusions

In this paper, a high-speed real-time communication architecture based on FPGA and DSP is proposed to enhance the efficiency and stability of real-time data acquisition and transmission in motion control process. FPGA integrates the peripherals such as the PCI interface controller and dual-clock FIFOs. As the core processor, DSP is responsible for performing data transfers and enhancing the throughput within the motion controller. The implementation of DMA transmission based on the PCI bus and the local bus enables data to be exchanged efficiently and frequently in real-time application. The adjustable size of data transferred under DMA mode also ensures the flexibility and the maneuverability of the communication architecture. Experiments are conducted on two different platforms and the results demonstrate that the proposed communication architecture is capable of achieving high-speed real-time data communication.

Acknowledgements. This research was supported in part by the National Basic Research Program of China (2013CB035804) and National Natural Science Foundation of China (U1201244).

References

1. Cho, J.U., Le, Q.N., Jeon, J.W.: An fpga-based multiple-axis motion control chip. IEEE Transactions on Industrial Electronics 6(3), 856–870 (2009)
2. He, S., Gao, X., Peng, C., Zhang, Y.: Development of dsp and fpga based 4-axis motion controller. In: Sixth International Symposium on Precision Engineering Measurements and Instrumentation, pp. 75441K–75441K. International Society for Optics and Photonics (2010)
3. Monmasson, E., Idkhajine, L., Cirstea, M.N., Bahri, I., Tisan, A., Naouar, M.W.: Fpgas in industrial control applications. IEEE Transactions on Industrial Informatics 7(2), 224–243 (2011)
4. Pu, D., Sheng, X., Zhang, W., Ding, H.: An application of real-time operating system in high speed and high precision motion control systems. In: International Conference on Automation Science and Engineering, CASE 2007, pp. 997–1001. IEEE (2007)
5. Wu, J., Pu, D., Ding, H.: Adaptive robust motion control of siso nonlinear systems with implementation on linear motors. Mechatronics 17(4), 263–270 (2007)
6. Zhao, H., Xiong, Z.: Scheduling scheme for networked motion controller in cnc system based on qos and qop. In: 2010 International Conference on Mechatronics and Automation (ICMA), pp. 979–984. IEEE (2010)
7. Zhao, H., Zhu, L., Xiong, Z., Ding, H.: Development of fpga based nurbs interpolator and motion controller with multiprocessor technique. Chinese Journal of Mechanical Engineering 26(5), 940–947 (2013)
8. Zhengbing, Z.: Design and implementation of signal generator based on software radio. In: Cross Strait Quad-Regional Radio Science and Wireless Technology Conference (CSQRWC), vol. 2, pp. 1017–1020. IEEE (2011)

Hydraulic Actuator with Power Line Communication for Valve Remote Control System

Seok-Jo Go[1], Jang-Sik Park[2], Min- Kyu Park[3], and Yong-Seok Jang[4]

[1] Division of Mechanical Engineering,
Dongeui Institute of Technology, Busan, 614-715, Korea
sjgo@dit.ac.kr
[2] Department of Electronic Engineering, Kyungsung University, Busan, 608-736, Korea
jsipark@ks.ac.kr
[3] School of Mechanical and Automotive Engineering Technology,
Yeungnam College of Science and Technology, Daegu, 705-703, Korea
mk_park@ync.ac.kr
[4] Marsen Co. LTD, Myeongji-dong, Busan, 618-815, Korea
softmac@naver.com

Abstract. Valve remote control system is a convenient system which it is possible to operate the valves installed in the cargo or ballast tanks from the remote wheelhouse. This paper is dealing with an integrated hydraulic actuator with a power line communication module for a valve remote control system. The integrated hydraulic actuator consists of a gear pump, integrated block, and controller with a power line communication module. We try to design and implement the hydraulic actuator and this tries to be verified through an experiment. The developed actuator has two advantages. One is that the pipelines between an actuator and a power unit are not needed since a solenoid valve rock and a hydraulic power unit used in the conventional control can be removed. The other is that the pressure loss can be minimized due to simplifying a pipeline and improving the power transmission ability.

Keywords: Valve remote control system (VRCS), Integrated hydraulic actuator, Power line communication module, Cargo tank, Ballast tank.

1 Introduction

In order to ensure stability of ships and offshore structures, various type valves have been installed at a pipeline for controlling a ballast system or a liquid cargo system. Valve remote control system (VRCS) is the key equipment in a shipboard monitoring and control system. VRCS enables valves to be operated by control signals from the remote wheelhouse [1-6]. According to the power sources, VRCS is divided into the hydraulic, pneumatic, and electric-powered. The hydraulic VRCS is mainly applied to ship and offshore structures because it can operate at a higher torque level compared to pneumatic and electric-powered VRCS. And, it has a fast-responsibility by using an incompressible fluid.

X. Zhang et al. (Eds.): ICIRA 2014, Part II, LNAI 8918, pp. 414–423, 2014.
© Springer International Publishing Switzerland 2014

Due to the scale-up of the shipbuilding and the additional safety functions, it has the problem of setting up the hydraulic power unit additionally as the amount used of the valve and actuator increases. And the shipbuilding cost is increased because the pipe length is lengthened and it is difficult to install a complex pipeline. Therefore, in the previous study, the hydraulic actuator was developed for solving the installation problem and reducing the length of pipeline and the loss of transmission power. And the developed hydraulic actuator stably operated at three state (Fully open, 50% open, fully close) of valve position [7].

In this study, to operate valves from remote site, we try to design a compact type hydraulic actuator with a power line communication. First, we describe the trend of VRCS of ship. And then, hydraulic actuator parts are developed and assembled. Power line communication network is also applied to the control system. Finally, the operating performance tests are carried out for verification of the developed hydraulic actuator with a power line communication.

2   Trend of Valve Remote Control System

In a conventional VRCS as shown in Fig. 1, the electric control signals from the valve control console which is installed at wheelhouse transfer to the various valves such as a solenoid valve, a proportional control valve, and so on. And the compressed hydraulic oil from a hydraulic power package transfers to the hydraulic actuator through a pump, a solenoid valve, a check valve, a relief valve, and so on as shown in Fig. 2.

There are some problems in a conventional VRCS [4,7]. First, the quantities of valves and actuators for controlling a ballast and a liquid cargo are incredibly increased as scale-up and specializing of shipbuilding. Thus, the additional or huge hydraulic power package is needed. It is difficult in configuration of a hydraulic power unit and a power transmission line. Secondly, in the case of using the long distance pipelines,

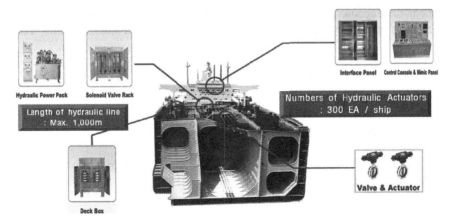

Fig. 1. Conventional valve remote control system

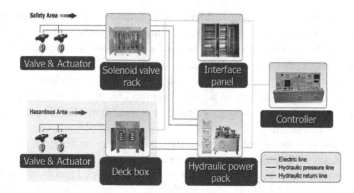

Fig. 2. Control flow diagram of a conventional valve remote control system

the incompressible fluid has compressible characteristics because moisture and bubble come up in the fluid. It is difficult to control the position of a hydraulic valve due to compressibility in a fluid. Thirdly, when the hydraulic power unit has a malfunction, the whole of the valve has to be passively operated. Finally, in case of the control system by using a centralized hydraulic power unit, it is difficult to individually control each actuator. In the previous study, the hydraulic actuator was developed for VRCS to solve these problems [7]. The actuator is composed of a gear pump, an AC motor, a check valve, a relief valve and a controller. And, this study proposes the new stand-alone type hydraulic actuator with the power line communication. The main idea is that actuator is integrated with the hydraulic power unit and the power line communication unit.

3   Integration of Hydraulic Actuator

We try to design and fabricate a compact hydraulic actuator with a power line communication network for solving the above problems. Figure 3 shows the schematic diagram of the integrated hydraulic actuator. The actuator consists of a bi-directional gear pump, an AC induction motor, an integrated block including a check valve and a relief valve, and a controller with a power line communication.

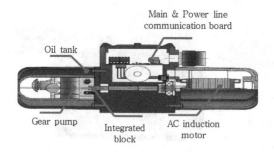

Fig. 3. Schematic diagram of the integrated hydraulic actuator

3.1   Integration of Hydraulic Actuator Unit

In the previous study, the compact type hydraulic actuator was fabricated as shown in Fig. 4 [7]. It is composed of the gear pump for generating a hydraulic power, the AC induction motor for rotating the gear pump, the micro check valve, the micro relief valve, and the oil tank.

The micro check valve and the micro relief valve are inserted into the integrated block which is connected with head of the gear pump. In case of the gear pump, the discharging flow rate is determined by the size of toothed gear. Table 1 shows the specifications of the gear pump. We selected a second column of Table 1 (displacement: 0.3cc/rev) and manufactured the gear pump as shown in Fig. 4.

As using the integrated hydraulic actuator, the solenoid valve rack and the hydraulic power package shown in Fig. 1 can be eliminated as shown in Fig. 5. It means that the integrated hydraulic actuator can provide the high quality of efficiency, profitability, and stability.

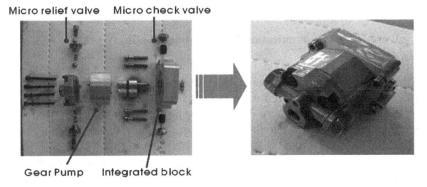

Fig. 4. Development of the gear pump and the integrated block

Table 1. Specifications of the gear pump

	Specifications				
Displacement (cc/rev)	0.2	0.3	0.5	0.7	1.1
Flow (Liter/min, 2500RPM) DC	0.5	0.7	1.2	1.7	2.7
Flow (Liter/min 1700RPM) AC	0.3	0.5	0.8	1.1	1.8
Max. Operating Pressure (bar)			190		
Max. Peak Pressure (bar)			230		
Max. Speed (rpm)			7,000		
Area (mm^2)	54	56	59	64	67

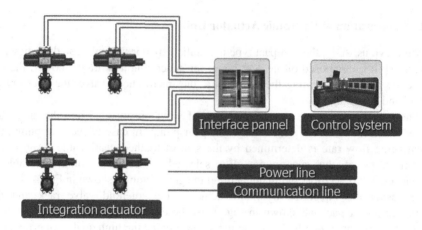

Fig. 5. Velve remote control system using the integrated hydraulic actuator and the power line communication

3.2   Power Line Communication Network Based on LonWorks

The power line communication is a system which uses the existing power line infrastructure to transmit and receive data. AC or DC power line can be used and a maximum communication speed is up to 400Mbps. The power line communication has great advantages such as a weight reduction, a cost reduction, and so on. Therefore, it has attracted attention as the next generation communication technology [8-11].

Figure 6 shows the schematic diagram of the power line communication module by using the PL3150 Transceiver (Echelon Co.) [12]. The PL3150 transceiver utilizes a dual carrier frequency signaling technology to provide superior communication reliability in the face of interfacing noise sources. In the case of acknowledged messaging, packets are initially transmitted on the primary frequency and if an acknowledgement is not received, the packet is retransmitted on the secondary frequency. In the case of repeated messaging, packets are alternately transmitted on the primary and secondary frequencies. Figure 6 includes the PLC coupler circuits. Injecting a communication signal into a power mains circuit is normally accomplished by capacitively coupling the output of a transceiver to the power mains. In addition to the coupling capacitor, an inductor or a transformer is generally present. The coupling capacitor and the inductor or the transformer together act as a high-pass filter when receiving the communications signal. The high-pass filter attenuates the large AC mains signal at 60 Hz, while passing the communication signal of the transceiver. The value of the capacitor is chosen to be large enough so that its impedance at the communication frequencies is low, yet small enough that its impedance at the mains power frequency 60 Hz is high. The value of the inductor is chosen to provide a relatively high impedance at the communication frequency of the PL3150 transceiver.

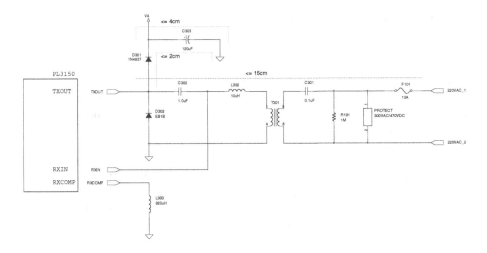

Fig. 6. Schematic diagram of the developed power communication module

3.3   Controller

The controller is composed of a MCU (PIC16F877A/I PT), a relay, a photo coupler, a 16-bit A/D converter, and a serial communication port. The Relay and the photo coupler are used for electrically isolating and controlling of the induction motor. The position of valve is measured from a valve position sensor and a 16-bit A/D converter. The controller is communicated with the other controller by using RS232 and RS422/485. Figure 7 shows the developed control module.

The controller has three function modes for monitoring status of a valve and a hydraulic actuator. In Manual Mode, an actuator is independently controlled as open and close. Dead band and delay time are adjusted by using Setting Mode. And Auto Calibration Mode offers to decrease error between an actuator with a valve.

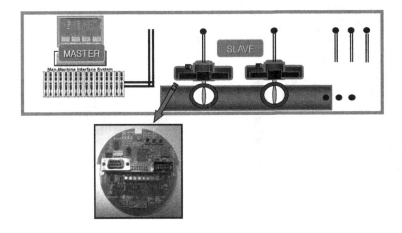

Fig. 7. Control module

4   Performance Evaluation

4.1   Power Line Communication Test

Figure 8 shows the developed hydraulic actuator with the power line communication module for VRCS. The gear pump, the AC induction motor, the controller, the integrated block and the power line communication module are together comprised for fabricating the actuator.

Figure 9 shows a communication check by setting of a communication voltage, a margin, and a speed. In the power line communication, voltage range of AC and DC is tested as shown in Figs. 10 and 11. The results show 5~30 V in DC range 110/220 V in AC range. The communication speed is set as 5,400 pbs, the frequency is used 132 kHz or 3.6 kHz. The communication test is executed as shown in Fig. 12. We confirmed a stable communication through Rx & Tx Flicker of LED.

Fig. 8. Developed hydraulic actuator with the power line communication module

Fig. 9. Power line communication using PLDSK

Fig. 10. DC vlotage in power line communication

Fig. 11. AC vlotage in power line communication

Fig. 12. Signal check

4.2   Performance Evaluation of the Developed Integrated Hydraulic Actuator with the Power Line Communication Module

The maximum working pressure is 150 bar, the maximum output torque of this actuator is 5,000 Nm, and it can be operated and monitored at remote wheelhouse in ship by power line communication. In order to evaluate performance of the developed actuator, experiments has been carried out for three state (Fully open, 50% open, Fully close) of valve position. And the valve monitoring program is implemented by using LabVIEW as shown in Fig. 13. The developed actuator is stably operated as shown in Fig. 14.

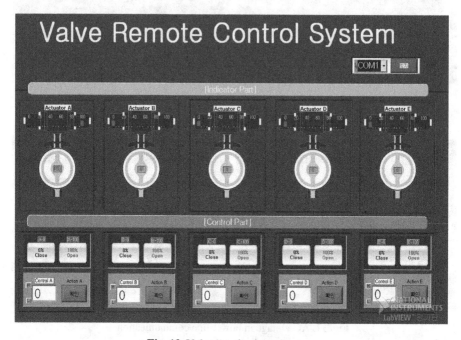

Fig. 13. Valve monitoring program.

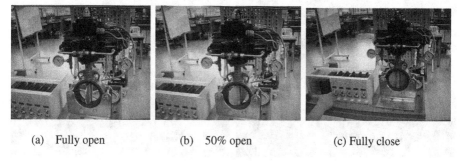

(a) Fully open   (b) 50% open   (c) Fully close

Fig. 14. Performance evaluation

5   Conclusion

This study shows the modular type hydraulic actuator in small size. An AC induction motor, a gear pump, and valves are integrated, and a power line communication is applied to the hydraulic actuator for remote control. The effectiveness of the developed system was shown through experiments. The experiment results show that the system can stably operate at three of valve position from remote site. The developed actuator has two advantages. One is that the pipelines between an actuator and a power unit are not needed since the solenoid valve rock and the hydraulic power unit used in the conventional control can be removed. The other is that the pressure loss can be minimized due to simplifying a pipeline and improving the power transmission ability.

Acknowledgments. This work was supported by Business for Cooperative R&D between Industry, Academy, and Research Institute funded Korea Small and Medium Business Administration.

References

1. Kim, K.: Strategy for Sustainable Growth of Marine Equipment Industry. In: 1st KOMEA FORUM (2008)
2. Kim, J.: Strategy for the Enhancement of the International Competitiveness of the Marine Equipment Industrym. Korea Marine Equipment Research Institute's Report (2008)
3. Shi, J., Yu, G., Zhang, W.: The ship monitoring and control network system design. In: 2011 International Conference on Management Science and Industrial Engineering, Harbin, China (2011)
4. Tomas, V., Kitarovic, J., Antonic, R.: The trend in integrated control and monitoring system for ships. In: 47th International Symposium ELMAR, Zadar, Croatia (2005)
5. Seil Serer Co. Ltd, http://www.seilseres.com
6. Nordic Flow Control Co. INC, http://www.nordicflowcontrol.com/
7. Park, M.K., Go, S.J., Park, J.S., Choi, M.H., Kim, C.D., Sung, K.G.: Development of Stand-Alone Type Hydraulic Actuator for Valve Remote Control System. In: Applied Mechanics and Materials, vol. 145, pp. 48–52. Trans Tech Publications, Switzerland (2012)
8. Kim, J.H., Lee, H.B.: Tends Research and Development View of Power Line Communication. In: Proceedings of the ITFE Summer Conference, pp. 573–578. Korea Institute of Information & Telecommunication Facilities Engineering (2008)
9. Choi, S.S., Oh, H.M., Kim, Y.S., Kim, Y.H.: Study on Very High-Rate Power Line Communications for Smart Grid. In: Trans. KIEE, vol. 60(6), pp. 1255–1260. The Korean Institute of Electrical Engineers (2011)
10. Kim, S.J., Lee, B.G., Kim, G.Y., Kim, S.H.: Field Testing of High-Speed Power Line Communications. In: Proceedings of the ITFE Summer Conference, pp. 227–230. Korea Institute of Information & Telecommunication Facilities Engineering (2009)
11. Jin, T.S.: Power-line Communication based Digital Home-Network Technology. In: Proceedings of the Korean Institute of Information and Communication Sciences Conference, pp. 992–995. The Korean Institute of Information and Communication Sciences (2010)
12. Echelon Co., http://www.echelon.com

Development of High Velocity
Flying Touch Control Algorithm
for Hot Rolling

Hyun-Hee Kim[1] , Sung-Jin Kim[1], Yong-Joon Choi[2], and Min-Cheol Lee[1]

[1] Pusan National Univercity, Jangjeon 2-dong,
Geumjeong-gu, Busan, Korea
[2] Posco, 6261, Donghaean-ro, Nam-gu, Pohang-si,
Gyeongsangbuk-do, Korea
sleepingjongmo@nate.com, jins2410@naver.com,
cyj@posco.com, mclee@pusan.ac.kr

Abstract. To improve a quality of the steel plates in the hot rolling process is an important objective of research. The flying touch hot rolling process is the method that controls the gap between the rolling rolls when the steel plate is transported through rolling rolls. If the steel plate is processed in hot rolling process when the gap between rolling rolls is fixed, the steel plate is damaged by friction and it occurs scratch. So it needs to control precisely the gap and angular velocity of rolling rolls to minimize the friction between steel plate and rolling rolls.

Keywords: hot rolling, flying touch, robotics lab, synchronization control.

1   Introduction

To improve a quality of the steel plates in the hot rolling process is an important objective of research. The steelworks' laboratory develops a control algorithm of hot rolling process to provide a good quality of coils to consumers, because the quality of steel plates is directly related to the value of products.

It needs proper thickness of slabs to start the hot rolling process. The slab is the processed steel which is made by continuous casting process, and the iron ore is changed to the steel in blast furnace by iron making process. Because the slab made by several processes is too thick to use in factories, it needs to modify the thickness of slab, and this procedure is a hot rolling process. The simple diagram of the hot rolling process is described in fig. 1.

The first procedure is a furnace process which heats cold slab. The goal temperature is generally higher than the melting point of the steel. The heated slab is transported to roughing mill to modify a thickness and width. The steel whose thickness and width are adjusted in roughing mill is transferred to crop shear which cuts the useless head part of the steel, after then, the finishing mill finally fine-tunes the steel. This finishing mill process makes the accurate size of consumer's request. In run out table, the heated steel is cooled to make recrystallization. Because the final

X. Zhang et al. (Eds.): ICIRA 2014, Part II, LNAI 8918, pp. 424–431, 2014.

steel plate's length is about 1000m, it needs coiling process to transport to factories. These coils are finally supplied to factories so that they can produce useful goods which are required by end-consumers.

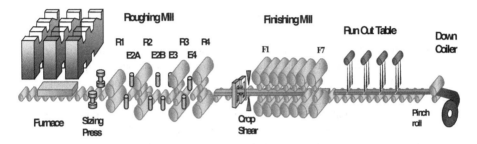

Fig. 1. The simple diagram of the hot rolling process. The slabs are transported to left-to-right direction[1].

1.1   Flying Touch

The flying touch hot rolling process is the method that controls the gap between the rolling rolls when the steel plate is transported through rolling rolls. If the steel plate is processed in hot rolling process when the gap between rolling rolls is fixed, the steel plate is damaged by friction and it occurs scratch. So it needs to control precisely the gap and angular velocity of rolling rolls to minimize the friction between steel plate and rolling rolls. The simple explanation about flying touch hot rolling process is described in fig. 2.

v_1 is a transportation velocity of slab, v_2 is an angular velocity of the upper rolling roll, and v_3 is an angular velocity of the lower rolling roll. v_2 and v_3 are connected by mechanical axes and joints, so upper and lower rolling rolls rotate at

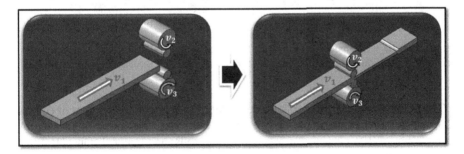

Fig. 2. The explanation picture about flying touch hot rolling process. The transportation and rolling direction should be matched to process the steel plate. The useless head part will be cut in crop shear.

same speed. v_1, v_2, and v_3 should be synchronized to minimize the scratch of the steel plate. In this paper, we propose the control algorithm to minimize impact on the steel plate by control the gap and velocity of rolling rolls.

2   The Design of Hot Rolling Model

Because the real hot rolling system is huge and dangerous, it needs miniature model to simulate flying touch control algorithm. The miniature model should apply the ratio of total size of real flying touch hot rolling system, and the velocity of steel plate should be coincided.

We can apply the gap control profile and the angular velocity control profile of rolling rolls of miniature model on the real flying touch hot rolling system. The 3D CAD model of miniature flying touch hot rolling model is described in fig. 3.

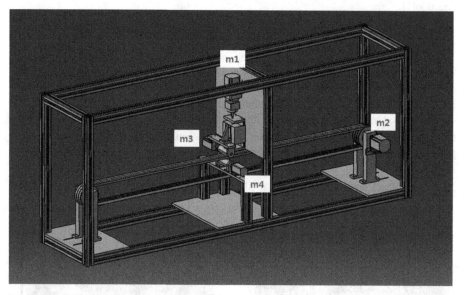

Fig. 3. The 3D CAD model of miniature flying touch hot rolling model. It is designed to simulate the position profile of the gap between rolling rolls and the angular velocity profile of the rolling rolls.

The flying touch hot rolling miniature model is 1/40 size of real system. It consists of 4 actuators which are controlled by micro controller. The motor 1 translate the upper rolling roll to control gap between rolling rolls, and motor 2 transport a belt. Because we cannot use a steel plate in miniature system, the steel plate is substituted to the rubber belt.

The motor 3 rotates the upper rolling roll, and the motor 4 rotates the lower rolling roll. In real system, the upper and lower rolling rolls should be rotated by only one motor because of its mechanical connection. But the miniature system is too small to design universal joints, so we used two motors at each rolling roll and control two motors by synchronization control.

3   Simulation

The 3D CAD model is designed to control the gap between the rolling rolls and to synchronize the velocity of rubber belt and rolling rolls. So we need the profile of the gap position data and the roller velocity data. Robotics Lab is used to simulate how the perturbation of motors is occurred when the rolling rolls contact the steel plate. The simple model of the flying touch hot rolling system is described in fig. 4.

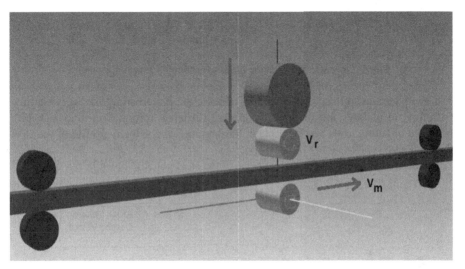

Fig. 4. The robotics lab modeling of flying touch hot rolling system. It is designed to research the influence of contact between the rolling rolls and the steel plate by computing the perturbation of each motor.

In the robotics lab, a model is formed. In the model, the metal is flowing between rolling rolls through left-to-right direction.

The rotate velocity of roller is V_r, and the metal flow velocity is V_m

In the simulation, the flying touch by rolling rolls was tested in three cases.

$$\text{Case 1} \quad V_r = V_m$$
$$\text{Case 2} \quad V_r = 1.1 \, x \, V_m$$
$$\text{Case 3} \quad V_r = 0.9 \, x \, V_m$$

For these three cases, the perturbation is estimated. It is assumed that, if the perturbation is increased, the probability of occurring scratches will increase. For all three cases, the plots look almost same.

The perturbation of motors from contacting rolling rolls with a steel plate is represented in fig. 5.

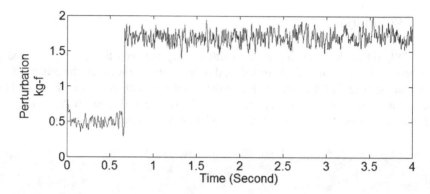

Fig. 5. Estimated Perturbation of metal at flying contact with roller

It can be assumed that, at the point of contact, the perturbation rises very sharply. Therefore, in real system, the dP/dT can give sufficient information to synchronize the speed of rolling rolls and steel plate. Furthermore, the velocity profile of the steel plate can give synchronizing information.

The desired velocity and actual velocity of the steel plate is shown in fig. 6

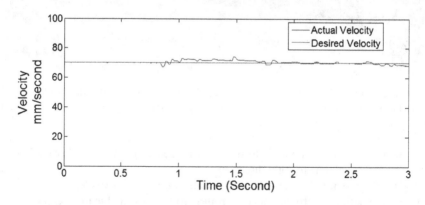

Fig. 6. The desired velocity and actual velocity of the steel plate.

4   Experiment

The manufactured 1/40 size miniature flying touch hot rolling machine is shown in fig. 7, and equipment for experiment is shown in fig. 8. Each AC servo motor needs servo driver to control the velocity and of rolling rolls and rubber belt. The purpose of this experiment is to synchronize rolling rolls' rotational velocity and rubber belt's velocity.

Fig. 7. 1/40 size miniature model of real flying touch hot rolling system for simulation

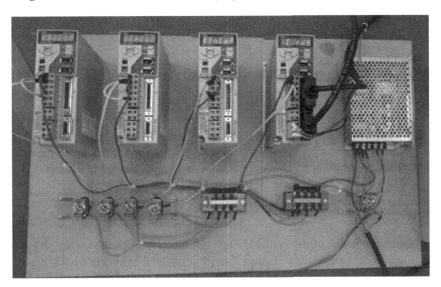

Fig. 8. Servo drivers, SMPS, switches to simulate flying touch hot rolling control algorithm

If the rolling rolls' rotational velocity and rubber belt's velocity are synchronized, we can assume that the scratches of processed steel plate are reduced. The rubber belt's velocity is measured at 0.455m/s, and the converted rotation velocity of rolling rolls should be 30.333 rad/s. The experiment result is shown in fig. 9.

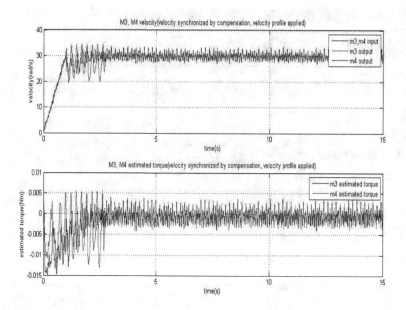

Fig. 9. Experiment result of 1/40 size miniature model of flying touch hot rolling system. The velocity profile(up) and torque profile(down) is shown.

In the experiment, the average rotational velocity of rolling rolls is 30.110 rad/s, so the accordance rate with rubber belt's velocity is about 99%. And the estimated torque profile of each motor is going to almost 0. Therefore, it can prove the rubber belt's velocity and rolling rolls' rotational velocity are synchronized.

5   Conclusion

The miniature model of flying touch hot rolling system is design by 3D CAD modeling and the perturbation of each motor is simulated by robotics lab to research the influence of contacting rolling rolls with a steel plate. The velocity profile of the Robotics Lab simulation is used to synchronize the rolling rolls' velocity.

In the experiment, the rubber belt's velocity and rolling rolls' rotational velocity were synchronized at 99%, and the torque profile of each motor was almost 0. Therefore, the velocity synchronization can reduce the scratches of processed steel plate in flying touch hot rolling process.

Acknowledgments. This research was supported by the MOTIE (Ministry of Trade, Industry & Energy), Korea, under the Industry Convergence Liaison Robotics Creative Graduates Education Program supervised by the KIAT (N0001126).

References

1. Choi, Y.-J., Lee, M.-C.: PID Sliding Mode Control for Steering of Lateral Moving Strip in Hot Strip Rolling. International Journal of Control, Automation, and Systems, 399–407 (2009)
2. Sansal, K., Yildiz, J., Fraser, F., Biao, H., Yale, Z., Fred, W., Vit, V., Mike, D.: Dynamic modeling and simulation of a hot strip finishing mill. Applied Mathmatical Modelling 33, 3208–3225 (2009)
3. Choi, Y.-J., Yoo, K.-S., Lee, M.-C.: Development of the Strip Off-center Meter Using Line Scan Camera in FM Line. Journal of Control, Automation, and Systems Engineering 11(6) (2005)
4. Rohde, W., Rosenthal, D.: High-tech rolling in hot strip mills – theory and practice. Metallurgical Plant and Technology 1, 48–55 (1990)

A New Dynamic Modelling Algorithm for Pneumatic Muscle Actuators

Jinghui Cao[1], Sheng Quan Xie[1,2], Mingming Zhang[1], and Raj Das[1]

[1] Department of Mechanical Engineering, The University of Auckland,
20 Symonds Street, Auckland City, New Zealand
{jcao027,mzha130}@aucklanduni.ac.nz,
{s.xie,r.das}@auckland.ac.nz
[2] State Key Laboratory of Digital Equipment and Technology,
Huazhong University of Science and Technology, Wuhan, China

Abstract. Pneumatic muscle actuators (PMAs) have been widely used in wearable robots due to its high power to weight ratio and intrinsic compliance. However, dynamic modelling of PMAs, which is important to control performance, has not been researched extensively. Hence, a testing device was designed and built to investigate PMA's dynamics. The device automates the experimental process by providing motions and recording pressure, force, position and velocity data. The gathered experimental data enable the authors to validate a previous PMA dynamic model. Meanwhile, new models are developed from the original model. Statistical analysis proves that the new models can better represent the PMA dynamics during the experiments.

Keywords: Pneumatic Muscle Actuator, Wearable Robotics, Rehabilitation Robotics.

1 Introduction

Pneumatic muscle actuators (PMAs) was firstly invented by Joseph L. McKibben for an orthotic application in 1950s [1]. There were not many further applications or development of PMAs until 1980s, when the original PMAs were redesigned to actuate robotic arms in Japan [2]. In the past two decades, PMAs have redrawn the attentions of researchers mainly due to their advantages their advantages of having high small weight, high strength and high power to weight ratio. PMA is also of controllable compliance, due to the compressibility of the gas [3].These advantages make the PMAs widely adopted in robot-human interaction applications such as robotic exoskeleton or orthosis [4-6].

Although there are various constructions of PMAs, the most common one used in research is the McKibben design which consists of a cylindrical flexible airtight tube that fits inside a sheath with braided thread. When a PMA is inflated, it widens and shortens to generate a contracting force and displacement [7]. In spite of the advantages of PMA, a major challenge of using PMAs as actuator is about controlling them precisely. Due to the compliance, elasticity of the material and friction, the PMA

X. Zhang et al. (Eds.): ICIRA 2014, Part II, LNAI 8918, pp. 432–440, 2014.

operation is highly nonlinear and subjects to hysteresis behaviour. Hence, a number of researches have been conducted to model PMA operations and thus control algorithms based on developed models. The essential predecessor of accurate control is the appropriate model of the PMAs.

A number of researchers have investigated different modelling method of PMAs. Chou and Hannaford [8] developed the simple but widely used geometrical model on McKibben PMA. Tondu and Lopez [9]further improved the model in [8] by compensating the end deformation (PMA is not a cylindrical shape when inflated) and attempting to offset the hysteretic behavior between inflation and deflation of a PMA. Another commonly adopted model of PMA is the phenomenological model developed by Reynolds el al. [10] . The PMA model is similar to the Hill's muscle model in biomechanics. It contains parallel force generation, spring and damping elements as shown in Figure 1. The dynamic force behavior is represented by Equation 1

$$F(P) + B(P)\dot{x} + K(P)x = L + M\ddot{x} \tag{1}$$

where: $F(P)$, $B(P)$, $K(P)$ are force, damping and spring parameters which are all linearly dependent on pressure (as indicated in Equation 2) and determined experimentally; M is the mass of the load; x is the contracting length of PMA and L is the force exerted on the load.

$$\begin{cases} F(P) = F_0 + F_1 P \\ K(P) = K_0 + K_1 P \\ B(P) = B_0 + B_1 P \end{cases} \tag{2}$$

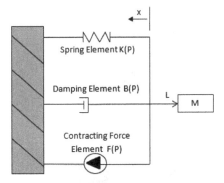

Fig. 1. The model used in [10] for the dynamic behavior of PMA

Compared to McKibben PMAs used in the previous mentioned modelling research, the PMAs manufactured by FESTO are able to generate much larger forces for the same amount of contraction.

PMA manufactured by FESTO is a special type of McKibben PMA actuators. Its pressure tight rubber tube and braided sheath are integrated into the contraction system. Due to the different in construction, PMAs manufactured by FESTO have different properties compared to the PMAs modelled in [8-10]. At the same pressure,

FESTO PMA can generate significant larger force than conventional McKibben PMAs with same size and contracting length. This thus makes PMA manufactured by FESTO increasing popular in wearable robotic research. However, modelling of FESTO muscle has not been extensively researched. Models developed for conventional muscles are adapted to model FESTO muscles with adjusted parameters. Choi et al.[11] adopted the dynamic model [10] to FESTO muscles with adjusted parameters; meanwhile, sliding mode control was implemented to cope with inaccuracy in modelling. Experiment based static models have also been developed to express the force-pressure-contraction relationship[12, 13]. However, to make better use of the FESTO PMAs in dynamic application, a more specific dynamic model for FESTO PMA is yet to be developed.

In this paper, an improved dynamic model to the one described in[10] will be investigated. Section 2 describes the setup of the testing device and the how the experiments are conducted. Section 3 is devoted to analyze the experimental results, propose the improved dynamic models, as well as validating the models. Section 4 concludes this paper and provides the direction for future research.

2   Experimental Setup and Procedures

The experiment setup (Figure 2) contains two PMAs. PMA_1 (FESTO DMSP-20-400N-RM-CM) at the bottom is the muscle to be modelled. PMA_2 (FESTO DMSP-40-300N-RM-CM) is to provide motions as required during the experiments. Both PMAs are pressure controlled with a pressure regulator (FESTO VPPM-6L-L-1-G18-0L6H-A4N). A pressure sensor (SPTE-P10R-S6-V2.5K) is also placed next to the

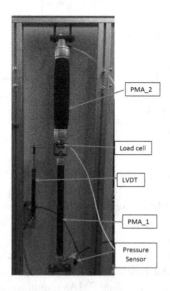

Fig. 2. Setup of the testing device

pneumatic connection to PMA_1 for more accurate pressure measurement. The linear potentiometer (Variohm VLP100) and the load cell (FUTEK LTH350) are used to measure displacement and exerted force respectively. The assumption of this testing setup is that mass of all moving components are negligible.

A National Instrument CompactRio is employed for hardware interface and control processing. The experimental procedures are implemented in LabVIEW program. At the beginning of the one experiment, the PMA_1 is inflated to a certain pressure. By adjusting the bolts at the bottom of the test rig along the thread rod, an appropriate initial contracting length (x_0) can be achieved, and at this instant the linear potentiometer has a reading of d_0. Being appropriate means the bolts are in contacted with the bottom plate of PMA_1. As the result, the contracting length during the experiment can be calculated using the following equation:

$$x = x_0 - (d - d_0) \tag{3}$$

where: x is the instantaneous contracting length of PMA_1; d is the instantaneous linear potentiometer reading.

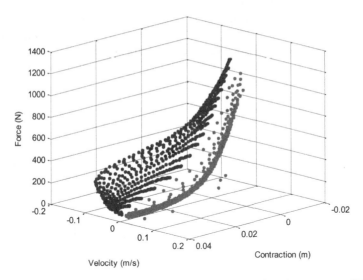

Fig. 3. The force verses contraction and velocity scatter plot, when the pressure of PMA_1 is 2.2bar. Red and blue dots represent data during the contraction and stretching of PMA_1 respectively

During one set of experiment, the pressure of PMA_1 is regulated to the fixed value, while the pressure of PMA_2 is manipulated to simulate different motions. Initially, PMA_2 is fully deflated. After the measurement of x_0 is completed, the press regulator of PMA_2 will receive a step input for a certain pressure. As a response, PMA_2 contracts to stretch PMA_1. After the movement is settled and a short wait, a control signal will be applied to bring the pressure of PMA_2 back to zero. Thus,

PMA_1 contract again back to x_0. Then, the set pressure of PMA_2 increases by 0.2 bar, and repeats the inflation-settling-deflation process. Higher set pressures to PMA_2 lead to larger displacements, larger forces and higher velocities during both the inflation and deflation of PMA_2. The set point for PMA_2 keeps increasing while the experimental steps are repeated, until a cap force is exceeded. Up to here, the experiment for the selected pressure of PMA_1 is completed. Fig. 3 is showing the gathered data for the set of experimental data when PMA_1 is regulated to 2.2 bar. Only dynamic data (velocity magnitude greater than 0.005 m/s) are gathered. The contracting and stretching data points are separated, so hysteresis can be considered in the proposed model. The same experiments are repeatedly conducted to various pressures of PMA_1 ranging from 0.6 to 5 bars, with three experiments for each pressure.

3   Result Analysis

Firstly, the validation of the existing dynamic model proposed in [10] with the newly gathered data is attempted. For a fixed pressure, the K, B and F parameters in Equation 1 can be calculated using a first-order polynomial curve fit for the coefficients. To cope with the hysteresis, two fitted surfaces shown in Fig. 4 are generated for one set of experimental results, one for the contracting and the other for the stretching process. The R^2 for the two fits are 0.941 and 0.913, which indicate good accuracy of the model of the specific pressure.

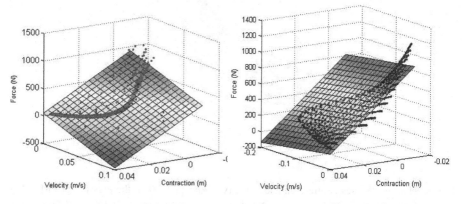

Fig. 4. The virtualization of curve fitting result for the same set of experimental data as in Figure 3. The left plot is for the contracting process and stretching data is shown on the right.

With two sets of force, spring and damping parameters acquired for each of the 56 experiments have been conducted. Relationships between pressure and each of the parameters can be visualized in Fig. 5. The first order linear relationships are built according to Equation 2 and the fitted coefficients are shown in the third column of Table 1. Relationships are represented by the green straight lines in Fig. 5. As can be seen from the figure, the green lines are well correlated to the data plots except the one for the spring element. To further quantify the accuracy of the model, the R^2 values (shown in the bottom row of Table 1) of the entire population of experimental data are calculated. For every valid data point, the force calculated via Equations 1 and 2 are compared with the measured force. A total of 29,811 and 43,682 data points are included in the analysis for stretch and contraction respectively. The result of regression analysis indicates that the model is not capable of representing the dynamic behavior of the PMA. Hence, there are demands for a new or modified dynamic model.

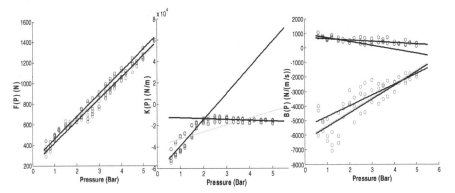

Fig. 5. From left to right: F(P), K(P) and B(P) versus pressure plot. Parameters for contraction are plotted in red and for stretch are in blue.

Table 1. The curve fitting results for different dynamic models

Factor		Coefficient	Model in [10]	Piecewise Model Below 2 bar	Above 2 bar
Force Element		F_0 (N)	254.4	239.5	207.4
		F_1 (N/Bar)	201.7	219.4	213.6
Spring Element		K_0 (N/m)	-39085	-64,104	-12,188
		K_1 (N/m/Bar)	24634	24,634	-767.5
Damping Element	Contraction	B_0 (N·s/m)	-5,178	-4,680	-4,429
		B_1 (N·s/m/Bar)	831.7	348.9	666.0
	Stretch	B_0 (N·s/m)	841.3	697.4	964.9
		B_1 (N·s/m/Bar)	-133.8	-12.41	-165.2
Regression Analysis	Contraction	R^2	0.0895	0.9679 (0.9671)	
	Stretch		0.6177	0.9746 (0.9742)	

Two approaches have been attempted with the same experimental data as describe previously. Firstly, by analyzing the plots in Fig. 5, it was realized that the trend of spring element is different when the pressure is below and above 2 Bar. Therefore, the original dynamic model is then modified by fitting separate force, spring and damping parameters for data below and above a pressure of 2 Bar, which is shown in black lines in Fig. 5. The fitting results as well as the R^2 values are also listed in the columns titled "Piecewise Model" of Table 1. The R^2 values are calculated in the same way as the ones for the original model. The R^2 values outsides the brackets are computed with all three elements from the proposed piecewise model; meanwhile, the ones inside the brackets are computed with only the spring element from the piecewise model and the other elements from the original model. There is essentially no difference between the R^2 values calculated with these two methods. Therefore, method only with piecewise spring element is preferred for the ease of further application.

Another method is to utilize a quadratic polynomial model as shown Equation 3 instead of the linear polynomial as in Equation 1.

$$F(P) + B_1(P)\dot{x} + K_1(P)x + K_2(P)x^2 + B_2(P)x\dot{x} = L + M\ddot{x} \tag{4}$$

where all five pressure dependent parameters can be expressed in the form of:

$$X(P) = X_2 P^2 + X_1 P + X_0 \tag{5}$$

The reason for proposing this model is that it is able to represent the dynamics at a fixed pressure. For example, in Fig. 4 the linear polynomial surfaces can only achieve R^2 values of 0.941 (contraction) and 0.913 (stretch); whereas, the fitted quadratic polynomials can bring these values to 0.989 and 0.991. Similar to previous linear models described previously, the pressure dependent quadratic curves can be generated for the five parameters. Hence, the overall R^2 values are calculated as 0.704 (contraction) and 0.906 (stretch), which are improved from the original model, but not as good as the proposed piecewise model. Furthermore, the nonlinearity of the quadratic model and added number of parameters significantly increase the complexity of further control implementation, as well as the computational power requirement.

During the data processing, the mass of the air muscle was also taken into consideration. However, in the worst scenario, the product of the entire mass (0.3kg) of the PMA and the maximum acceleration (3m/s) is less than the noise of the load cell reading. Therefore, the mass of the PMA is treated as negligible.

To exam the generalization of the proposed piecewise dynamic model, the identical experimental process was repeated on a different PMA with 290mm length and 20mm diameter (FESTO DMSP-20-290N-RM-CM). A total of 37 experiments were conducted with the new targeted PMA and the results are presented with Fig. 6 and Table 2, which are in the same formats as Fig. 5 and Table 1 for the 400mm PMA. By analyzing the results, it can be summarized that the proposed piecewise dynamic model is capable in representing different sizes of PMA.

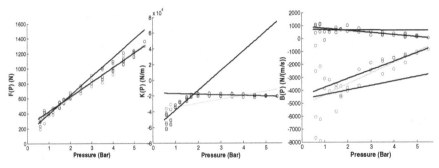

Fig. 6. The experimental results for the 290mm PMA. From left to right: F(P), K(P) and B(P) versus pressure plot. Parameters for contraction are plotted in red and for stretch are in blue. The black lines indicate the fitting of the proposed piecewise model and the green lines are fitted from the model developed in [10].

Table 2. the curve fitting results for different dynamic models for the 290mm PMA

Factor		Coefficient	Model in [10]	Piecewise Model Below 2 bar	Piecewise Model Above 2 bar
Force Element		F_0 (N)	194.8	140.7	229.9
		F_1 (N/Bar)	204.0	250.3	194.9
Spring Element		K_0 (N/m)	-39458	-63,382	-16,352
		K_1 (N/m/Bar)	5,027	25,085	-787.9
Damping Element	Contraction	B_0 (N·s/m)	-6,275	-6,381	-5,463
		B_1 (N·s/m/Bar)	946.4	948.2	737.4
	Stretch	B_0 (N·s/m)	763.9	961.6	709.3
		B_1 (N·s/m/Bar)	-91.04	-268.5	-91.04
Regression Analysis	Contraction	R^2	0.4871	0.9724	
	Stretch		0.6986	0.9765	

4   Conclusion and Future Work

New dynamic models of PMA have been developed based on the original model developed in [10]. Only experiments with pressure 2 Bar and above were reported by Reynolds el al. [10]. Whereas, the models proposed in this article are based on extensive experiments with pressures ranging from 0.6 to 5.0 Bar. The piecewise linear model is preferred by the authors, due to its high correlation to measured data, meanwhile still being in a simple form. The proposed models are experimental based and muscle specific. However, the experimental procedures and data processing have been implemented to be fully automated. As a result, the models can be easily transferred to PMAs with different sizes.

Currently, the author are working with trajectory controlling lower limb exoskeleton actuated by antagonistic pairs of PMA with the developed piecewise linear dynamic model, as well as a parallel ankle rehabilitation robot driven by four PMAs. In the future, it is aimed to compare the effect on control performance of the new model with other PMA models.

References

1. Nickel, V.L., Perry, J., Garrett, A.L.: Development of useful function in the severely paralyzed hand. The Journal of Bone & Joint Surgery 45, 933–952 (1963)
2. Inoue, K.: Rubbertuators and applications for robots. In: Proceedings of the 4th International Symposium on Robotics Research, pp. 57–63. MIT Press (1988)
3. Caldwell, D.G., Tsagarakis, N., Medrano-Cerda, G.: Bio-mimetic actuators: Polymeric pseudo muscular actuators and pneumatic muscle actuators for biological emulation. Mechatronics 10, 499–530 (2000)
4. Ferris, D.P., Czerniecki, J.M., Hannaford, B.: An ankle-foot orthosis powered by artificial pneumatic muscles. Journal of Applied Biomechanics 21, 189 (2005)
5. Hussain, S., Xie, S.Q., Jamwal, P.K., Parsons, J.: An intrinsically compliant robotic orthosis for treadmill training. Medical Engineering & Physics 34, 1448–1453 (2012)
6. Beyl, P., Knaepen, K., Duerinck, S., Van Damme, M., Vanderborght, B., Meeusen, R., Lefeber, D.: Safe and Compliant Guidance by a Powered Knee Exoskeleton for Robot-Assisted Rehabilitation of Gait. Advanced Robotics 25, 513–535 (2011)
7. Kelasidi, E., Andrikopoulos, G., Nikolakopoulos, G., Manesis, S.: A survey on pneumatic muscle actuators modeling. In: 2011 IEEE International Symposium on Industrial Electronics (ISIE), pp. 1263–1269. IEEE (2011)
8. Chou, C.-P., Hannaford, B.: Measurement and modeling of McKibben pneumatic artificial muscles. IEEE Transactions on Robotics and Automation 12, 90–102 (1996)
9. Tondu, B., Lopez, P.: Modeling and control of McKibben artificial muscle robot actuators. IEEE Control Systems 20, 15–38 (2000)
10. Reynolds, D., Repperger, D., Phillips, C., Bandry, G.: Modeling the dynamic characteristics of pneumatic muscle. Annals of Biomedical Engineering 31, 310–317 (2003)
11. Choi, T.-Y., Lee, J.-J.: Control of manipulator using pneumatic muscles for enhanced safety. IEEE Transactions on Industrial Electronics 57, 2815–2825 (2010)
12. Sarosi, J.: New approximation algorithm for the force of Fluidic Muscles. In: 2012 7th IEEE International Symposium on Applied Computational Intelligence and Informatics (SACI), pp. 229–233. IEEE (2012)
13. McDaid, A., Kora, K., Xie, S., Lutz, J., Battley, M.: Human-inspired robotic exoskeleton (HuREx) for lower limb rehabilitation. In: 2013 IEEE International Conference on Mechatronics and Automation (ICMA), pp. 19–24. IEEE (2013)

Multi-channel Transmission Mechanism Based on Embedded Motion Control System

Hui Wang, Chao Liu, Jianhua Wu, Xinjun Sheng, and Zhenhua Xiong

State Key Laboratory of Mechanical System and Vibration, School of Mechanical
Engineering, Shanghai Jiao Tong University, Shanghai 200240, China
{351582221,aalon,wujh,xjsheng,mexiong}@sjtu.edu.cn

Abstract. A multi-channel Direct Memory Access (DMA) transmission
mechanism based on embedded multi-axis motion system is proposed in
this paper. The motion control system comprises of a computer and a
motion controller based on DSP and FPGA. DSP is vital important in
transferring data sets among servo driver modules, FPGA module and
itself. The PCI protocol module and DMA buffer pool are integrated
in FPGA to set up DMA channels between the computer and the motion controller, which simplifies the address mapping of local bus and
increases the efficiency of data transmission. Experiments are conducted
on the self-designed motion control system and the results demonstrate
that the DMA transmission mechanism has distinct advantages, such
as simple hardware structure, excellent low transmission fluctuation and
high hard real-time performance.

Keywords: motion control system, DSP, FPGA, multi-channel DMA,
real time.

1 Introduction

The typical PC-based motion control system contains a computer and a motion controller. The computer is responsible for path planning and interpolation
points transmitting to the motion controller. The points are acquired by the real-
time control program implemented in motion controller[11]. Since much time is
required for the computer to run the system program, parse command-line ar-
guments, control human-machine interface (HMI), and carry out other computational tasks, the computer can not meet the requirements of real-time control
at the microsecond level. For instance, the Vxworks operation system which is
famous for its hard real-time feature still has poor performance in motion control
even though its time slice is no less than 10ms[1]. Since the time-slice length of
the computer is too long to realize real-time control at the microsecond level,
hard real-time interrupt mechanism is generally adopted in motion controller.
In this case, bus type and bus protocol implemented in motion controller are
very important to the improvement of real-time data interaction. The computer,
accordingly, is connected with the motion controller via industrial buses, such
as PCI, real-time ethernet, industrial field bus and so on; the memory-sharing
technique is applied in data exchanging over the mentioned bus[3].

X. Zhang et al. (Eds.): ICIRA 2014, Part II, LNAI 8918, pp. 441–452, 2014.

Generally, the hardware architecture of the motion controller is " bus protocol chip + FPGA + DSP " [6,12]. When the working frequency of PCI bus protocol chip, FPGA, and DSP are set as 50 MHz, 50 MHz, and 150 MHz respectively, the transfer rate under DMA channel can reach 35.2 MB/s after strict mathematical analysis [8]. The DMA channel of DSP allows zero-overhead transfers at the rate of no less than 4 clock/word without processor intervention[7]. For data exchange between local bus and internal bus through DMA channel, DSP-Cache method is widely applied which can avoid bus confliction and can make full use of DMA channel resources. However, in this method, only two channels are employed for data exchange[4]. While, when different data sets are exchanged at one time, the DSP-Cache method is unable to satisfy the requirements, so a multi-channel DMA transmission mechanism is proposed in this manuscript. The mechanism is also capable of improving the performance of multi-channel data transmission, especially for real-time data exchanging via DMA channel.

In the motion control field, there are several types of buses transferring data between the computer and the motion controller, whose qualities affect the system reliability. One of the buses is the industrial ethernet which is generally applied and is advantageous in high-speed data transmission and multi-node system construction. Moreover, the real-time performance and the stability of the industrial ethernet are enhanced after several years of research. However, no uniform standard exists for each company has its own copyright and license, which makes it difficult to implement those protocols in FPGA[9]. Another popular bus is industrial field bus. Based on distributed control technology, it can easily build a real-time serial network, where different devices can be linked together. But its transmission rate is closely associated with the transmission distance. For example, the maximum rate of Controller Area Network (CAN) bus is about 0.125 MB/s within a distance of 100 meters, but it decreases to approximately 0.625 KB/s when the transmission distance extends to 10000 meters[5]. The Peripheral Component Interconnect (PCI) bus is another widely used bus. Although multiple devices are hard to share information on normal PCI bus, it is still preferred due to its significant advantages, such as strict specification, easy implementation and stable transmission[2]. For instance, its burst transfer rate can be up to 132MB/s according to the PCI local bus specification V2.1. Meanwhile, DMA communication based on PCI bus is usually applied in the data acquisition field[10]. In order to decrease fluctuating transmission rate and offer an efficient connection between the computer and the motion controller, a hardware architecture" FPGA + DSP " is proposed in this manuscript to accomplish data sets interaction which embedded PCI bridge protocol into FPGA.

In this paper, a novel real-time data-interaction mechanism is presented, and a " FPGA+DSP " architecture is designed for motion controller. In addition, according to priorities of data sets, a multi-channel DMA mechanism is utilized to transfer data sets among the computer, FPGA, DSP, and DAC. Meanwhile, the connection implemented in FPGA is established by the internal PCI protocol module and first-in-first-out (FIFO) buffer module, which is used to carry out the handshake with DMA channels of compter and DSP. In this case, the mechanisms is dominated and executed by DSP, which greatly reduce the real-time demands for the computer. Moreover, the processors of the computer and DSP can be liberated from the complicated Input/Output (I/O) affairs. So the bottleneck of the traditional transmission mechanism is solved properly.

This article is organized as follows. In section 2, the architecture of the system is introduced. Section 3 describes DSP program design, and then the DMA channel parameters are settings are given. The buffer pool modules and PCI protocol module are also proposed. Furthermore, data transmissions among the computer, DSP and servo driver modules both in I/O mode and DMA mode are given in detail. The COM-to-LPT module is constructed in Section 4. In Section 5, experiments are carried out, and results are presented which validate the effectiveness of the transmission mechanism. We conclude this article with some final remarks in section 6.

2   System Description

The communication architecture of motion control system is shown in Fig. 1, in which PCI protocol module is implemented in FPGA. Both the I/O and DMA transmission mode are used to exchange data between the computer and FPGA. What's more, the channels of computer and DSP are connected by using data buffer pools and data request-response units in FPGA. Thus, the small number of data such as simple commands and parameters uses the I/O transmission mode to transfer, while the data packet carrying information like actual positions or interpolation points can be adopted DMA transmission mode to transfer. Meanwhile, the DMA controller manipulates the transfer operations of interpolated points from PC to FPGA. Similarly, the DSP controller guides the transmission of bulk data from FPGA to PC, such as a packet of actual positions. Through the proposed mechanism, the data transmission procedure is detailed as follows. Once the packet of interpolated points is transferred entirely from the DMA data buffer pool to internal RAM of DSP, the DSP processor is used to subdivide those points into fine interpolation points. And then the closed-loop calculation is carried out in each servo cycle, which fetches the

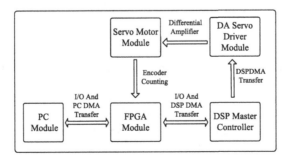

Fig. 1. Communication architecture of motion control system

data — one is from those points in order and the other is the actual position of motor acquired from optical encoder. After that, the calculation results are transmitted to the COM-to-LPT module located in FPGA. And the COM-to-LPT module converts the results from the serial data to multiple streams of data afterwards. At last, the data is handed over to the servo driver module.

3   Design and Implementation of Multi-channel Structure

In this paper, the self-designed four-axis motion controller is used as the experiment platform, whose architecture is " FPGA+DSP ". The FPGA chip with type of Cyclone II, made by ALTERA company, has 8256 lE logic units, 125 general-purpose I/O pins, 36 M4K storage units and up to 167 MHz I/O frequency. The DSP chip TMS320F28335 has a high-performance 32-bit CPU, up to 150 MHz clock frequency and a six-channel DMA controller for ADC, MCBSP, EPWM, XINTF, and SARAM. Besides the FPAG and the DSP, two AD1866 chips are implemented in the motion controller for DA conversion. The chips comprise four DAC channels. Each chip is a complete dual 16-bit DAC, which offers excellent performance while requiring a single +5 V power supply, and each DAC is equipped with a high performance output amplifier. The employed amplifier can achieve fast settling and high slew rate, since its ability of producing ±1 V signals at load currents up to ±1 mA and its fast CMOS logic element which allows for an input clock rate up to 16 MHz.

3.1   Design and Configuration of DSP Module

The External Memory Interface (EMIF) and Multichannel Buffered Serial Port (MCBSP) of DSP are adopted to exchange data with FPGA. Therefore, the 150 MHz operation frequency of DSP is provided to FPGA through XCLKOUT pin, which ensures that the clocks can be synchronized between them. Meanwhile, DSP supervises the data transmissions among chips on the motion controller and makes a closed-loop calculation to control motor according to commands from computer.

As shown in Fig. 2, DSP has six DMA channels. The rate of transmissions are configured as follows. In order to ensure the rate of 37.5 MB/s between FPGA and itself, 8 clocks are taken to transfer one 16 bit data through EMIF port. And considering the maximum input clock rate of AD1866, 256 clocks are taken to transfer a 16-bit data through MCBSP. Namely, the transfer data is 1.17 MB/s. Thus, in each servo cycle, four axis velocity values, stored in the relevant address space located in the front of RAM 4, are updated and transferred to MCBSP in real time through DSP DMA bus. In addition, the RAM 5 and RAM 6 can hold totally 4096 32bit-wide data in all, which connect Zone 6 respectively through DMA bus and DSP bus. In this case, two different data packets of interpolation points from PC can be transferred twice to the RAMs in DMA mode. In each interval of 40 servo cycles, DSP processor fetches the desired four-axis interpolation points out one after another and subdivides

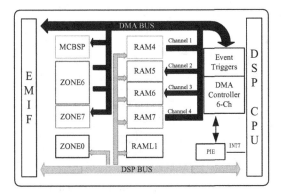

Fig. 2. DSP configuration block diagram

Table 1. DSP multi-channel DMA parameters settings

Channel No.	Priority	Data flow direction	Description	Interrupt source
1	1	RAM 4 to MCBSP	motors control	MCBSP-A Tx
2	2	FIFO 0 to RAM 5	PC data transfer	Software
3	2	FIFO 0 to RAM 6	PC data transfer	Software
4	2	RAM 7 to FIFO 1	DSP data transfer	Software

each point into fortieths to make closed-loop calculation. At the same time, since RAM 7 can store two thousand and forty-eight 32bit-wide data, four-axis actual position points from optical encoder are stored in RAM 7 sequently in each servo cycle. Once RAM 7 is filled with required data, the packet of actual positions is shifted from RAM 7 to the buffer through DMA bus. After the buffer is full, the request-response unit notifies the computer to read the data packet. From the above analysis, During the data exchange process between DSP and FPGA, the four-axis velocity values need to be transferred to MCBSP in real time via DMA bus. The unavoidable conflict occurs on bus occupancy. To deal with the problem, the multi-channel transmission mechanism is proposed. Data sets, accordingly, are allotted into different DMA channels and time slices are assigned to each set by PIE arbiter.

Table 1 shows DSP multi-channel DMA parameters settings, including channels settings and their priorities. The DMA channel 1 is set higher priority, other channels have the same priorities. Therefore, in channel 1, there is one data transfer from RAM 4 to MCBSP for real-time motion control in each servo cycle. Meanwhile, other transfers are assigned regular channels. Thus, the channels are serviced in round-robin fashion as " CH2 → CH3 → CH4 → CH2 → · · ·". Once any channel is serviced, the event trigger receives the transmission request and sends it to PIE arbiter. And then the processor notifies the DMA controller to master data flows. When channel 1 receives an interrupt trigger

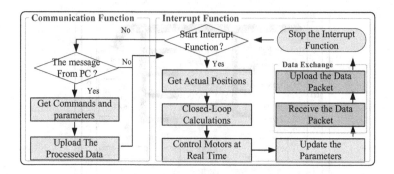

Fig. 3. DSP software's schematic chart

from its peripheral before the former channel completes, the transmission in former channel is suspended and channel 1 is serviced until the data word transfer in current channel is completed. After the data transfer in Channel 1 is completed, the former channel resumes to transfer data. As a result, the four-axis velocity values transfer can offer a real-time guarantee for motion control.

The programs implemented in the motion controller is written in C language, and the code is stored in RAML1. As shown in Fig. 3, the programs contain two parts: one is the interrupt handler function, the other is the communication function. The interrupt handler function is responsible for the closed-loop control, the data packet transmission, the control parameter update and the velocity command transfer. The other function is responsible for the command and parameter interaction between the computer and motion controller. Due to the reasonable time distribution feature of DSP, both of them can work cooperatively.

3.2   Design and Configuration of FPGA Module

The motion controller mainly comprises of a computer, FPGA, DSP and servo driver modules. The FPGA serves as a bridge to connect them together, which contains a internal PCI protocol module, a data buffer pool module, a request-response unit module, a COM-to-LPT module and a frequency-division-and-phase-detection module. The PCI protocol module maps the memory of motion controller into the computer RAM and guides data exchange between the controller and computer. a data buffer pool module contains the DMA data buffer pool and the I/O buffer pool. The DMA data buffer pool establishes a channel for data packet exchange, while the I/O data buffer pool serves for a small amount of data transmission. The request-response unit module manages access rights of buffer pools assigned to the computer and DSP. The COM-to-LPT module converts the four-axis velocity values from the serial form to the multiple-stream form to match with servo driver module. The frequency-division-and-phase-detection module, which acquires differential signals from

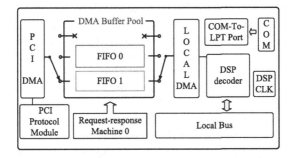

Fig. 4. The DMA channel connection between the computer and DSP

optical encoders and transforms them into single-ended signals, is responsible for signal filtering, frequency multiplication, phase detection and index signal acquisition.

The DMA data buffer pool, composed of two FIFOs, is instantiated in the Quartus II software using the MegaWizard Plug-in Manager. The depth of DMA data buffer pool is 8192 bytes. As shown in Fig. 4, FIFO 0 stores interpolation point data that comes from the computer buffer and FIFO 1 stores actual position data acquired by DSP. One side of the pool is linked with DSP decoder module, and the other side is connected with PCI protocol module. The DSP decoder module allows other modules to map their address regions for XINTF Zones as RAM in DSP. And the PCI protocol module is implemented according to the local bus specification V2.1. It contains a set of well-defined configuration registers used to identify devices and supply device configuration, which ensures the realization of DMA access to motion controller. Since these two modules work with different frequency, the communication between the DMA channel on PCI bus and that on local bus is asynchronous. Therefore, it requires for a reasonable clock and FIFO depth configuration.

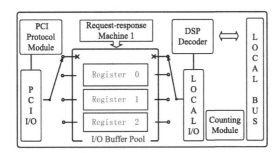

Fig. 5. The I/O channel connection between the computer and DSP

The request-response unit 0 serves as a transponder to notify the computer and motion controller to fetch the data. Once the computer finishes the transition of a data packet to FIFO 0, the request signal is activated. DSP monitors the signal in every servo cycle. The embedded DMA controller implements the request after the signal changes and consequently initiates the transmission on channel 2 and channel 3 in turns. When RAM 5 or RAM 6 is full, DSP processes the data after sending a control instruction to deactivate the request signal. When DSP transfers a feedback data packet from RAM7 to FIFO 1 via channel 4, a trigger signal is generated by DSP after data transfer is finished. After that, the request-response unit 0 receives the signal and produces the other request signal to notify the PCI Protocol module. The module initiates a DMA transfer and the data is transmitted from FIFO 1 to the computer over PCI Bus.

The I/O data buffer pool, instantiated in the Quartus II software, is shown in Fig. 5. One side of the pool is linked with DSP decoder module, and the other side is connected with PCI protocol module. Furthermore, Three 16-bit-wide registers are built in the buffer pool to store a small amount data temporarily. The data contains a 16-bit computer command and a 32-bit parameter. After the buffer pool receives data from the computer, the request-response unit 1 activates the request signal. Once the signal is captured, the DSP processor moves the data from buffer pool to RAML1. When DSP finishes the calculation and transfers the feedback data form RAML1 to buffer pool over DSP bus, a trigger signal is generated by DSP afterwards. And then, the request-response unit 1 receives the signal and produces a request signal to notify the PCI protocol module. The module initiates a I/O transfer and transmits the data from the buffer pool to the computer over PCI Bus.

4   The COM-to-LPT Module Design

The transmission of four-axis velocity values are initiated in series format after transmit control registers (XCR), McBSP bit clock (ClKX), frame period (FPER) and frame width (FWID) are steed. XCR stores the data stream temporarily and sends the stream to data transmit (DX) pin on the rising edge of ClKX. The frame sync generator (FSG) signal is capable of distinguishing each 16-bit datum in the data stream. The data stream transfers from DX pin to COM-to-LPT module through Channel 1 at the rate of 1.17 MB/s in each servo cycle. The COM-to-LPT module is written in Verilog language, which transforms FSG signal into four-chip-select signals. After AD1866 chips receive the data stream, ClKX signal and four-chip-select signals simultaneously for DA conversion. The actual effect of the COM-to-LPT module is shown in Fig. 6.

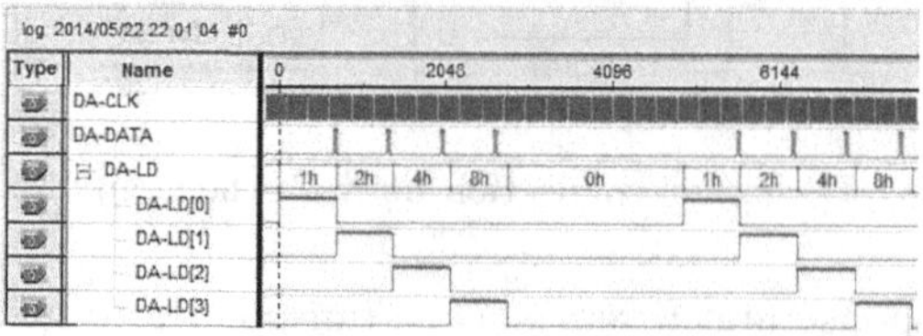

Fig. 6. The actual effect of the COM-to-LPT module

5 Experimental Studies

5.1 Experimental Setup

As shown in Fig. 7, the multi-channel DMA motion control system comprises of a self-designed four-axis motion controller, a AC servo driver unit, and a computer. The AC servo driver unit contains four YASKAWAY SGDV-2R8A01A series AC Servo Drivers and four SGMJV-04ADA21 series AC Servo Motors with the maximum speed of 3000 rpm. The computer is the ADVANTECH 610L series industrial PC with E5300 2.6G CPU, 1G memory, 250G hard disk and Windows XP SP3 operation system. In order to verify the priority of the proposed transmission mechanism in this manuscript, the corresponding experiments are carried out in the following parts.

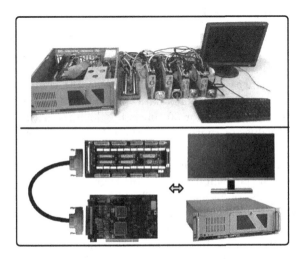

Fig. 7. Experimental setup

5.2 Feasibility Analysis of the Proposed Mechanism

The feasibility analysis of the multi-channel DMA transmission mechanism is discussed in this subsection. The four-axis motion controller exchanges data among the computer, the servo driver module and itself in DMA mode. The parameters are given as: data packet size: 2048/32 bit, PCI transmission rate: 132 MB/s, PCI bus width: 32 bit, DSP bus width: 16 bit. As to the DSP XINTF setting, zone 6 and zone 7 are selected. The read and write access parameters of which are as follows: lead period: 2 cycles, active period: 4 cycles, trail period: 2 cycles. Thus, the transfer rate of zone 6 and Zone 7 is 37.5 MB/s. The settings of MCBSP are shown as follows: CLKG frequency: 16 cycles, frame-synchronization period pulse width: 256 cycles, transmit word length: 16 bit.

Packet exchanges are carried out between the computer and motion controller in DMA mode. Meanwhile, DSP processes the data in the on-chip memory and then transfers the results into COM-to-LPT module in real time through DMA channel 1. At the same time, the computer monitors packet traffic on PCI bus and counts the number of packet transited. Moreover, the SignalTap II Embedded Logic Analyzer in Quartus II is used to monitor the transmission of the four-axis-velocity values. The experiment is implemented for 30 times, in which the actual running times are recorded. The values of runtime of data transmission between the computer and motion controller are given in Fig. 8. Besides that, according to the parameter settings shown in Table 1, 30 experiments for each channel are carried out to prove the feasibility of the mechanism. During the operation of each channel, the start and end time of packet transmission are recorded by the timer of DSP. And the values of runtime per channel under the multi-channel DMA transmission mechanism are given in Fig. 9.

5.3 Performance Comparison

In this subsection, experiments are carried out to verify the effectiveness of the multi-channel DMA transmission mechanism. As to the DSP XINTF settings, the Zone 0 is selected. The other setting parameters are the same as the above. Thus, the data exchanges are carried out between the computer and motion controller in I/O mode. The number of data transmitted each time is 4096 words, the experiment is implemented for 30 times, in which the actual running times are recorded. The values of runtime with the I/O transmission mechanism are given in Fig. 8.

Compared with the I/O transmission mechanism, the performance of data transmission between the computer and DSP is greatly improved by the multi-channel DMA transmission mechanism. The time consumed ranges from 250 ms to 360 ms in the I/O transmission mode. The average value is 263 ms and the mean square error (MSE) is 24.6 ms. Meanwhile, no data loss occurs during the transmission experiment. In contrast, under the DMA transmission mode, the time consumed ranges from 15 ms to 16 ms. The average value is 15.6 ms and the MSE is 0.48 ms. Compared with I/O mode, the multi-channel DMA transmission mode is more than 16 times as transmission rate, and the MSE

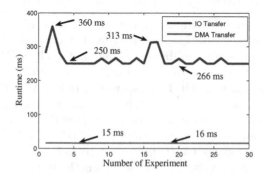

Fig. 8. Results of comparative experiments

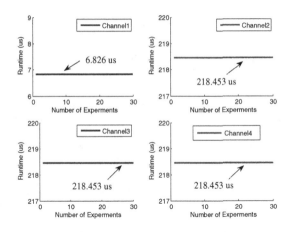

Fig. 9. Running times of data transmissions via four channels

of the transmission rate in DMA mode drops to nearly one-fiftieth of that in I/O mode. Meanwhile, Due to the real-time transmission ability of DSP and the sharing clock between DSP and FPGA, the rapidity and stability of the motion controller is improved. Therefore, the running times of each channel remain the same in the experiments, and are 6.826 μs, 218.453 μs, 218.453 μs, and 218.453 μs respectively. Furthermore, The data exchange in I/O mode consumes a lot of computing resources of processor compared to DMA mode. Therefore, the stability and hard real-time ability of the motion control system are enhanced by the application of the multi-channel DMA transmission mechanism.

6   Conclusions

In this paper, a multi-channel management mechanism with a simple hardware realization is proposed which can increase the data transfer rate and decrease the transmission fluctuation rate. The data sets, dominated and executed by DSP, can be transferred into different DMA channels in priority mode. In addition, the internal PCI protocol module and the two data buffer pool modules are implemented in FPGA, which are responsible for the connection of the PCI bus and DSP DMA channels. Experiments are conducted on the self-designed motion control system and the results verify that the proposed transmission mechanism is capable of achieving better reliability and higher transmission rate than traditional I/O transfer mode. Our future research will integrate local rational B-spline curve interpolation technology in the motion controller, which expects to achieve real-time curve interpolation.

Acknowledgements. This research was supported in part by the National Basic Research Program of China (2013CB035804) and National Natural Science Foundation of China (U1201244).

References

1. Barbalace, A., Luchetta, A., Manduchi, G., Moro, M., Soppelsa, A., Taliercio, C.: Performance comparison of vxworks, linux, rtai and xenomai in a hard real-time application. In: 15th IEEE-NPSS Real-Time Conference 2007, pp. 1–5. IEEE (2007)
2. Bureš, P.: The problems of continuous data transfer between the pc user interface and the pci card control system. In: Mechatronics, pp. 483–487. Springer (2012)
3. Dozio, L., Mantegazza, P.: Linux real time application interface (rtai) in low cost high performance motion control. Motion Control 2003 (2003)
4. Guo, Q., Zhang, B.: Dsp-cache optimization based on dma. Journal of Electrical & Electronic Education 2, 021 (2009)
5. Hsieh, C.C., Hsu, P.L.: The can-based synchronized structure for multi-axis motion control systems. In: 2005 IEEE International Conference on Systems, Man and Cybernetics, vol. 2, pp. 1314–1319. IEEE (2005)
6. Huang, F.Q., Lin, W.P., Liang, J.C.: Design of laser processing system of five-axis motion controller based on dsp and fpga. Advanced Materials Research 846, 98–102 (2014)
7. Instruments, T.: Tms320f28335 digital signal controllers data manual (2007)
8. Jia, Q., Huang, Z., Liu, Y.: Study on performance of pci interface for embedded system. Future Communication Technology (2 Volume Set) 51, 3 (2014)
9. Jung, I.K., Lim, S.: An ethercat based control system for human-robot cooperation. In: 2011 16th International Conference on Methods and Models in Automation and Robotics (MMAR), pp. 341–344. IEEE (2011)
10. Yan, J.F., Wu, N.: High speed dma data transfer system based on pci bus. Journal-University of Electronic Science and Technology of China 36(5), 858 (2007)
11. Yeh, S.S., Hsu, P.L.: Analysis and design of integrated control for multi-axis motion systems. IEEE Transactions on Control Systems Technology 11(3), 375–382 (2003)
12. Yolacan, E., Aydin, S., Ertunc, H.M., et al.: Real time dsp based pid and state feedback control of a brushed dc motor. In: 2011 23rd International Symposium on Information, Communication and Automation Technologies, vol. 201, pp. 1–6 (2011)

A Method to Construct Low-Cost Superficial Tactile Array Sensors

Thomas Cobb[1], Muhammad Sayed[2], and Lyuba Alboul[2]

[1] Faculty of Arts, Computing, Engineering and Sciences, Sheffield Hallam University
[2] Centre for Automation and Robotics Research, Sheffield Centre for Robotics,
Sheffield Hallam University, UK
a8042763@my.shu.ac.uk, muhammad.b.h.sayed@gmail.com, l.alboul@shu.ac.uk

Abstract. This paper describes the research, development, testing and initial implementation of the necessary environment interaction tactile sensors in order to give appropriate feedback on a humanoid robot hand. Steps of the method to construct a tactile array sensor are presented including testing of individual components and their assembly.

1 Introduction

Recent years of technological advances, a growing hobbyist interest and the advent of CAD/CAM and 3D printing have made more complex mechanical designs for humanoid robot hands available to a larger audience[1]. Although mechanical designs are now widely and cheaply available, the sensing and control systems are not. This paper describes the research, development, testing and initial implementation of environment interaction sensors in order to give feedback of a humanoid robot hands contact forces necessary for any interaction with the environment[2].

The aim of this paper is to develop a low-cost superficial tactile pressure distribution sensor array that can be constructed in any arbitrary shape for installation on anthropomorphic robotic hands. The paper describes main tasks to achieve this; tactile sensors providing feedback from the environment, a method of attaching the sensors, and a data acquisition system to provide feedback from sensors.

2 Sensing Solution

Studies on human and animal sensing show that the most important aspects of interaction sensing are the abilities to sense pressure and temperature. Of these two parameters, pressure was considered the most significant in contribution to the dexterity of the hand and therefore the primary concern with regard to sensor development and therefore this project.

In order to fully optimise the dexterity of a hand, the pressure sensor solution is required to have a resolution that can detect multiple contact points and if possible indicate the shape of the contact areas.

X. Zhang et al. (Eds.): ICIRA 2014, Part II, LNAI 8918, pp. 453–462, 2014.

A resistive pressure sensor array has been chosen to simplify the processing part of interpreting the measured values. This is an advantage over capacitive pressure sensing as it reduces the time required to poll an array for feedback.

A common solution amongst developers for collecting information about force is to use Force Sensitive Resistors (FSRs). A variety of FSRs are available in different shapes and configurations. FSRs are simple and easy to use but the main drawback is that they cannot detect the difference between a single contact point and multiple contact points. In order to determine the exact shape and position of a contact area, a tactile array sensor is required.

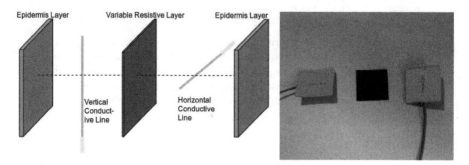

Fig. 1. Left: construction of pressure sensor showing two orthogonal crossover lines forming a single taxel, Right: an implementation of the single taxel sensor using copper lines

A standard structure of two orthogonal arrays of conductive lines sandwiching a variable resistive layer has chosen to keep the design as simple as possible. The conductive lines need to be enclosed in an insulating epidermis layer to protect the sensor array. Electronics attached to the sensor measures the resistance at the points of intersection of the vertical and horizontal lines forming a tactile pixel (taxel). Figure 1 shows the layer construction of the sensor using only two orthogonal lines forming a single taxel at the crossover point.

2.1   Related Work

There is a plethora of scientific papers on tactile sensors and, specifically, on robot tactile sensing. However, most of those papers are related to industrial robotics. One of the first books on tactile sensing to be worth mentioned, is the book of R. A. Russell [3]. In this book various functions of tactile sensing in both robots and living creatures are described, various types of tactile and touch sensors are surveyed, as well as applications in industrial robotics, manipulation and assembly tasks, are listed. A reader can find recent developments together with further references in the Handbook of Robotics [4]. However, applications listed there concern industrial robotics.

There are a couple of works that tackle some aspects of tactile sensing, similar to the ones proposed in this work. In [5] the development and applications of an artificial, flexible, force sensitive skin are presented, but the sensor is large and is not applicable to small areas such a humanoid robot finger. The tactile sensing system, presented in [6], is similar in its structure, to our system. However, their solution is, again, developed for sensing large areas, and the lattice of their conductive layer is too sparse.

2.2 Design Criteria

Several key factors can be identified that can be used to evaluate the constructed sensor and guide the construction process itself.

Size is a key requirement, the sensor must be thin enough to allow for superficial installation on existing humanoid robot hands. A simple measure to asses sensor size is to consider the possibility to "wear" the sensor on a human hand; if the size does not hinder normal hand functionality then it is an acceptable size.

Flexibility is another key factor that is closely related to the first one; a sensor that is small but stiff at the joints will hinder the hand's dexterity. This factor is affected by material selection for the components of the sensor.

Durability is required for extended use, especially with superficial installation; areas covering the joints may suffer from fatigue after many cycles of bending stresses, and eventually be failing.

Sensitivity of the sensory element should produce a sensor of an overall sensitivity that approximates the human hand. This could be as low as 0.055gm for men and 0.019gm for women [7]. The maximum force generated by the average human hand is slightly less than 12N [8].

Spatial resolution should also approximate the human hand; high resolution sensory receptors in the skin - mechanoreceptors - have densities between 70 to 140 units/cm^2 at the fingertips and 30 to 40 units/cm^2 on the phalanges [7]. Two-point threshold studies suggests that humans can differentiate between 2 points as close as 0.9mm apart [9].

Cost and feasibility of construction should be low enough to allow robotics researchers with limited experience of material science and sensor construction to build and customise their own systems.

Temporal resolution - how many usable readings per unit time - and **restoration time** - the time taken to go back to normal value after the load removed - were not considered in this paper for simplicity.

2.3 Variable Resistive Layer

Electrically conductive packaging materials made of polymeric foil impregnated with carbon black are used to protect electronic components from electrostatic discharge. Such materials can be easily obtained in sheet forms and are commercially available under different trademarks such as Velostat and Linqstat. The conductivity - and hence resistivity - of the sheets vary when the material is subjected to compressing pressure.

Table 1. Resistive characteristics of Velostat when force is applied

Weight (Kg)	Force (N)	Resistance (ohm)
0.1	0.98	460
0.2	1.96	230
0.3	2.94	130
0.4	3.92	80
0.5	4.90	69
0.6	5.88	40
0.7	6.86	32
0.8	7.85	30
0.9	8.83	27
1	9.81	25
1.1	10.79	22
1.2	11.77	20
1.3	12.75	18
1.4	13.73	16
1.5	14.71	16

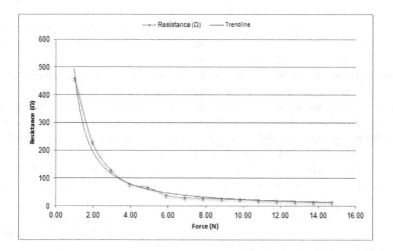

Fig. 2. Resistance of Velostat plotted against the force applied

Products produced by brand-name manufacturers have standard dimensions and resistance characteristics published in the product datasheet. However these datasheets are not concerned with the variable resistance characteristics. Variable resistive characteristics of Velostat has been investigated experimentally.

An experiment was conducted using a 20X20mm piece of Velostat sandwiched between two copper plates attached to a digital multi-meter. Mass increasing at regular intervals was then placed on the test piece. The range of mass applied started at 0.1Kg and increased at intervals of 0.1Kg up to 1.5Kg. This range of mass was chosen as the maximum force generated by the average human hand is slightly less than 12N [8].

Results shown in Table 1 and Figure 2 shows that resistance across the Velostat decreases in a very predictable way as the force applied increases; therefore the behaviour of the sensor should be predictable. The resistance is within a practical range that can be easily integrated in the development of the sensor. The change in resistance is of greater magnitude at the lower end of the range of force applied. This could be useful when attempting to control light touch manipulations with a humanoid robot hand.

Using $F = ma$, where a is acceleration due to gravity $= 9.80665$m/s, 1.5Kg is chosen as the maximum mass as it would generate a force F $= 1.5*9.80665 = 14.71$N. This easily encompasses the range of the average human hand force, the maximum force generated from a low-cost robotic hand will most likely fall within this region.

It is worth noting that there may be differences in the behaviour of one piece of Velostat to another and it may be worthwhile testing several samples. In conclusion, these results and observations supports the choice of Velostat as a suitable material for the variable resistive layer.

2.4   Conductive Lines

The use of copper wires for conductive lines has its benefits when connecting the sensor to a prototype board, however long vertical lines made of copper wires result in a loss of flexibility. This therefore results in a loss in dexterity of the humanoid robot hand it is working in conjunction with. Durability of the wire is also an issue; after a number of manipulations the wire could break as a result of bending stresses applied. This could be critical as the hand would lose any pressure feedback which could result in damage.

Also there is an issue of the resolution achievable by using copper wires. The close proximity of the wires has an adverse effect on the resistance readings as one line could be supporting the line immediately around it and reducing contact with the variable resistive layer. A flatter line would offer less support to the line around it therefore a point contact would have more effect on a single taxel thus allowing a higher resolution to be achieved. Therefore, a flatter, flexible material was proposed to alleviate the issues highlighted to a great extent, which is conductive thread. Conductive thread has very different resistive characteristics to a pure copper wire.

Two experiments were performed in order to compare the different resistive characteristics of 0.2mm diameter copper wire and conductive thread before the thread was utilised in a sensor. The first experiment was to measure the resistivity of both the thread and the wire over a length of 300mm. The second experiment was to measure the resistance of a contact crossover point on both materials as illustrated in Figure 3.

2.5   Epidermis Layer

From the structure of the sensor shown in Figure 1, it can be seen that the resistive layer and conductive lines need to be enclosed between insulating epidermis

Fig. 3. Crossover contact points of copper wire lines and conductive thread lines

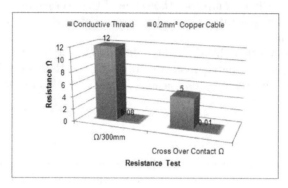

Fig. 4. Resistive characteristics comparison of copper wire and conductive thread

layers to protect the sensor array. After the study of a few prospective materials, neoprene foam was deemed a suitable material as it is durable and insulating. Cotton fabric was used as the insulating epidermal layers with conductive thread multi-point sensor.

2.6   Initial Prototype Sensor

Figure 1 shows components of a single taxel sensor constructed using 0.2mm diameter copper wire for the conductive lines with Velostat as the variable resistive layer and neoprene foam as the insulating layers. Figure 6 shows a closed single taxel sensor and it's components using conductive thread for the conductive lines.

Figure 5 shows experimental results. Separate tests were conducted for resistance and voltage, for voltage the sensor is used in a potential divider circuit with a $10k\Omega$ pull-down resistor. Predicted voltage is calculated using resistance results. From these results it can be seen that the sensor is predictable in its performance.

2.7   Prototype Multi-point sensor

The sensor layout and its dimensions are based on the average human index finger[10]; the sensor is split into three main sensing areas that correspond with

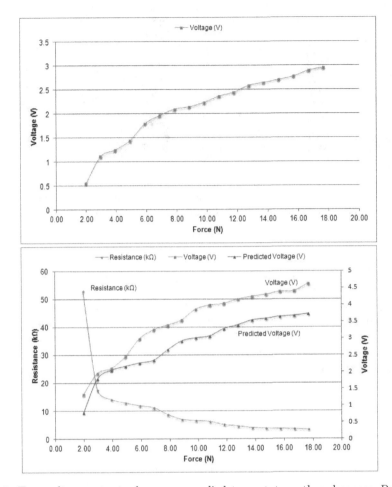

Fig. 5. Top: voltage output when mass applied to prototype thread sensor, Bottom: resistance and voltage of prototype copper sensor

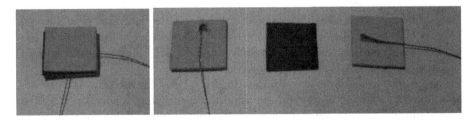

Fig. 6. Single-taxel conductive thread sensor, Left: closed, Right: components

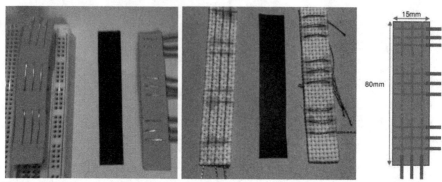

Fig. 7. Left: components of prototype multi-point sensor using copper wire and neoprene, Centre: components of prototype multi-point sensor using conductive thread and cotton fabric, Right: arrangement of conductive lines and occurrences of taxeks

the phalanxes of the finger. Each sensing area is comprised of 3 horizontal and 3 vertical lines giving 27 crossover points (taxels) in total (Figure 7).

The sensor has been tested with both an evenly distributed force and point force applied at different regions. The results of tests show that distribution and change in output voltages are in an approximate agreement of what was recorded with a single crossover taxel. However the voltages recorded were lower than that of a single point. This could be due to the structure of the sensor and the close proximity of the taxels.

2.8   Sensor Attachment Solution

It was decided that the preferred solution during development was to use a latex palm glove. The glove was slightly larger than desired however it provided clearly defined flat areas for sensor implementation and due to its structure presented a limited amount of stretch. An example of how the sensors are placed on the glove is shown in Figure 8.

2.9   Data Acquisition Solution

Each vertical line is used in a potential divider format with a $10k\Omega$ resistor. Outputs from potential dividers are fed directly to three analogue inputs on an Arduino Uno board. Each horizontal line is attached to a digital output of the Arduino board. A program loops through switching each digital output high in turn and reads the associated analogue values.

The project used Processing 2 in conjunction with the Arduino IDE to acquire and visualise tactile pressure information. Processing 2 is an open source programming language that allows real time visualisation of the data collected using the microcontroller.

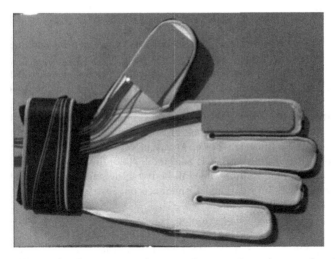

Fig. 8. Example of sensor attachment solution using a latex palm glove

3   Conclusions and Future Work

A prototype multi-point sensor was built and tested using the conductive thread for the lines, Velostat as the variable resistive layer and cotton fabric as the insulating epidermal layers. The testing was conducted identically to the method described above to ensure a fair comparison of the results could be achieved. The multi-point sensor predominantly worked as predicted. As with a wire conductive lines sensor the distribution and change in output voltages gained were as predicted based on what was recorded with a single crossover taxel. However this time the taxels appeared to have less interference with each other when point pressure was applied. This leads to the conclusion that this sensor is an improvement on the wire multi-point sensor.

Overall the sensor system worked as it was designed to. From the results of testing and development it can be concluded that a conductive thread based sensor has significantly more promise than a copper wire based sensor as it allows better flexibility and durability. The overall sensor system solution was still successful at interpreting the pressure into a voltage with sufficient accuracy.

Nevertheless, there are still issues to be addressed. Experiments were performed with one piece of Velostat under an assumption that thickness is evenly distributed. Further experiments need to be conducted with several pieces of material and from different manufactures.

The sensor could provide a template for controlling relatively affordable humanoid robot hands. This would be of great benefit to robotics developers and people within the robotics education sector.

References

1. Langevin, G.: Here is "InMoov", The robot hand you can print and animate (Blog) (2012), http://www.inmoov.fr (accessed March 29, 2014)
2. Biagiotti, L., Lotti, F., Melchiorri, C., Vassura, G.: How Far Is the Human Hand? A Review on Anthropomorphic Robotic End-effectors. University of Bologna, Bologna (2004)
3. Andrew Russell, R.: Robot Tactile Sensing. Prentice-Hall, Englewood Cliffs (1990)
4. Sicilliano, B., Khatib, O.: Handbook of Robotics. Springer (2008)
5. Papakostas, T.V., Lima, J., Lowe, M.: A large area force sensor for smart skin applications. Proceedings of IEEE Sensors 2, 1620–1624 (2002)
6. Pan, Z., Cui, H., Zhu, Z.: A flexible full-body tactile sensor of low cost and minimal connections. IEEE International Conference on Systems, Man and Cybernetics, 2003 3, 2368–2373 (2003)
7. Jones, L.A., Lederman, S.J.: Human Hand Function. Oxford University Press (2006)
8. Radwin, R., Oh, S., Jensen, T., Webster, J.: External finger forces in submaximal five-finger static pinch prehension. Ergonomics 35(3), 275–288 (1992)
9. Goethals, P.: Tactile Feedback for Robot Assisted Minimally Invasive Surgery: An Overview. University of Leuven (2008)
10. Agnihotri, A., Purwar, B., Jeebun, N., Agnihotri, S.: Determination of Sex by hand Dimensions. The Internet Journal of Forensic Science (2005)

Enhanced Positioning Systems
Using Optical Mouse Sensors

Mingxiao He[1,*], Xuemei Guo[1], and Guoli Wang[1,2]

[1] School of Information Science and Technology
Sun Yat-Sen University, Guangzhou 510006, China
[2] SYSU-CMU Shunde International Joint Research Institute, Foshan 5280000, China
isswgl@mail.sysu.edu.cn

Abstract. This paper explores the use of optical mouse sensors for building positioning systems. A low-cost, customized optical sensor positioning system has been developed, with desirable high performance. Unlike the traditional dead-reckoning method using rotary encoders, optical mouse sensors are unacted on the errors induced by wheel slippage. However, some research has indicated that height variations can bring errors in 2D displacement measurements when using optical mouse sensors. What's more, the so-called error cumulative phenomenon is an inevitable problem to every motion positioning system. To overcome these weaknesses, our main contributions are in two aspects. One is that a new adaptation mechanism is presented to enhance the positioning robustness to surface height variations. This performance enhancement attributes to the height-variation awareness of our presented system, in addition to the required positioning capability. The other is that a novel calibration paradigm with the assisted landmarks is proposed to eliminate the error cumulative phenomenon. This induced performance enhancement benefits from the high-speed optic flow sensing capability of optical mouse sensors. Experimental results are reported to validate the enhancement capability of the presented positioning system.

Keywords: Optical flow, Mouse sensor, Position sensing, Height variation, Landmark.

1 Introduction

Nowadays more and more movable devices with intelligence can be found in our life, such as a domestic robot, an electric wheelchair, an autonomous vacuum cleaner and ect. A key problem for these movable devices is how to get their position information. Only by knowing the position of the movable devices, can these devices be able to avoid obstacles and plan path, given the importance many approaches have been proposed. Most of the position sensing strategies fall into three categories: beacon-based, inertial-based and vision-based. These categories each have their own advantages and disadvantages. However, when

---

* Corresponding author.

X. Zhang et al. (Eds.): ICIRA 2014, Part II, LNAI 8918, pp. 463–474, 2014.
© Springer International Publishing Switzerland 2014

some researchers turn their eyes on the optical flow method, a new way for positioning is found.

Employing optical flow sensors for horizontal 2D displacement measurement has been widely applied in the field of UAV (unmanned aerial vehicle) for many years[1], [2]. Sophisticated methods[3] have been proposed to make custom-built optical sensor positioning systems for UAVs. Since the success of using optical flow method in the area of UAV, more researchers focus on applying it to ground devices. The idea of using optical mouse sensors to estimate the position of movable devices such as mobile robots has been proved to be feasible. Sungbok Kim[4], [5] and J. Palacin[6] place optical mouses on the mobile robot to estimate the position and velocity. But the applicabilities of optical mouses are sorely limited, conventional optical mouses need to be placed quite near to the work surface. One aspect that uneven surfaces make the optical mouses losing accuracy, another aspect that the work surfaces of movable devices are not flat usually. In order to enlarge the reliable working interval between the optical mouse sensor and its work surface to overcome this weakness, Robert Ross[7], A.M. Harsha S. Abeykoon[8], etc. assemble the optical mouse sensor with new lens which has lager focal length. Their prototype shows that the surface height variations during the movement of optical sensors can be a crucial effect factor. Although Robert Ross gives two approaches to mitigate the influence induced by height variations[9], he does not propose a way to calibrate his prototype to avoid the error accumulation. In this paper, a positioning system has been developed to provide robust odometry for movable devices. For enhancing the reliability of our system, artificial landmarks and multiple isomorphic optical sensor units are used.

The remainder of this paper is structured as follows: Section 2 introduces the implementation of a positioning system with optical mouse sensor units. In section 3, two methods has been proposed to enhance the reliability of our positioning system. Section 4 gives experimental testing of the proposed positioning system and the conclusion goes in the Section 5.

2   Positioning System

This positioning system has a number of optical mouse sensor units, which are basic and vital. Every unit uses the optical flow theory to calculate the horizontal displacement and return the displacement increments to a microcontroller. Then the positioning system will output reliable position measurements by using the microcontroller to process and fuse the data from all the units.

2.1   Optical Mouse Sensor Unit

To make the mouse sensor working stably at a distance from its work surface, a new optical system with larger focal length lens and high illumination light need to be developed. What's more, the resolution of the restructured sensor unit needs to be recalibrated.

New Optical System

The working principle of mouse sensor chips is that: a CMOS camera is inbuilt in the sensor chip to take photos of the surface under the sensor at a very high rate; then the DSP(Digital Signal Processor) with firmware algorithms, which are based on comparing the difference in successive imaging arrays, infers the horizontal displacement. So the new optical system should guarantee that clear surface pictures the with enough features can be captured. Like a camera system, the lens needs to be carefully selected. How a lens affects the output of an optical mouse sensor chip is shown by equation (1):

$$n_{x(y)} = \frac{n_p}{2Htan(\frac{\alpha}{2})}v\Delta t .$$ (1)

Where $n_{x(y)}$ is the value of the register Delta_X(Y) in the mouse sensor, which represents the counts of unit displacement in X(Y) direction, n_p is the number of pixels in the imaging array in the direction of movement of the sensor, v is the velocity in X(Y) direction, H is the distance between the lens and the work surface under the sensor, α is the FOV(field of view) of the lens.

Sensor with lens which has larger FOV require more features in the images of the work surface be available. And the distance(H) at which the sensor can output reliable measurements will be shorter. But an advantage is that the sensor is less susceptible to the noise induced by angular oscillations during motion. While sensor with small FOV lens is the opposite[10]. Another important consideration is the light intensity which the sensor requires (80 to 100 W/m^2). Choosing a lens with the larger f-stop means the illumination light of the sensor unit will have high power consumption to provide enough light intensity.

Through experiments, the lenses which are used for cameras in video monitoring system are chosen for the restructured optical mouse sensor units. With the proprietary designed base, the lens can be easily assembled and adjusted. In order to keep the working distance between the surface and the lens at about 30mm, the lens with 8mm focal length is chosen. The FOV of the lens is 40° and the f-stop is 2.0. A Laser Diode with the condenser lens removed is used for illuminating, the wavelength of the laser is 650nm. The optical mouse sensor chip for the restructured sensor unit is ADNS-2610.

Fig. 1 (a) shows the optical mouse sensor unit which can output reliable displacement measurements over the surface at a distance of 26-43mm. A microcontroller has been used to read out the data in the registers of the sensor. A testing paper with specific mark is put under the unit. In Fig. 1 (b), the pixel data from the mouse sensor chip shows that clear images of the surface are captured, namely the sensor unit works in a valid state.

Recalibrating the Sensor Unit

Conventional optical mouses sense the position by using the value of registers Delta_X and Delta_Y. To convert the increments of the register values into real displacement, the resolution factor is needed. For the conventional optical mouse with sensor chip ADNS-2610, the resolution is 400 CPI(count per inch)[11].

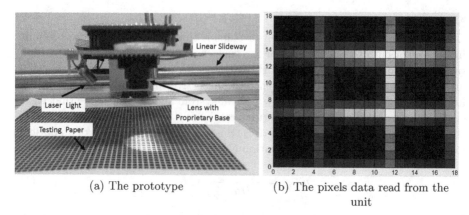

(a) The prototype

(b) The pixels data read from the unit

Fig. 1. Optical Mouse Sensor Unit

Here the resolution of the restructured optical mouse sensor unit is unknown. Another way to calculate the displacement is using the optical flow theory. Equation (2) shows the relation between the registers value and the horizontal displacement:

$$v\Delta t = \frac{Hn_{x(y)}}{R_d h} .$$
(2)

Where the h is the distance between the lens and the optical mouse sensor, R_d is the designed resolution for the optical mouse, the other notations are same as those in equation (1). And for conventional optical mouse H and h are equal. Actually, equation (2) can be deduced from equation (1). Thus, by definition, the resolution R of the developed unit can be calibrated by equation (3):

$$R = \frac{h}{H}R_d .$$
(3)

According to the imaging principle of convex lens, h is greater than the lens focal length but less than twice of it, H is greater than twice of the lens focal length.

2.2   Framework of Positioning System

Fig. 2 shows the framework of a system which is developed for positioning. This positioning system can be attached to the movable devices which need to be localized. Here the Freescale microcontroller (xs128) is used for processing data from the developed sensor units. And the position data can be sent to an external device for further use via a bluetooth serial port module (HC-06). More than one optical mouse sensor unit are used in the positioning system for providing reliable measurements.

3   Methods for Reliability Enhancement

For every position sensing system, the so-called error cumulative phenomenon is a tough issue. What is even more unfortunate, is that the height variations

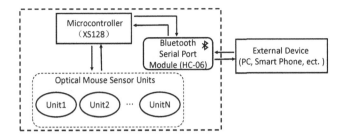

Fig. 2. The Framework of Positioning System

can also induce errors for the system using optical sensors. In order to enhance reliability of the positioning system mentioned above, two methods are proposed in this section: calibration with assisted landmarks and height adaptive recalibration.

3.1 Method 1: Calibration with Assisted Landmarks

Using artificial landmarks are an effective way to get absolute displacement to eliminate the error cumulative phenomenon for calibrating the position of movable devices. Some researchers use RFID tags for mobile robot positioning and which has been shown as a good way to lower the cumulative errors[12]. To design artificial landmarks for calibrating measurements of the system proposed in this paper, the main problem is how the landmarks can be recognised by the optical mouse sensor units. Here an important value, which can be read out from the Pixel_Sum register of the optical mouse sensor, is used as a feature for recognition. The value of Pixel_Sum reflects the average pixel value of the optical flow images. That means landmarks with different colours can be recognised by the optical mouse sensor units.

Landmarks Selection. Experiments have been done to test the values of Pixel_Sum when the sensor units move over different surfaces. Here the blue and the black mouse mats are chosen as artificial landmarks for giving a paradigm. Fig. 3 shows the result. Obviously, the mats can be recognised by setting different threshold values of Pixel_Sum. SQUAL is a parameter of surface quality, which is provided by the optical mouse sensor. The sensor units work better over surfaces with high SQUAL. As shown in Fig. 3, the SQUAL values of the mats are high enough to make sure the sensor units work reliably.

Landmarks Placement. The two kinds of mats are placed on the work surface of the positioning system in such a way, the mats are distributed at a unit interval between each other. And the same kind of mats are in a nonadjacent format. More specifically, as shown in Fig. 4, the space between mats is 100mm and the size of mats is 20*20mm^2.

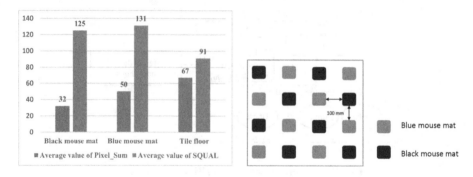

Fig. 3. Pixel_Sum and SQUAL Values over Different Surfaces

Fig. 4. The Placement of Artificial Landmarks

Measurements Calibration. The cumulative errors could be corrected in such a way, when the positioning system senses the fist landmark, the microcontroller will identify the landmark by the value of Pixel_Sum and record the colour of the landmark. When the device senses a landmark again (we assume the condition, which the device misses all the 8 landmarks adjacent to the last landmark, does not exist), the landmark should be identified once more. By comparing the two landmarks a conclusion can be draw, if their colours are different the device must move a unit of distance(100mm) in X or Y direction, else if their colours are same the device must move a unit of distance in both X and Y direction. Combining with the original displacement information from the positioning system, the measurements of the system can be calibrated to eliminate the error cumulative phenomenon.

3.2   Method 2: Height Adaptive Recalibration

In equation (2), to measure the horizontal 2D displacement, the distance between the lens and the sensor work surface H needs to be known. Although H can be measured right after the positioning system is attached to the movable device, H will change when the movable device moves on a uneven surface. If take the initial measurement of H into equation (2) for calculation, errors will be induced. In order to alleviate the errors induced by height variations, 3 or more optical sensor units, which are placed at different height, are used in the proposed system for sensing displacement as well as providing a height adaptation mechanism. In this paper 3 sensor units are used as an example. Fig. 5 (a) shows the relationship between the SQUAL values (S_{squal_i}, i=1,2,3) which are read from a sensor unit and the H. And the relation curves of 3 isomorphic sensor units at different height are shown in Fig. 5 (b). The vertical interval between the adjacent sensor units is 5mm and Sensor Unit 1 is the highest Sensor Unit 3 is the lowest. The curve in Fig. 5 (a) is fitted by two lines. So that the height H_i (i=1,2,3) can be calculated by S_{squal_i}. More specifically, as shown in equation (4):

$$H_i = \begin{cases} c \cdot S_{squal_i} + e & (H_i > H_{max}, \; c < 0) \\ d \cdot S_{squal_i} + f & (H_i \leq H_{max}, \; d > 0) \end{cases} . \tag{4}$$

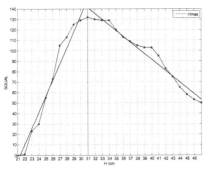
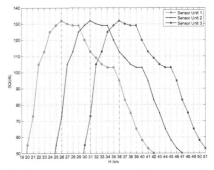

(a) The SQUAL of a sensor unit at different height (work surface: black mouse mat)

(b) The relation curves of 3 isomorphic sensor units

Fig. 5. The Relationship Between the SQUAL and H

Where H_{max} as shown in Fig.5 (a) is the distance between the lens and the work surface of the sensor unit, which can make the SQUAL value reaching the peak. H_{max} is also the initial interval between the lens of Sensor Unit 2 and the work surface, when the positioning system is attached to the movable device. The values of c, d, e, f are connected with the work surface type. To compare H_i with H_{max}, the state of S_{squal_i} during the height changing can be used. As shown in Table 1:

Table 1. The relationship table of S_{squal_i} and H_i

The state of			H_i
S_{squal_1}	S_{squal_2}	S_{squal_3}	
Reduces	Reduces	Rises	$H_1 > H_{max}$
or			$H_2 > H_{max}$
Rises	Rises	Reduces	$H_3 \leq H_{max}$
Reduces	Rises	Rises	$H_1 > H_{max}$
or			$H_2 \leq H_{max}$
Rises	Reduces	Reduces	$H_3 \leq H_{max}$
Rises	Rises	Rises	The surface quality changes, parameters
or			(c, d, e, f) in equation (4) should be
Reduces	Reduces	Reduces	changed according to the surface type.

Furtherly, take H_i and the Delta_X(Y) register value ($n_{x(y)}$) of Sensor Unit i into equation (5), the displacement of the positioning system in X(Y) direction can be deduced:

$$D_{x(y),i} = \frac{2.54H_i}{400h}n_{x(y)} \quad (cm).$$ (5)

Where, $D_{x(y),i}$ is the displacement in X(Y) direction measured by Sensor Unit i.

When using this method, we assume the distance H never exceeds the range of 26-36mm. Namely, if the type of surface does not change, the SQUAL values of all the sensor units will not have an accordant variation.

To fuse the measurements of every sensor units, a weighted fusion method has been shown with good results[9]. Here, the results from all the sensor units are weighted by a matrix $\mathbf{W}$ to get a reliable measurement $D_{x(y)}$. More specifically, as shown in equation (6):

$$D_{x(y)} = \mathbf{WD}$$ (6)

where,

$$\mathbf{W} = \left[\frac{S_{squal_1}}{\sum_{i=1}^{3} S_{squal_i}} \quad \frac{S_{squal_2}}{\sum_{i=1}^{3} S_{squal_i}} \quad \frac{S_{squal_3}}{\sum_{i=1}^{3} S_{squal_i}} \right] ,$$ (7)

$$\mathbf{D} = \left[D_{x(y),1} \ D_{x(y),2} \ D_{x(y),3} \right]^{\mathrm{T}} .$$ (8)

When the device moves on an uneven surface, the interference from those sensor units which are not working at a suitable height can be minimise by using the weighted matrix. In the following section experimental testings are used to assess the methods proposed above.

4   Experimental Testing

For improving accuracy, a linear slideway with a proprietary holder on it is used as the experimental equipment.

4.1   Analysis of the Optical Mouse Sensor Units' Performance

To analyze the performance of the restructured sensor units prototype, a single unit was attached to the holder on a linear slideway, as shown in Fig. 1 (a). The unit was moved back and forth over a 150mm long black mouse mat in the X(Y) direction of the sensor, with the values of SQUAL and Delta_X(Y) recorded then repeated 20 times. Then the steps above were repeated over a green mouse mat and the tile floor. The results are shown in Table 2:

The largest relative error is 13.4mm over a 150mm track and the largest RMSE (root mean square error) is 10.914mm. With the SQUAL value of the work surface decreasing, the measurement errors increase. On the whole, the sensor unit has a similar performance with the conventional optical mouse.

Table 2. Performance test for the optical sensor unit

$H=31$, $h=9.5$ (mm)	Surfaces (The real displacement is 150mm)		
	Black mouse mat	Blue mouse mat	Tile floor
SQUAL	130	132	78
$\overline{Delta\_X}$	714.30	722.55	684.90
$\overline{D_x}$ (mm)	148.0	149.7	141.9
RMSE_X (mm)	2.469	1.008	8.258
$\overline{Delta\_Y}$	696.75	718.60	671.45
$\overline{D_y}$ (mm)	144.3	148.9	139.1
RMSE_Y (mm)	6.004	1.426	10.914

4.2 Experimental Testing for Method 1

To validate the effectiveness of using landmarks for measurement calibration, the experiment was implemented in such a way: a single unit was attached to the holder on a linear slideway , as shown in Fig. 6 (a), the mouse mats as landmarks were placed on the ground under the sensor unit, the space between mats is 100mm and the size of mats is 20*20mm²; the whole length of the track is 240mm; the unit was moved back and forth for 30 times, every time when the sensor reached at both ends of the track the absolute relative errors were calculated and recorded. For the first time, no algorithms of landmarks recognition and errors correction were used. For the second time, algorithms of Method 1 were used. the results are shown in Fig. 6 (b). As shown, the measurements were calibrated by using Method 1 so that the absolute relative errors did not increase all the time.

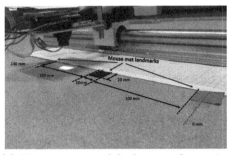

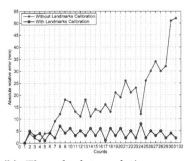

(a) The experimental deployment for testing Method 1

(b) The absolute relative errors which are calculated while the sensor unit moving back and forth along a 240mm track

Fig. 6. Experimental Testing for Method 1

4.3   Experimental Testing for Method 2

To test the efficacy of the proposed positioning system, a prototype system consists of 3 optical mouse sensor units was placed on a linear slideway. To make sure the surface has enough features, a special paper was used as the work surface under the sensor units. Obstructions were placed on the surface, each obstruction has a certain thickness to make the surface under the sensor uneven. The sensor units were placed at different height and the vertical interval between the adjacent sensor units is 5mm. Sensor Unit 1 is the highest and Sensor Unit 3 is the lowest. More specifically, as shown in Fig. 7. The device was moved along a 450mm track for 20 times. Each time the measurements from all the sensor units and the measurements by using weighted fusion method(equation (6)) were recorded. What is different is the first 10 times we did not use Method 2 to recalibrate the H of each sensor units, the results are shown in Fig. 8 (a). The last 10 times SQUAL values were used for height recalibration, the results are shown in Fig. 8 (b). More numerical results are given in Table 3.

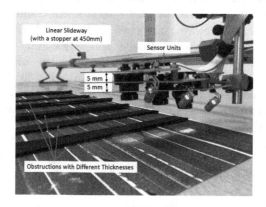

Fig. 7. The experimental deployment for testing Method 2

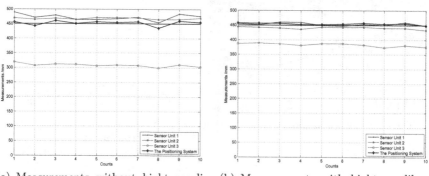

(a) Measurements without hight recalibration

(b) Measurements with hight recalibration

Fig. 8. Experimental results by using Method 2

Table 3. Experimental results of the positioning system

Measurement unit is mm		Average measurement	RMSE	Maximum relative error
Without hight recalibration	Sensor Unit 1	472.20	24.23	39
	Sensor Unit 2	468.20	18.37	21
	Sensor Unit 3	308.30	141.79	151
	Positioning System	453.00	8.01	14
With hight recalibration	Sensor Unit 1	453.10	5.47	10
	Sensor Unit 2	440.30	10.23	17
	Sensor Unit 3	382.60	67.60	77
	Positioning System	454.10	4.89	7

As shown in Fig. 8 (a), although the measurements by fusing the data from all the sensor units are close to the real value, the measurements from each units are shown to be not reliable. As shown in Fig. 8 (b), by using Method 2, the measurements from every sensor units become closer to the real value. Naturally, the fusion results become more reliable with smaller fluctuations around the real value.

5   Conclusion

In this paper, using optical mouse sensors for positioning is discussed. The way of using optical mouse sensor to make a high-performance sensor unit has been proposed. The sensor unit prototype has been shown to have a wider applicability in movable device positioning than the conventional optical mouse. The experimental testing results show that the unit can work at 26-43mm from its work surface with accurate measurements generated. Then a positioning system consists of the sensor units, a microcontroller and a bluetooth serial port module is introduced. For enhancing the reliability of this system, 2 methods are proposed to mitigate errors. Assisted artificial landmarks can be used to calibrate the positioning system with the error cumulative phenomenon being eliminated. Fusing the data of 3 sensor units, which are put at different height, helps to mitigate the errors induced by height variations. This low-cost, compact positioning system can be using for any application where the position detection is needed. In practical use, attaching the device to objects for position sensing is underway. A smart environment can be made by everything is able to sense its position in the environment.

References

1. Aubepart, F., El Farji, M., Franceschini, N.: FPGA implementation of elementary motion detectors for the visual guidance of micro-air-vehicles. In: IEEE International Symposium on Industrial Electronics, vol. 1(1), pp. 71–76. IEEE Press (2004)

2. Kim, J., Brambley, G.: Dual Optic-flow Integrated Inertial Navigation. In: Australasian Conference on Robotics and Automation (2005)
3. Honegger, D., Meier, L., Tanskanen, P., Pollefeys, M.: An open source and open hardware embedded metric optical flow CMOS camera for indoor and outdoor applications. In: IEEE International Conference on Robotics and Automation (ICRA), vol. (1), pp. 1736–1741. IEEE Press (2013)
4. Kim, S.: Isotropic optical mouse placement for mobile robot velocity estimation. In: IEEE/ASME International Conference on Advanced Intelligent Mechatronics (AIM), vol. 1(1), pp. 7–12. IEEE Press (2013)
5. Kim, S., Kim, H.: Optimal optical mouse placement for mobile robot velocity estimation. In: IEEE International Conference on Mechatronics (ICM), vol. 1(1), pp. 310–315. IEEE Press (2013)
6. Palacin, J., Valganon, I., Pernia, R.: The optical mouse for indoor mobile robot odometry measurement. Sensors and Actuators A 126(1), 141–147 (2006)
7. Ross, R., Devlin, J.: Analysis of real-time velocity compensation for outdoor optical mouse sensor odometry. In: The 11th International Conference on Control Automation Robotics & Vision (ICARCV), vol. 1(1), pp. 839–843. IEEE Press (2010)
8. Harsha, S., Abeykoon, A.M., Udawatta, L., Dunuweera, M.S., Gunasekara, R.T., Fonseka, M., Gunasekara, S.P.: Enhanced position sensing device for mobile robot applications using an optical sensor. In: IEEE International Conference on Mechatronics (ICM), vol. 1(1), pp. 597–602. IEEE Press (2011)
9. Ross, R., Devlin, J., Wang, S.: Toward Refocused Optical Mouse Sensors for Outdoor Optical Flow Odometry. IEEE Sensors Journal 12(6), 1925–1932 (2012)
10. Griffiths, S.R.: Remote terrain navigation for unmanned air vehicles. Master Thesis, Department of Mechanical Engineering, Brigham Young University (2006)
11. Agilent ADNS-2610 Optical Mouse Sensor Data Sheet. Agilent Technologies, Santa Clara, CA (2004)
12. Guo, Y., Guo, X., Wang, G.: Using RFID in localization for indoor navigation of mobile robots. In: International Conference on Mechatronics and Automation (ICMA), vol. 1(1), pp. 1892–1897. IEEE Press (2012)

A Novel Silicon Based Tactile Sensor
on Elastic Steel Sheet for Prosthetic Hand

Chunxin Gu, Weiting Liu[*], and Xin Fu

The State Key Lab of Fluid Power Transmission and Control, Zhejiang University,
310027 Hangzhou, China
{cxgu,liuwt,xfu}@zju.edu.cn

Abstract. This paper focuses on developing a novel tactile sensor to be used in a prosthetic hand. A novel structure of combining an elastic steel sheet and a piezoresistive gauge is designed and fabricated by the hybrid method of the traditional machining process and the MEMS technology. Normal force loading tests by applying force from 0 to 15N and 0 to 0.1N are performed respectively to obtain the preliminary characterization of the designed sensor. The results of the testing experiments show the good linearity, sensitivity and hysteresis of the sensor. The sensor is mounted on the fingertip of a commercial prosthetic hand and covered by polydimethylsiloxane (PDMS)，which can be tested for further application.

Keywords: Tactile sensor, prosthetic hand, piezoresistive, hybrid, MEMS.

1   Introduction

Prosthetic hand is widely investigated by different research groups and at the same time commercial prostheses are also available in the market. However, most of the commercial prostheses have no or just simple on/off tactile sensor, which makes the manipulation less flexible and the function less reliable. Besides, the tactile sensors in the past researches usually have complex mechanical structures and rigorous fabrication process, which limit their further applications on the prosthetic hand.

So far, several kinds of tactile sensors have been developed based on the principles of piezoresistive [1-3], capacitive [4-5], piezoelectric [6-7], optical [8] methods and etc. The silicon based piezoresistive sensors have the high sensitivity and small sizes with well established design and fabrication techniques. However, the sensor elements (especially the silicon elements) are usually fragile [9] which decreases the reliability of the sensory system. Hence, taking advantages of the good qualities, it is meaningful to improve the robustness of the silicon based tactile sensors for their integration on prosthetic hands.

Considering the requirements of commercial prosthetic hands such as high reliability, low expense and good static force sensing ability, this paper presents an

---

[*] Corresponding author.

X. Zhang et al. (Eds.): ICIRA 2014, Part II, LNAI 8918, pp. 475–483, 2014.
© Springer International Publishing Switzerland 2014

easily fabricated silicon piezoresistive sensor, combining with a simple mechanical structure, which performs in a reliable way under comparatively large load in addition to high sensitivity, high linearity and low hysterisis of force measurement. As the force sensing range for human hand perception is about 0.1N-10N [10], the maximum indenting load is set as 15N for the safety of potential overloading and the minimum one as 0.1N in the testing experiments.

2   Sensor Design

The sensor consists of three parts as illustrated in Fig. 1, the metal framework, the silicon gauge and the flexible printed circuit board (FPCB). The metal framework, which is U-shaped at the y-z cross section, has a 0.3mm thick square elastic sheet with two 0.9mm thick supporting braces. The sheet's width is 15mm which is designed to fit the thumb of the prosthetic hand (SJQ18-G, Danyang, China). The metal framework's material is a kind of stainless steel, 17-4ph with Young's modulus as 197Gpa and Poisson ratio as 0.3.

The metal framework has two braces used to be fixed in the thumb of the prosthetic hand, which ensures the reliable mount and the proper space between the rear surface of the steel sheet and the thumb, protecting the fragile silicon gauge. As the other two sides of the metal sheet are free, the sheet is prone to deform along the y direction.

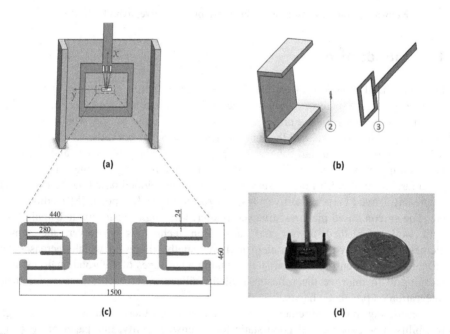

Fig. 1. Sensor structure. (a) A schematic of the sensor structure. (b) An exploded view diagram of the sensor (①the steel sheet, ②the silicon gauge, ③the FPCB). (c) Dimensional drawing of the silicon gauge. (d) Photograph of the sensor.

On the center of the rear surface of the steel sheet is the piezoresistive silicon gauge with the main mechanical features and dimensions shown in Fig. 1(c). The whole silicon gauge is symmetrical about the vertical centerline which can be divided into two independent parts with the initial resistance as 3.4 kΩ for each side. The grey parts are the silicon piezoresistors all of which align in the y direction in order to make the piezoresistors gain the maximum sensitivity. The width of the each piezoresistor is 24μm and the length has two values, 280μm and 440μm respectively. The yellow parts are the pads, among which the three one in the center are defined for welding the golden wires and the others mainly as the joints between the nearby parallel silicon piezoresistors.

In order to reserve the mounting space for the silicon gauge, the FPCB is hollow in the center with a tailor-like strip containing three leads. Thanks to the golden wires between the pads on the silicon gauge and the wire leads on the FPCB, the gauge is easily connected to the external processing circuitry.

By means of the finite element analysis (FEA) simulation tool (ANSYS Workbench 14.5), the mechanical stress in the silicon device is simulated. The stress is evaluated to prove the safety of the silicon gauge after deformation. Therefore, the force is only set at the maximum value as 15N (redundant compared to 10N) to obtain the maximum stress. As the silicon gauge is fixed on the steel sheet, its deformation is mainly up to the steel sheet deformation. Thus, the model of the simulation can completely ignore the influence of the silicon gauge and the FPCB. The model also simplifies the structure as a total flat sheet and fixes the five degrees of freedom (DOF) in two opposite edges except the DOF rotating around each edge. The force is uniformly applied on the rod indenter that contacts the surface of the sheet and transfers the force. Since the main deformation is along the y-axis direction in which the piezoresistors longitudinally align, only the y-axis normal stress is cared about. The FEA outcomes are shown in Fig. 2. On the rear surface of the strain steel, the stress is tensile when the force is applied. From Fig. 2(b), the area around the center (especially the red parts) is quite different from the regions nearby due to the singularity region in which the edge of the indenter base contacts the steel sheet. The maximum y-axis normal stress (though related to the singularity region) around the center is about 380.89 MPa which is far less than the silicon yield stress (2.8-7 GPa) [11], thus the silicon gauge is quite safe under this design.

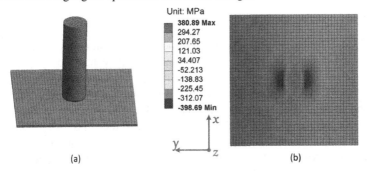

Fig. 2. FEA outcomes with an applied force of 15N. (a) The FEA model. (b) The distribution of the y-axis normal stress on the rear surface of the steel sheet.

3   Sensor Fabrication

The metal framework is grinded from a steel block before being polished to get a 0.3mm thick sheet and two 0.9mm thick braces. And the silicon gauge is fixed on the rear surface of the steel sheet by the technology of glass sintering. Around the gauge, the FPCB is glued on the surface. Using the ultrasonic bonding machine, the golden wires are connected to the FPCB and the silicon gauge respectively.

The sensing elements have been fabricated with an SOI wafer based on the dry etching technology. The SOI wafer is p-type, doped with boron with a <110> surface orientation. Fig. 3 illustrates a SEM image of the fabricated sensor.

Fig. 3. SEM picture of the silicon gauge structure

4   Preliminary Characterization Experiments

In order to investigate the measuring ability of the silicon sensor, preliminary characterization experiments have been performed. The signal conditioning circuitry and the testing platform have been set up; the sensor mounting and packaging on the prosthetic hand have been tried.

4.1   Read out Electronics

A Wheatstone bridge circuit is set up for each piezoresistor (initial resistance about 3.4 kΩ) independently, which produces an output voltage proportional to the resistance change $\Delta R/R$ in the piezoresistors. In the Wheatstone bridge, there are two resistors R1, R2 with precise resistance of 3.4 kΩ and a variable resistor Rvar used to adjust the initial offset level to zero. The bridge output signal is then led to an operational amplifier (AD620, Analog Devices) with the adjustable amplifier gain A from 1 to 10000. The final output signal is measured by a data acquisition (DAQ) card (NI 6343, National Instruments) which is connected to a computer. The input voltage is Vin =5.0 V while the output voltage Vout is controlled under 5.0 V to fit the measurement range of the DAQ card. Considering a balanced bridge where R1=R2=Rvar=R, the output voltage Vout is

$$V_{out} = AV_{in} \frac{\Delta R}{4R + 2\Delta R} \approx AV_{in} \frac{\Delta R}{4R}$$

(1)

4.2   Experimental Set-Up

The characterization apparatus is shown in Fig. 4. It consists of a three-axial motion platform, a sensor mount platform, a link block, a three-axial force sensor, a signal conditioning electronic circuit board, a data acquisition card and a computer for data acquisition and analysis.

The three-axial force sensor (FC3D50-50N, Force China) is fixed on the three-axial motion platform (M-VP-25XA-XYZL, Newport) through a link block. And a rod indenter is also screwed into the force sensor. Thus, the indenter can move precisely in three-axial direction, by which a force can be loaded at a proper location of the designed sensor. The force value is obtained by the three-axial force sensor whose measurement range is from -50N to 50N with precision as 0.01N. The three-axial motion platform can drive the indenter at the max range of 25 mm in each direction and the lowest speed at 1 μm/s. Thus, a precise force variation can be obtained by the motion control of the platform.

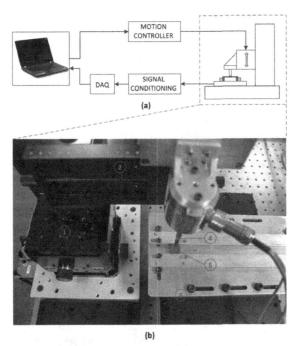

Fig. 4. (a) A schematic of the testing system; (b) Photograph of the force loading system (①the three-axial motion platform, ② the link block, ③ the three-axial force sensor, ④ the rod indenter, ⑤the designed sensor, ⑥the sensor mount platform).

The diameter of the rod indenter is 3mm. A force is loaded when the indenter contacts the surface of the steel sheet. It can be assumed that the force loaded on the metal sheet is uniformly distributed at the circle contact area.

Normal load tests will be performed to analyze the mechanical properties of the sensor. In order to get the maximum deformation of the steel sheet under the same load, the load area is confirmed to be near the center of the upper surface of the metal sheet. As both the axis and the motion direction of the indenter are perpendicular to the surface of the metal sheet, the shear force can be ignored and only the normal force is considered.

Before the loading of the force, the target position of the indenter should be determined. Firstly, the three-axial motion platform is manually driven to locate the center of the indenter base at one of the vertices of the square steel sheet and only a small offset vertically. Secondly, the motion platform is automatically driven at the displacement of 7.5 mm (half of the width of the steel sheet) in both x and y directions. Thirdly, the motion platform is automatically driven again, but only in the vertical direction, at the speed of 0.01 mm/s to load the force. The measurement value of the loading force from the three-axial force sensor can be monitored in the interface of the SignalExpress (Version 2011, National Instrument). If the force value is beyond the planned one, the motion can be aborted immediately through the motion control interface in the PC. Therefore, the automatic motion of the platform can be easily controlled to obtain a proper force value. Considering the requirements of the contact force in the human hand of grasping objects, the maximum force to load is set at 15N and 0.1N respectively.

Afterwards, the sensor is tried to be mounted on the commercial prosthetic hand for further testing experiments such as grasping tests. The inside of the two braces of the sensor is glued to the lateral sides of the thumb of the prosthetic hand. In addition, the rear side of the steel sheet keeps a normal distance from the fingertip of the thumb to protect the silicon gauge. Then a flexible layer packaging follows by using the packaging material polydimethylsiloxane (PDMS) at 10:1 of monomer to curing agent and a designed mould with the chamber shaped like a humanoid thumb. Fig. 5 shows the final image of the sensor mounted on prosthetic hand with the elastic cover.

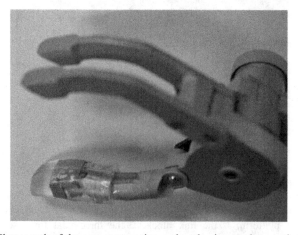

Fig. 5. Photograph of the sensor mounting and packaging on the prosthetic hand

5   Experimental Results and Discussion

The continuous normal load tests have been performed when an applied force is increasing from 0 to 15N and decreasing back to zero at a constant translational speed of 0.01mm/s along the z direction. These tests have been repeated five times to make sure the reliability of the system.

Typical sensor outputs $\Delta R/R$ in response to the force loading and unloading are plotted in Fig. 6. Using the linear regression technology, the linear fitting lines are also plotted in Fig. 6. The resulting sensitivity is $0.0047N^{-1}$ for loading and $0.0046N^{-1}$ for unloading. The coefficient that quantifies the linearity of the curve is 0.999 for both loading and unloading, very close to one, which means good linearity. The maximum hysteresis is 0.0015 happened at the zero force loading moment, which is also low compared to the sensitivity.

The set-up of the minimum normal load tests is the same as the tests mentioned above except the applied force from 0 to 0.1N. In Fig. 7, typical sensor resistance response to the force loading and unloading is plotted. There is an instant response when the applied force steps up to 0.1N or down to 0. The resistance change value to the force step is about 5Ω which is not large enough for reliable force discrimination but expected to be improved following the sensitivity improvement.

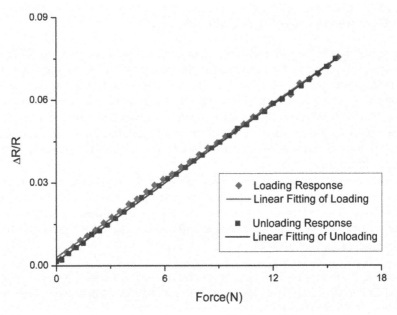

Fig. 6. Hysteresis cycle for one piezoresistive gauge. The linear fittings of loading response and unloading response are also plotted.

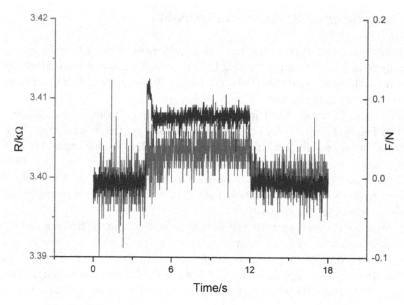

Fig. 7. Minimum load tests results with applied force as low as 0.1N

6   Conclusions and Future Work

A silicon based tactile sensor with a novel structure has been designed and fabricated. The testing experiments by indenting at the center of the steel sheet has been performed which proves the high reliability of this sensor. Meanwhile, the tests indicate the good measurement ability that the force range effectively measured is from 0.1N to 15N with the high linearity, high sensitivity and low hysteresis. In the last, the sensor is well mounted on the commercial prosthetic hand and packaged by the PDMS layer, which will be used for further application of the sensor integration on the prosthetic hand.

Future work will consist in the optimization of the geometric structure of the sensor to improve the sensitivity more and the development of the silicon gauge array integrated on the steel sheet. Moreover, the testing experiments of indenting on different regions of the steel sheet and the influence of the PDMS layer covered on the sensor will also be performed.

Acknowledgements. This work was supported by National Basic Research Program (973) of China (No.: 2011CB013303) and International S&T Cooperation Program of China (No.: 2012DFG71860).

References

1. Beccai, L., Roccella, S., Arena, A., Valvo, F., Valdastri, P., Menciassi, A., Carrozza, M.C., Dario, P.: Design and fabrication of a hybrid silicon three-axial force sensor for biomechanical applications. Sens. Actuators A: Physical 120(2), 370–382 (2005)

2. Noda, K., Hoshino, K., Matsumoto, K.: A shear stress sensor for tactile sensing with the piezoresistive cantilever standing in elastic material. Sens. Actuators A: Phys. 127(2), 295–301 (2006)
3. Kim, K., Lee, K.R., Kim, W.H.: Polymer-based flexible tactile sensor up to 32×32 arrays integrated with interconnection terminals. Sens. Actuators A: Phys. 156, 284–291 (2009)
4. Lee, H., Chung, J., Chang, S.: Normal and shear force measurement using a flexible polymer tactile sensor with embedded multiple capacitors. J. Microelectromech. Syst. 17, 934–942 (2008)
5. Lee, H.K., Chang, S.I., Yoon, E.: A flexible polymer tactile sensor: fabrication and modular expandability for large area deployment. J. Microelectromech. Syst. 15(6), 1681–1686 (2006)
6. Hosoda, K., Tada, Y., Asada, M.: Anthropomorphic robotic soft fingertip with randomly distributed receptors. Robot Auton. Syst. 54, 104–109 (2006)
7. Dargahi, J., Parameswaran, M., Payandeh, S.: A micromachined piezoelectric tactile sensor for an endoscopic grasper-theory, fabrication and experiments. J. Microelectromech. Syst. 9(3), 329–335 (2000)
8. Heo, J.S., Chung, J.H., Lee, J.J.: Tactile sensor arrays using fiber Bragg grating sensors. Sens. Actuators A: Physical 126(2), 312–327 (2006)
9. Yousef, H., Boukallel, M., Althoefer, K.: Tactile sensing for dexterous in-hand manipulation in robotics–A review. Sens. Actuators A: Physical 167(2), 171–187 (2011)
10. Jones, L.A., Lederman, S.J.: Human Hand Function. Oxford University Press (2006)
11. Madou, M.: Fundamentals of Microfabrication. CRC Press (1997)

Development of a Compliant Magnetic 3-D Tactile Sensor with AMR Elements

Ping Yu, Xiaoting Qi, Weiting Liu[*], and Xin Fu

The State Key Lab of Fluid Power Transmission and Control, Zhejiang University,
310027 Hangzhou, China
{yuping55,liuwt,xfu}@zju.edu.cn

Abstract. In this paper, a novel compliant magnetic tactile sensor which can measure both normal and shear force is proposed. The sensor is composed of an elastic dome with a cylindrical permanent magnet bonded to the internal surface and a 3-axis anisotropic magneto resistive(AMR) chip fixed on the rigid substrate. Output voltages of AMR chip change with the displacement of magnet when the elastic dome deforms by contact. The measurement of a prototype sensor shows that the full-scale ranges of detectable forces are about 4N, 4N and 20N for the x-, y-, and z-directions, respectively. In this ranges, output voltages change linearly with the contact forces.

Keywords: 3-D tactile sensor, magnetic, AMR, compliant.

1 Introduction

The tremendous development of robotic and prosthetic hand has proposed great demands for tactile sensing[1]. Tactile sensors that can be mounted on the curved, deformable fingertips and capable of multi-axis force measuring are paid more and more attentions. By providing information about the both normal and shear components of the contact forces at the digit-object interface, contact conditions could be gathered in detail and used to improve manipulation[2-3].

So for, several kinds of multi-axis force tactile sensors have been developed based on the principles of piezoresistive [4-5], capacitive [6], piezoelectric [7], optical [8] methods and etc . Most of the previous multi-axis force sensors have been fabricated using MEMS. They provide good resolution and sensitivity, but they are not suitable for mounting on the curved, deformable fingertips because of the rigid substrate. Recently, flexible sensors have been developed to allow the mounting of the sensor on curved surfaces[9-10].However, these sensors cannot detect shear forces and their delicate electronic components are vulnerable to damage when mounted directly on gripping surfaces.

---

[*] Corresponding author.

X. Zhang et al. (Eds.): ICIRA 2014, Part II, LNAI 8918, pp. 484–491, 2014.

Here we describe a magnetic type tactile sensor that is sensitive to the range of normal and shear forces encountered in robotic and prosthetic hand applications. Because of the characteristic of magnetic field, the sensitive element (permanent magnet) and signal detecting chip of the tactile sensor can be separately mounted on the elastic layer and rigid substrate. In this way, the tactile sensor will be robust and easy to repair.

2   Sensor Design

Fig.1 illustrate the conceptual diagram of the proposed structure of 3-D tactile sensor. It is mainly composed of an elastic dome, a cylindrical permanent magnet and a 3-axis anisotropic magneto resistive(AMR) chip .The dome is made of elastic material so that the tactile sensor could have compliance and softness characteristics. In parallel, dome structure with curved geometry can increase the portion of load gathered by the tactile sensor in case of contact with a planar surface[11]. The elastic dome has a flat top face in inner surface for the magnet to be bonded to. The 3-axis AMR chip is fixed on the rigid substrate, sharing the same axis with the magnet.

When a force is applied on the top, the elastic dome deforms causing the displacement of the permanent magnet. The resistance of AMR chip in X, Y and Z axis change with the displacement of the magnet. By detecting the variation of the resistance, normal and tangential forces applied can be obtained.

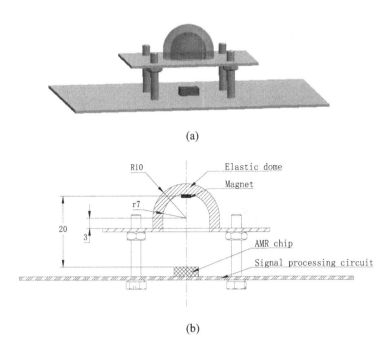

(a)

(b)

Fig. 1. (a) 3D design of the tactile sensor. (b) Schematic showing a cross section of the sensor.

3   Sensor Fabrication

A prototype of the proposed 3-D tactile sensor is developed based on the structure shown in Fig.1. The elastic dome with dimensions of R=10mm, r=7mm, is made of Polydimethylsiloxane(PDMS). The cylindrical neodymium permanent magnet (HXN3-1, MISUMI) has a diameter of 3mm and a thickness of 1mm .Its magnetic flux density at surface is 2000~2400Guass. The 3-axis AMR chip (HMC1053, Honeywell) contains three magnetoresistive chips positioned for orthogonal three-axis sensing. It has field range of ±6 Gauss for all x-, y- and z- axes.

Casting technique was used to make the elastic PDMS (Sylgard 184, Dow Corning) dome. The liquid part A and part B were mixed at the ratio of 10:1. In order to fabricate bubble free sample, the mixture was thoroughly degassed in a vacuum desiccator at low pressure for 30 min. Then the degassed mixture was poured into an aluminium mold (Fig.2a,b) and moved into an incubator for curing. Curing temperature was set to 80°C while curing time was set to 3 hours. The fabricated elastic PDMS dome is shown in Fig.2c.

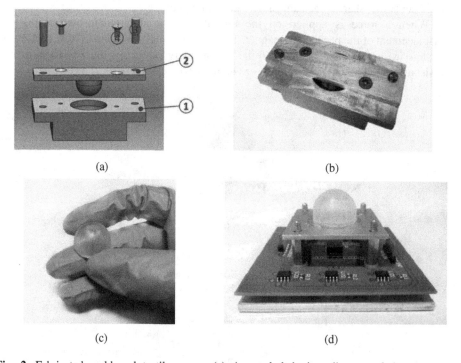

(a)   (b)

(c)   (d)

Fig. 2. Fabricated mold and tactile sensor (a) An exploded view diagram of the sensor aluminum mold(①Core mold, ②Outer mold, ③Location pin, ④ Screw) (b) Photograph of the assembled aluminum mold. (c) Photograph of the PDMS dome (d) Photograph of the sensor tactile sensor prototype.

The assembling process of the tactile sensor is comprised of four steps. Firstly, the AMR chip was fixed on a PCB. Secondly, an aluminum plate was placed parallel to the

PCB by bolt. Thirdly, the permanent magnet was bonded to the top face in inner surface of the elastic dome by glue. Fourthly, the elastic dome with magnet was glued to the aluminum plate. In assembling, the magnet and the AMR chip should be mounted in-line to reduce initial offset of output signals. Location holes in PCB and aluminum plate were used for aligning. The assembled prototype is shown in Fig.2d.

4   Preliminary Characterization Experiments

In order to investigate the ability of the sensor to measure both normal and shear forces, preliminary characterization experiments have been performed. The signal conditioning circuitry and the testing platform have been set up.

4.1   Signal conditioning

The 3-axis AMR chip contains three orthogonal arranged 4-element Wheatstone bridges(Fig.3). The output voltage of every bridge is

$$V_{out1} = S \times B \times V_b . \tag{1}$$

where S is the sensitivity of   HMC1053, B is the magnetic flux density, V_b is the input voltage.

The bridge output signal is then led to an instrumentation amplifier(AD623, Analog Devices, USA).The amplifier gain G can be adjusted from 1 to 1000. Thus the output of the circuit is

$$V_{out} = G \times (S \times B \times V_b) + V_{ref} \tag{2}$$

where G is the amplifier gain ,V_{ref} is the reference voltage.

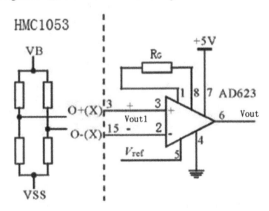

Fig. 3. Diagram of the signal conditioning circuit

4.2   Experiment Set-Up

The characterization apparatus is shown in Fig.4. It consists of a three-axial motion platform(M-VP-25XA-XYZL, Newport, USA), a link block, a three-axial force sensor(FC3D50-50/50/50N,Shanghai Forcechina Measurement Technology Co., Ltd, China), a rod indenter with flat head, a data acquisition card and a PC for data acquisition and analysis.

The three-axial force sensor is fixed on the three-axial motion platform through a link block. The indenter is screwed into the three-axial force sensor. Thus, the indenter can move precisely in three-axial direction by which vertical force and tangential force can be applied to the proposed tactile sensor. The three-axial force sensor whose measurement range is from -50N to 50N with precision as 0.01N for all x-,y-,and z-axes is used to monitor the applied force vector precisely. The three-axial motion platform can drive the indenter at the max range of 25 mm in each direction and the lowest speed at 1 μm/s. Thus, a precise force variation can be obtained by the motion control of the platform. The diameter of the rod indenter head is 20mm.

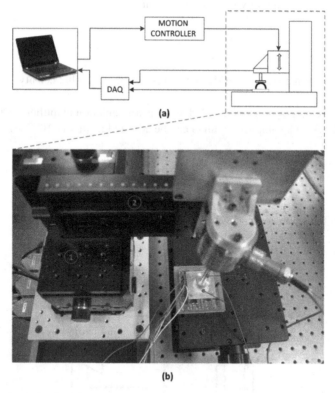

Fig. 4. (a) A schematic of the testing system; (b) Photograph of the force loading system (①the three-axial motion platform, ②the link block, ③the three-axial force sensor, ④the rod indenter, ⑤the designed sensor, ⑥the sensor mount platform)

Normal load tests and Shear load tests will be performed to calibrate the proposed tactile sensor. In the normal load tests, the three-axial motion platform is firstly manually driven in all x-, y-, and z-directions to locate the indenter directly over the elastic dome and then automatically driven in the vertical(z axis) direction only at the speed of 0.01mm/s to load normal force. The loading forces from the three-axial force sensor can be monitored in the interface of the SignaExpress (Version 2011, National Instrument). The motion is manually stopped immediately through the motion controller when the maximum force is reached. The maximum force load is set at 20N, 4N,4N for the x-,y-,and z- directions. In the shear load tests, the normal force is kept in constant and then the motion platform is automatically driven in the horizontal (x or y axis) direction to apply tangential force.

5   Experimental Results and Discussion

5.1   Normal Load Tests

The continuous normal load tests have been performed when an applied force is increasing from 0 to 20N at a constant translational speed of 0.01mm/s along the z direction.

Typical sensor outputs Vx, Vy and Vz in response to the force loading are plotted in Fig.5. The slope of curve Vz is 0.058V/N while the coefficient of determination is 0.997, showing a high linearity of the sensor repose.

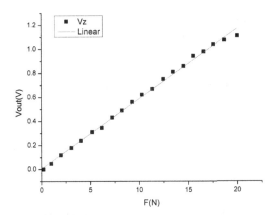

Fig. 5. Normal loading response of the tactile sensor

5.2   Shear Load Tests

Preliminary tangential tests have be carried out by keeping normal force Fz at 15N and moving the motion platform in x axis direction to continuously increase the tangential force Fx from 0N to 4N.

Fig.6 shows the sensor outputs Vx, Vy and Vz in response to the force loading. The slope of curve Vx is 0.078V/N while the coefficient of determination is 0.98, showing a high linearity of the sensor repose.

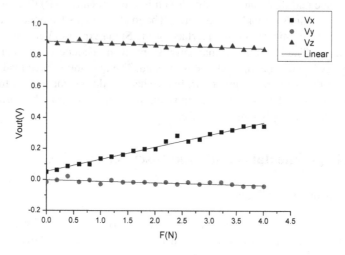

Fig. 6. Tangential loading response of the tactile sensor

6   Conclusions and Future Work

A magnetic 3-D tactile sensor has been designed and fabricated. The normal load and shear load tests has been performed. Tests results indicated that the fabricated sensor can separate the normal and shear force. For an applied normal force between 0 and 20N, the sensitivity is 0.058V/N with a linearity of 99.7%. For an applied shear force between 0 and 4N, preliminary tangential tests show a shear sensitivity of 0.078V/N with a linearity of 98%.

Due to the assembling error and magnetic flux density distribution character of the magnet, tactile sensor exhibit coupling characteristic for three dimensional forces, as shown in Fig.6. It is necessary to perform uncoupling treatment, like magnet location, to solve this problem in future. Moreover, future work will consist in the integration with a multi-fingered robot or prosthetic hand.

Acknowledgements. This work was supported by National Basic Research Program (973) of China (No.: 2011CB013303) and International S&T Cooperation Program of China (No.: 2012DFG71860).

References

1. Dahiya, R.S., Valle, M.: Robotic Tactile Sensing. Springer, New York (2013)
2. Johansson, R.S., Flanagan, J.R.: Coding and use of tactile signals from the fingertips in object manipulation tasks. Nat. Rev. Neurosci. 10, 345–359 (2009)

3. Beccai, L., Roccella, S., Ascari, L., Valdastri, P., Sieber, A., Carrozza, M.C., Dario, P.: Development and experimental analysis of a soft compliant tactile microsensor for anthropomorphic artificial hand. IEEE/ASME Transactions on Mechatronics 13, 158–168 (2008)
4. Beccai, L., Roccella, S., Arena, A., Valvo, F., Valdastri, P., Menciassi, A., Dario, P.: Design and fabrication of a hybrid silicon three-axial force sensor for biomechanical applications. Sensor. Actuat. A-Phys. 120(2), 370–382 (2005)
5. Kim, K., Lee, K.R., Kim, W.H., Park, K.B., Kim, T.H., Kim, J.S., Pak, J.J.: Polymer-based flexible tactile sensor up to 32× 32 arrays integrated with interconnection terminals. Sensor. Actuat. A-Phys. 156, 284–291 (2009)
6. Tiwana, M.I., Shashank, A., Redmond, S.J., Lovell, N.H.: Characterization of a capacitive tactile shear sensor for application in robotic and upper limb prostheses. Sensor. Actuat. A-Phys. 165, 164–172 (2011)
7. Hosoda, K., Tada, Y., Asada, M.: Anthropomorphic robotic soft fingertip with randomly distributed receptors. Robot. Auton. Syst. 54, 104–109 (2006)
8. De Maria, G., Natale, C., Pirozzi, S.: Force/tactile sensor for robotic applications. Sensor. Actuat. A-Phys. 175, 60–72 (2012)
9. Shamanna, V., Das, S., Çelik-Butler, Z., Butler, D.P., Lawrence, K.L.: Micromachined integrated pressure–thermal sensors on flexible substrates. J. Micromech. and Microeng. 16, 1984–1992 (2006)
10. Lowe, M., King, A., Lovett, E., Papakostas, T.: Flexible tactile sensor technology: Bringing haptics to life. Sensor Rev. 24, 33–36 (2004)
11. Vásárhelyi, G., Fodor, B., Roska, T.: Tactile sensing-processing: Interface-cover geometry and the inverse-elastic problem. Sensor. Actuat. A-Phys. 140, 8–18 (2007)

Elastic Dynamic Analysis on the Scraping Paste Mechanism of the Printing Equipment for Silicon Solar Cells[*]

Jinglun Liang[1], Xianmin Zhang[2], Liangchun Cui[2],
Jiapeng Han[2], and Yongcong Kuang[2]

[1] School of Mechanical Engineering, Dongguan University of Technology
Dongguan, Guangdong Province, China
[2] School of Mechanical & Automotive Engineering, South China University of Technology
Guangzhou, Guangdong Province, China

Abstract. Elastic-dynamic analysis of a scraping paste mechanism for screen printing is carried out in this paper to improve the printing quality during the deposition process. Based on finite element method, Kineto-Elasto dynamic(KED) assumption and Lagrange equation, the elastic dynamic model of the scraping paste mechanism is established firstly. Comparing the natural frequencies of the system obtained by two methods, the effectiveness of the proposed modeling method is verified. Moreover, the sensitivity of dynamic displacement is also obtained by the direct differential method. Then the elastic deformation subjected to the length of the effective sliding shaft, the section diameter and the printing pressure is investigated by numerical simulation in Matlab. The results show that, the elastic deformation of the sliding shaft is not a constant during the printing process, but increases in non-linear along with the increase of the length of the effective sliding shaft, decreases along with the increase of the section diameter. As the printing pressure rises, its elastic deformation also increases in non-linear. The results provide necessary information and guidance for process parameters optimization, structure improvement to achieve high accuracy printing.

Keywords: screen printing, scraping paste mechanism, dynamic analysis, KED method.

1   Introduction

Silicon solar cell is a semiconductor electron device, which can effectively absorb solar radiation and convert it into electrical energy, widely applied in all kinds of lighting and power-generation system. Screen printing is an important process for

[*] This work was supported by the National Science Foundation for distinguished young scholars of China under Grant No. 50825504, National advanced energy technology research program 863 (Grant No. 2012AA050302) and Guangdong Hong Kong Technology Cooperation Funding (Dongguan Project 2012205122).

X. Zhang et al. (Eds.): ICIRA 2014, Part II, LNAI 8918, pp. 492–503, 2014.

silicon solar cells production and the printing quality (thickness, width, accuracy, shape, etc) influences the quality of silicon solar cell. The scraping paste mechanism is one of the key mechanisms of screen printing equipment for silicon solar cells and its dynamic performance has direct impact on the printing quality. Scraping paste mechanism is usually designed or improved only according to suppressing the paste scraped from the front to the back or from the back to the front during the deposition process, while the influence of the elastic deformation of the mechanism is usually ignored. So there are some differences between the ideal model and the actual situation. Therefore, in order to improve the quality of the printed paste on silicon solar cell and realize high accuracy and robust printing, it is necessary to analyze the elastic deformation of scraping paste mechanism according to various factors during the paste deposition process.

Kineto-Elasto dynamic (KED) analysis [1] is used to solve the mechanical system dynamics problems, especially the problems of the elastic deformation of the mechanism and joint. One of main applications of KED is to analyze the elastic motion response of a mechanical component, while its motion law is determined [2]. In recent years, some explorations were done in this area. Hu [3] has presented dynamic analysis of a novel 2-DOF high-speed parallel manipulator and used KED method to establish the dynamic equation. Zhang [4, 5] has researched the elastic dynamic of linkage mechanism, considering the coupling between the rigid body motion and the elastic motion. Using finite element method and KED method, dynamics analysis of the flexible crane system of the heavy travelling overhead working truck has been carried out by Du [6].

In this paper, the KED analysis of the scraping paste mechanism of the printing equipment for silicon solar cells is carried out. Firstly the mechanism model is simplified and its elastic dynamic model is established by finite element method, KED method and Lagrange equation. And the effectiveness of the proposed modeling method is verified by comparing the natural frequencies of the model obtained by two methods. The sensitivity of the dynamic displacement is obtained by the direct differentiation method. With the actual structure and motion parameters, the elastic deformation of the scraping paste mechanism is investigated by numerical simulation in MATLAB.

2   Simplification of Mechanism Model

The scraping paste mechanism of the printing equipment for silicon solar cells is mainly composed of a servo motor, a ball screw, tow sliding shafts and squeegee parts, etc, as shown in Fig. 1.

During scraping the paste and the deposition process, according to the snap-off distance (the distance between the screen and the substrate) and calibrated pressure, the squeegee is driven to press the screen down into line-contact with the substrate (silicon wafer) by the motor or the air cylinder mounted internally. Then the sliding shafts and the squeegee are driven to traverse across the screen, pushing the paste through the screen image onto the substrate. After the squeegee separates from the screen, a metal flood blade travels back over the screen, spreading the paste ready for the next print cycle.

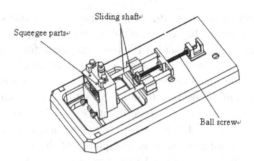

Fig. 1. Diagram of the scraping paste mechanism

During scraping the paste process, the squeegee is driven to press the screen with a force downward   usually named printing pressure and larger than 60 N. Due to the reaction, the squeegee and the sliding shaft receive a force upward, which is equal to the printing pressure. As a result, elastic deformation of the sliding shaft like a cantilever beam occurs, which affects the stability of the printing quality in practice during the deposition process. Obviously, the maximum deformation of the sliding shaft occurs at the end point of the sliding shaft. To simply the model, the sliding shaft can be treat as a cantilever beam with a lumped mass on the terminal, as shown in Fig. 2. According to theoretical mechanics [7] knowledge, it is evident that the terminal of the sliding shaft is acted by a force F upward and an anticlockwise moment M. As the sliding shaft moves in horizontal, the length of the effective sliding shaft (cantilever beam) increases or decreases.

3   Establishment of the Elastic Dynamic Equation

The elastic dynamic equation of the sliding shaft of scraping paste mechanism is established by finite element principle, KED method. Its procedures are as follows: First of all, the space beam element is used to simulate the sliding shaft of scraping paste mechanism. Then, choosing displacement modes and establishing the generalized coordinates, dynamic equations of the beam element are derived by Lagrange equation. Finally, the total dynamic equations are formed by assembling the entire elements according to the compatibility at the nodes [8-9].

3.1   Spatial Finite Element Model

The sliding shaft of scraping paste mechanism is simulated by the beam element. In this paper, space beam elements with circular cross section are chosen as the basic element model, as shown in Fig. 3. The space beam element has two nodes with i, j respectively. i-xyz is the element coordinate system. The axial forces which exert at nodes i and j are denoted as F_{Ni} and F_{Nj} respectively. $F_{Qyi}, F_{Qyj}, F_{Qzi}, F_{Qzj}$ stand for the shear forces in y and z direction. M_{xi}, M_{xj} denotes the torques. $M_{yi}, M_{yj}, M_{zi}, M_{zj}$ denotes the bending moments around y and z axis.

Fig. 2. Force diagram of the sliding shaft

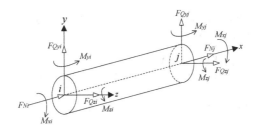

Fig. 3. Model of space beam element

Assuming that the axial, lateral (two direction), torsion deformations of the space beam element occur, then $\delta = [u_1 \ v_1 \ w_1 \ \theta_{x1} \ \theta_{y1} \ \theta_{z1} \ u_2 \ v_2 \ w_2 \ \theta_{x2} \ \theta_{y2} \ \theta_{z2}]^T$ can be denoted the vector of the beam element in the generalized coordinate, where u_1, u_2 is the axial deformation displacements of nod i and nod j respectively, v_1, v_2 are the lateral deformation displacements of two endpoints along y axis, while w_1, w_2 along z axis, and θ_{x1}, θ_{x2}, θ_{y1}, θ_{y2}, θ_{z1}, θ_{z2} are the elastic angular displacements of two endpoints around the x-axis, y-axis and z-axis. The elastic displacements along the x-axis, y-axis and z-axis and the elastic angular displacements around the x-axis, y-axis and z-axis are denoted as u, v, w, θ_x, θ_y, θ_z respectively. It is assumed that the axial deformation displacement u and the torsion angle θ_x are linear functions of x, and the lateral deformation displacements v and w are expressed by cubic polynomial about x, using the boundary condition of elements.

When the sliding shaft is moving in horizontal, because elastic displacement of the element is very small, the coupling effect of rigid motion and elastic deformed motion can be neglected. So, the absolute speed is superposition of rigid motion speed and elastic deformed speed at any point on the element. Let $\delta_r=[x_i \ y_i \ z_i \ \theta_x \ \theta_y \ \theta_z x_j y_j \ z_j \ \theta_x \ \theta_y \ \theta_z]^T$, and it represents rigid motion displacement of the element, where x_i, y_i and z_i are rigid body displacements of the element at node i along the x-axis, y-axis and z-axis respectively, θ_x, θ_y and θ_z are rigid body angular displacements of the element around the x-axis, y-axis and z-axis respectively, x_j, y_j and z_j are rigid body displacements of the element at node j along the x-axis, y-axis and z-axis respectively. Then, the speed of any point on the element can be expressed as:

$$
\begin{cases}
\dot{u}_a = \dot{u}_r + \dot{u} & \dot{v}_a = \dot{v}_r + \dot{v} \\
\dot{w}_a = \dot{w}_r + \dot{w} & \dot{\theta}_{ax} = \dot{\theta}_{rx} + \dot{\theta}_x \\
\dot{u}_r = N_u \dot{\delta}_r & \dot{v}_r = N_v \dot{\delta}_r \\
\dot{w}_r = N_w \dot{\delta}_r & \dot{\theta}_{rx} = N_{\theta_x} \dot{\delta}_r
\end{cases}
\tag{1}
$$

where $\dot{u}_a$, $\dot{u}_r$ and $\dot{u}$ are the absolute speed, rigid body speed and elastic speed in the x direction respectively, while $\dot{v}_a$, $\dot{v}_r$ and $\dot{v}$ in the y direction and $\dot{w}_a$, $\dot{w}_r$ and $\dot{w}$ in the z direction, $\dot{\theta}_{ax}$, $\dot{\theta}_{rx}$ and $\dot{\theta}_x$ are absolute angular velocity, rigid body angular velocity and elastic angular velocity around the x direction, respectively.

3.2   Dynamic Equations of the Element

Before dynamic equations of the beam element are derived by Lagrange equation, it is necessary to calculate kinetic energy and strain energy of the element.

3.1.1   Kinetic Energy of the Element

Assuming that the mass of the element gathers at the neutral line, namely x axis. Then, kinetic energy of the element includes its translational kinetic energy and rotational kinetic energy around its axis can be written as [8]:

$$T = \frac{1}{2}\int_0^l \rho A\left(\dot{u}_a^2 + \dot{v}_a^2 + \dot{w}_a^2\right)dx + \frac{1}{2}\int_0^l \rho I_p \dot{\theta}_{ax}^2 dx \tag{2}$$

where l is the element length, ρ is the element density, A is the element cross sectional area, I_p is the element cross sectional polar moment of inertia.

Substituting (1) into (2), we have

$$T = \frac{1}{2}\left(\dot{\delta} + \dot{\delta}_r\right)^{\mathrm{T}} M_e \left(\dot{\delta} + \dot{\delta}_r\right) \tag{3}$$

where M_e is mass matrix of the element, and $M_e = \rho A \int_0^l NN^{\mathrm{T}} dx + \rho I_p \int_0^l N_{\theta_x}^{\mathrm{T}} N_{\theta_x} dx$.

3.1.2   Strain Energy of the Element

Taking account of bending strain energy, tensile strain energy and torsion strain energy, and ignoring the buckling strain energy, strain Energy of the element can be expressed as [9]:

$$V = \frac{1}{2}E\int_0^l\left[A\left(\frac{\partial u}{\partial x}\right)^2 + I_z\left(\frac{\partial \theta_z}{\partial x}\right)^2 + I_y\left(\frac{\partial \theta_y}{\partial x}\right)^2\right]dx + \frac{1}{2}\int_0^l GI_p\left(\frac{\partial \theta_x}{\partial x}\right)^2 dx \tag{4}$$

where E is elastic modulus, G is shear modulus, I_y is cross sectional moment of inertia to y-axis, I_p is cross sectional moment of inertia to z-axis.

Substituting the vector of the beam element into (4), we have

$$V = \frac{1}{2}\delta^{\mathrm{T}} K_e \delta \tag{5}$$

where K_e is the element stiffness matrix.

Substituting (3) and (4) into Lagrange equation

$$\frac{d}{dt}\left(\frac{\partial(T-V)}{\partial\dot{\delta}}\right) - \frac{\partial(T-V)}{\partial\delta} = F \tag{6}$$

Dynamics equation of the element can be written as

$$M_e\left(\ddot{\delta} + \ddot{\delta}_r\right) + K_e\delta = F_e \tag{7}$$

where F_e is the generalized force array of the element applied load.

Let U^e be the global generalized coordinate vector, δ_i is the local generalized coordinate vector of element i. Defining the coordinate transformation matrix R_i, yields

$$\delta_i = R_i U^e \ (i = 1, 2, \ldots\ldots, n) \tag{8}$$

where $U^e = [U_1, U_2, \ldots\ldots, U_{12}]^T$, R_i is the conversion matrix of element i, n is the number of elements.

Using coordinate transformation matrix and taking the global generalized coordinates vector U^e as variable, based on (7), the motion differential equation of the element can be rewritten as

$$\left(M_{e\text{-}s}\right)_i\left(\ddot{U}^e + \ddot{U}^e_r\right) + \left(K_{e\text{-}s}\right)_i U^e = \left(F_{e\text{-}s}\right)_i \tag{9}$$

where $\left(M_{e\text{-}s}\right)_i = R_i^T\left(M_e\right)_i R_i$, $\left(K_{e\text{-}s}\right)_i = R_i^T\left(K_e\right)_i R_i$, $\left(F_{e\text{-}s}\right)_i = R_i^T\left(F_e\right)_i$.

3.1.3   Dynamic Equation of the System

In order to form the system equation, the generalized coordinates array $U = [U_1, U_2, \ldots\ldots, U_N]^T$ can be defined, where $N=6\times(n+1)$. The elastic acceleration array $\ddot{u}$ and the rigid acceleration array $\ddot{u}_r$ can be determined too. So the relationship between U^e and U can be written as

$$U^e = B_i U \tag{10}$$

where B_i is a coordinates coordination matrix with $12\times N$, and its elements are 0 or 1.

Substituting (10) into (9), and dot-multiplying both sides of the result by B_i^T, yields

$$M_i\left(\ddot{U} + \ddot{U}_r\right) + K_i U = F_i \tag{11}$$

where $M_i = B_i^T \left(M_{e\text{-}s}\right)_i B_i$, $K_i = B_i^T \left(K_{e\text{-}s}\right)_i B_i$, $F_i = B_i^T \left(F_{e\text{-}s}\right)_i$.

According to (11), assembling all the elements, the elastic dynamics equation of the system can be easily obtained as

$$M\ddot{U} + KU = F - M\ddot{U}_r \tag{12}$$

where $M = \sum_{i=1}^{n} M_i, K = \sum_{i=1}^{n} K_i, F = \sum_{i=1}^{n} F_i$

M, K are the general mass matrix and the general stiffness matrix respectively, while U, F are the elastic displacement array and the generalized force array, and the second term on the right side of (12) is the rigid inertia force array. In addition, the general mass matrix M also includes the lumped mass. In other words, the kinetic energy of the lumped mass is taken into account.

4   Frequency Characteristic Analysis

Natural frequency and vibration mode of the system are determined by its own stiffness matrix K and mass matrix M.

$$(K - w^2 M)\phi = 0 \tag{13}$$

where w is the natural frequency, ϕ is vibration mode matrix.

To obtain natural frequency is obviously to get generalized eigenvalues of K relative to M. The solution can be derived as follows.

$$\det(M^{-1}K - w^2 E) = 0 \tag{14}$$

In order to verify the validity of the modeling method in this paper, the natural frequencies of the sliding shaft of the scraping paste mechanism are compared by two methods. The first one is that the natural frequencies are calculated in MATLAB based on (13). The other one is to perform the mode analysis of the sliding shaft and get its natural frequencies in ANSYS software. When sliding shaft locate at the position of the maximum stretching-length, the former three orders natural frequencies of the sliding shaft are acquired by two methods. The results are shown in Table I. It is evident that that the relative errors of the natural frequencies by two methods are less than 5%. So the proposed modeling method is effective and accurate.

Table 1. The comparison results of natural frequency

Natural frequency	The results from MATLAB (Hz)	The results from ANSYS (Hz)	Relative error (%)
The first frequency	355.80	342.9	3.76
The second frequency	1533.81	1523.5	0.68
The third frequency	2188.17	2127.6	2.85

5   Dynamic Displacement Response Sensitivity

The response sensitivity of dynamic displacement is the change ratio of displacement as to building design parameters. Design parameters of the sliding shaft include geometric size parameters, material characteristic parameters, etc. Differentiating (12) with respect to parameter b, we have

$$M\frac{\partial \ddot{U}}{\partial b} + K\frac{\partial U}{\partial b} = \frac{\partial F}{\partial b} - \frac{\partial M}{\partial b}(\ddot{U} + \ddot{U}_r) - M\frac{\partial \ddot{U}_r}{\partial b} - \frac{\partial K}{\partial b}U \qquad (15)$$

It is evident that (12) and (15) have the same structural form. As a result, they have the same solving method.

6   Example Analysis

The parameters of the sliding shaft of the scraping paste mechanism are as follow: The material is mild steel, the density ρ is 7.5×10^3 kg/m^3, elastic modulus E is 20.6×10^{10} pa, and shear modulus G is 79.4×10^9 pa. The section of bar is circular. The number of elements is 4. The section diameter d is 0.025 m. According to the actual printing conditions, usually, the original effective length of the sliding shaft is 0.051 m, the printing stroke is 0.18 m, and the printing pressure is 75 N. The force simplified to the sliding shaft F is 37.5 N, the torque M is 1.757 N·m, and the lumped mass m is 3.17 kg.

The velocity of the sliding shaft is as follows:

$$V = \begin{cases} 2\times10^3 t & 0 \leq t < 0.1s \\ 200 & 0.1 \leq t < 0.9s \\ 200 - 2\times10^3(t-0.9) & 0.9 \leq t \leq 1s \end{cases} \qquad (16)$$

The effective length of the sliding shaft as a function of the time is shown in Fig. 4. It is obvious that the length of bar increases in non-linear during the printing process.

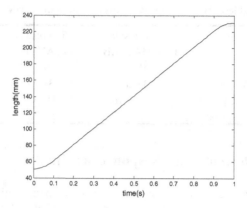

Fig. 4. The plot of the effective length as a function of the time

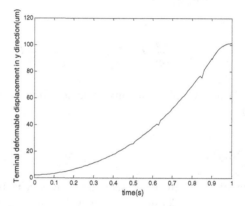

Fig. 5. The deformable displacement in y direction

The deformable displacements in x, y direction and the deformable angular displacement around z direction of the terminal of the sliding shaft as functions of the time can be obtained by numerical simulation in MATLAB. It is obvious that the elastic deformation displacement of the sliding shaft along the x direction and the elastic deformation angular displacement around the z direction are very small. So the terminal deformation displacement of the sliding shaft in y direction (elastic deformation for short) will be only discussed in the following papers. As shown in Fig. 4 and Fig. 5, the elastic deformation of the sliding shaft increases in non-linear along with the increase of time (due to the increase of the effective length). So it is easily understood that as the increase of the effective length of the sliding shaft during the actual printing process, the distance between the screen and the wafer will be larger unpredictably and the angle between the squeegee and the screen will be changed too, which will lead to different thickness of the printed paste on the same silicon wafer and then influence the quality of the silicon cell.

The elastic deformation of the sliding shaft on the terminal subjected to the section diameter will be discussed in this segment. Supposing the effective length of the sliding shaft is 231 mm (the maximum length), according to the actual situation, the elastic deformation subjected to the section diameter is calculated within the range from 15 mm to 35 mm, and the result is shown in Fig. 6(a). The elastic deformation of the sliding shaft decreases in non-linear along with the increase of the section diameter. Moreover, the shaft deformation changes greatly when the diameter is less than 20 mm, and the plot of the shaft deformation becomes flat when the diameter is greater than 30 mm. As shown in Fig. 6(b), when the section diameter is less than 20 mm, the deformable displacement in y direction is more sensitive for its change, and the sensitive basically has no variation when the section diameter is greater than 30 mm, which also proves the change tendency of the curve in Fig. 6(a). As a result, when the cross-section diameter of the sliding shaft is designed or improved, the elastic deformation caused by its section diameter and the material cost should be comprehensively considered.

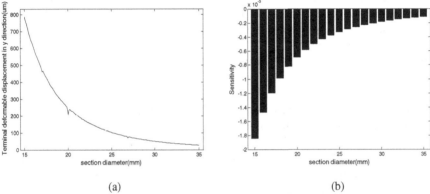

(a)   (b)

Fig. 6. (a) The deformable displacement in y direction as a function of the section diameter. (b) The sensitivity of terminal deformable displacement in y direction subjected to the section diameter.

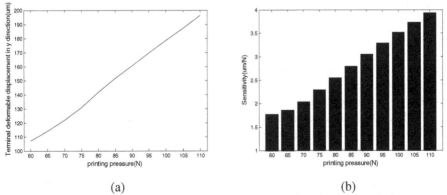

(a)   (b)

Fig. 7. (a) The deformable displacement in y direction as a function of the printing pressure value. (b) The sensitivity of terminal deformable displacement in y direction subjected to the printing pressure value.

The deformations of the sliding shaft shown in Fig. 5 and Fig. 6(a) are obtained when the printing pressure is 75 N. But during the production process, the printing pressure will be modified with a range of 60 N to 110 N, due to the difference of screen templates. So the elastic deformation of the sliding shaft as a function of the printing pressure is also investigated, as shown in Fig. 7(a), supposing the effective length of the sliding shaft is also the maximum. As to the plot, as the printing pressure is larger, the elastic deformation of the sliding shaft also becomes larger in non-linear. The plot of the sensitivity of the deformable displacement in y direction as a function of the printing pressure is shown in Fig. 7(b). When the printing pressure is larger than 80 N, the deformable displacement is more sensitive as to its change. The results shown in Fig. 7(a) and Fig. 7(b) are coincident.

During the printing process, the printing pressure is a key parameter for screen printing. If the pressure is too large, the silk screen is easy to be destroyed by the squeegee, and the silicon wafer is easily broken. If the pressure is too small, the paste will remain on the screen and the paste cannot be deposited on the wafer. Therefore, a proper printing pressure should be determined by many factors, including the deformable displacement.

In conclusion, the elastic deformation of the sliding shaft in printing process will increase along with the increase of the effective length of the sliding shaft, the decrease of the section diameter and the increase of the printing pressure. The elastic deformation of the sliding shaft makes the distance between the screen and the silicon wafer different from the setting value, which leads to different thickness of the paste printed on silicon wafers and influences the printing quality. Therefore, the elastic deformation error of the sliding shaft subjected to the effective length, the section diameter and the printing pressure should be considered in the design or optimization of the scraping paste mechanism of the printing equipment of the silicon solar cells.

7   Conclusion

In this paper, the KED analysis of the scraping paste mechanism of the printing equipment for silicon solar cells is carried out. Firstly the mechanism model is simplified and its elastic dynamic model is established by finite element method, KED method and Lagrange equation. On this basis, comparing the natural frequencies of the model obtained by two methods, the effectiveness of the modeling method is verified. Using the direct differentiation method, the expression of the response sensitivity of dynamic displacement is constructed. By numerical simulation in MATLAB, the elastic deformation of the sliding shaft of scraping paste mechanism is gained during the printing process. There are some conclusions. (1) The elastic deformation of the sliding shaft is not a constant during the printing process, but increases in non-linear along with the increase of the length of the effective sliding shaft. (2) Along with the increase of the section diameter, its elastic deformation decreases in non-linear. (3) As the printing pressure increases, its elastic deformation becomes larger in non-linear. The abovementioned results provide necessary information and guidance for process parameters optimization, structure improvement to achieve high accuracy printing.

References

1. Songhui, N., Xiaolong, L., Aihong, Q.: The method of the stress analysis in elastic linkages. Natural Science Journal of Xiangtan University 23(1), 79–83 (2001)
2. Shanzeng, L., Yue-qing, Y., Jianxin, Y.: Dynamic analysis of a 3-RRC parallel flexible manipulator. Journal of Vibration and Shock 27(2), 157–161 (2008)
3. Junfeng, H., Xianmin, Z.: Elastodynamic analysis of a novel 2-DOF high-speed parallel manipulator. Journal of South China University of Technology (Natural Science) 37(11), 123–128 (2009)
4. Xianmin, Z., Jike, L., Yunwen, S.: A high efficient frequency analysis method for closed flexible mechanism systems. Mech. Mach. Theory 33(8), 1117–1125 (1998)
5. Xianmin, Z., Hongzhao, L., Yunwen, S.: Finite dynamic element analysis for high-speed flexible linkage mechanisms. Computers & Structures 60(5), 787–796 (1996)
6. Liang, D.: The dynamic analysis for the flexible crane system of the heavy traveling overhead working truck. Harbin Institute of Technology, Harbin (2010)
7. Harbin Institute of Technology Theoretical mechanics Department: Theoretical mechanics (I), 6th edn., pp. 13–18. Higher Education Press, Beijing (2004)
8. Ce, Z., Yongqiang, H., Ziliang, W.: The analysis and design of elastic linkage, 2nd edn., pp. 89–93. Mechanical Industry Press, Beijing (1997)
9. Shanzeng, L.: Dynamics of 3-DOF spatial flexible parallel robots. Beijing University of Technology, Beijing (2009)

Covariant Decomposition
of the Three-Dimensional Rotations

Clementina D. Mladenova[1], Danail S. Brezov[2], and Ivaïlo M. Mladenov[3]

[1] Institute of Mechanics, Bulgarian Academy of Sciences, Sofia
[2] University of Architecture, Civil Engineering and Geodesy, Sofia
[3] Institute of Biophysics, Bulgarian Academy of Sciences, Sofia
clem@imbm.bas.bg, danail.brezov@gmail.com, mladenov@bio21.bas.bg

Abstract. The main purpose of this paper is to provide an alternative representation for the generalized *Euler* decomposition (with respect to arbitrary axes) obtained in [2,3] by means of vector parameterization of the *Lie* group SO(3). The scalar (angular) parameters of the decomposition are explicitly written here as functions depending only on the contravariant components of the compound vector-parameter in the basis, determined by the three axes. We also consider the case of coplanar axes, in which the basis needs to be completed by a third vector and in particular, two-axes decompositions.

Keywords: Rotations, vector-parameterization, group decompositions, Lie algebras, Lie groups.

1 Introduction

Determination of the compound rotation which is equivalent to some sequence of given rotations is straightforward but the inverse problem of finding a set of rotational matrices such that their composition is equivalent to a given rotation is a challenging problem. In particular this is extremely difficult problem in the cases when the axes of the chosen rotations are not orthogonal [5]. The necessity of such decomposition of the arbitrary rotational motions into two or three successive rotations is dictated by the practical needs in industry and engineering sciences. It is worth to mention that the factorizations of orthogonal matrices play an important role in modern navigation and control of aircrafts, submarines, and communication satellites crystallography and diffractometry, nuclear magnetic resonance, or digital image processing and optics. In any of these areas it is necessary to perform several successive displacements in order to obtain the desired setting. The present paper is a natural continuation of our long research program concerning rotation decompositions which started with [7].

The real generalizations in *Euler* decomposition for axes in general position appear hardly at the end of last century, probably stimulated from development of the computational techniques. In the general case, the matrix coefficients do not depend in a simple way on the angles of the decomposition. For this

X. Zhang et al. (Eds.): ICIRA 2014, Part II, LNAI 8918, pp. 504–511, 2014.

purpose the *Rodrigues* formula, or some algebraic structure like quaternions or vector-parameters can be used to give the connections for the unknown angles. Nowadays, the most successful methods are: quaternion methods, methods using vector-parameter, and methods based on the *Rodrigues* formula.

2   Vector-Parameters in the Euler Decomposition

Vector-parameters, also known as *Rodrigues'* or *Gibbs'* vectors, are naturally introduced via stereographic projection. For the rotation group in $\mathbb{R}^3$ we consider the spin cover $SU(2) \cong \mathbb{S}^3 \longrightarrow SO(3) \cong \mathbb{RP}^3$ and identify $\mathbb{S}^3$ with the set of unit quaternions (cf. [8])

$$\zeta = (\zeta_0, \boldsymbol{\zeta}) = \zeta_0 + \zeta_1 \mathbf{i} + \zeta_2 \mathbf{j} + \zeta_3 \mathbf{k}, \quad |\zeta|^2 = \zeta \bar{\zeta} = 1, \quad \bar{\zeta} = (\zeta_0, -\boldsymbol{\zeta}), \ \zeta_\alpha \in \mathbb{R}. \tag{1}$$

The corresponding group morphism is given by the adjoint action of $\mathbb{S}^3$ in its *Lie* algebra of skew-*Hermitian* matrices, in which we expand vectors $\mathbf{x} \in \mathbb{R}^3 \to x_1 \mathbf{i} + x_2 \mathbf{j} + x_3 \mathbf{k} \in \mathfrak{su}(2)$. The resulting $SO(3)$ matrix transforming the *Cartesian* coordinates of $\mathbf{x}$ has the form

$$\mathcal{R}(\zeta) = (\zeta_0^2 - \boldsymbol{\zeta}^2)\mathcal{I} + 2\,\boldsymbol{\zeta} \otimes \boldsymbol{\zeta}^t + 2\,\zeta_0 \boldsymbol{\zeta}^\times \tag{2}$$

where $\mathcal{I}$ and $\boldsymbol{\zeta} \otimes \boldsymbol{\zeta}^t$ denote the identity and the tensor (dyadic) product in $\mathbb{R}^3$ respectively, whereas $\boldsymbol{\zeta}^\times$ is the skew-symmetric matrix, associated with the vector $\boldsymbol{\zeta}$ via *Hodge* duality. The famous *Rodrigues'* rotation formula then follows directly with the substitution

$$\zeta_0 = \cos\frac{\varphi}{2}, \qquad \boldsymbol{\zeta} = \sin\frac{\varphi}{2}\,\mathbf{n}, \qquad (\mathbf{n}, \mathbf{n}) = 1.$$

On the other hand, we may choose to get rid of the unnecessary fourth coordinate by projecting $\boldsymbol{\zeta} \to \mathbf{c} = \frac{\boldsymbol{\zeta}}{\zeta_0} = \tan\left(\frac{\varphi}{2}\right)\mathbf{n}$ and thus obtain the entries of the rotation matrix (2) expressed as rational functions of the *vector-parameter* $\mathbf{c}$

$$\mathcal{R}(\mathbf{c}) = \frac{(1 - \mathbf{c}^2)\mathcal{I} + 2\,\mathbf{c} \otimes \mathbf{c}^t + 2\,\mathbf{c}^\times}{1 + \mathbf{c}^2}. \tag{3}$$

Quaternion multiplication then gives the composition law of vector-parameters as

$$\langle \mathbf{c}_2, \mathbf{c}_1 \rangle = \frac{\mathbf{c}_2 + \mathbf{c}_1 + \mathbf{c}_2 \times \mathbf{c}_1}{1 - (\mathbf{c}_2, \mathbf{c}_1)}, \qquad \mathcal{R}(\mathbf{c}_2)\mathcal{R}(\mathbf{c}_1) = \mathcal{R}(\langle \mathbf{c}_2, \mathbf{c}_1 \rangle) \tag{4}$$

and in the case of three rotations $\mathbf{c} = \langle \mathbf{c}_3, \mathbf{c}_2, \mathbf{c}_1 \rangle$ we have

$$\mathbf{c} = \frac{\mathbf{c}_3 + \mathbf{c}_2 + \mathbf{c}_1 + \mathbf{c}_3 \times \mathbf{c}_2 + \mathbf{c}_3 \times \mathbf{c}_1 + \mathbf{c}_2 \times \mathbf{c}_1 + (\mathbf{c}_3 \times \mathbf{c}_2) \times \mathbf{c}_1 - (\mathbf{c}_3, \mathbf{c}_2)\mathbf{c}_1}{1 - (\mathbf{c}_3, \mathbf{c}_2) - (\mathbf{c}_3, \mathbf{c}_1) - (\mathbf{c}_2, \mathbf{c}_1) + (\mathbf{c}_3, \mathbf{c}_2, \mathbf{c}_1)}. \tag{5}$$

It is not difficult to see that the operation is associative and constitutes a representation of $SO(3)$, since the identity and inverse elements are also well-defined by $\langle \mathbf{c}, 0 \rangle = \langle 0, \mathbf{c} \rangle = \mathbf{c}$, $\langle \mathbf{c}, -\mathbf{c} \rangle = 0$. Among the advantages of this representation are more economical calculations, rational expressions for the matrix entries of $\mathcal{R}(\mathbf{c})$ and a correct description of the topology of $SO(3) \cong \mathbb{RP}^3$. For applications to rigid body mechanics, we refer to [6,9]. The reader can consult also [10] for the inverse kinematics aspects and [1] for the singular axes constellations.

As for the generalized *Euler* decompositions, we start with the much simpler two axes setting $\mathcal{R}(\mathbf{c}) = \mathcal{R}(\mathbf{c}_2) \mathcal{R}(\mathbf{c}_1)$, where $\mathbf{c}_k = \tau_k \hat{\mathbf{c}}_k$ and $\mathbf{c} = \tau \mathbf{n}$ ($\hat{\mathbf{c}}_k^2 = \mathbf{n}^2 = 1$) are the corresponding vector-parameters. We also denote $(\hat{\mathbf{c}}_j, \mathcal{R}(\mathbf{c})\,\hat{\mathbf{c}}_k) = r_{jk}$ and $(\hat{\mathbf{c}}_j, \hat{\mathbf{c}}_k) = g_{jk}$. Taking an appropriate scalar product provides the necessary and sufficient condition for the existence of the above decomposition in the form $r_{21} = g_{21}$. Next, multiplying $\mathbf{c} = \langle \mathbf{c}_2, \mathbf{c}_1 \rangle$ on the left with $\mathbf{n}^\times$ and projecting along $\hat{\mathbf{c}}_1$ and $\hat{\mathbf{c}}_2$ respectively, we obtain

$$\tau_1 = \frac{\tilde{v}_3}{g_{12}v_1 - v_2}, \qquad \tau_2 = \frac{\tilde{v}_3}{g_{12}v_2 - v_1} \qquad (6)$$

where we make use of the notations

$$v_k = (\hat{\mathbf{c}}_k, \mathbf{n}), \qquad \tilde{v}_1 = (\hat{\mathbf{c}}_2 \times \hat{\mathbf{c}}_3, \mathbf{n}), \qquad \tilde{v}_2 = (\hat{\mathbf{c}}_3 \times \hat{\mathbf{c}}_1, \mathbf{n}), \qquad \tilde{v}_3 = (\hat{\mathbf{c}}_1 \times \hat{\mathbf{c}}_2, \mathbf{n}) \cdot$$

Note that vanishing denominators in the above expressions are related to *half-turns*, i.e., rotations by a straight angle [4]. In particular, if $\mathbf{n} \perp \hat{\mathbf{c}}_{1,2}$ ($v_1 = v_2 = 0$), we have a decomposition into a pair of reflections, which is a well-known result in elementary geometry.

In the case of three axes $\mathcal{R}(\mathbf{c}) = \mathcal{R}(\mathbf{c}_3)\mathcal{R}(\mathbf{c}_2)\mathcal{R}(\mathbf{c}_1)$, such that $\hat{\mathbf{c}}_2$ cannot be parallel to $\hat{\mathbf{c}}_1$ or $\hat{\mathbf{c}}_3$, we use the scalar product $(\hat{\mathbf{c}}_3, \mathcal{R}(\mathbf{c})\,\hat{\mathbf{c}}_1) = (\hat{\mathbf{c}}_3, \mathcal{R}(\tau_2\hat{\mathbf{c}}_2)\,\hat{\mathbf{c}}_1)$ to obtain

$$(r_{31} + g_{31} - 2g_{12}g_{23})\,\tau_2^2 + 2\omega\,\tau_2 + r_{31} - g_{31} = 0, \qquad \omega = (\hat{\mathbf{c}}_1, \hat{\mathbf{c}}_2 \times \hat{\mathbf{c}}_3).$$

The above quadratic equation has real roots, given by

$$\tau_2^\pm = \frac{-\omega \pm \sqrt{\Delta}}{r_{31} + g_{31} - 2g_{12}g_{23}} \qquad (7)$$

as long as its discriminant is non-negative

$$\Delta = \begin{vmatrix} 1 & g_{12} & r_{31} \\ g_{21} & 1 & g_{23} \\ r_{31} & g_{32} & 1 \end{vmatrix} \geq 0 \qquad (8)$$

which plays the role of a necessary and sufficient condition for the existence of the decomposition. In order to find the remaining two scalar parameters, we use the composition

$$\mathbf{c}_1 = \langle -\mathbf{c}_2, -\mathbf{c}_3, \mathbf{c} \rangle, \qquad \mathbf{c}_2 = \langle -\mathbf{c}_3, \mathbf{c}, -\mathbf{c}_1 \rangle, \qquad \mathbf{c}_3 = \langle \mathbf{c}, -\mathbf{c}_1, -\mathbf{c}_2 \rangle. \qquad (9)$$

Namely, multiplying with $\hat{\mathbf{c}}_k^\times$ on the left and projecting over $\mathbf{n}$, we obtain the linear-fractional relations between τ_k, which yield the solutions for the general case in the form

$$\tau_1^\pm = \frac{g_{32} - r_{32}}{(g_{32} + r_{32})\tau v_1 - (g_{31} + r_{31})\tau v_2 + (r_{31} - g_{31})/\tau_2^\pm}$$

(10)

$$\tau_3^\pm = \frac{g_{21} - r_{21}}{(g_{21} + r_{21})\tau v_3 - (g_{31} + r_{31})\tau v_2 + (r_{31} - g_{31})/\tau_2^\pm}$$

while in the symmetric one we consider the limit $\tau \to \infty$ and thus obtain

$$\tau_1^\pm = \frac{g_{23} - v_2 v_3}{v_1 \tilde{v}_1 + v_2 \tilde{v}_2 + (v_1 v_3 - g_{13})/\tau_2^\pm}, \qquad \tau_3^\pm = \frac{g_{12} - v_1 v_2}{v_2 \tilde{v}_2 + v_3 \tilde{v}_3 + (v_1 v_3 - g_{13})/\tau_2^\pm}.$$

In the three axes setting we may also have degenerate solutions, related to a singularity of the map $\mathbb{RP}^3 \to \mathbb{T}^3$, known as *gimbal lock*, which is given by the condition

$$\hat{\mathbf{c}}_3 = \pm \mathcal{R}(\mathbf{c})\,\hat{\mathbf{c}}_1.$$

(11)

In that case the parameters τ_1 and τ_3 cannot be determined independently. Instead, we have the effective two-axes decomposition $\mathcal{R}(\mathbf{c}) = \mathcal{R}(\tau_2 \hat{\mathbf{c}}_2)\mathcal{R}(\tilde{\tau}_1 \hat{\mathbf{c}}_1)$, where the solutions

$$\tilde{\tau}_1 = \frac{\tau_1 \pm \tau_3}{1 \mp \tau_1 \tau_3} = \frac{\tilde{v}_3}{g_{12} v_1 - v_2}, \qquad \tau_2 = \frac{\tilde{v}_3}{g_{12} v_2 - v_1}$$

(12)

form a one-parameter set, expressed in terms of the generalized *Euler* angles as

$$\varphi_1 \pm \varphi_3 = 2 \arctan\left(\frac{\tilde{v}_3}{g_{12} v_1 - v_2}\right), \qquad \varphi_2 = 2 \arctan\left(\frac{\tilde{v}_3}{g_{12} v_2 - v_1}\right).$$

3   Covariant Form of the Solutions

First, we consider the simpler case of two axes $\mathbf{c} = \langle \tau_2 \hat{\mathbf{c}}_2, \tau_1 \hat{\mathbf{c}}_1 \rangle$, in which it is necessary to complete the basis with a third vector

$$\mathbf{c} = \xi_1\, \hat{\mathbf{c}}_1 + \xi_2\, \hat{\mathbf{c}}_2 + \xi_3\, \hat{\mathbf{c}}_1 \times \hat{\mathbf{c}}_2.$$

(13)

If we denote the adjoint matrix of g with γ, we have $|\,\hat{\mathbf{c}}_1 \times \hat{\mathbf{c}}_2\,|^2 = 1 - g_{12}^2 = \gamma^{33}$. Note that in formula (6) we use the covariant components of $\mathbf{c}$ in the same basis

$$\tau v_1 = \xi_1 + g_{12}\xi_2, \qquad \tau v_2 = \xi_2 + g_{12}\xi_1, \qquad \tau \tilde{v}_3 = \gamma^{33}\xi_3.$$

(14)

Thus, by direct substitution, we obtain the decomposability condition $r_{21} = g_{21}$ as

$$\xi_1 \xi_2 + (1 - g_{12}\xi_3)\xi_3 = 0$$

(15)

and the solutions themselves are given by the expressions

$$\tau_1 = -\xi_3/\xi_2, \qquad \tau_2 = -\xi_3/\xi_1. \tag{16}$$

One peculiar symmetry becomes immediately apparent from the above formula, namely $\tau_1\xi_2 - \tau_2\xi_1 = 0$.

In the three axes setting we consider first the case, in which $\{\hat{c}_k\}$ constitutes a basis

$$\mathbf{c} = \xi_1\,\hat{c}_1 + \xi_2\,\hat{c}_2 + \xi_3\,\hat{c}_3 = \langle \tau_3\hat{c}_3, \tau_2\hat{c}_2, \tau_1\hat{c}_1 \rangle.$$

By substituting the matrix entries r_{ij}, calculated according to (3) in the solutions (7), (10) and using the inverse metric tensor $g^{-1} = \omega^{-2}\gamma$ for lifting the indices of $\mathbf{c}$ we obtain

$$\tau_2^{\pm} = \frac{-\omega \pm \sqrt{\omega^2 - \sigma^2 + 2\gamma^{13}\sigma}}{\sigma - 2\gamma^{13}}$$

$$\tag{17}$$

$$\sigma = 2\frac{\gamma^{13}\xi_2^2 - \gamma^{23}\xi_1\xi_2 - \gamma^{12}\xi_2\xi_3 + \gamma^{22}\xi_1\xi_3 - \omega\xi_2}{\xi_1^2 + \xi_2^2 + \xi_3^2 + 2g_{12}\xi_1\xi_2 + 2g_{23}\xi_2\xi_3 + 2g_{13}\xi_1\xi_3}$$

for the middle parameter, and respectively for the other two

$$\tau_1^{\pm} = \frac{\gamma^{13}\xi_1\xi_2 + \gamma^{12}\xi_1\xi_3 - \gamma^{11}\xi_2\xi_3 - \gamma^{23}\xi_1^2 - \omega\xi_1}{\omega\,(\xi_1^2 + \xi_2^2 + 2g_{12}\xi_1\xi_2 + g_{13}\xi_1\xi_3 + g_{23}\xi_2\xi_3) - \gamma^{23}\xi_1 + \gamma^{13}\xi_2 + \kappa_2/\tau_2^{\pm}}$$

$$\tag{18}$$

$$\tau_3^{\pm} = \frac{\gamma^{13}\xi_2\xi_3 + \gamma^{23}\xi_1\xi_3 - \gamma^{33}\xi_1\xi_2 - \gamma^{12}\xi_3^2 - \omega\xi_3}{\omega\,(\xi_2^2 + \xi_3^2 + g_{12}\xi_1\xi_2 + g_{13}\xi_1\xi_3 + 2g_{23}\xi_2\xi_3) + \gamma^{13}\xi_2 - \gamma^{12}\xi_3 + \kappa_2/\tau_2^{\pm}}.$$

Here we have used the notation $\kappa_2 = \gamma^{13}\xi_2^2 - \gamma^{23}\xi_1\xi_2 - \gamma^{12}\xi_2\xi_3 + \gamma^{22}\xi_1\xi_3 - \omega\xi_2$. In the case $\omega = 0$ we use expansion in the basis (13) and the explicit relations (14) between the covariant and contravariant components of $\mathbf{c}$ in order to obtain

$$\tau_2^{\pm} = \pm\sqrt{\frac{\mathring{\sigma}}{2\gamma^{13} - \mathring{\sigma}}}, \qquad \mathring{\sigma} = 2\frac{\gamma^{13}\xi_2^2 - \gamma^{23}(\xi_1\xi_2 + \xi_3) - g_{13}\gamma^{33}\xi_3^2}{1 + \xi_1^2 + \xi_2^2 + \gamma^{33}\xi_3^2 + 2g_{12}\xi_1\xi_2}. \tag{19}$$

Denoting $\mathring{\kappa}_2 = \gamma^{13}\xi_2^2 - \gamma^{23}(\xi_1\xi_2 + \xi_3) - g_{13}\gamma^{33}\xi_3^2$, we have for $\tau_{1,3}$ the expressions

$$\tau_1^{\pm} = \frac{\gamma^{13}(\xi_1\xi_2 - \xi_3) - \gamma^{23}\xi_1^2 + g_{23}\gamma^{33}\xi_3^2}{(\gamma^{13} + g_{12}\gamma^{23})\xi_1\xi_3 + (\gamma^{23} + g_{12}\gamma^{13})\xi_2\xi_3 + \gamma^{13}\xi_2 - \gamma^{23}\xi_1 + \mathring{\kappa}_2/\tau_2^{\pm}}$$

$$\tag{20}$$

$$\tau_3^{\pm} = \frac{g_{12}\xi_3^2 - \gamma^{33}(\xi_1\xi_2 + \xi_3)}{(g_{12}\gamma^{23} + g_{13}\gamma^{33})\xi_1\xi_3 + (\gamma^{23} + g_{23}\gamma^{33})\xi_2\xi_3 + \gamma^{13}\xi_2 + \mathring{\kappa}_2/\tau_2^{\pm}}.$$

If the compound rotation is symmetric, i.e., $\varphi = \pi$ and $\mathcal{R}(\mathbf{c}) = \mathcal{O}(\mathbf{n}) = 2\mathbf{n}\otimes\mathbf{n}^t - \mathcal{I}$, we consider the limit $\tau \to \infty$ in the solutions after substitution of the coordinates

ξ_k with the contravariant components η_k in the expansion of the unit vector $\mathbf{n}$ ($\xi_k = \tau \eta_k$) and dropping all linear and constant terms in the so obtained expressions. For example, in the case $\omega = 0$ we have

$$\tau_1^{\pm} = \frac{\gamma^{13}\eta_1\eta_2 - \gamma^{23}\eta_1^2 + g_{23}\gamma^{33}\eta_3^2}{(\gamma^{13} + g_{12}\gamma^{23})\eta_1\eta_3 + (\gamma^{23} + g_{12}\gamma^{13})\eta_2\eta_3 + m}$$

$$\tag{21}$$

$$\tau_3^{\pm} = \frac{g_{12}\eta_3^2 - \gamma^{33}\eta_1\eta_2}{(g_{12}\gamma^{23} + g_{13}\gamma^{33})\eta_1\eta_3 + (\gamma^{23} + g_{23}\gamma^{33})\eta_2\eta_3 + m}$$

where

$$m = (\gamma^{13}\eta_2^2 - \gamma^{23}\eta_1\eta_2 - g_{13}\gamma^{33}\eta_3^2)/\tau_2^{\pm} \tag{22}$$

$$\tau_2^{\pm} = \pm\sqrt{\frac{\mathring{\sigma}}{2\gamma^{13} - \mathring{\sigma}}}, \qquad \mathring{\sigma} = 2\frac{\gamma^{13}\eta_2^2 - \gamma^{23}\eta_1\eta_2 - g_{13}\gamma^{33}\eta_3^2}{\eta_1^2 + \eta_2^2 + \gamma^{33}\eta_3^2 + 2g_{12}\eta_1\eta_2}. \tag{23}$$

The case when $\omega \neq 0$ and the decomposition with respect to two given axes are treated similarly.

As for the degenerate case (11), if $\omega = 0$ we may use the result obtained in the two axes setting combined with (12) in order to express

$$\tilde{\tau}_1 = -\xi_3/\xi_2, \qquad \tau_2 = -\xi_3/\xi_1. \tag{24}$$

If $\omega \neq 0$ on the other hand, the solutions are given by

$$\tilde{\tau}_1 = \frac{\tau_1 \pm \tau_3}{1 \mp \tau_1\tau_3} = \frac{\omega\,\xi_3}{\gamma^{23}\xi_3 - \gamma^{33}\xi_1}, \qquad \tau_2 = \frac{\omega\,\xi_3}{\gamma^{13}\xi_3 - \gamma^{33}\xi_1}. \tag{25}$$

In both cases we may use η_k instead of ξ_k so that the expressions are valid when $\tau \to \infty$.

If we need to express ξ_k on the other hand, it is straightforward to use the composition law (5) and then take the correct scalar products. Thus, in the case $\omega \neq 0$ we obtain

$$\xi_1 = \frac{(1 - g_{23}\tau_2\tau_3)\tau_1 + \omega^{-1}(\gamma^{12}\tau_1\tau_3 - \gamma^{13}\tau_1\tau_2 - \gamma^{11}\tau_2\tau_3)}{1 - g_{12}\tau_1\tau_2 - g_{13}\tau_1\tau_3 - g_{23}\tau_2\tau_3 + \omega\tau_1\tau_2\tau_3}$$

$$\xi_2 = \frac{(1 + g_{13}\tau_1\tau_3)\tau_2 + \omega^{-1}(\gamma^{22}\tau_1\tau_3 - \gamma^{12}\tau_2\tau_3 - \gamma^{23}\tau_1\tau_2)}{1 - g_{12}\tau_1\tau_2 - g_{13}\tau_1\tau_3 - g_{23}\tau_2\tau_3 + \omega\tau_1\tau_2\tau_3} \tag{26}$$

$$\xi_3 = \frac{(1 - g_{12}\tau_1\tau_2)\tau_3 + \omega^{-1}(\gamma^{23}\tau_1\tau_3 - \gamma^{13}\tau_2\tau_3 - \gamma^{33}\tau_1\tau_2)}{1 - g_{12}\tau_1\tau_2 - g_{13}\tau_1\tau_3 - g_{23}\tau_2\tau_3 + \omega\tau_1\tau_2\tau_3}.$$

For $\omega = 0$ the corresponding result is

$$\mathring{\xi}_1 = \frac{(1 - g_{23}\tau_2\tau_3)\tau_1 - \gamma^{13}(1 - g_{12}\tau_1\tau_2)\tau_3/\gamma^{33}}{1 - g_{12}\tau_1\tau_2 - g_{13}\tau_1\tau_3 - g_{23}\tau_2\tau_3}$$

$$\mathring{\xi}_2 = \frac{(1+g_{13}\tau_1\tau_3)\tau_2 - \gamma^{23}(1-g_{12}\tau_1\tau_2)\tau_3/\gamma^{33}}{1 - g_{12}\tau_1\tau_2 - g_{13}\tau_1\tau_3 - g_{23}\tau_2\tau_3}$$

$$\mathring{\xi}_3 = \frac{\gamma^{23}\tau_1\tau_3/\gamma^{33} - \gamma^{13}\tau_2\tau_3/\gamma^{33} - \tau_1\tau_2}{1 - g_{12}\tau_1\tau_2 - g_{13}\tau_1\tau_3 - g_{23}\tau_2\tau_3}.$$

(27)

Likewise, in the case of two axes we have a linear system for $\xi_{1,2}$ with solutions, given by the formulas

$$\xi_1 = \frac{\tau_1}{1 - g_{12}\tau_1\tau_2}, \qquad \xi_2 = \frac{\tau_2}{1 - g_{12}\tau_1\tau_2}.$$

(28)

Since the above expressions are rational in terms of the parameters τ_j, if any of these diverges, i.e., $\varphi_k = \pi$, we can still obtain the correct formulae, applying *l'Hôpital's* rule.

Similar expressions hold for the relations between the scalar parameters and the covariant components of **c** in the corresponding basis. However, these are almost straightforward to write considering the results obtained in [2,3]. Another possible generalization involves the hyperbolic case, i.e., the three-dimensional *Lorentz* group $SO(2,1)$, which can be treated in an analogous way. Some of the advantages of this new representation for the numerous applications of the generalized Euler decomposition (cf. [6,8,9]) are quite obvious. The explicit dependence only on the contravariant components allows, apart from purely geometric considerations, for straightforward differentiation, as well as for obtaining the decomposition in a different basis from one that has been given.

4 Conclusion

The well known theoretical and practical problem of finding two or three rotational matrices such that their composition is equivalent to a given rotation is extremely difficult in the particular cases when the chosen axes of the rotations are not orthogonal. Solution of the problem is quite sensitive on the chosen parameterization of the rotation group. The present paper is a part of a long-standing research program about rotation decompositions and their applications. Here we present an alternative representation for the generalized *Euler* decomposition (with respect to arbitrary axes) by means of vector parameterization of the *Lie* group $SO(3)$. The advantages of this parameterization turn out to be indispensable in such situations.

References

1. Bongardt, B.: Geometric Characterization of the Workspace of Non-Orthogonal Rotation Axes. J. Geom. Mechanics 6, 141–166 (2014)
2. Brezov, D., Mladenova, C., Mladenov, I.: Vector Decompositions of Rotations. J. Geom. Symmetry Phys. 28, 67–103 (2012)

3. Brezov, D., Mladenova, C., Mladenov, I.: Some New Results on Three-Dimensional Rotations and Pseudo-Rotations. In: AIP Conf. Proc., vol. 1561, pp. 275–288 (2013)
4. Brezov, D., Mladenova, C., Mladenov, I.: Quarter Turns and New Factorizations of Rotations. Comptes Rendus de l'Académie Bulgare des Sciences 66, 1105–1114 (2013)
5. Davenport, P.: Rotations About Nonorthogonal Axes. AIAA Journal 11, 853–857 (1973)
6. Mladenova, C.: Group Theory in the Problems of Modeling and Control of Multi-Body Systems. J. Geom. Symmetry Phys. 8, 17–121 (2006)
7. Mladenova, C., Mladenov, I.: Vector Decomposition of Finite Rotations. Rep. Math. Phys. 68, 107–117 (2011)
8. Kuipers, J.: Quaternions and Rotation Sequences. Geometry, Integrability and Quantization 1, 127–143 (2000)
9. Piña, E.: Rotations with Rodrigues' Vector. Eur. J. Phys. 32, 1171–1178 (2011)
10. Rull, A., Thomas, F.: On Generalized Euler Angles. Mechanisms and Machine Science 24, 61–68 (2015)

Structure Performance Improvement of Parallel Mechanisms by Optimizing Topological Configurations

Jiaming Deng, Huiping Shen[*], Ju Li, Xiaomeng Ma, and Qingmei Meng

School of Mechanical Engineering,
Changzhou University
Changzhou 213016, China
{czdydjm,shp65}@126.com

Abstract. This paper deals with structure performance improvement of parallel mechanisms by optimizing topological configurations based coupling-reducing principle. Taking three-translation parallel mechanism 3-$R//R//C$ as the example, a coupling-reducing configuration is designed by only optimizing its branch chain architecture between the moving platform and the static platform without changing the structure of the branch chain itself. This leads that the coupling degree of the configuration is reduced from $k = 1$ to $k = 0$. So it is easy to get its forward position kinematics analytical solution. After comparing and analyzing input-output decoupling characteristic, the singularity, workspace, dexterity of the optimization configuration with that of typical configuration, it is found that these performance are obviously better than the typical configuration before optimization. This paper provides an effective method of topology configurations optimization design.

Keywords: Parallel mechanism, Coupling degree, Configuration, Structure Performance.

1 Introduction

Topological structure of the spatial parallel mechanism is not only complex but also has its own particularity. It includes two meanings, (1) The topological structure of the branched chain itself, i.e., joint type of the branched chains, the connection relation between the links and dimensional constraint which refers to the constraint type of the relative orientation between the link and joint axis, such as parallel, coaxial, concurrent, coplanar, vertical and arbitrary cross and so on, in all six basic types[1]; (2) The topological architecture of the branched chain relative to the moving and the fix platform, i.e., the geometrical arrangement relation of the branched chain

---

[*] Corresponding author.

X. Zhang et al. (Eds.): ICIRA 2014, Part II, LNAI 8918, pp. 512–521, 2014.

between the moving platform and the fix platform. Because of the above two reasons, there are a lot of "configurations" for a parallel mechanism. At the same time, the topological structure of parallel mechanism exercises a great influence on the structure, kinematic and dynamic performance. Once the mechanism is designed and assembled, the mechanism topology structure features full-cycle invariance (excluding singular positions), which has nothing to do with the motion positions.

On the one hand, the existing main theories and methods for topological structure design of the parallel mechanism, e.g., screw theory based method[2], displacement subgroup based method[3], POC set based method[4, 5,6], are all focused on the topological structure design of the branched chain itself. In general, one or more topological structures can be gotten by structure synthesis of parallel mechanisms based on the output motion type or specific function. Then get its derivative mechanism by changing the topological structure of the branched chain itself (e.g., joint type and number, the connection relation between the links). The kinematic and dynamic performance analysis for a given topological structure are basic work for product design, which are studied by many scholars, involving the input-output (i.e.,I-O)decoupling [7], the singular analysis [8], performance judgment analysis and the global performance index of acceleration [9,10], dexterity and isotropic [11], etc. However, there is less research on the topological architecture of the branched chain relative to the moving and the fix platform.

On the other hand, these is also less research on the influence of the topological structure of the parallel mechanism, especially the topology structure optimization on its performance (including structure, kinematics and dynamics performance), but which is of great importance to the structure synthesis, the dimension synthesis[12].Obviously, a parallel mechanism which has the same branched chain structure may has different topological structure configurations, performance difference of which is certainly larger. Revealing such mapping regularity has more practical value on the mechanism performance improvement and optimization.

This paper deals with structure performance improvement issues of parallel mechanisms by optimizing topological configurations based coupling-reducing principle. First, taking the typical 3-DOF translational parallel mechanism 3-R//R//C as an example, we design a special configurations without by changing its topology structure of branched chain itself, only by optimizing its topology architecture between the moving and the fix platform, the results of which is that the coupling degree of the configuration is reduced from $k = 1$ to $k = 0$ .We called it as coupling-reducing configuration for short or optimization configuration. Thus, it is easy to get its analytical forward position kinematics solution. Then after analyzing the I-O decoupling, singularity, workspace, dexterity of the coupling-reducing configuration, we found performance of the coupling-reducing configuration is obviously improved and better than the typical configuration before optimization.

2   The Structure Coupling-Reducing Principle of Parallel Mechanisms

2.1   Definition of Mechanism Coupling Degree

Mechanism coupling degree k ($k{\geq}0$) which reflects the complexity of mechanism topological structure refers to the coupling complexity between each loop position variables of basic kinematic chain (BKC) of mechanism[13]. It is proved that the bigger the k is, the stronger and higher the mechanism coupling and complexity will be; and the analytical forward position solution can be got directly only when $k{=}0$; If $k{>}0$, the analytical forward position solution can not be obtained directly and have to be solved by simultaneous position equations solution of multiple loops, while k is the minimum dimension of simultaneous equations. For sake of system, calculation formula of coupling degree is given as follows[1,4,13]:

(1) Constraint Degree of Single Opened Chain

The constraint degree of a single opened chain (SOC) is defined as

$$\Delta_j = \sum_{i=1}^{m_j} f_i - I_j - \xi_{L_j} = \begin{cases} \Delta_j^- = -5,-4,-3,-2,-1. \\ \Delta_j^0 = 0 \\ \Delta_j^+ = +1,+2,+3,\cdots \end{cases} \tag{1}$$

Where, m_j is the number of motion pairs of first SOC_j; f_i is the degree of freedom of first i joint (except local degrees of freedom); I_j is the number of drive pairs of first SOC_j; ξ_{L_j} is the number of independent displacement equation of first j independent loop. For a BKC, there is

$$\sum_{j=1}^{v} \Delta_j = 0 \tag{2}$$

(2) Coupling Degree of Basic Kinematic Chain

The definition of coupling degree of BKC is defined as

$$k = \frac{1}{2}\min\{\sum_{j=1}^{v} |\Delta_j|\} \tag{3}$$

Where, $\min.\{\bullet\}$ is the of $SOC(\Delta_j)$ whose is minimum when BKC split into $SOC(\Delta_j)$.

The number of mechanism BKC and value of coupling degree can be found from the above equation, and mechanism coupling degree k is the maximum among BKC coupling degree, i.e.

$$k = \max\left\{k_1, k_2, \cdots, k_i, \cdots\right\} \tag{4}$$

Where, v is the number of independent loops; k_i is the first i coupling degree of BKC.

2.2   Structure Coupling-Reducing Principle and Coupling-Reducing Design Method

Forward position solution is one of the most important and basic problems in analysis and design of parallel mechanisms. By Section 2.1, we know that mechanism coupling degree value should be reduced to make sure forward position solution is easily solved, which is called as mechanism structure coupling-reducing principle because its value reflects the ability to describe the complexity of mechanism structure.

According to the formula (1), (3) and (4) ,we find the value of coupling degree k is completely depends on the constraint degree Δ value of first independent loop. Therefore, in order to realize structure coupling-reducing which means reducing the coupling degree k, the constraint degree Δ value of first loop of the mechanism should be reduced. Further, we will propose two corresponding coupling-reducing design methods based on reducing the number of degree of freedom of joints in the first loop as follows.

① Paralleling and coinciding the axis of the joints(including joints with multiple degree of freedom, such as, spherical pair (S), cylindrical pair(C), universal pair(T) axis) to reduce the number of the degree of freedom of joints in the first loop, which can turn the partial degree of freedom into the negative degree of freedom of first loop that should be calculated in the second loop not in the first loop. This is known as the reducing-coupling design method based on the shift of negative freedom degree;

② Combining joints or reducing the number of edges of the moving platform to reduce the number of degree of freedom of joints in the first loop.This is known as the reducing-coupling design method based on joints combining.

3   Typical Configuration of Three-Translation Parallel Mechanism

Typical configuration of three-translation 3-R//R//C parallel mechanism as shown in Fig.1 [4,14,15]. It consists of the triangular moving platform and the fix platform connected by three same structure branched chains, whose joint axis are all parallel to each other, known as $SOC\{-R_{i1}//R_{i2}//C_{i3}-\}, i=1,2,3$ (R is represent revolute pair and C represent cylindrical pair). Moreover, the joint axis on the moving platform and the fix platform is coincided with each edge of the triangle platform respectively.

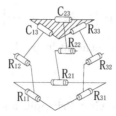

Fig. 1. Typical configurations of three-translation parallel mechanism

3.1   The Degrees of Freedom and the Output Motion Type

Select any point on the moving platform as the basis point, thereupon

1)•Determine POC set of the branched end link

$$M_{b_i} = \begin{bmatrix} t^3 \\ r^1 \left(// R_{i1} \right) \end{bmatrix}, i = 1, 2, 3$$

2) The three branched chains are arranged between the moving and the fixed platforms triangularly, therefore, any two branched chains can constitute the first loop, and the number of independent displacement equation ξ_{L_1} is

$$\xi_{L_1} = \dim.(M_{b_1} \cup M_{b_2}) = \dim.(\begin{bmatrix} t^3 \\ r^1 (// R_{11}) \end{bmatrix} \cup \begin{bmatrix} t^3 \\ r^1 (// R_{21}) \end{bmatrix}) = \dim.(\begin{bmatrix} t^3 \\ r^2 (// (R_{11}, R_{21})) \end{bmatrix}) = 5$$

$$F_{(1 \sim 2)} = \sum_{i=1}^{m} f_i - \sum_{j=1}^{1} \xi_{L_j} = 8 - 5 = 3$$

$$M_{pa(1 \sim 2)} = \begin{bmatrix} t^3 \\ r^1 (// R_{11}) \end{bmatrix} \cap \begin{bmatrix} t^3 \\ r^1 (// R_{21}) \end{bmatrix} = \begin{bmatrix} t^3 \\ r^0 \end{bmatrix}$$

3) Determine the number of independent displacement equation ξ_{L_2} of second independent loop

$$\xi_{L_2} = \dim.(M_{pa(1 \sim 2)} \cup M_{b_3}) = \dim.(\begin{bmatrix} t^3 \\ r^0 \end{bmatrix} \cup \begin{bmatrix} t^3 \\ r^1 (// R_{31}) \end{bmatrix}) = \dim.(\begin{bmatrix} t^3 \\ r^1 (// R_{31}) \end{bmatrix}) = 4$$

4) Determine degree of freedom of the mechanism

$$F = \sum_{i=1}^{m} f_i - \sum_{j=1}^{2} \xi_{L_j} = 12 - (5 + 4) = 3$$

5) Determine POC set of the moving platform

$$M_{pa} = M_{pa(1\sim2)} \cap M_{b_3} = \begin{bmatrix} t^3 \\ r^0 \end{bmatrix} \cap \begin{bmatrix} t^3 \\ r^1(//R_{31}) \end{bmatrix} = \begin{bmatrix} t^3 \\ r^0 \end{bmatrix}$$

Therefore, when R_{11}, R_{21}, R_{31} in the fix platform are chosen as active joint, the moving platform can achieve three-translation output.

3.2   The Calculation of Coupling Degree k

1) Determine the first single opened chain SOC_1 and its constrained degree Δ_1

$$SOC\{-R_{11}//R_{12}//C_{13} - C_{23}//R_{22}//R_{21}-\}$$

$$\Delta_1 = \sum_{i=1}^{m_1} f_i - I_1 - \xi_{L_1} = 8 - 2 - 5 = +1$$

2) Determine SOC_2 and its constrained degree Δ_2

$$SOC\{-R_{31}//R_{32}//C_{33}-\}$$

$$\Delta_2 = \sum_{i=1}^{m_2} f_i - I_2 - \xi_{L_2} = 4 - 1 - 4 = -1$$

3) Determine BKC and coupling degree κ of mechanism

Therefore, there is only one BKC in this mechanism, and the coupling degree is

$$\kappa = \frac{1}{2}\sum_{j=1}^{v} |\Delta_j| = \frac{1}{2}(|+1| + |-1|) = 1$$

We analyze and find that 3-CRR mechanism [16] 3-PRRR mechanism [17] and Isoglide3-T3 parallel mechanism (i.e. 3-PRRPR or 3-CRC [18]) have the same coupling degree, i.e., $k=1$. Therefore, the numerical forward position solution can be obtained by one-dimensional searching method.

4   Optimization of Three-Translation Parallel Mechanism Configuration

4.1   Optimization Design of Topological Configurations

According to the reducing-coupling design method based on the shift of negative freedom degree in Section 2.2, the number of degree of freedom of joints in the first loop can be reduced by making the revolute joint (or cylindrical joint) axis in any a pair of branched chains are parallel or coincident to each other, and this two branched chains constitute the first loop. Therefore, a coupling reducing configurations is designed this way.

Make the revolute joint R(or cylindrical joint C) axis in the fix platform(or the moving platform) of the first and third branched chains terminal parallel to each other, as shown in Fig.2; Obviously, the POC set of each branched chains end link are invariant.

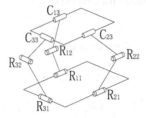

Fig. 2. The coupling-reducing configurations of 3-R//R//C three-translation parallel mechanism

4.2   Structural Analysis of the Optimization Configuration

Here, we analyze only the topological structure of coupling-reducing configuration showed in Fig.2 as follows.

(1) Calculation of Degree of Freedom

1) The first loop consists of the first and third branched chains, whose rotation joint R axis are parallel to each other, and the number of independent displacement equation ξ_{L_1} is

$$\xi_{L_1} = \dim.(M_{b_1} \cup M_{b_2}) = \dim.\left(\begin{bmatrix} t^3 \\ r^1(//R_{11}) \end{bmatrix} \cup \begin{bmatrix} t^3 \\ r^1(//R_{21}) \end{bmatrix}\right) = \dim.\left(\begin{bmatrix} t^3 \\ r^1(//R_{11}) \end{bmatrix}\right) = 4$$

$$F_{(1-2)} = \sum_{i=1}^{m} f_i - \sum_{j=1}^{1} \xi_{L_j} = 8 - 4 = 4$$

$$M_{pa(1-2)} = \begin{bmatrix} t^3 \\ r^1(//R_{11}) \end{bmatrix} \cap \begin{bmatrix} t^3 \\ r^1(//R_{21}) \end{bmatrix} = \begin{bmatrix} t^3 \\ r^1(//R_{11}) \end{bmatrix}$$

2) Determine the number of independent displacement equation ξ_{L_2} of second independent loop

$$\xi_{L_2} = \dim.(M_{pa(1-2)} \cup M_{b_3}) = \dim.\left(\begin{bmatrix} t^3 \\ r^1(//R_{11}) \end{bmatrix} \cup \begin{bmatrix} t^3 \\ r^1(//R_{31}) \end{bmatrix}\right) = \dim.\left(\begin{bmatrix} t^3 \\ r^2(//\lozenge(R_{11}, R_{31})) \end{bmatrix}\right) = 5$$

3) Determine degree of freedom of the mechanism

$$F = \sum_{i=1}^{m} f_i - \sum_{j=1}^{2} \xi_{L_j} = 12 - (4+5) = 3$$

4) Determine POC set of the moving platform

$$M_{pa} = M_{pa(1\sim2)} \cap M_{b_3} = \begin{bmatrix} t^3 \\ r^1(//R_{11}) \end{bmatrix} \cap \begin{bmatrix} t^3 \\ r^1(//R_{31}) \end{bmatrix} = \begin{bmatrix} t^3 \\ r^0 \end{bmatrix}$$

(2) Calculation of Coupling Degree k

1) Determine SOC_1 and its constraint degree

$$SOC\{-R_{11}//R_{12}//C_{13} - C_{23}//R_{22}//R_{21} -\}$$

$$\Delta_1 = \sum_{i=1}^{m_1} f_i - I_1 - \xi_{L_1} = 6 - 2 - 4 = 0$$

Here, it is easy to judge that translational degree of freedom $P^{(C13)}$ along C_{13} axis and translational degree of freedom $P^{(C23)}$ along C_{23} axis is negative degree of freedom, so it should not be included in the first loop. But for SOC_2, they are the effective degrees of freedom, which should be included in.

2) Determine SOC_2 and the constraint degree Δ_2

$$SOC\{-P^{(C_{13})} - P^{(C_{23})} - R_{31}//R_{32}//C_{33} -\}$$

(The translational degrees of freedom, i.e., $P^{(C13)}$, $P^{(C23)}$ in SOC_1 has been included)

$$\Delta_2 = \sum_{i=1}^{m_2} f_i - I_2 - \xi_{L_2} = 6 - 1 - 5 = 0$$

3) Determine BKC and coupling degree k of mechanism

This mechanism has two BKC, and their coupling degree is

$$\kappa = \frac{1}{2} \sum_{j=1}^{v} |\Delta_j| = 0$$

Obviously, the coupling degree is reduced to zero. This shows that structural complexity of three configurations has been reduced to a minimum, and forward position solutions are very easily to obtain.

Literature[19] has analyzed and compared the other kinematic performance of typical configuration as shown in Figure 1 with that of the optimization configuration as shown in Figure 2, which includes mechanism I-O decoupling, singularity, workspace and dexterity for the two configurations. Further, it is found that kinematics performances of the optimization configuration are obviously better than the typical configuration before optimization. For example, strong I-O decoupling of the typical configuration is changed into partial I-O decoupling. So it is easier to control motion of the optimization configuration. For saving space, the details are omitted here and can be found in *Ref.*[19].

5   Conclusions

The issue on structure performance improvement of parallel mechanisms by optimizing topological configurations based coupling-reducing principle is studied in this paper. A coupling-reducing configurations of three-translation *3-R//R//C* parallel mechanisms is designed. Main conclusions can be drawn as follows:

(1) Parallel mechanisms which have the same topological structure of branch chain may have different branch chain architecture between the moving platform and the fix platform, i.e., different configurations. But the structural performance, kinematic and dynamic performance of these configurations are quite different.

(2) Optimizing the branch chain architecture between the moving and the fix platform can not only reduce the coupling degree from larger value to smaller value (for the example in this paper, i.e., from $k=1$ to $k=0$, and easier to obtain analytical forward position solution), but make change strong I-O decoupling into partial I-O decoupling, which make be so easier for kinematic control. Other kinematic and dynamic performance of optimization configurations are also obviously better than the typical configuration before optimization.

This paper expands the issue and direction of topological structure optimization of parallel mechanism and provides valuable reference for configuration optimization design of the other parallel mechanisms.

Acknowledgments. This project is supported by National Natural Science Foundation of China (Grant No.51375062, 51475050, 51405039).

References

[1] Yang, T.: Topology Structure Design of Robot Mechanisms. Machinery Industry Press, Beijing (2004)
[2] Zhen, H., et al.: Higher spatial mechanism. Higher Education Press, Beijing (2006)
[3] Herve, J.M.: Design of Parallel Manipulators via the Displacement Group. In: Proc. of the 9th World Cong. on Theory of Machine and Mechanisms, Milan, pp. 2079–2082 (1995)
[4] Yang, T., Liu, A., Luo, Y., Shen, H., et al.: Topology design of robot mechanism. Science Press, Beijing (2012)
[5] Yang, T.-L., Liu, A.-X., Shen, H.-P., et al.: On the Correctness and Strictness of the Position and Orientation Characteristic Equation for Topological Structure Design of Robot Mechanisms. ASME Journal of Mechanical Design 5, 021009-1 18 (2013)

[6] Gao, F., Yang, J., Ge, Q.: Parallel robot type synthesis of the GF set theory. Science Press, Beijing (2011)

[7] Wang, X., Dai, Y., Li, S.: Decoupling motion of general mechanism. Journal of National University of Defense Technology 24(2), 85–90 (2002)

[8] Ma, L., Yin, X., Yang, T.: Special configuration analysis model three translational parallel robot. Journal of Jiangsu University 23(2), 43–45 (2002)

[9] Liu, X.-J., Wang, J.: Parallel Kinematics- Type, Kinematics, and Deisgn. Springer (2014)

[10] Guo, X.: Research on dynamics theory of parallel robotic mechanism. Yan Shan University (2002)

[11] Tian, H., Jinsong, W.: Stewart parallel robot of local dexterity and isotropic conditions of analytic. Chinese Journal of Mechanical Engineering 35(5), 41–46 (1999)

[12] Melert, J.P.: Parallel Robots, 2nd edn. Springer (December 27, 2005)

[13] Yang, T.: The basic theory of mechanical system – mechanism, kinematics, dynamics. Mechanical Industry Press, Beijing (1996)

[14] Yin, H.: Research on topology and kinematics of the parallel mechanism. Changzhou University (2014)

[15] Tsai, L.W.: Multi-Degree-of-Freedom Mechanisms for Machine Tools and the Like. Patent No.5656905. U.S (1997)

[16] Kong, X.W., Gosselin, C.: A class of 3-DOF translational parallel manipulators with linear input-output equations. In: Proc. of the Workshop on Parallel Mechanisms and Manipulators, Quebec, Canada, October 3-4 (2002)

[17] Kim, H.S., Tsai, L.W.: Evaluation of a Cartesian parallel manipulator. In: Lenarcic, J., Thomas, F. (eds.) Advances in Robot Kinematics, pp. 21–28. Kluwer Academic Publishers, Dordrecht (2002)

[18] Gogu, G.: Structural Synthesis of Parallel Robotic Manipulators with Decoupled Motions. Report ROBEA MAX-CNRS (2003)

[19] Zhen, W.: Study on the relationship between structure and properties of the topology of parallel mechanism. Changzhou University (2013)

Towards a Dynamic Based Agents Architecture for Cellular Networks Optimisation: Cell Breathing

Dahi Zakaria Abd El Moiz and Mezioud Chaker

Modeling and Implementation of Complex Systems Laboratory
Dept. of New Technologies of Information and Communication
Constantine 2 University, Constantine City, Algeria
{zakaria.dahi,chaker.mezioud}@univ-constantine2.dz

Abstract. One of the major optimisation issues of today's most advanced cellular phone networks (2G, GPRS, EDGE, 3G, LTE, 4G) is the "Network Congestion" (NC). This issue is resulting generally from an unfair distribution of the traffic between antennas. Moreover, this problem causes the increasing of dropped calls, which is unacceptable regarding today's high standards of the mobile phone industry. To recover this traffic balance, an optimisation process is performed called "Load Balancing" (LB). Most of the works done in this area focus more on the efficiency of the optimisation techniques rather than the dynamic and automatic aspects of these last ones. Knowing that radio telephony networks are real life applications, makes that the real time and the dynamical resolving as much important as the efficiency of the techniques used. That's why, in this paper we have tackled for the first time the network congestion issue from a software engineering point of view. A high level modeling based on a multi-agent system and algerbra process Π-calculus is used in order to design self-adaptative, dynamical system that can respond and cope wih the network congestion issue.

Keywords: Cellular Phone Network, Dynamic Optimisation, Network Optimisation Problems, Multi Agent Systems, Π-calculus, Cell Breathing, Load Ballancing.

1 Introduction

During planning and managing a cellular phone system, engineers have to face many challenging optimisation problems [11]. In addition, cellular phone networks like other real life applications rely on perpetual cycles of *request⌣respond* tasks where costumers are constantly establishing, suspending, dropping (calls, text messages, internet services). Reason that made radio mobile networks subject to unpredictable variation of traffic. Therefore, the whole difficulty lies in the dynamic aspect of their development, which adds more complexity to the design task.

X. Zhang et al. (Eds.): ICIRA 2014, Part II, LNAI 8918, pp. 522–534, 2014.
© Springer International Publishing Switzerland 2014

In fact, any design process that can't handle or cope with this dynamic evolution of traffic may cause many issues. One of them is the "*Network Congestion*" (*NC*)[16]. Technically, this problem is caused by the fact that each antenna has a given amount of traffic that it can handle. Once this amount exceeded the upcoming calls to this antenna will be ignored. To cope with this issue, traffic has to be distributed fairly between cells. This balance recovery is known also as "*Load Balancing*" (*LB*)[17]. It is performed through a process called "*Cell Breathing*" (*CB*) which corresponds to a constant "*shrinking*" and "*expanding*" of the cell coverage of the saturated antenna and its neighbouring antennas.

The most recent theoretical works that have been done in this field [13],[16], [17],[18],[19], focused more on the efficiency of the resolving techniques rather than the dynamical and automatic aspects of the optimisation process. In fact, cellular phone networks are real life systems and open public which makes the need for reactivity, real time adjusting, dynamical evolution, robustness against system crashes as much important as the efficiency of the optimisation techniques themselves.

Therefore, the system must not only provide a solution (*efficiency of resolving*), but also react to failures (*reactivity*) and hazards (*adaptivity*) by adjusting the solution (*real time reacting*), or be able to deal with problems whose data are only partially known (*intelligence*), and where the information arrives at over time. Reason why, the whole optimisation process has to be seen as a system where each component ensure one of these requirements.

Based on these observations, in this paper we tackled theoretically from a software engineering point of view the *load balancing process*. Trying to design an application architecture that can handle efficiency, failures, reactivity and adaptivity required for an industrial mobile phone system.

The resolving of the network congestion does not rely only on the congested antenna but also implies its neighbouring antennas. In reality, a cell coverage "*shrinks*" or "*expands*" according to the load of traffic of all the antennas and not only one. So, the system has to be designed as a set of agents that can interact, cooperate, and exchange necessery data of all the cells of the network. Reason why, the need for involving a multi-agent system seemed promising.

The proposed system, is based on a set of agents of different types. Each agent is responsible of performing a particular task. Afterwards, the need for the integration of the concept of "*process mobility*" or "*interaction mobility*" seemed a promising solution to application failures. The choice was made for the Π-calculus process, for its ability to dynamically change the topology of applications.

2   Mobile Network

A typical radio mobile network is a set of several subsystems. Each subsystem is responsible of performing a particular task such radio transmission, network management, data storing and so on. All along these last two decades, several multi technological radio networks have been deployed (2G[1], GPRS[2] ,EDGE[3], LTE[4], 3G[5],4G[6]).

Each time a new generation of radio telephony network is implanted, the radio network witnesses three major enhancement of the previous technology. Firstly, the enhancement of the data flow. Secondly, updating the previous multiplexing[7] techniques by new adequate ones, or the commutation techniques. Thirdly, the use of a different bandwith (frequencies). Besides the enhancement of the electrical and electronical hardware from one generation to another to go with the mentioned technological enhancements.

On the other hand, despite all the differences between each generation and the other (2G, GPRS, EDGE, LTE, 4G), the radio transmission core especially antenna remains the same, with identical types of anetnnas, wave propagation models and parameters. So, any optimisation of the radio transmisson core will be totally workable (with slight diffrences) for any generation of the radio mobile network. Reason why, our work will mainly focus on the radio subsystem and especially antennas.

2.1 Base Station Tranceiver (BTS)

The BTS (Antenna) has as function the management of radio transmissions (modulation, demodulation, equalisation, coding and error correcting). It also manages the data link layer for the signal exchange between the mobile and the network operator's infrastructure. Actually, a BTS can handle a maximum of one hundred simultaneous calls (in 2G networks for example) [8], which corresponds to a capacity of 43 Erlang [8]. Technially to handle such amount of traffic a BTS has to be equiped with at least 7 TRX card [9] in 2G networks and a certain number of "Channel Element"[10] or "WBBP Board" [11] in 3G networks.

Each antenna has a specific type of coverage, and have specific characteristics. Generally the antenna is characterized by:

- Height (h_{ant}) : is the height of the antenna. It expresses the elevation of the antenna above the ground level.
- Transmission gain (G_{ant}) : relates the intensity of an antenna in a given direction to the intensity that would be produced by a hypothetical ideal antenna that radiates equally in all directions (isotropically) and has no losses.

---

[1] 2G : Second generation network or so-colled GSM.
[2] GPRS : General Packet Radio Service so-called 2.5G.
[3] EDGE : Enhanced Data Rates for GSM Evolution so-called 2.75G.
[4] LTE : Long Term Evolution Networks so-called 3.9G.
[5] 3G : Third generation network.
[6] 4G : Fourth generation network.
[7] M.T : Modulate and combine several given data on a shared medium.
[8] Erlang : Unit of measurement of traffic.
[9] TRX Card : Tranceiver-receiver, device responsible of traffic handling in 2G.
[10] CE : Channel element, device responsible of traffic handling in 3G.
[11] WBBP : WCDMA Base Band Process.

- Transmission loss (L_{ant}) : is often due to the electronics and materials that surround the antennas. These tend to absorb some of the radiated power (converting the energy to heat), which lowers the efficiency of the antenna.
- Transmission power (P) : expresses the amount of power needed for its operability.
- Vertical deviation (T_{ant}) : called also *tilt*, is defined as the angle between the main beam of the antenna and the horizontal plane [12] (see figure 1).
- Horizantal deviation (A_{ant}) : called also *azimuth*, is the projected vector from the antenna (origin) to a point of interest (mobile) perpendicularly onto a reference plane. The angle between the projected vector and a reference vector on the reference plane is called the azimuth (see figure 1).
- Propagation diagram : it determines how the antenna is irradiating the environnement vertically and horizantally with its wave signal. A_{ant} and T_{ant} determine two type of diagrams : Vertical Diagram ($VDIAG$) and Horizantal Diagram ($HDIAG$).

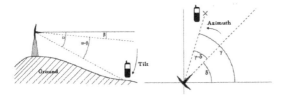

Fig. 1. Tilt and Azimuth

3 Multi Agent Systems

A growing number of researchers define an agent as a computer system located in an environment [5], where it is able to act in full autonomy over its actions in order to reach the objectives of its designing.

Artificial intelligence researchers also agree on the necessity of the existence of some characteristics so that we can speak of an agent. These are the most essential : the autonomy (*acts without outside intervention*), the interaction (*communicates with other agents*), the proactivity (*pursues a goal (goal oriented)*), the cooperation (*capable of coordinating with other agents to achieve a common goal*), the mobility (*able to move to another environment* [6]), the rationality (*capable of acting according to its internal goals and knowledge*), the intelligence (*aware of its reasoning : beliefs, goals, plans, assumptions are explicit*).

Generally agents communicate via interaction. The interaction is a mechanism that makes all existing agents in the system more dynamic, by highlighting the mechanisms of communication and cooperation [7]. By the means of interactions, all the entities immersed in an environment can interact in various ways : the interaction without communication (*based on the inference of the actions of other agents*), the interaction via communication (*based on a finite set of signals without interpretation or fixed syntax*), the interaction via sending messages and plans (*based on sending messages as method calls of the object-oriented languages, or sending plans*), and finally the interaction via the blackboard (*based on putting the information in a common memory space (shared) called "blackboard"*).

4 Π-Calculus

To Develop secured and reliable software, we must first formalize it well [12]. Which is why we use formal languages such the Π-calculus. This last one allows us to build a mathematical model to verify some properties expected from the software (security, non-blocking, quickness, dynamical evolution, robustness againt failure). Algebra processes are appropriate frameworks for the specification and verification of reactives systems. This area has experienced different approaches: CSP [13], CCS [14], Π-Calculus [1],[2].

In these algebras , any well-formed term denotes a process. The fundamental abstraction is that we are not interested in the process behavior only through a number of interactions points called "channels". Synchronization and communication are expressed by the laws of internal and external composition. Note that as well in CCS than in the Π-calculus, we make no assumptions about executions's speeds that are in different processes. So, these last ones are presumed to move at different speeds. Π-Calculus is a formal language for reasoning about distributed communicating systems. It allows describing a systems as set of mobile processes. It means, that the processes and communication links between processes can change their places at any time in the system. The use of Π-Calculus will introduce the concept of mobility, which was absent in CCS. The objective is to allow a dynamic reconfiguration of the topology of applications in case of dynamic evolution or failure of the network.

4.1   Polyadic Π-Calculus

The polyadic Π-Calculus allows the transmission of a tuple of values during an interaction, to benefit from an important property which is the parameterization, through the simultaneous emission of a series of values in a particular order to a process. Many software programs exceed a million lines of code. It is difficult to take into account such size with the old formal methods. By the use of some formal techniques such the Π-Calculus and more specifically the Π-Calculus polyadic, the abstraction of many details through parameterisation will reduce the system specification to an acceptable size. The syntax used is as follows:

$$P ::= O | \alpha P | P | P | P + P | [x + y] P | [x/ = y] P | \overrightarrow{(u, v)} P | \overrightarrow{A(u)} \tag{1}$$

Where the prefixes α(actions) are defined as follows:

$$p ::= \Pi | \overrightarrow{\alpha(u)} | \overrightarrow{\overline{\alpha} < u >} \tag{2}$$

---

[12] Formalize : Mathematical description in term of process.
[13] CSP : Communicating Sequential Processes.
[14] CCS : Calculus of Communicating Systems.

- $\overrightarrow{u}$: Denotes a list of variables called "names".
- $\alpha(\overrightarrow{u})$: Reception of a list of variables $\overrightarrow{u}$ on the canal α.
- $\alpha<\overrightarrow{u}>$: Transmission of a list of variables $\overrightarrow{u}$ on the canal α.

Table 1. Operators of Π-Calculus

Π-Calculus	Description
$p.P$	Prefixing
$P + Q$	Choice
$P \mid Q$	Parallelism
O	Process Nil
$\sum P_i$	Generalized choice
$!P$	Replication
$A(x)$	Abstraction
$[x = y]$	Matching

4.2 Higher Order Π-Calculus

It does not only allow to pass variables and channels but also processes [3].
Example:

$$P|Q \tag{3}$$

with :

$$P \equiv \alpha(x, y, z) and Q \equiv \alpha(u, v, R).v(u).R \tag{4}$$

$$P|Q \equiv \overline{\alpha}(x, y, S).O|\alpha(u, v, R).\overline{v}(u).R \tag{5}$$

In the example above, the process S was transmitted from the process P to the process Q.

5 Problem Description

A technical constraint of the antenna used for radio transmission makes that each one of them can't handle generally above a given amount of traffic in Erlang. So, any excess of this mesure will result in losses and communications drops.

A more efficient, economical and green way to reduce and prevent this kind of issue is firstly to use the same hardware which will be less hardware consuming. Secondly, a saturated antenna don't imply automatically that the neighbouring antennas are congested too. So, to recover this load ballance the idea is to perfrom dynamically what we call "cell breathing" [14] [15].

The idea is that each congested antenna and its neighbouring uncongested antenna dynamically "shrink" and "expand" their coverage (see figure 2) so the traffic will be fairly distributed.

The typical manual and static resolving cost in time, human, and hardware. Since, the adding of antenna is a way to expand original coverage of the network, the idea was to optimise the coverage of the already existing anetannas to fulfill dynmically each time this requirement.

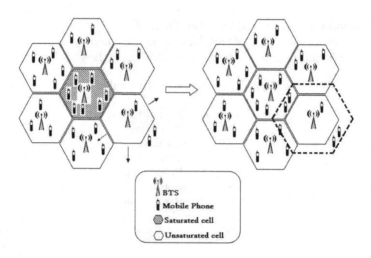

Fig. 2. Cell Breathing Phenomenon

The issue that remains is to : identify which neighbouring cell shall expand and which ones shall shrink according to their own traffic load and the traffic load of the saturated antenna too. Reason why, we've proposed a dynamic architecture that can handle and perform all these computations to cope dynamically with this optimisation issue. Our approach is aiming to acheive a green (*respectfull of environnement*), less time, human and hardware consuming network optimisation.

5.1   Cell Definition

Technically a "*cell*" can be seen as a set of points from the environement where each point receives a given antenna's signal above a threshold. Threshold depends on the type of service proposed by the operator (outdoor, indoor, deep-indoor , incar). So, any adjustment of the coverage area of a given antenna will be technically expressed by the modification of the strength of signal received in this area. Mathematically, the strength of a signal received is expressed by the "*field strength*".

Generally field strentgh is computed using several costumized formulas such in [9], [10] :

$$FS = P + G_{ant} - L_{ant} - PL(x_{RP}, y_{RP})$$
$$- VDIAG(AI - T_{ant}) - VDIAG(\gamma - A_{ant}) \qquad (6)$$
$$+ G_{mob} - L_{mob}$$

with :

$$\gamma = (180 \setminus \Pi)atan2 * (y_{RP} - y_{ant}, x_{RP} - x_{ant}) \qquad (7)$$

where (P) is the transmission power of the antenna (mesured in dB)[15] , (G_{ant}, L_{ant}) and (G_{mob}, L_{mob}) are respectively the propagation gain and loss of the antenna and the mobile (mesured in dB). (PL) is the propagation path loss computed at the specefic point in the cell (mesured in dB). (T_{ant}, A_{ant}) are respectively tilt and azimuth of the antenna (mesured in degrees). $(VDIAG, HDIAG)$ are respectively the vertical and horizontal propagation diagram of the antenna. (x_{ant}, y_{ant}) and (x_{RP}, y_{RP}) are respectively the cartesians coordiantes of the antenna and the concerned receiving point in the cell.

After analysing this formula, technically three parameters are the principal ones customizable in real time and able of adjusting field strength instantly : the power of transmission (P), tilt and azimuth (T_{ant}, A_{ant}). Whereas other variables such antenna and mobile loss and gain (G_{ant}, L_{ant}) and (G_{mob}, L_{mob}) are specific to each device and fixed by industrial producer. Propagation diagrams are related to the type of antenna $(VDIAG, HDIAG)$ (Omnidirectionnel or Directive antenna). Finally path loss (PL) are computed each time using empirical models such Okumura-Hata model or Walfish-Ikagami model.

So, in other words the *"cell breathing"* can be formulated as an optimisation problem where the resolving techniques have to find optimal parameters of the : *power, tilt, and azimuth* of each antenna.

6   Proposed Architecture

The BTS is a component of transmission and reception, with a minimum of intelligence. Its function is the management of radio transmissions. This is the first network component responsible of supporting the request of a subscriber. A cell that serves many mobiles sees its coverage area shrinks due to unaffordable load of traffic. So, coverage holes appear and calls will be rejected. The load on a cell corresponds to the ratio between the demand for traffic on that cell and its actual capacity. This exceeding of load is a major issue in radio network design. To avoid this problem, the BTSs must operate reliably on the ressources (increase their intelligence). Things that lead us to think that cells with a load approaching 100 % (called *"requesting cells"*) can get the support from other neighboring BTS with no load (called *"candidate cells"*). This may lead to an automatic reorganization of the frequency plan, without the need of new *"TRX cards"* for 2G networks or *" WBBP Board"* in 3G networks or inserting new BTS.

A mobile is attributed to the cell that offers the best radio quality on its canal. When a mobile moves from one cell to another, it sees the *"field strentgh"* of the first cell decreasing and the one of the second increasing progressively. Assigning a call from a mobile to a BTS is a process that involves many parameters (eg : azimuth, tilt, power, etc.) [4].

The objective of the proposed architecture is to provide an automatic, dynamic resolving of the netwrok congestion without the need of manual intervention. Thus, the system receives as input an initial network configuration and

---

[15] dB : decibel.

develops an optimized reconfiguration of the network. Saving a collection of diverse solutions (history) allows a better and faster adaptation of the network according to the registered changes.

The adjusting of the solution is not performed till recording the changes of the environment. The approach is defined by a set of state of the environment $E = \{state_1, state_2, ..., state_n\}$. Each state is characterized by :

- Set of demanding cells.
- Set of candidate cells.
- Covered area.

6.1   General Architecture of the System

The general architecture of the optimisation system includes different types of agents. Each agent is responsible of resolving a problem or executing a specific task.

- Cell agent : this agent is responsible of detecting the overloaded cell (reactive agent), through the computing of the following function :

$$L = \frac{A}{C} \tag{8}$$

$$With \ : \ L \ : cell \ load.$$
$$A \ : traffic \ request \ on \ the \ cell.$$
$$C \ : effective \ capacity \ of \ the \ cell.$$

  If the cell load reaches 100 %, the cell is declared "requesting" else it is declared "candidate" with a degree of participation (number of communications which it could be able to support).
- Supervisor Agent: the role of this agent (deliberative agent) is to identify each time which cells are of type "requesting" and which one are of type "candidate". After working with the GIS agent, supervisor agent will decide what is the next cell (closest to the location of the subscriber) to cover the overloaded area. Then, it delegates to the evaluator agent the task of computing the necessary parameters (azimuth, tilt, power). After optimisation, these parameters will be sent back to the supervisor agent to ensure customisation of the BTS (adjusting antenna) of the concerned cell. Then, requests the history agent to make a backup of this network configuration (solution) according to recorded state of the environment.
- GIS Agent: this agent (cognitive agent) has a geographical representation of the area of the cellular network (line, point, surface, size, etc), with various changes (in case of occurrence of obstacles or new constructions).
  Note: The GIS agent has a 3D geographic representation of the cell [22], for each instant t. So-called 3D-T cells.

- Evaluator Agent: depending on the size of the area to cover, this agent takes the initiative to calculate the necessary parameters: *azimuth*, *tilt* and *power*. It sends these parameters to the supervisor agent. Since *"cell breathing"* can be formulated as an optimisation problem which its aim is to find optimal parameters. The goal of the evaluator agent is to accomplish this task in a polynomial time(*efficiency*). So, evaluator agent can include exact (*heuristics*) or approached (*metaheuristics*) algorithms such Genetic Algorithm (GA) or Particle Swarm Optimisation (PSO) for their ability to tackle optimisation problem with high complexity.
- Historic Agent: this agent maintains the historical of previous solutions according to the events. This will allow the reuse of good solutions for a better and faster adaptation of the network in future phases.

  Note: Once the overload resolved (the subscriber has moved to another cell, or decreases its communication), the network returns again to its original configuration.

6.2   Interactions between Agents

1 : The cell agent state the situation of the "requesting ", "candidate" cell, and the degree of participation in the supervisor agent.

2 : The supervisor agent asks the GIS agent about the geographical description of the specified area.

3 : The GIS agent delivers the requested information to the supervisor agent.

4 : The supervisor agent asks the historic agent if there's an earlier solution to the present case in the base of solutions .

5 : The historic agent gives the solution (if it exists) to the supervisor agent.

6 : If there is an earlier solution to the current case, the supervisor agent delivers it to the cell agent.

7 : Otherwise it returns the settings for the current case to the evaluator agent.

8 : The evaluator agent delivers the new solution to the supervisor agent.

9 : The supervisor agent delivers the solution to the cell agent. And calls the historic agent of this backup solution.

7   Towards a Dynamic Architecture Based on Π-Calculus

Considering that the radio telephony networks are real life systems. So, a particular intrest has to be given to the dynamical and real-time aspect of these kind of systems. Crashes and system failures have to be strongly considered. Regarding that, many issues still existing. Two of them are:

1. In case of the loss of a link between two agents. How can we intervene ?
2. If we talk about real-time optimisation, is there a way to reduce the communication time between these different agents ?

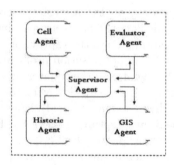

Fig. 3. General System Architecture

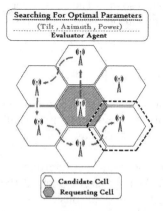

Fig. 4. Movement of The Evaluator Agent

Based on Π-calculus, as formal language during the specification of the agents of the previous system. It will allow to prevent the system from the failures (crashes) or the dynamic evolution of the network. For example: if the supervisor agent loses its communication link with the GIS agent, the expected goal of the optimisation process will not be achieved. The answer to this concern is the following: the Π-calculus as a formal language provides a dynamic reconfiguration of the topology of the applications. Which means that the evaluator agent may delegate its link to one of its agents to complete the requested task.

This mobility link in Π-calculus can be defined as follows :

$$S \equiv ba.S' \text{ and } G \equiv a(x).G' \text{ and } E \equiv \bar{b}(x).ax.E'$$
$$With \ : \ S \ : \ Supervisor \ Agent.$$
$$G \ : \ GIS \ Agent.$$
$$E \ : \ Evaluator \ Agent.$$

The second benefit is that the evaluator agent can move to the cells surrounding the overloaded cell to respond to each detected saturation. The neighboring cells deliver their three parameters: *azimuth, tilt, power*, according to their current load of traffic. In this case, the evaluator agent has the updated parameters that can be applied. It will move from one cell to another (by browsing neighbouring cells), and retrieves the three parameters of each cell. As shown in the figure 4.

8   Conclusion and Perspective of Research

This paper has two main purposes. Firslty, it introduces a model for the network congestion problem. Secondly, it presents a new dynamical solution to the problem of network congestion. An agent-based solution is presented, which provides the redesigning of the network through effective cooperation between various agents. Each agent is responsible of solving a particular problem or performing a specific task. The solution includes also the algebra process Π-calculus, for its ability to dynamically change the topology of applications and cope with any system crashes or network dynamic evolution.

As perspective, we seek to test our architecture in real life scenarios. Secondly, enhance the design of our architecture with more complex backup plans to cope with system failures and dynamic evolution of the network. Thirdly, we will focus on the design of the core of the optimisation process of the proposed architecture which is the evaluator agent. Finally, we think of including industrial automata.

References

[1] Attiogbe, C.: Introduction to Algebras Process and LOTOS (January 2004)

[2] Krivine, J.: Reversible Process Algebras and Concurrent Declarative Programming (November 2006)

[3] Lhouassaine, C.: Mobility and Π-Calculus (June 2002)

[4] Boyer, A.: Antenna (October 2011)

[5] Wooldridge, M.J.: An Introduction to Multi-Agent Systems (2002)

[6] Gherbi, T., Borne, I.: A Meta-Model for Applications Based on Mobile Agents. In: Conference on Software Engineering (CIEL) (June 2012)

[7] Labrou, Y., Fenin, T., Peng, Y.: Agent Communication Langages: The Current Landscape. IEEE Intelligent Systems, 45–52 (March-April 1999)

[8] Girodon, S.: GSM, GPRS and UMTS Networks (June 2002)

[9] Zimmermann, J., Hons, R., Muhlenbein, H.: ENCON: An evolutionary algorithm for the antenna placement problem. Computers and Industrial Engineering Journal (2003)

[10] Zimmermann, J., Hons, R., Muhlenbein, H.: From Theory to Practice: An Evolutionary Algorithm for the Antenna Placement Problem. Advances in Evolutionary Computing (2003)

[11] Resende, M.G.C., Pardalos, P.: Handbook of Optimization in Telecommunications. Springer (2008)

[12] Bratu, V.-I.: Self-optimization of Antenna Tilt in Mobile Networks, Master of Science Thesis Stockholm, Sweden (2012)

[13] Rivera, L.A.S., Nuaymi, L., Bonnin, J.M.: Analysis of a Green-Cell Breathing Technique in a hybrid access network environment. In: WD 2013: Wireless Days IFIP and IEEE International conference, pp. 1–6 (2013)

[14] Bhaumik, S., Narlikar, G., Chattopadhyay, S., Kanugovi, S.: Breathe to Stay Cool: Adjusting Cell Sizes to Reduce Energy Consumption. In: Green Networking Conference (August 30, 2010)

[15] Abdul-Rahman, A., Pilouk, M.: Spatial data modelling for 3D GIS. Springer (2008)

[16] Ekpenyong, M.E.: Managing Cell Congestion in Broadband Wireless Networks: A Comprehensive Simulation Approach. Arabian Journal for Science and Engineering 37(3), 631–645 (2012)

[17] Yang, Z., Niu, Z.: Load Balancing by Dynamic Base Station Relay Station Associations in Cellular Networks. IEEE Wireless Communications Letters 2(2), 155–158 (2013)

[18] Byun, H., Yu, J.: Automatic handover control for distributed load balancing in mobile communication networks. Journal Wireless Networks 18(1), 1–7 (2012)

[19] Wang, H., Liu, N., Li, Z., Wu, P., Pan, Z., You, X.: GA unified algorithm for mobility load balancing in 3GPP LTE multi-cell networks. Science China Information Sciences Journal 56(2), 1–11 (2013)

Modified Formula of Mobility for Mechanisms

Wenjuan Lu, Lijie Zhang[*], Yitong Zhang, Yalei Ma, and Xiaoxu Cui

Key Laboratory of Parallel Robot and Mechatronic System,
Yanshan University, Qinhuangdao 066004
Key Laboratory of Advanced Forging & Stamping Technology and Science,
Yanshan University, Qinhuangdao 066004

Abstract. How to determine the degree of freedom (DOF) of parallel mechanisms (PMs) and multiloop spatial mechanisms has been a long standing problem, and it still is an active field of research with plenty open questions. Firstly, three categories formulas expressed with loop-unit are introduced from a new perspective of their treatment for the motion parameter of each independent closed loop(ICL). Secondly, based on some concepts and theory of link group, the analysis provides insight into why the previous formulas do not work for some multiloop mechanisms. Then a modified formula is presented by adding an addition index called loop-redundant-rank, meanwhile the steps of estimate of mobility property are presented. Finally a typical examples is selected to calculate the DOF with the formula mentioned above. And the results coincide with the prototype data, which shows the effectiveness of the proposed method to a certain extent. It can be seen that the formula presented in this paper can do quick mobility calculation and mobility property analyses of parallel mechanisms with elementary legs.

Keywords: mobility, loop-unit, addition item, independent closed loop.

1   Introduction

Mobility or the DOF of a mechanism is the first consideration in the kinematic and the dynamic modelling of mechanisms and robots. IFToMM terminology defines the mobility as the number of independent co-ordinates needed to define the configuration of a kinematic chain or mechanism [1].

Earlier works on the mobility of mechanisms go back to the second half of the 19th century to Chebychev, Grübler [2]. The development of a criterion for mobility calculation has a history of about 150 years. Prof. Gogu [3], Huang [4] et. al have summarized dozens of typical formulas in detail from different point of view. With the evolution of mechanisms from plane to space, single-degree to multi-freedom, serial-robot to parallel-robot, especially the advent of parallel mechanisms with overconstraints, it is found that a lot of drawbacks of the traditional formulas gradually show, and the results obtained by traditional mobility formulas are often in coincide

---

[*] Corresponding Author.

X. Zhang et al. (Eds.): ICIRA 2014, Part II, LNAI 8918, pp. 535–545, 2014.
© Springer International Publishing Switzerland 2014

with the prototype data. As Tsai [5] said that "the value given by the above equations will be the lower bound of degree of freedom." Rico [6] said that "there are many examples of parallel manipulators where the criterion given by the equation provides the incorrect number of degrees of freedom." and Zhao [7] claimed "Up to now, we have not found a suitable method to calculate the DOF of the mechanisms(3-PTT and 4-PTT) correctly." Ouyang [8] presented a new formula in his paper and side: "The new formula and Kutzbach-Grubler Formula work for all the DOF calculation in plane and space mechanisms, while others are not refined."

All of the above is to say: many traditional methods and formulas are not suited to some modern or classical mechanisms, the formulas have to be reviewed and the flaw of which needs to be considered. For the unit of topological structure can be divided into: elements-unit, loop-unit and limb-unit [9], therefore the mobility calculation can be considered from all kinds of structural unit. In this paper, selecting a group of mobility formulas expressed with loop-unit, the problems existing in these traditional theories are analyzed. On this basis, a modified new formula is put forward.

2   Traditional DOF Formulas Based on Loop Rank

For the mobility formulas expressed with loop-unit, in addition to the number of joints, the DOFs of joints, other key controlling parameters are generally used: the motion parameter(b) (also named as "rank") and the constraint parameter(d) of each ICL of the mechanism, the number of overclosing constraints and so on. However the motion parameters of ICL in original mobility formulas expressed with loop-unit have not been considered comprehensively.

2.1   Original Mobility Formula

In 1963, Ozol [3] proposed the following mobility equation thinking in terms of closed loop in mechanism:

$$M = \sum_{c=1}^{5}(6-c)C_c - 6q \tag{1}$$

where $\sum_{c=1}^{5}(6-c)C_c$ is the total number of scalar kinematic parameters of the mechanism, q is the number of ICL. In this formula, the coefficient of q is 6, which represents that the formula is only applicable to the mechanism with all the rank of ICL being 6. Therefore, for the overconstrainted mechanism PCM [10] shown in Fig. 1, of which the ICL rank is not 6. So we can say that Eq. (1) does not work for this mechanism and an erroneous result will be got: $M = (6-5)\times 12 - 6\times 2 = 0$.

Analogously, there are still formulas for mobility calculation presented by Koenigs [3], Rossner [3] and Boden [3] i.e.., they are only applicable to the mechanism whose rank of ICL is 6.

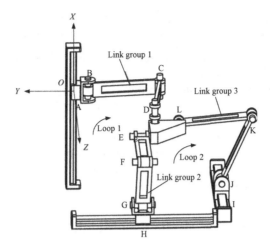

Fig. 1. PCM Mechanism

2.2   The Formula Not Differentiating the Loop Rank

Take into consideration the fact that the rank (or motion parameter) of ICL will be reduced and less than 6, as the overconstraints appearing in mechanisms. In 1968, Manolescu [3] extends the Eq. (1) proposed by Ozol by considering the overconstraints in ICL, and presents a new formula expressed with loop-unit:

$$M = \sum_{c=d+1}^{5} (6-c)C_c - (6-d)q \qquad (2)$$

where d is the constraint parameter of each ICL, which shows that the overconstraints in closed loop is considered. However, the formula does not differentiating the ranks of different chosed loops. Therefore, this formula is not applicable to the mechanism with different values of constraint parameter for ICLs. In fact, even if for PCM having the same d for all the closed loops as Gogu said, the result obtained by this formula is:

$M = (6-5) \times 12 - (6-1) \times 2 = 2$, which is quite different from the reality.

Hochman [3], Dobrovolsk [3] i.e. have put forward the similar formulas which have considered the rank of loop but not differentiated the loop rank in a mechanism.

2.3   The Formula Differentiating the Loop Rank

Sustained efforts have been made with the development of history. In 1999, considering the mechanism with different ranks for different closed loops, a more perfect equation was proposed by Antonescu [3]:

$$M = \sum_{i=1}^{5} iF_i - \sum_{j=2}^{6} jR_j \qquad (3)$$

where F_i represents the number of joints having the degree of connectivity $f_i = i$ and R_j represents the number of closed loops having the motion coefficient $b_j = j$. Thus $\sum_{j=2}^{6} jR_j$ represents the total number of ranks of all the closed loops. The values of j range from 2 to 6, which show that different closed loops will have different ranks. However by using Eq. (3), for PCM mechanism, the same erroneous results are obtained as Eq. (2): $M = 1 \times 12 - 5 \times 2 = 2$.

Moreover, Voinea and Atanasiu [3], Manolescu and Manafu [3], Gronowicz [3] i.e. have presented mobility formulas differentiating the loop rank in different ways.

Since the case that different ranks for different loops has already been considered, why is the error result still got?

3   Some Concepts for Mechanism Mobility

To simplify analysis, firstly, some concepts and theory, which are called theory of link group, needed for DOF calculation should be recalled and presented.

(1) General link group (or simply link group) [11]: a general link group is regarded as the combination of non-coincident links between any independent loop and its adjacent loop. The mobility of a link group can be equal to 0, or less than 0, or more than 0.

(2) Displacement parameters of link group [12]: the displacement parameters are referred to linear and angular displacement parameters. It is the total displacement parameters of link group before its end part is restricted by other link groups at a given coordinate. It is expressed as G_k^{gz}, which is named displacement vector of link group.

$G_k^{gz}\,(\alpha \quad \beta \quad \gamma, \quad x \quad y \quad z)$, where $\alpha\,\beta\,\gamma, x\,y\,z$ are formal parameters. The value of the formal parameter can be either 0 or 1. 0 denotes nonexistence of the movement while 1 denotes existence. $\alpha\,\beta\,\gamma$ named angular displacement parameters, represent the rotations around X-, Y-, Z-axes, respectively. $x\,y\,z$, called as linear displacement parameters, represent the translations along X-, Y-, Z-axes, respectively.

(3) Displacement parameters of ICL: It is the total displacement parameters of ICL before it is restricted by other ICLs at the given coordinate. It is expressed as L_k^{hl}, which is named displacement vector of ICL. $L_k^{hl}\,(\alpha \quad \beta \quad \gamma, \quad x \quad y \quad z)$, where $\alpha\,\beta\,\gamma, x\,y\,z$ are formal parameters, and represent the rotations around and translations along X-, Y-, Z-axes, respectively. The value of the formal parameter can be either 0 or 1. 0 denotes nonexistence of the movement while 1 denotes existence. Dimension of L_k^{hl} is expressed by d_k^{hl}, namely, $d_k^{hl} = \dim(L_k^{hl}) = \dim\,(\alpha \quad \beta \quad \gamma, \quad x \quad y \quad z)$, which is named as the rank of the kth ICL. It can also be written as $d_k^{hl}\,(\alpha \quad \beta \quad \gamma, \quad x \quad y \quad z)$ for short.

(4) Base point [12]: In the reference system, the point, determined by the intersection of the effective parameters of link group, is named base point. Base point parameters can be expressed with a vector $^N\!O_B^{M,J}\,(\alpha\,\beta\,\gamma, x\,y\,z)$. Dimension of $^N\!O_B^{M,J}$ is expressed by $^N\!d_B^{M,J}$, namely $^N\!d_B^{M,J} = \dim\,(^N\!O_B^{M,J})$.

4   Cause Analysis

Based on the concepts mentioned above, below, we will analyze the reasons that cause the above problems mentioned in section 2. Without loss of generality, the cause analysis will be done combined with the PCM mechanism shown in Fig. 1. The PCM mechanism is formed by two ICLs. Suppose loop 1 contains link groups 1 and 2, at the given coordinate, displacement parameters of loop 1 can be obtained:

$$L_1^{hl} = (\alpha \quad 0 \quad 0, \quad x \quad y \quad z) \cup (0 \quad \beta \quad 0, \quad x \quad y \quad z) = (\alpha \quad \beta \quad 0, \quad x \quad y \quad z),$$

the rank of which is $d_1^{hl} = \dim (\alpha \quad \beta \quad 0, \quad x \quad y \quad z) = 5$. Loop 2 contains link groups 2 and 3. Therefore, displacement parameters of loop 2 are: $L_2^{hl} =$

$(0 \quad \beta \quad 0, \quad x \quad y \quad z) \cup (0 \quad 0 \quad \gamma, \quad x \quad y \quad z) = (0 \quad \beta \quad \gamma, \quad x \quad y \quad z)$, the rank of which is $d_2^{hl} = \dim (0 \quad \beta \quad \gamma, \quad x \quad y \quad z) = 5$. When there is only the first independent loop consists of link group 1 and link group 2 connected in parallel between the fixed platform and the movable platform, the motion of movable platform is the intersection of the displacement parameters of the two link groups: $(\alpha \quad 0 \quad 0, \quad x \quad y \quad z) \cap (0 \quad \beta \quad 0, \quad x \quad y \quad z) = (0 \quad 0 \quad 0, \quad x \quad y \quad z)$. It can be seen that the rotation around Y-axis of movable platform has been constrained, which represents that the rank of ICL before it is restricted by former ICL. Actually, when a parallel mechanism formed by closing the two loops, that is to say when adding the third link group to loop 1, some motion of loop 2, rotation around Y-axis, will be restrained, and the rank of loop 2 will be reduced. The motion of movable platform will be: $(0 \quad 0 \quad 0, \quad x \quad y \quad z) \cup (0 \quad 0 \quad \gamma, \quad x \quad y \quad z) = (0 \quad 0 \quad \gamma, \quad x \quad y \quad z)$, rather than $(0 \quad \beta \quad \gamma, \quad x \quad y \quad z)$. At this time, if the rank of the loop 2 is calculated by 5, the result of mobility using formula will be smaller than the reality.

In short, the essence is that the rank of the closed loop will be changed after it is restricted by other ICLs, which is ignored in the process of mobility calculation.

5   New Theory of DOF

5.1   A Modified Formula Based on Loop Rank

The development of freedom theory is a process of going for perfect step by step. The formula presented by Moroskine [3]: $M = N - r$, which is valid without exception in theory, but with poor feasibility. To solve this problem, in 2005, Gogu [13] extended the formula, proposed a new method for calculating "r" by the method of linear transformation, and put forward a new formula for quick mobility calculation without the need to develop the kinematic equations.

The traditional formula for mobility analysis is the Grübler-Kutzbach(G-K) criterion [10], however it has been documented for failure in many classical or modern mechanisms. The most core reason is that not all the overconstraints have been considered. Seizing the nature for mobility calculation, Prof Huang [14] developed

"G-K formula" and presented a "modified G-K criterion", which is obtained by adding "parallel-redundant-constraint" to "G-K formula". On the basis of screw theory, he formed the mobility principle based on reciprocal screw, with which almost all the "paradoxical mechanisms" summarized by Gogu have been solved [10].

Similarly, having analyzed the reason for the error in the process of mobility calculation by Eqs. (1)~(3), an addition item should be added to modify the formulas. The value of addition item is equal to the reduced rank of loop when adding a link group to the former closed loop. This item is called "loop-redundant-rank" here, and expressed as η: $\eta = \sum_{j=1}^{q}\left[d_j^{hl} - \dim\left(\bigcap_{i=1}^{i=j} G_i^{gz} \cup G_{j+1}^{gz}\right)\right]$. Therefore, a modified formula based on loop rank can be obtained:

$$M = \sum_{i=1}^{5} iF_i - \sum_{j=1}^{q} d_j^{hl} + \eta \tag{4}$$

$$\eta = \sum_{j=1}^{q}\left[d_j^{hl} - \dim\left(\bigcap_{i=1}^{i=j} G_i^{gz} \cup G_{j+1}^{gz}\right)\right] \tag{5}$$

where, M denotes the DOF of a mechanism; F_i is the number of kinematic joints with i DOF; q is the number of independent loops; η is the number of loop-redundant-rank; d_j^{hl} is the rank of the jth independent loop; G_i^{gz} denotes the displacement parameters vector of the ith link group.

Substitute Eq. (5) into Eq. (4):

$$M = \sum_{i=1}^{5} iF_i - \sum_{j=1}^{q} d_j^{hl} + \eta = \sum_{i=1}^{5} iF_i - \sum_{j=1}^{q} d_j^{hl} + \sum_{j=1}^{q}\left[d_j^{hl} - \dim\left(\bigcap_{i=1}^{i=j} G_i^{gz} \cup G_{j+1}^{gz}\right)\right] =$$

$$\sum_{i=1}^{5} iF_i - \sum_{j=1}^{q} d_j^{hl} + \sum_{j=1}^{q} d_j^{hl} - \sum_{j=1}^{q} \dim\left(\bigcap_{i=1}^{i=j} G_i^{gz} \cup G_{j+1}^{gz}\right) =$$

$$\sum_{i=1}^{5} iF_i - \sum_{j=1}^{q} \dim\left(\bigcap_{i=1}^{i=j} G_i^{gz} \cup G_{j+1}^{gz}\right) \tag{6}$$

Eq. (6) is the final modified formula based on loop rank.

5.2   Estimate of Mobility Property

With a rapid development of mechanical engineering over the past half century after World War II, a great number of multi-DOF over-constrained new mechanisms have appeared. To apply mechanisms into practice, it is not enough to only calculate the number of mobility. We also need to know the mobility characteristics. That is to say, the issue of estimate of mobility property should be addressed. However there is few research in these mobility theories from the perspective of loop-unit. Based on the mobility theory of link group, it is easy to judge the mobility property.

The DOF of the output member is [15]:

$$F' = {}^N d_B^{M,J} + \sum F_{k'} \tag{7}$$

where $F_{k'}$ is the DOF of the link group whose DOF, F_k, is less than zero. F_k can be calculated by:

$$F_k = \sum_{i=1}^{P_k} f_i - \dim\left(G_k^{gz}\right) \tag{8}$$

It is presented in previous paper [12]. If $F' > 0$, it means that the output member is movable. And now the displacement parameters of base point, ${}^N O_B^{M,J} = \bigcap_{k=1}^{q+1} G_k^{gz}$, can mean the mobility property of output member. That is to say that the estimate of mobility property is done just with intersection operation for displacement parameters of link group [16]. In general, procedure for estimate of mobility property is summarized as follows:

1. *Determine the displacement parameters of each link group, G_k^{gz}.*

2. *Determine the DOF of each link group, F_k, using Eq. (8).*

3. *Seclect the DOF of the link group whose DOF is less than zero, that is $F_{k'}$.*

4. *Determine base point parameters, ${}^N O_B^{M,J}$, with intersection operation. Meanwhile dimension of ${}^N O_B^{M,J}$ can be determine.*

5. *Determine the DOF of the output member, F', using Eq. (7). If $F' = 0$, it means the output member is fixed. While $F' > 0$ it means the output member is movable, at this moment ${}^N O_B^{M,J}$ means the mobility property of output member, including both independend motion and parasitic motion.*

5.3   The Rules of Union Operation for Displacement Parameters of Link Group

The determination of displacement parameters of link group, G_k^{gz}, is an important prerequisite for the mobility calculation and estimate of mobility property. In a certain coordinate system, displacement parameters of link group depend on some key controlling factors: type of kinematic pair, relative position and orientation of pair axes and so on. For the topological structure of link group composed with only translational pairs and kinematic elements connected in series, there are only linear displacement parameters. The displacement parameters of link group can be obtained by superposition of each pair's displacement parameter, and can be described as follows:

$$(0 \; 0 \; 0, \; x \; 0 \; 0) \cup (0 \; 0 \; 0, \; x \; 0 \; 0) = (0 \; 0 \; 0, \; x \; 0 \; 0);$$

$$(0 \; 0 \; 0, \; 0 \; y \; 0) \cup (0 \; 0 \; 0, \; x \; 0 \; 0) = (0 \; 0 \; 0, \; x \; y \; 0);$$

$$(0 \; 0 \; 0, \; x \; y \; 0) \cup (0 \; 0 \; 0, \; x \; 0 \; 0) = (0 \; 0 \; 0, \; x \; y \; 0); \text{ and so on.}$$

The displacement parameters of link group composed of prismatic pairs can be obtained relatively simply. It just needs to determine the linear displacement parameters according to the axial direction of prismatic pairs.

However, for the link group composed of multiple revolute pairs, the determination of displacement parameters is generally no longer a case of simple superposition of the angular displacement parameters, that is $(\alpha \quad 0 \quad 0, \quad 0 \quad 0 \quad 0) \cup$ $(\alpha \quad 0 \quad 0, \quad 0 \quad 0 \quad 0) \neq (\alpha \quad 0 \quad 0, \quad 0 \quad 0 \quad 0)$.

There may be also some parasitic linear displacement parameters when the rotation axes meet a certain geometric condition. For convenience of description, we have introduced ten types of basic link group and the displacement parameters of which are listed in Ref [15].

A relatively complicated link group can be considered as a connection by kinematic pair according to certain ways. That is to say a complicated link group can be divided into some basic link group types. Therefore the displacement parameters of the link group will be obtained by a superposition of which of basic link groups. The proposal of displacement parameters of basic link group types establishes the foundation of the determination of the displacement parameters of link groups.

6   Mobility Analysis of a Typical Mechanisms

In ref [3], as Gogu said the rank of the two ICLs of the PCM mechanism shown in Fig.1 are both equal to 5, the right result cannot be obtained with the traditional formulas . The mobility analysis of this mechanism will be done next with the new formula.

By the assumptions, loop 1 is denoted by ABCDEFGH and loop 2 is denoted by EFGHIJKL. The displacement vectors of each link group are:

$$G_1^{gz} = (\alpha \quad 0 \quad 0, \quad x \quad y \quad z),$$
$$G_2^{gz} = (0 \quad \beta \quad 0, \quad x \quad y \quad z);$$
$$G_3^{gz} = (0 \quad 0 \quad \gamma, \quad x \quad y \quad z).$$

Therefore, the displacement vector of loop 1 including link group 1 and link group 2 is obtained by union operation:

$$L_1^{hl} = G_1^{gz} \cup G_2^{gz} = (\alpha \quad 0 \quad 0, \quad x \quad y \quad z) \cup (0 \quad \beta \quad 0, \quad x \quad y \quad z)$$
$$= (\alpha \quad \beta \quad 0, \quad x \quad y \quad z)$$

And $G_1^{gz} \cap G_2^{gz} \cup G_3^{gz} = (\alpha \quad 0 \quad 0, \quad x \quad y \quad z) \cap (0 \quad \beta \quad 0, \quad x \quad y \quad z)$
$$\cup (0 \quad 0 \quad \gamma, \quad x \quad y \quad z)$$
$$= (0 \quad 0 \quad \gamma, \quad x \quad y \quad z).$$

By Eq. (6), the DOF of the PM in Fig.1 can be obtained:

$$M = \sum_{i=1}^{5} iF_i - \sum_{j=1}^{q} \dim\left(\bigcap_{i=1}^{i=j} G_i^{gz} \cup G_{j+1}^{gz}\right) =$$

$$(1 \times 12) - [\dim\left(G_1^{gz} \cup G_2^{gz}\right) + \dim\left(G_1^{gz} \cap G_2^{gz} \cup G_3^{gz}\right)] =$$

$$12 - [\dim(\alpha \quad \beta \quad 0, \quad x \quad y \quad z) + \dim(0 \quad 0 \quad \gamma, \quad x \quad y \quad z)] = 12 - (5+4) = 3.$$

On the other hand, with Eq. (5), the loop-redundant-rank can be calculated:

$$\eta = \sum_{j=1}^{q}\left[d_j^{hl} - \dim\left(\bigcap_{i=1}^{i=j} G_i^{gz} \cup G_{j+1}^{gz}\right)\right] = d_2^{hl} - \dim\left(G_1^{gz} \cap G_2^{gz} \cup G_3^{gz}\right) =$$

$$5 - \dim(0 \quad 0 \quad \gamma, \quad x \quad y \quad z) = 5 - 4 = 1 \ .$$

Adding Eq. (3) with the addition item, a right result will be obtained:

$$M = \sum_{i=1}^{5} iF_i - \sum_{j=2}^{6} jR_j + \eta = 1 \times 12 - 5 \times 2 + 1 = 3.$$

Mobility property analyses:

(1) Determine the DOF of each link group:

$$F_1 = F_2 = F_3 = \sum_{i=1}^{P_k} f_i - \dim\left(G_k^{gz}\right) = 4 - 4 = 0 ;$$

(2) Determine $F_{k'} : \sum F_{k'} = 0$

(3) Determine base point parameters:

$$^N O_B^{M,J} = \bigcap_{k=1}^{q+1} G_k^{gz} = G_1^{gz} \cap G_2^{gz} \cap G_3^{gz} =$$

$$(\alpha \quad 0 \quad 0, \quad x \quad y \quad z) \cap (0 \quad \beta \quad 0, \quad x \quad y \quad z) \cap (0 \quad 0 \quad \gamma, \quad x \quad y \quad z)$$

$$= (0 \quad 0 \quad 0, \quad x \quad y \quad z).$$

And, $^N d_B^{M,J} = \dim(^N O_B^{M,J}) = \dim((0 \quad 0 \quad 0, \quad x \quad y \quad z)) = 3$

(4) Determine the DOF of the output member:

$$F' = {}^N d_B^{M,J} + \sum F_{k'} = \dim(^N O_B^{M,J}) + 0 = 3 > 0$$

Therefore $^N O_B^{M,J} = (0 \quad 0 \quad 0, \quad x \quad y \quad z)$ means that the moving platform has three independent translation along X-axis, Y-axis and Z-axis respectively.

7   Conclusion

The development of freedom theory is a process of going for perfect step by step. The development history of a group of mobility formulas expressed with loop-unit is introduced with their consideration of the motion parameter of ICL. To modify these defective traditional mobility formulas, new concepts including the displacement parameters of ICL and loop-redundant-rank are presented. Then a new formula for mobility calculation by adding a correction item called loop-redundant-rank is put forward.

In the new formula presented in this paper, intersection operation is done to avoid constraint calculation. Meanwhile, taking a link group and an ICL as a unit instead of a pair, which can makes the mobility calculation simple.

Although there have been many formulas for mobility calculation in different forms, the authors from a new point of view, apply the theory of link group and intersection rule presented by authors to the traditional mobility formulas expressed with loop-unit. As Prof Yang Tingli said, it is very meaningful for applying different methods to solve the same problem.

Acknowledgements. The authors acknowledge the financial support from the Natural Science Foundation of China (51275438).

References

1. Ionescu, T.G.: Terminology for mechanisms and machine science. Mechanism and Machine Theory 38, 597–901 (2003)
2. Huang, Z., Liu, J.F., Li, Y.W.: 150-year unified mobility formula issue. Journal of Yanshan University 35(1), 1–14 (2011)
3. Gogu, G.: Mobility of mechanisms: A critical review. Mechanism and Machine Theory 40(9), 1068–1097 (2005)
4. Huang, Z., Li, Q.C., Ding, H.F.: Theory of Parallel Mechanisms. Springer (2012)
5. Mruthyunjaya, T.S.: Kinematic structure of mechanisms revisited. Mechanism and Machine Theory 38, 279–320 (2003)
6. Rico, J.M., Aguilera, L.D., Gallardo, J., et al.: A More general mobility criterion for parallel manipulators. ASME Journal of Mechanical Design 128, 207–219 (2006)
7. Zhao, J.S., Zhou, K., Feng, Z.J.: A theory of degrees of freedom for mechanisms. Mechanism and Machine Theory 39(6), 621–643 (2004)
8. Ouyang, F., Cai, H.Z., Liao, M.J.: Comparable study on new and old formulas for calculating DOFs of mechanisms and structures. China Mechanical Engineering 24(21), 2942–2947 (2013)
9. Yang, T.L., Liu, A.X., Luo, Y.F., et al.: Theory and application of robot mechanism topology. Science Press, Beijing (2012)
10. Huang, Z., Liu, J.F., Li, Y.W.: On the degree of freedom-the general formula of the degree of freedom which has been searched for 150 years. Science Press, Beijing (2010)
11. Zhang, Y.T., Mu, D.J.: New concept and new theory of mobility calculation for multi-loop mechanisms. Sci. China Tech. Sci. 1604, 1598–1604 (2010)
12. Zhang, Y.T., Lu, W.J., Mu, D.J., et al.: A novel mobility formula for parallel mechanisms expressed with mobility of general link-group. Chinese Journal of Mechanical Engineering 26(6), 1082–1090 (2013)
13. Gogu, G.: Mobility and spatiality of parallel robots revisited via theory of linear transformations. European Journal of Mechanics A/Solids 24, 690–711 (2005)

14. Huang, Z., Li, Q.C.: General methodology for type synthesis of lower-mobility symmetrical parallel manipulators and several novel manipulators. International Journal of Robotics Research 21(2), 131–145 (2002)
15. Lu, W.J., Zhang, L.J., Zeng, D.X., Zhang, Y.T.: Research on Application of GOM Formula—A Novel Mobility Formula. China Mechanical Engineering 25(17), 2283–2289 (2014)
16. Lu, W.J., Zhang, L.J., Zhang, Y.T.: Fast solution method of mobility calculation based on list intersection. China Mechanical Engineering (2014)

A Long Distance Polar Rover
with Multifunctional Wind Energy Unit

Jicheng Liu[*], Lingyun Hua, Jie Ma, Zili Xu, and Pange Jiang

School of Mechatronics Engineering and Automation, ShangHai University,
Shanghai 200072, China
{liujicheng,hualingyun,robomark,zilixu,jpg123}@shu.edu.cn

Abstract. Aimed at making full use of the Antarctic renewable wind energy, this paper presents a long distance polar rover installed a multifunctional wind energy unit. The multifunctional wind energy unit integrates the wind power driving and wind power generation, which can convert the force acting on the unit caused by the wind energy into the rover's driving force and the wind power at the same time. Meanwhile, assisted by the track diving subsystem, the polar rover can change the rover's motion direction and cross the obstacle. The track diving subsystem can move back and forth, which can change the center of mass of the rover and improve the mobility of the polar rover. Furthermore, on the basis of the aerodynamics and mechanics theory, the motion performance and automatic reset capability of the polar rover was investigated.

Keywords: Antarctic, polar rover, wind energy, changing mass center, Long distance exploration.

1   Introduction

The environment of the Antarctic inland poses numerous challenges for mobile robots. As the important moving carrier for the Antarctic exploration, the polar rover must face the limited energy stored in the polar rover which greatly restricts the polar rover to execute a long distance exploration.

Many research works have been done successfully to investigate how to reasonably utilize the Antarctica's renewable energy, such solar power and wind energy in order to enlarge the rover's exploration field.

Nomad rover designed by Carnegie Mellon University was applied to Antarctic meteorite exploration, a wind power generation unit installed on Nomad rover was used to supplement electric power for this rover [1]. Ref. [2] designed an Antarctic robot with vertical axis wind turbine. This robot had been tested in Zhongshan Station in Antarctic, the results shows that the renewable wind energy system can provide continuous power supply for the rover. But the results also show that the steering mode, energy capture efficiency, control mode are not meet the requirements of the environment. Meanwhile, the wind-driven spherical robots, for example JPL tumbleweed rover and the Tumble Cup, were tested in the Antarctic environment.

---

[*] Corresponding author.

X. Zhang et al. (Eds.): ICIRA 2014, Part II, LNAI 8918, pp. 546–554, 2014.
© Springer International Publishing Switzerland 2014

Experiments show that the motion is strongly episodic, once the rover gets moving quickly for a while, then stop or control the rover's direction was difficult [3,4]. Moreover, in the aspect of using renewable solar energy, Cold Regions Research and Engineering Laboratory developed first ground vehicle specifically designed to conduct long-duration autonomous science campaigns over terrestrial ice sheets. A five sided box of solar panels surrounds the chassis of the four-wheel mobile robot, the collecting solar power reflected from the snowfield as well as directly incident to the panels. A unique power management system controls the operating point of each panel to match collective power input to power demand [5,6]. Furthermore, the flying robot and ice surface mobile robot were also developed for the Antarctic exploration [7-10], and the motion performance for these robots also was researched [11-13]. Therefore, how to make full use of the Antarctic renewable energy sources so that enlarges the polar rover's exploration field is one of main study topics.

A long distance polar rover installed a multifunctional wind energy unit was introduced in this paper. The multifunctional wind energy unit integrates the wind power driving and wind power generation, which can convert the force acting on the unit generated by the wind energy into the rover's driving force and the wind power at the same time. Meanwhile, the motion performance of the polar rover was investigated. The characteristics of changing mass center and its influence on polar rover was also analyzed. This research provides a mechanical structure for the unmanned polar rover to complete tasks driven by the renewable wind energy and interactive track driving sub-system.

2   The Structure of the Long Distance Polar Rover

2.1   Polar Rover Structure

The structure of the long distance polar rover is shown in Fig.1. The detecting instruments are fixed in the capsule of polar rover. For ensuring the polar rover have the automatic reset capability, six sledges is connected respectively with capsule by landing legs and the revolute joint whose rotation angle is limited. And the sledges distribute around the axis of capsule by 120°. Track driving subsystem can provides driving force for polar rover. The track driving subsystem can realize the action respectively by lifting the track unit, moving along the lead screw and rotating around the capsule's axis. And under these different action modes of the track unit, the polar rover can change its mass center for restoring the driving capacity to well-balanced state from unsteady state and crossing the ice valley.

The multifunctional wind energy unit is set at the rear of the polar rover, whose blades can be rotated so that the different blade wind angle can be obtained. The multifunctional wind energy unit can convert the force acting on the blade generated by the wind energy into the rover's driving force by adjusting the blade wind angle, and as well as convert the wind energy into the electric energy by the blades' rotation. Therefore, the characteristics of the polar rover ensure the polar rover to execute the long distance exploration. Meanwhile, assisted by the track diving subsystem, the polar rover can change the rover's motion direction, cross the obstacle and possess the automatic reset capability.

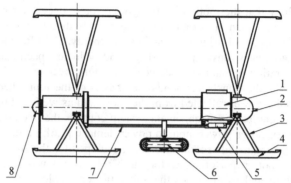

1. transmission mechanism   2. capsule   3. landing leg   4. Sledge   5. motor
6. track unit   7. lead screw   8. multifunctional wind energy unit

Fig. 1. Structure of the polar rover

2.2   The Multifunctional Wind Energy Unit

The multifunctional wind energy unit is shown in Fig.2. The blades are the key parts of the polar rover, and the blade wind angle can be adjusted by driving the motor. Under the force acting on the blades can be divided into driving force and lift force. The driving force pushes the polar rover to move forward. In contrast, the lift force can rotate the blades. Furthermore, the rotational motion of the blades is transmitted from output shaft to reduction gearbox of power generation unit, which can convert surplus wind energy into wind power. When the resistance acting on the blade,

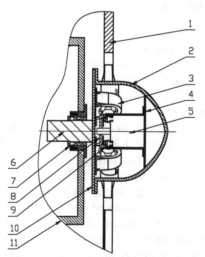

1. blade   2. outside protective cover   3. bearing pedestal   4. motor frame 5. motor
6. output shaft   7. bearing   8. big bevel gear 9. small bevel gear 10. end cover   11. capsule

Fig. 2. Section view of the multifunctional wind energy unit

namely the driving force of the polar rover becomes too large or too small so that the polar rover is in the unstable motion stage, the motor drive the bevel gear pair to adjust blade deflection angle and change blade windward area ultimately in order to ensure the polar rover to obtain the controllable driving force.

The multifunctional wind energy unit integrates the wind power driving and wind power generation. Thus, the polar rover can make full use of the Antarctic renewable wind energy sources, which enlarge the polar rover's exploration field.

3   Motion Performance of the Polar Rover Driven by Wind Energy

When the force caused by the wind energy acts on the blades, the force becomes the pressure force acting on the blades and then product the thrust force so that push the polar rover to move forward. The reasonable thrust force can be achieved by changing the appropriate blades' windward area.

To facilitate study, this paper assumes the blades are in the ideal fluid filed as shown in Fig. 3.

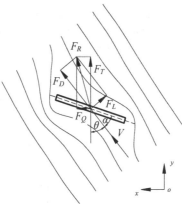

Fig. 3. Air force acting on a blade

In Fig.3, V is the wind velocity, α and θ is the angle between the direction of wind velocity and plate surface and the angle between wind velocity and rover forward direction, respectively.

On the basis of the Bernoulli's principle, the aerodynamic force F_R acting on the blade can be expressed as:

$$F_R = \frac{1}{2} \rho C_r S V^2 \qquad (1)$$

where ρ is the air density (kg/m³), C_r is the thrust force coefficient, S is the wing span area (m²), $S=R{\times}c$, R is the blade's wing span length (m), c is the blade chord length (m).

Aerodynamic force F_R can be divided into Lift force F_L and resistance F_D (as shown in Fig.3). F_L and F_D is vertical and along with the direction of wind velocity respectively, which can be described as:

$$F_L = \frac{1}{2}\rho C_L S V^2 \tag{2}$$

$$F_D = \frac{1}{2}\rho C_D S V^2 \tag{3}$$

where C_L is the lift coefficient, C_D is the resistance coefficient.

To project the lift force F_L and resistance F_D into two components on x and y axis. F_Q is parallel to the rotating plane and F_T is perpendicular to the rotating plane, which can be written as the following relation:

$$F_T = F_L \sin\theta + F_D \cos\theta \tag{4}$$

$$F_Q = F_L \cos\theta - F_D \sin\theta \tag{5}$$

Polar rover is pushed to move forward by aerodynamic thrust force F_T generated by airflow. In contrast, the torque caused by radial force F_Q can rotate the blades.

When there are N blades in the multifunctional wind energy unit, then the total thrust force T and the total torque Q can be gotten:

$$T = \frac{1}{2}N\rho V^2 S \left(C_L \sin\theta + C_D \cos\theta\right) \tag{6}$$

$$Q = \frac{1}{2}N\rho V^2 S R \left(C_L \cos\theta - C_D \sin\theta\right) \tag{7}$$

According to the Eq. (6) and Eq. (7), when external condition is certain, total thrust force T and the total torque Q are in direct proportion to N, S and V^2. With the change of wind velocity or wind direction, the polar rover adjusts the windward area of blades so that obtain the controllable thrust force. Meanwhile, the multifunctional wind energy unit can generate the wind power under the driving of torque Q.

4   Motion Performance of the Polar Rover Driven by Track Unit

The long distance polar rover can change the position of the track unit along the capsule's axis by rotating the lead screw. The lead screw can drive the track unit to move forward or backward so that the mass center of the polar rover is adjusted (as shown in Fig. 4). When the polar rover are moving on the relatively even ice or snowfield, the mass center O_2 of the track unit and the mass center O_1 of rover except the track unit should pose on the same vertical line in order to ensure the uniformity of the force acting on the sledges. However, when the polar rover is driven by the track unit, track unit contact the ice or snowfield and the rear two sledges are raised. Therefore, the total weight of the polar rover is exerted on the track unit and the front two sledges. Because the lift height of rear two sledges is less than the length of polar rover, for simplifying the analysis, this paper assumes that the distance l between O_1 and the contact force point on front sledges, and the distance e between O_1 and O_2 is equal to the corresponding geometric length of polar rover.

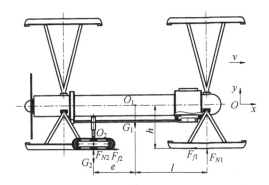

Fig. 4. The sketch of changing mass center

In Fig. 4, v is the polar rover's translational speed, G_2 is the weight of the track unit, G_1 is the polar rover's weight except the track unit weight. F_{N1} and F_{N2} is the normal force exerted on the sledge and tract unit respectively, F_{f1} is the friction force acting on the sledge, F_{f2} is the driving force of the track unit.

Then, according to the geometric model as shown in Fig.4, F_{N1} and F_{N2} can be described as:

$$\begin{cases} F_{N1} = \dfrac{G_1 e}{l+e} \\ F_{N2} = \dfrac{G_1 l + G_2(l+e)}{l+e} \end{cases} \qquad (8)$$

Therefore, the driving force F_{f2} generated by the track unit can be expressed as:

$$F_{f2} = \dfrac{\mu G_1}{\left(\dfrac{l}{e}+1\right)} \qquad (9)$$

where μ is the friction coefficient between sledge and ice or snowfield

To define $\overline{F}_{f2} = F_{f2}/(\mu G_1)$ and $\delta = l/e$, then the normalized function $\overline{F}_{f2}$ can be deduced that:

$$\overline{F}_{f2} = \dfrac{1}{\delta+1} \qquad (10)$$

According to the Eq. (3), The Fig. 4 reveals the dependence of $\overline{F}_{f2}$ on δ.

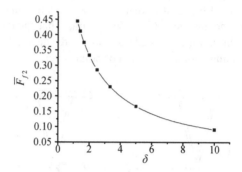

Fig. 5. The dependence of normalized function $\overline{F}_{f2}$ and δ

It can be seen from the Fig. 5, $\overline{F}_{f2}$ decreases with the increase of δ. In other words, the driven force F_{f2} increases with the moving of the track unit far away from the O_1. However, it can also be observed that the changing range of the driven force F_{f2} is different with the changing range of moving distance e.

Meanwhile, the track unit can decrease the velocity of the rover, and the track unit also plays an important role in changing the motion direction of the polar rover.

5   Automatic Reset Capability of the Polar Rover

Renewable wind energy can be used to drive the polar rover. On the contrary, the wind can also induce the overturn of polar rover. Therefore, the polar rover must possess the automatic reset capability.

The cylindrical capsule of the long distance polar rover can reduce effectively the wind resistance and decrease the risk of overturn. Especially, the track unit can rotate around the capsule's axis. After the overturn of the polar rover, another two sledges will contact the ice or snowfield, and the track unit will be parallel with the surface of ice or snowfield or keep a certain angle with the ground. Then, the polar rover will recover to the normal stage with the rotation of track unit even the polar rover overturns under the action of an external force, as shown in Fig.6.

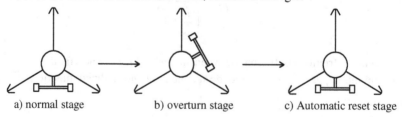

a) normal stage   b) overturn stage   c) Automatic reset stage

Fig. 6. Automatic reset capability of the polar rover

6   Conclusions

A long distance polar rover with a multifunctional wind energy unit was developed. The multifunctional wind energy unit integrates the wind power driving and wind power generation, which can convert the force acting on the unit generated by the wind energy into the rover's driving force and the wind power at the same time. It can make full use of the Antarctic renewable wind energy and enlarge the polar rover's exploration range. Meanwhile, the polar rover can be driven by the track driving system so that improve its tractive and steering performance. Furthermore, the track unit can ensure the polar rover recover to the normal stage even the polar rover overturn under the environment of the Antarctic inland.

Acknowledgments. This research was supported by National Natural Science Foundation of China (Grant No. 61005073) and Creative Research Fund of Shanghai Municipal Education Commission (14YZ009) and Natural Science Foundation of Shanghai City (Grant No. 14ZR1415700).

References

1. Wagner, M.D., Pedersen, L., Whittaker, W.L.: Technology and Field Demonstration of Robotic Search for Antarctic Meteorites. International Journal of Robotics Research 19(11), 1015–1032 (2000)
2. Wang, T.M., Zhang, T.Y., Liang, J.H., Chen, J.: Design and Field Test of a Rover Robot for Antarctic Based on Renewable Energy. Journal of Mechanical Engineering 49(19), 21–30 (2013)
3. Hajos, G.A., Jones, J.A., et al.: An Overview of Wind-driven Rovers For Planetary Exploration. In: 43th AIAA Aerospace Sciences Meeting and Exhibit, Reno, Nevada, pp. 1–13 (January 2005)
4. Xie, S.R., Chen, J.Q., Luo, J., Li, H.Y., Yao, J.F.: Gu, Jason.: Dynamic Analysis and Control System of Spherical Robot for Polar Region Scientific Research. In: Proceeding of the IEEE International Conference on Robotics and Biomimetics, Shenzhen, China, pp. 2540–2545 (December 2013)
5. Ray, L., Price, A., Streeter, A., Denton, D., Lever, J.H.: Design of a Mobile Robot for Instrument Network Deployment in Antarctica. In: Proceedings of the 2005 IEEE International Conference on Robotics and Automation, Barcelona, Spain, pp. 2123–2128 (2005)
6. Lever, J.H., Ray, L.E.: Revised Solar-power Budget for Cool Robot Polar Science Campaigns. Cold Regions Science and Technology 52, 177–190 (2008)
7. Liang, J.H., Lei, X.S., Wang, S., Wu, Y.L., Wang, T.M.: A Small Unmanned Aerial Vehicle for Polar Research. In: Proceedings of the IEEE International Conference on Automation and Logistics, Qingdao, China, pp. 1178–1183 (September 2008)
8. Li, H.B., Xiao, Y.L.: Singularity-free Azimuth/heading Error Algorithm of Lateral Guidance for Lifting Re-entry Vehicle. Journal of Astronautics 29(3), 901–906 (2008) (Chinese)
9. Li, B.R., Qin, W.J., Guo, J.X., Zhao, Y.W., Liang, J.H.: The Experiment and Application of Intelligent Robot Techniques in the Antarctic Expedition. Chinese Journal of Polar Research 21(4), 336–343 (2009) (Chinese)

10. Chen, C., Bu, C.G., He, Y.Q., Han, J.D.: Environment Modeling for Long-range Polar Rover Robots. Science China 58, 75–82 (2013) (Chinese)
11. Williams, R.M., Ray, L.E., Lever, J.H.: Autonomous Robotic Ground Penetrating Radar. In: Proceeding of the IEEE International Conference on Imaging Systems and Techniques (IST), pp. 7–12 (2012)
12. Faieghi, M.R., Delavari, H., Baleanu, D.: A Novel Adaptive Controller for Two-degree of Freedom Polar Robot with Unknown Perturbations. Commun Nonlinear Sci Numer Simulat 17, 1021–1030 (2012)
13. Lever, J.H., Delaney, A.J., Ray, L.E., et al.: Autonomous GPR Surveys Using the Polar Rover Yeti. Journal of Field Robotics 30(2), 194–215 (2013)

DSE Stacker: Close-Packing Stacker
with Double-Screw Elevating for Reagent Dispenser

Dan Wang, Wenzeng Zhang, and Zhenguo Sun

Key Lab for Advanced Materials Processing Technology (Ministry of Education),
Dept. of Mechanical Engineering, Tsinghua University, Beijing 100084, China
danwang1230@gmail.com, wenzeng@mail.tsinghua.edu.cn

Abstract. Traditional stackers for reagent dispensers are complex and cumbersome in structure since the transmission systems in the chambers are as tall as the chambers storing separated microplates. This paper designs a novel close-packing stacker system with function of elevating and separating of microplates, called DSE stacker. The DSE stacker, driven by two motors, can manipulate close-packed microplates one by one in a tall chamber. The DSE stacker consists of two symmetrical screws, two stepper motors, a chamber, a carrier, and transmissions. In a DSE output stacker, the lowest microplate can be extracted, separated from upper microplates and then transferred automatically to a carrier under the chamber. In a DSE input stacker, the process is reversed. This paper gives framework on kinematics of the DSE stacker system in detail. The results of analyses show that the stacker system is valid.

Keywords: Reagent dispenser, stacker system, close-packing, microplate manipulation, screw mechanism.

1   Introduction

Microplates including filter microplates, PCR micro-plates and deep well plates [1, 2], are commonly used in medical, chemical and biological laboratories, especially for High Throughput Screening (HTS) [3-5]. Compared with manually dispensing, reagent dispenser, applied to automatically fill microplate with media, is high efficient and labor saving.

Efficiency and stability of stackers are important for bio-reagent dispenser, which is still a research focus since there are still some difficult problems for a long time, among which elevating and separating mechanism of stacker is key technology. Stacker system can be used to store microplates, extract empty microplates from the stacker, transfer microplates to the dispensing position, and insert filled ones into the stacker. In recent years, a lot of achievements have been made in this area. According to arrangement mode of microplates, stacker system with elevating and separating of microplates has two main kinds: gap-packed (GP) stacker and close-packed (CP) stacker.

X. Zhang et al. (Eds.): ICIRA 2014, Part II, LNAI 8918, pp. 555–566, 2014.
© Springer International Publishing Switzerland 2014

Traditional gap-packed (GP) stacker for reagent dispenser driven by one motor is cumbersome and too complex in structure. For example, a GP stacker [6] includes two motor and two transmission chains with equally spaced teeth to separate microplates one from another, which leads to low space utilization since there is inevitable gaps between two adjacent microplates. What is more, the transmission system in the chambers is as tall as the chamber of separated microplates, leading to complicated structure and high demand of installation and machining accuracy.

In addition, traditional close-packed (CP) stackers driven by more than three motors are smaller and complicated in control. Many expensive commercial products, such as the Viafill stacker, the BioStack stacker and the Connect stacker [7], are developed and improved in recent decades. The CP stacker is driven by at least three motors: the first one for grabbing or releasing the upper microplates, the second one for pushing a plate vertically to fetch the lowest microplate, and the third one for transferring the microplate on the plate to the dispensing position. The motor action process of CP stacker is complex, which results in a complicated control system.

This paper designs a novel close-packing stacker system with function of elevating and separating of microplates, called DSE stacker.

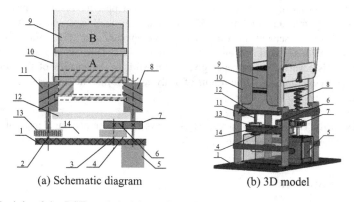

(a) Schematic diagram   (b) 3D model

Fig. 1. Principle of the DSE stacker. 1-base; 2-passive shaft; 3-active shaft; 4-active pulley; 5-motor; 6-second gear; 7-first gear; 8- active screw; 9-microplate; 10-chamber; 11-passive screw; 12-carrier; 13-passive pulley; 14-belt.

2   Design of DSE Stacker

2.1   Components of DSE Stacker

DSE stacker, shown in Fig. 1, consists of two symmetrical screws, two reduced stepper motors, transmission mechanisms, a chamber, a carrier and close-packed microplates. The shaft of the active screw is sleeved inside the first gear; the shaft of the passive screw is sleeved inside the passive pulley. The motor is fixed in the base. The first gear sleeves the motor output shaft, while the second gear sleeves the active shaft. The active pulley sleeves the active shaft and similarly, the passive pulley sleeves the passive shaft. A transmission mechanism connects the active and passive pulley. The passive shaft and the active shaft are sleeved inside the base. The active pulley, passive pulley and transmission mechanism can be selected as teeth belt, flat belt, tendon or chain driving [8]. The chamber installed on the base can be moved by a technician after the whole dispensing process has been completed. Microplates are close packed in the chamber and furthermore the bottom surface of the lowest microplate is in contact with the active screw and passive screw. Driven by the second motor and second transmission mechanism, the carrier is installed just under the chamber to move the microplate to the dispensing position.

2.2 Motion Process of DSE Stacker

The motion process of DSE stacker is like this. First of all, the motor rotates, driving the first gear to rotate, and further driving the second gear and the active pulley to rotate reversely. At the same time, the passive pulley is driven through the belt transmission, and then the passive screw rotates. Because the rotation directions and the handedness of the active and passive screws are inverse, they support and drive the lowest microplate moving along the spirals synchronously. As a result, the lowest microplate goes upward or downward according to the rotary direction of the motor.

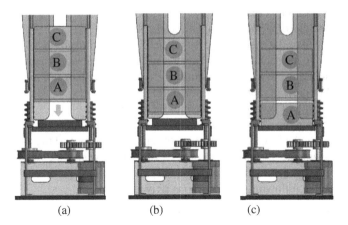

Fig. 2. Dropping process of microplate in DSE stacker

When the motor rotates in the direction as indicated by the arrow shown in Fig. 1(a), microplate A goes downward along the spirals of the screws, as shown in Fig. 2. While microplate A reaches to the endings of the spirals, it is out of touch with the active and passive screws, leaving the chamber to the carrier. At this time, the top surfaces of the screws support the projecting edges (namely, flange) of microplate B, and consequently support the whole upper close-packed microplate stack. Till now, microplate B takes the place of microplate A and the next circulation begins.

When the motor rotates the other way around, the active and passive screws pick up the microplate, which has been dispensed with reagent and momentarily stored in the carrier, from the carrier to the stacker. Finally, all microplates are packed one above another in the chamber.

3   Analysis of DSE Stacker

This section gives framework on kinematics of DSE stacker. Contact force [9] during elevating and separating process is analyzed firstly. After that, the simulation of the contact force of the screws is given. The results show that the screws can lift the lowest microplate upward and downward and meanwhile support the upper close-packed microplate stack stably. The contact force is related to the number of the microplates and the mechanical structure of the screws. At the end of this section, the total time of fetching out the lowest microplate is analyzed in detail.

Let:

N_1, N_2- supporting force of the active and passive screws caused by the microplate stack respectively, N;

f_1, f_2- friction force exerted on the active and passive screws by the microplate stack respectively, N;

Fs_1, Fs_2- resultant force of the active and passive screws, N;

ψ_1, ψ_2 - lead angle of the active and passive screws, rad;

H_1, H_2- axial length of the active and passive screws, mm;

P_1, P_2- thread pitch of the active and passive screws, mm;

μ_1, μ_2 - friction coefficient respectively;

G- gravity of the close-packed microplate stack, N;

n- rotational speed of the motor, rpm; h- height of one microplate, mm;

G_m- gravity of one microplate, N; n_m- number of microplates supported by the screws;

h_e- height of the flange of microplate, mm; w_e- width of the flange of microplate, mm;

w_s- width of the screw in radical direction, mm;

h_t- least axial length of two adjacent teeth of the screw, mm;

d- distance between the screw and the upper portion of the microplate, mm;

t_1- time of moving the lowest microplate between the chamber and the carrier, s.

In order to support the microplate stack stably and evenly, the active screw and passive screw must have identical materials and structure parameters except the handedness, so the following relationship is arrived at:

$$\begin{cases} P_1 = P_2 = P \\ \psi_1 = \psi_2 = \psi \\ H_1 = H_2 = H \\ \mu_1 = \mu_2 = \mu \end{cases} \tag{1}$$

Fig. 3 shows the contact force analysis of the DSE stacker. For the materials and structure parameters of the two screws are identical except the handedness, and the microplates are put in the middle of the two screws, one gets:

$$\begin{cases} N_1 = N_2 = N \\ f_1 = f_2 = f \\ F_{s_1} = F_{s_2} = F_s \end{cases} \tag{2}$$

O stands for the gravity center of the microplate stack in the chamber. When the lowest microplate is supported by the screws, the whole system relative to point O is balanced according to the force analysis. The following relationship is arrived at:

$$\begin{cases} G = n_m G_m \\ F_s = G/2 \\ f = F_s \sin\psi \\ N = F_s \cos\psi \\ f = \mu N \end{cases} \tag{3}$$

Combining the relationships above, the expressions of f and N is arrived at:

$$\begin{cases} f = \dfrac{1}{2} n_m G_m \sin\psi \\ N = \dfrac{1}{2} n_m G_m \cos\psi \end{cases} \tag{4}$$

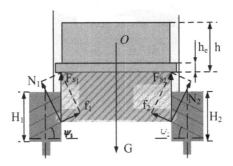

Fig. 3. Contact force analysis of DSE stacker

Set f as functions of ψ and n_m. After the microplate has been dispensed with reagent, the maximum mass of one microplate is about 285g, so G_m=2.85 N. The range of ψ is from 0~90° and the range of n is from 1 to 50. The relationships among f, n_m and ψ are shown in Fig. 4(a). Set N as functions of ψ and n_m. The relationships of N, n_m and ψ are shown in Fig. 4(b).

The conclusions from these figures are listed below:

1) When the number of microplates n_m increases, f and N will respectively increase distinctly. In order to improve the lifetime of the microplates and the screws, the number of microplates should be limited according to f and N.

2) When the lead angle of the screws ψ increases, f will increase while N decreases obviously, which means the lead angle of the active and passive screws cannot be too large, otherwise the supporting force is too little to brace the microplate stack stably.

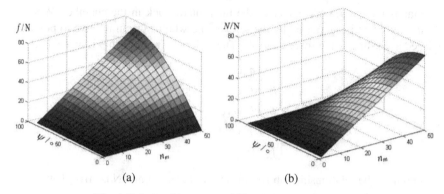

Fig. 4. Relationships among f, N and the other variables

Aiming to realize the function of elevating the microplate upward and downward, and supporting the upper close-packed microplate stack, the structure parameters of the screws are strictly restricted.

When the microplate stack goes downward along the spirals, as shown in Fig. 5, the screws cannot meddle with the operated microplate stack and should support the lowest microplate through the flange, so one gets:

$$\begin{cases} h_t > h_e \\ w_s > w_e \\ d > 0 \end{cases} \tag{5}$$

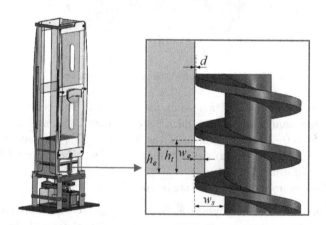

Fig. 5. Structure of the screws

As the lowest microplate is in the critical condition of leaving the screws, the screws should support the upper close-packed microplate stack in the chamber. In this way, the movements are continuous and vibrations are reduced as far as possible, so the following relationship is arrived at:

$$H_1 = H_2 = H \geq h \tag{6}$$

If $H < h$, when the lowest microplate gets to the ending of the spirals and is about to be released, the upper microplate stack has not come into contact with the screws, and inevitably the stack has to freely fall to the top of the screws, resulting in vibrations and impulsions. The DSE stacker mentioned in this paper can manipulate any microplates compliant to ANSI and furthermore any object that has structure similar to microplate. The key point is that there is a flange, like that of microplate, which can be supported by suitable screws. This kind of escapement mechanism is of great importance to every walk of life that requires extraction or insertion of one object from or to the chamber of the objects respectively.

The speed of elevating the microplate upward and downward is determined by the rotational speed of the motor n and the thread pitch of the active and passive screws P. The time of moving the lowest microplate from the chamber to the carrier and the time of moving the microplate in the carrier to the lowest position in the chamber is equal, denoted by t_1. The expression of t_1 is arrived at:

$$t_1 = \frac{60H}{n \cdot P} \tag{7}$$

4   Design of Reagent Dispenser with DSE Stacker

At the beginning of this section, peristaltic and syringe pumps are introduced. Then there comes the components of the reagent dispenser with DSE stacker. Thirdly, working process is talked about in detail. Fourthly, the kinematics analyses are given. This section ends up with reagent dispenser specifications.

4.1   Peristaltic Pump and Syringe Pump

Peristaltic pumps [10], world-widely used in chemical analysis and drug screening, deliver fluid by exerting force on flexible tubes. Because the flexible tubes containing fluid are disposable, the pump mechanisms do not get in touch with the fluid, ensuring sterility and preventing contamination. Typically, the flexible tubes are engaged with some rollers, which squeeze the tubes lengthwise. As a result, the fluid in the flexible tubes are forced to move. Peristaltic pumps are chosen for positive displacement and flow metering characteristic [11]. In addition, this kind of pumps have an external reservoir and therefore can transfer bulk liquids, which is necessary for HTS.

Syringe pumps [12], injecting liquid in a piston motion, are more accurate and more expensive compared with the syringe pumps. Most importantly, syringes are applied. The motions of syringe pumps are like this: a pump motor moves a pusher forward; the syringe plunger is compressed resulting in liquid dispensing; If the rotation direction of the motor is reversed, the plunger is moved back to fill the syringe with fluid.

The comparison of parameters of peristaltic and syringe pumps is shown in Table 1.Taking into consideration all the advantages and disadvantages of the two kinds of pumps, the reagent dispenser developed in this paper applies the 8-channel peristaltic pump.

Table 1. Parameters of peristaltic pump and syringe pump

	Peristaltic pumps	Syringe pumps
Precision	<0.1%	1~3%
Price	High	Low
Operation channels	1~10	1~16
Volumes	Finite	Bulk
Pulsing flow	No	Yes
Operation pressure	Up to 3000 psi	30 psi or less

4.2   Components of the DSE Reagent Dispenser

The vital functions of the DSE reagent dispensers [13] are shown in Fig. 6. The fundamental function of the DSE reagent dispensers is to fulfill the automated dispensing. In other words, reagent is transferred from reservoir to microplates automatically; furthermore, the operation objectives of the DSE reagent dispenser are not limited to one microplate but tens of microplates. In order to improve the space utilization, these microplates are closed-stacked in the vertical direction; moreover, in order to be injected with reagent, microplates must be extracted out of the microplate stack one after another; in addition, the microplates should be transported from one station to another and the liquid should be dispensed from the reservoir to the microplate; finally, the filled microplate should be inserted to the microplate stack.

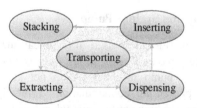

Fig. 6. Functions of the DSE reagent dispenser

The structure design of the reagent dispenser is to realize all the functions aforementioned, as shown in Fig. 7(a). First, the peristaltic pump can deliver the goal liquid from reservoir to microplates. Second, there are two DSE stacker systems: the output DSE stacker is used to store the close-packed microplate stack, the microplate of which is empty; the input DSE stacker is used to store the close-packed filled microplate stack; the input and output DSE stacker are applied to fulfill the functions of extraction and insertion, which means fetching one microplate out of the output chamber and reversely putting one microplate into the input chamber. Third, the carrier is applied to transport the microplate plate between the output DSE stacker, the peristaltic pump and the input DSE stacker.

Fig. 7(a) indicates the schematic diagram of the components of the DSE reagent dispenser. Fig. 7(b) shows the 3D model of the components and flexible tubes are not shown. A coordinate system is established: the plane XY coincides with the upper plane of the carrier; the center O coincides with the center line of the output DSE stacker; the positive direction of the x-axis is coincided with the direction from the output DSE stacker to the input DSE stacker; the positive direction of the z-axis coincides the upward vertical direction.

The DSE input and output stackers are installed on the base. To ensure the microplates are transported in the plane YZ, the four side faces of the two DSE stackers are parallel respectively. The carrier driven by the transmission mechanism, and mounted on the base, can move along y-axis. The peristaltic pump is purchased and integrated with the whole device.

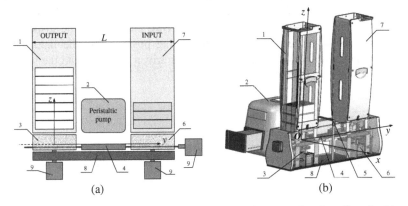

(a)   (b)

Fig. 7. Components of the DSE reagent dispenser. 1-output chamber; 2-peristaltic pump 3-output DSE stacker; 4-carrier; 5-pipe holder; 6-input DSE stacker; 7-input chamber; 8-base; 9-motor

4.3   Working Process of the DSE Reagent Dispenser

The working process of the DSE reagent dispenser is explained as below. At the beginning, the carrier, driven by the motor, moves to the position just under the output DSE chamber. Then the double-screw of the output DSE stacker works, resulting in the extraction of the lowest microplate to the carrier and the support of the upper microplate stack. Next, the carrier moves the microplate to the peristaltic pump station, where the microplate is dispensed with reagent. Afterwards, the microplate, carried by the carrier, is transferred to the position just under the input DSE chamber, and then the double-screw of the input DSE stacker works, resulting in the insertion of the microplate to the lowest position of the microplate stack, which indicates the ending of one cycle.

4.4   Mechanical Analysis of Reagent Dispenser

The dimensions of the DSE reagent dispenser (without the peristaltic pump) is as follows: the length L is 583mm; the width W is 150mm; the height H_d is 665mm. The number of the wells of microplates is diverse, such as 48, 96, 384, and so on. In this paper, the 96-well microplates are taken as an example. There are 12 wells per column and 8 wells per row.

Let t_2 represent time of moving one microplate between the output DSE stacker and the peristaltic pump, s; t_3 stands for time of moving one microplate between the peristaltic pump and the input DSE stacker, s; t_p represents time of dispensing reagent into one microplate, s; t_r stands for time of the carrier transferred from the input DSE stacker to the output DSE stacker, s; t stands for time of completing one cycle, s;

So one gets:

$$t = 2t_1 + t_2 + t_3 + t_p + t_r \tag{8}$$

According to the using condition, and the type of the motor and transmission mechanism, set $n=400r/min$, $P=7.5mm$, $h=39mm$, $H=40mm$, $n_m=15$, $t_r=10$s and then one gets $t_1=0.8$s. The distance between the output DSE stacker and the dispensing position l_1 , and the distance between the dispensing position and the input DSE stacker l_2 are equal, and $l_1=l_2=250$mm. The carrier is driven by screw and the lead P_h is 2mm, so:

$$t_2 = \frac{60 l_1}{n \cdot P_h} = 18.75s \tag{9}$$

In the moving direction of the carrier, the distance between the center of the first and last well of the microplate l_m is 100mm, so the following relationship is arrived at:

$$t_3 = \frac{60(l_2 - l_m)}{n \cdot P_h} = 11.25s \tag{10}$$

According to specifications of the peristaltic pump, it is reasonable to set $t_p=15$s, so according to (8),

$$t = 56.6s \tag{11}$$

At different time, the operated microplate is in different position, as shown in Fig. 8. It is obvious that the lowest microplate is moving on the y-z plane.

When the microplate is extracting from the output chamber to the carrier by the double-screw of the output DSE stacker, the y value remains unchanged and the z value decreases. The speed in z direction changes according to the rotation rate of the motor that drives the screws.

While the microplate moving to the dispensing station, the z value does not change while the y value increases. The speed in y direction changes according to the rotation rate of the motor that drives the carrier. As mentioned, the 8-channel peristaltic pump and 96-well microplates are chosen. At the time of dispensing reagent to the microplate, the longitudinal direction of the microplate is coincided with the direction from the output DSE stacker to the input DSE stacker. Wells in the same row are dispensed at the same time and all the 12 rows of wells are filled with water one after another, so the Fig. 8(a) shows stair-like pattern at the time segment t_p.

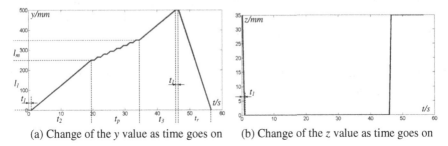

(a) Change of the y value as time goes on   (b) Change of the z value as time goes on

Fig. 8. Position of the lowest microplate at different time

As time goes on, the carrier transports the microplate from the dispensing station to the input DSE stacker, the y value increases and the z value is constant, as shown in Fig. 8(b). When the microplate is inserted from carrier to the input chamber of the output DSE stacker by the double-screw, the y value remains unchanged and the z value increases. The speed of inserting changes according to the rotation rate of the motor that drives the screws. At the time segment t_r, the carrier moves back to the output DSE stacker and another cycle starts.

4.5   Features of the DSE Reagent Dispenser

There are some vital features of the DSE reagent dispenser, which overcomes the shortages of the traditional dispensers. The advantages of the DSE reagent dispenser mentioned in this paper are presented as follows:

The number of motors used in the DSE stacker is less than that of motors used in the CP stacker, resulting in the reduction of the complexity of the control system. The double-screw applied in the DSE stacker is installed between the chamber and the carrier, and does not need extra space. However, the push-plate applied in the CP stacker has to move below the carrier to let the carrier move freely, and thus the upward and downward lifting of the push-plate wastes time, reduces efficiency and increases energy consumption. Compared with the GP stacker, the microplates are close-packed in the DSE stacker, which improves the space utilization. Besides, the double-screw is shorter than the tall transmission system used in the GP stacker. Therefore, the DSE stacker is simple in structure and easy to install.

In order to improve the stability, the number of the positive and passive screws of the DSE stacker can be increased, so the supporting force of each screw decreases, which means long lifespan or more microplates in the chamber.

The DSE reagent dispenser, or to be more specific, the DSE stacker can manipulate all ANSI-compliant microplates, and even all objects having the similar structure, which means the dispenser can be used in any situation that has the similar functional requirement.

5   Conclusion

In this paper, a novel close-packing stacker system with function of elevating and separating of microplates (DSE stacker) is designed. The DSE stacker can be designed as an output stacker or an input stacker for reagent dispenser. The DSE stacker is high efficient, simple in structure, control and compatible for different kinds of microplates. The results of analyses show that the stacker system is valid.

Acknowledgement. This paper was supported by the National Key Scientific Instrument/Equipment Development Project of China (No.2012YQ150087).

References

1. Tan, H.Y., Ng, T.W., Neild, A., et al.: Point spread function effect in image-based fluorescent microplate detection. Analytical Biochemistry 397(2), 256–258 (2010)
2. Mayr, L.M., Fuerst, P.: The future of high-throughput screening. Journal of Biomolecular Screening 13(6), 443–448 (2008)
3. Rhode, H., Schulze, M., Renard, S., et al.: An improved method for checking HTS/uHTS liquid-handling systems. Journal of Biomolecular Screening 9(8), 726–733 (2004)
4. Johnston, P.A., Johnston, P.A.: Cellular platforms for HTS: Three case studies. Drug Discovery Today 7(6), 353–363 (2002)
5. Archer, J.R.: History, evolution, and trends in compound management for high throughput screening. Assay and Drug Development Technologies 2(6), 675–681 (2004)
6. Hitch, J.R.: Automated labware storage system: U.S. Patent 6,099,230 (2000)
7. Felton, M.J.: Product Review: Liquid handling: dispensing reliability. Analytical Chemistry 75(17), 397-A (2003)
8. Li, G., Li, B., Sun, J., et al.: Development of a directly self-adaptive robot hand with pulley-belt mechanism. International Journal of Precision Engineering and Manufacturing 14(8), 1361–1368 (2013)
9. Zhang, W., Chen, Q., Sun, Z., Zhao, D.: Passive adaptive grasp multi-fingered humanoid robot hand with high under-actuated function. In: Proc. of IEEE Inter. Conf. on Robotics and Automation, pp. 2216–2221 (April 2004)
10. Shkolnikov, V., Ramunas, J., Santiago, J.G.: A self-priming, roller-free, miniature, peristaltic pump operable with a single, reciprocating actuator. Sensors and Actuators A: Physical 160(1), 141–146 (2010)
11. Rhie, W., Higuchi, T.: Design and fabrication of a screw-driven multi-channel peristaltic pump for portable microfluidic devices. Journal of Micromechanics and Microengineering 20(8), 085036 (2010)
12. Li, Z., Mak, S.Y., Sauret, A., Shum, H.C.: Syringe-pump-induced fluctuation in all-aqueous microfluidic system implications for flow rate accuracy. Lab on a Chip 14(4), 744–749 (2014)
13. Lu, G., Tan, H.Y., Neild, A., et al.: Liquid filling in standard circular well microplates. Journal of Applied Physics 108(12), 124701 (2010)

Topological Design of Hinge-Free Compliant Mechanisms Using the Node Design Variables Method

Jinqing Zhan[*], Kang Yang, and Zhichao Huang

School of Mechanical and Electrical Engineering, East China Jiaotong Uinveristy,
Nanchang 330013, China
zhan_jq@126.com, {yktianshi,hzcosu}@163.com

Abstract. The designs of compliant mechanisms using topology optimization typically lead to de facto hinges in the created mechanisms which can cause high stress concentration and are difficult to manufacturing. Topology optimization of hinge-free compliant mechanisms using the node design variables method is proposed. Within defined sub-domain, the projection function independent on element mesh is adopted to represent the relationship of node design variables and node density variables, which can achieve the minimum length scale constraint of the topological solution to avoid generating the de facto hinges. The method of moving asymptotes is adopted to solve the topology optimization problem. The numerical examples are presented to show the feasibility of the approach. It can obtain hinge-free compliant mechanisms which is convenient for manufacturing.

Keywords: compliant mechanisms, hinge-free, topology optimization, node design variables method.

1 Introduction

With the development of the micro and nanotechnology piezoelectric actuating technique becomes one of the most important techniques in the fields of micro-manipulation and micro-positioning. Compliant mechanisms are flexible structures which can generate some desired motions by undergoing elastic deformation instead of through rigid linkage as in rigid body mechanisms. It is suitable to be used in the fields of precision engineering. The suitability is mainly due to the monolithic nature of compliant mechanisms with less or no rigid body joints like reduced friction and backlash losses, part assembly costs, noise and vibrations, and easy unitization [1-3].

Topology optimization of compliant mechanisms has been drawn more and more attentions because it only needs to designate a design domain and the positions of the inputs and outputs, without a known rigid-link mechanism. One significant challenge when applying the continuum topology optimization method to compliant mechanism

---

[*] This work was financially supported by the National Natural Science Foundation of China (51305136) and the Science and Technology Plan Projects of Jiangxi Provincial Education Department (GJJ13319).

designs is the de facto hinges. How to avoid generating the de facto hinges during the topology optimization of compliant mechanisms has been paid more attention in recent years. Sigmund [4] introduced a new class of morphology-based restriction schemes that work as density filters to avoid the de facto hinges in the optimal mechanisms. Zhou [5, 6] proposed the hybrid discretization model for topology optimization of compliant mechanisms which can eliminate one-node connecting hinges as well as checkerboard patterns. Luo [7] and Chen [8] employed a quadratic energy functional in the objective to solving the de-facto hinge problem in the topological design of compliant mechanisms. Wang [9] developed an intrinsic characteristic stiffness method to design hinge-free compliant mechanisms. Takezawa [10] proposed a new level set method based on the phase field method and sensitivity analysis to design distributed compliant mechanisms. Zhu [11, 12] developed several new objective functions using level set method to avoid generating the de facto hinges in the created mechanisms. Despite all mentioned attempts, an efficient and effective scheme is still needed.

In this study, we proposed a new method for topology optimization of hinge-free compliant mechanisms using node design variables method in order to avoid generating the de facto hinges in the created mechanisms. Within defined sub-domain, the projection function independent on element mesh is adopted to represent the relationship of node design variables and node density variables, which can achieve the minimum length scale constraint of the topological solution to avoid one-node connecting hinges. The method of moving asymptotes is adopted to solve the topology optimization problem. The numerical example is presented to show the feasibility of the proposed approach. It can obtain hinge-free compliant mechanisms.

2   Optimization Formulations of Hinge-Free Compliant Mechanisms

2.1   The Node Design Variables Method

At present, the element densities are usually used to approximate the displacement field in a topology optimization problem. The displacement field and element density of each point in the domain is denoted as Q4/U implementation, as shown in Fig.1(a), where Q4 is the element used to approximate the displacement field using bi-linear interpolation method and U refers to uniform material density inside each element. The formulation is developed by

$$u_e = \sum_{i=1}^{m} N_i u_i \quad \rho_e = \rho .$$

(1)

where u_e is displacement of any point in an element, N_i is the shape function associated with node i , u_i is node displacement vector, m is the number of nodes of each element, ρ_e is material density of each point in an element, and ρ is element density .

One of the problems with the Q4/U implementation is that it only can ensure C_0 continuity of the displacement field. However, the element density field has C_{-1} continuity which frequently results in the phenomenon of numerical instability such as checkerboard patterns.

The node densities are used as the design variables to approximate the material density field. The displacement field and element density of each point is denoted as Q4/Q4 implementation, as shown in Fig.1b. Material density of each point inside an element is not uniform. The displacement vector and material densities is approximated using bi-linear interpolation method, which is expressed by

$$u_e = \sum_{i=1}^{m} N_i u_i \quad \rho_e = \sum_{i=1}^{m} N_i \rho_i .$$ (2)

where ρ_i is the node density associated with node i.

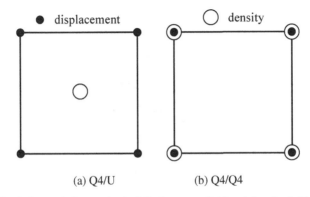

(a) Q4/U   (b) Q4/Q4

Fig. 1. Interpolation method of displacement field and density field

The Q4/Q4 implementation can ensure C_0 continuity of both displacement field and density field, which can avoid the phenomenon of numerical instability [13]. However, the implementation method can not be adopted for topological design of hinge-free compliant mechanisms.

The node design variables method adopts the projection function independent on element mesh to represent the relationship of node design variables and node density variables. It can achieve the minimum length scale constraint of the topological solution which avoid generating the de facto hinges. The projection function is defined as [14-15]

$$\rho_i = \max_{j \in \Omega_i} (d_j) .$$ (3)

where ρ_i is the material density of node i, d_j is the design variable located at node j, and Ω_i is the sub-domain corresponding to node i.

The sub-domain corresponding Ω_i to node i is defined as a circle with its center located at the node i, and the radius $r_{\min}$ equals half of the minimum allowable size of structural members in the optimized compliant mechanisms, as shown in Fig.2. The distant between any node j in the sub-domain and node i must satisfy the equation

$$r = \left| r_i - r_j \right| < r_{\min} \quad j \in \Omega_i . \tag{4}$$

where r is the distant between node i and node j, r_i and r_j are the distant between node i and the reference point and the distant between node j and the reference point, respectively.

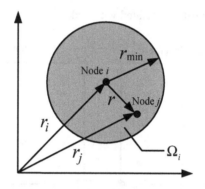

Fig. 2. Sub-domain corresponding to node

The designer-defined parameter $r_{\min}$ determines the minimum size of the topological solution. The sub-domain and related entities is illustrated in Fig.3.

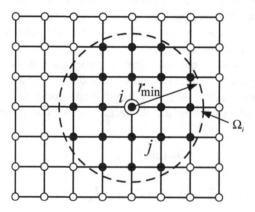

Fig. 3. Sub-domain corresponding to node

2.2   Optimization Model

The objective of the optimization problem is to maximize the output displacement at the output point and the volume is used as the constraints. The projection function independent on element mesh is adopted to avoid generating the de facto hinges in the optimized compliant mechanisms. The optimization model can be established as

$$\underset{d_1, d_2, \ldots, d_M}{Maximize}\ u_{out}(\mathbf{d})$$

$$Subject\ to:$$

$$\begin{cases} \rho_i = f(\mathbf{d}) \\ \rho_e = \sum_{i=1}^{m} N_i \rho_i \\ \mathbf{F} = \mathbf{KU} \\ \sum_{e=1}^{N} V_e \rho_e - V^* \leq 0 \\ 0 < d_{min} \leq d_j \leq 1, \quad j = 1, 2, \ldots, M \end{cases} \tag{5}$$

where u_{out} is the output displacement, $\mathbf{d}$ is the design variables, f is the projection function, ρ_i and ρ_e are the node variable and the element density, respectively. $\mathbf{F}$ represents the applied load, $\mathbf{K}$ is the global stiffness matrix, and $\mathbf{U}$ is the nodal displacement vector. V_e is the element material volume, and V^* refers to the prescribed total volume in the design. d_{min} is the lower bound on design variables and is taken to be a small positive value in order to avoid singularity of the global stiffness matrix. N and M denote the total number of finite elements and the total number of node, respectively.

2.3   Sensitivity Analysis and Solution Technique

To apply the gradient-based algorithm, the sensitivities of objective and constraint with respect to the design variables are needed. For topology optimization problem, there are typically two methods which can be applied to perform sensitivity analysis: the direct differentiation method and the adjoint method. In this work, the adjoint method is used to perform sensitivity analysis.

The sensitivity of the output displacement can be expressed by

$$\frac{\partial u_{out}}{\partial d_i} = \sum_{j \in S_i} \frac{\partial u_{out}}{\partial \rho_j} = -\sum_{j \in S_i} \boldsymbol{\lambda}^T \frac{\partial \mathbf{K}}{\partial \rho_j} \mathbf{U}. \tag{6}$$

where S_i is the set of nodes mapped form the node design variable d_i, $\boldsymbol{\lambda}$ is the adjoint displacement vector and can be found as the solution to the adjoint load problem.

$$\mathbf{K}\boldsymbol{\lambda} = \mathbf{L} . \tag{7}$$

where $\mathbf{L}$ is a unit dummy load applied at output point.

According to definition of finite element matrices, we got

$$\frac{\partial \mathbf{K}}{\partial \rho_j} = \sum_{e=1}^{N} \int_{\Omega_e} \frac{\partial \mathbf{K}_e}{\partial \rho_e} \frac{\partial \rho_e}{\partial \rho_i} d\Omega_e = \sum_{e=1}^{N} \int_{\Omega_e} \frac{\partial \mathbf{K}_e}{\partial \rho_e} N_i d\Omega_e = \sum_{e=1}^{N} \int_{\Omega_e} p\rho_e^{p-1} \mathbf{K}_e N_i d\Omega_e . \tag{8}$$

where $\mathbf{K}_e$ is the element stiffness matrix, P is the penalization power which is chosen equal to 3 in this study .

The sensitivity of the material volume is expressed by

$$\frac{\partial V}{\partial d_i} = \sum_{j \in S_i} \frac{\partial V}{\partial \rho_j} = \sum_{j \in S_i} \sum_{e=1}^{N} \int_{\Omega_e} N_i d\Omega_e . \tag{9}$$

The topology optimization problems are solved typically using two classified approaches: Optimality Criteria (OC) method [16] and mathematical programming method. The OC method has been successfully applied to topology optimizations, and it is simple and easy to be implemented. But, it is difficult to construct the explicitly expressed heuristic updating schemes for complicated objectives. In mathematical programming method, sequential convex programming approaches include the Method of Moving Asymptotes (MMA) [17], Sequential linear programming (SLP)[18] and others. MMA is more flexible and theoretically well founded to deal with complicated optimization problem. MMA is used to update the design variables in this research.

3   Numerical Example

This example demonstrates the design of a compliant inverting mechanism. The design domain, boundary conditions, and the prescribed input and output motion are sketched in Fig. 4. The size of the design domain is 80 mm by 80 mm. The elasticity modulus of the structural material is set to 1GPa and its Poisson ratio is 0.3. The input and output springs stiffness are set to k_{in} = 0.001 mN/μm and k_{out} = 0.001 mN/μm, respectively. The applied force is F_{in} = 10 mN and the maximal material volume usage is restricted to 30%. Due to the symmetric nature of this problem, only the half of the design domain is used. The design domain is discretized into 40×80 four node plane finite elements.

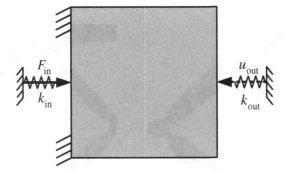

Fig. 4. Design domain of compliant inverter mechanism

The optimal topology is obtained by variable density method with sensitivity filter as shown in Fig. 5. From Fig. 5, one finds that de facto hinges occurs in the optimal mechanism. The de facto hinges can result in a sharp stress concentration which causes the mechanisms to be not useful in most real applications. Meanwhile, the optimal compliant mechanism contains much undesirable gray areas which cause it to be not convenient for manufacturing.

Fig. 5. Optimal mechanism obtained by the variable density method with filter technique

The optimal compliant mechanisms are obtained by the using node design variables method without filter technique as shown in Fig. 6. Compared Fig. 6 with Fig. 5, one can find that de facto are completely eliminated in the created mechanisms. It shows that the projection function method achieve the minimum length scale constraint of the topological solution to avoid generating the de facto hinges in the optimal compliant mechanisms. The hinge-free compliant mechanisms can be obtained using the node design variables. Moreover, the created mechanisms have distinct topology which is meaningful and economical for fabrication.

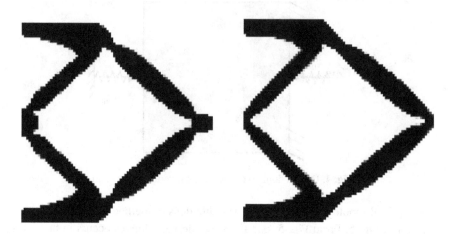

(a) The minimum allowable radius $r_{min} = 1$ (b) The minimum allowable radius $r_{min} = 2$

Fig. 6. Optimal mechanism obtained by the node design variables method

Fig. 6 shows optimal topologies for the minimum allowable radii of 1 and 2. It is clear that decreasing minimum allowable radii can result in thinner structural members in the optimal hinge-fee compliant mechanism. The node design variables method provides direct control over member sizes through the minimum allowable radius.

4   Conclusions

In this study, a new method for topology optimization of hinge-free compliant mechanisms using the node design variables method is presented.

(1) The projection function independent on element mesh is adopted to represent the relationship of node design variables and node density variables in the node design variables method. The minimum length scale constraint can be achieved using the node design variables method to avoid generating the de facto hinges in the created mechanism.

(2) The optimal hinge-free compliant mechanisms obtained using the node design variables method have distinct topology which is meaningful and economical for manufacturing.

(3) The node design variables method can provide direct control over member sizes through varying the minimum allowable radius. Decreasing the minimum allowable radius can lead to finer member in the final optimal mechanisms.

References

1. Howell, L.L.: Compliant mechanisms. John Wiley & Sons, New York (2001)
2. Tian, Y., Shirinzadeh, B., Zhang, D., Zhong, Y.: Three flexure hinges for compliant mechanisms designs based on dimensionless graph analysis. Precis. Eng. 34, 92–100 (2010)
3. Zhan, J.Q., Zhang, X.M.: Topology optimization of compliant mechanisms with geometrical nonlinearities using the ground structure approach. Chin. J. Mech. Eng. 24, 257–263 (2011)
4. Sigmund, O.: Morphology-based black and white filters for topology optimization. Struct. Multidisc. Optim. 33, 401–424 (2007)
5. Zhou, H.: Topology optimization of compliant mechanisms using hybrid discretization model. J. Mech. Design. 132, 111003(1–8) (2010)
6. Zhou, H., Mandala, A.R.: Topology optimization of compliant mechanisms using the improved quadrilateral discretization model. J. Mech. Robot. 4, 021007(1–9) (2012)
7. Luo, J.Z., Luo, Z., Chen, S.K., Tong, L.Y., Wang, M.Y.: A new level set method for systematic design of hinge-free compliant mechanisms. Comput. Methods. Appl. Mech. Eng. 198(2), 318–331 (2008)
8. Chen, S.K., Wang, M.Y., Liu, A.Q.: Shape feature control in structural topology optimization. Comput. Aided. Design 40, 951–962 (2008)
9. Wang, M.Y., Chen, S.K.: Compliant mechanism optimization: analysis and design with intrinsic characteristic stiffness. Mech. Based. Design. Struct. 37, 183–200 (2009)
10. Takezawa, A., Nishiwaki, S., Kitamura, M.: Shape and topology optimization based on the phase field method and sensitivity analysis. J. Comput. Phy. 229, 2697–2718 (2010)
11. Zhu, B.L., Zhang, X.M., Wang, N.F.: A new level set method for topology optimization of distributed compliant mechanisms. Int. J. Numer. Methods. Eng. 91, 843–871 (2012)
12. Zhu, B.L., Zhang, X.M., Wang, N.F.: Topology optimization of hinge-free compliant mechanisms with multiple outputs using level set method. Struct. Multidisc. Optim. 47, 659–672 (2013)
13. Matsui, K., Terada, K.: Continuous approximation of material distribution for topology optimization. Int. J. Numer. Methods. Eng. 59, 1925–1944 (2004)
14. Guest, J.K., Prevos, J.H., Belytschko, T.: Achieving minimum length scale in topology optimization using nodal design variables and projection functions. Int. J. Numer. Methods. Eng. 61, 238–254 (2004)
15. Chau, H.L.: Achieving minimum length scale and design constraint in topology optimization: A new approach. Master's thesis, University of Illinois at Urbana-Champaign, Urbana (2006)
16. Zhou, M., Rozvany, G.: The COC algorithm, part II: topological, geometry and generalized shape optimization. Comput. Methods. Appl. Mech. Eng. 89, 197–224 (1991)
17. Svanberg, K.: The method of moving asymptotes: A new method for structural optimization. Int. J. Numer. Methods. Eng. 42, 359–373 (1987)
18. Nishiwaki, S., Frecher, M.I., Min, S., Kikichi, N.: Topology optimization of compliant mechanisms using the homogenization method. Int. J. Numer. Methods. Eng. 42, 535–559 (1998)

Research on Lightweight Optimization Design for Gear Box

Guoying Yang, Jianxin Zhang, Qiang Zhang[*], and Xiaopeng Wei

Key Lab of Advanced Design and Intelligent Computing, Ministry of Education,
Dalian University, Dalian, China, 116622
zhangq26@126.com

Abstract. In order to meet the requirements of saving and environmental protection, the development of automotive lightweight is imminent. Through the finite element analysis of transmission housing, a method based on improved genetic algorithm is proposed for lightweight automobile transmission housing. To improve the efficiency and accuracy, the Latin method is used in the experimental design to generate test sample points, which combined with the response surface of polynomial technology to create an approximate model. Finally, an improved GA is applied to resolve the optimal parameters. The results show that the mass of gear box is reduced under the premise such as strength, stiffness and vibration resistance. This method can provide certain engineering guidance for the lightweight optimization of gear box.

Keywords: gear box, approximation model, GA, lightweight optimization.

1 Introduction

Yearly highlight environmental, energy and safety issues as the car brought, automotive lightweight has become an important direction on the development of automobile industry. Currently, there are three major automotive lightweight approaches [1-3]. The use of lightweight automotive technology to achieve lightweight vehicles, such as TWB, thermoforming technology, etc; The use of new lightweight materials to achieve lightweight vehicles, such as aluminum, magnesium alloy, plastic, etc; the other is to optimize the car's structure to achieve lightweight, such as making its parts become hollow, thin-walled and miniaturization, which includes size optimization and structural optimization.

Automotive companies from the developed world had begun to study automotive lightweight technology in the 1980s, which is late relatively in china. There is a big gap between domestic and foreign countries on automobile lightweight technology at this stage. The application of new materials requires a longer development cycle and higher research and development costs. In comparison, automotive structural optimization is not only cost less, but also achieving better results on lightweight.

---

[*] Corresponding author.

X. Zhang et al. (Eds.): ICIRA 2014, Part II, LNAI 8918, pp. 576–585, 2014.

In this paper, we select an automatic housing as a structural model for an automobile. On the basis of meeting the mechanical and vibration characteristics, we adopt experimental sampling techniques, which combine with approximate model method and evolutionary algorithm to optimize the design of the thickness of the transmission housing, and achieved satisfactory results. The whole process of the flowchart is shown in Figure 1.

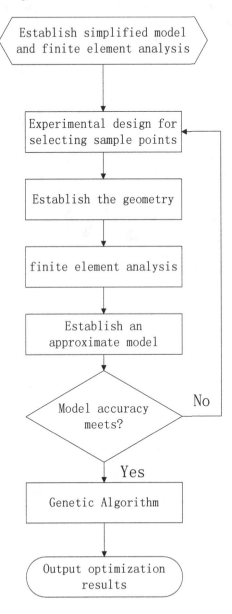

Fig. 1. Whole process of the flowchart

2   Establish a Simplified Model and Finite Element Analysis

2.1   Build Three-Dimensional Model

A three-dimensional model is built according to the model of automatic transmission housing from a company as it shown in Figure.2 (a). In this paper, in order to facilitate the next step for finite element analysis, the structure of models have little impact should be simplified. It simplify as follows: simplified fillets, holes, thread, chamfer, etc. based on the fact that there is no prejudice to the finite element analysis.

(a)   Entity model of automatic gearbox

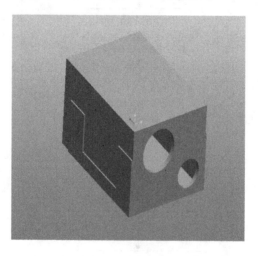

(b) Simplified three-dimensional model of automatic gearbox

Fig. 2. Build three-dimensional model

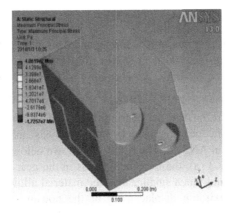

(a) The analysis of stress

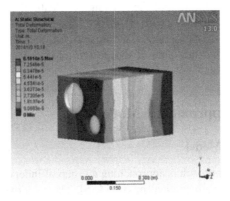

(b) The analysis of displacement

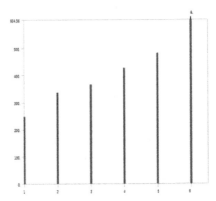

(c) The analysis of frequency

Fig. 3. Finite element analysis results

2.2   Finite Element Analysis

The three-dimensional is stored in IGS format and then imported in ANSYS software. Upon inquiry, the shell material is gray cast iron HT200, the manual of Mechanical Design shows that its modulus of elasticity is $E = 130GPa$, Poisson's ratio is $\mu = 0.25$ and density is $\rho = 7.2g/cm^3$. Considering the discontinuity of the housing, the mesh is refined in the bearing, bolts, etc. A total of 37,992 nodes and 18,715 units are in the gearbox model. Static analysis and dynamic analysis are carried out separately in ANSYS. The results are shown below.

As shown in Figure 3 (a), the maximum stress in the gear box is 48.62MPa at the Bolt hole in housing and much smaller than the material allowable stress 200MPa. In Figure 3 (b), the maximum displacement of the housing is 0.081614mm, which appears in the axial wall and around the two bearing holes, less than its static stiffness 0.095mm. Figure 3 (c) is modal analysis of the housing, which shows the front six natural frequencies of vibration of the housing, first natural frequency of vibration is 247.77Hz, greater than its system resonance frequency 160Hz. By analysis, the margin respectively exists between the results of the strength, static stiffness, vibration frequency and the value of the design allows. Therefore, the case can be lightweight optimization.

3   Establish Approximate Model

3.1   Approximate Model

Approximate model approach is approaching a group of independent variables and the response variable by mathematical model [4, 5]. The relationship between the input variables and output response can be described by the following formula:

$$y(x) = \tilde{y}(x) + \varepsilon \tag{1}$$

$y(x)$ is the actual value of the response, the function is unknown; $\tilde{y}$ is the response approximation, and we know its polynomial. ε is the random error of approximation and the actual value, usually obey the standard normal distribution $(0, \sigma^2)$. In this paper, the optimization target is the quality of the housing, main influence factors is wall thickness of the housing shell, therefore, so select the housing bottom wall thickness, axial and lateral wall thickness as design variables. The objective function can be described as:

$$\min m(x_1, x_2, x_3) \tag{2}$$

$$x_1 \in [3,6], x_2 \in [14,20], x_3 \in [3,8]$$

$$s.t. \quad K \le 0.1\,\text{mm}, \quad \sigma \le 200\text{MPa}, \quad f \ge 160\text{Hz}$$

In the function, m is the quality of the housing; K is the maximum allowable amount of displacement; σ is the intensity value of the housing; f is the first order frequency. Wall thickness is in millimeters.

The advantage of approximate model is that it can establish empirical formula, get quantitative relationship between input and output; Reduce call the time-consuming simulation program, improve efficiency of optimization, the actual computation time is usually several orders of magnitude shorter; Smooth the response function and reduce the "noise value", conducive to faster convergence to the global optimum. Establish approximate model includes sample data collection, approximate model type selection and the approximate model validation. Establish approximate model includes sample data collection, approximate model type selection and the approximate model validation.

3.2   Sample Collection

The prerequisite for the establishment of the approximate model is to sample the points of variables. Appropriateness of the sample points' selection directly determines the correctness of the approximate model. In order to ensure the efficiency and accuracy of the approximate model, you need to select the appropriate sample point, if the sample point selected properly, it will result in an approximate model to get the wrong result.

1	18	9	4	11	10	30
2	25	26	22	13	7	3
3	19	28	11	14	29	9
4	5	10	27	20	14	28
5	21	1	17	28	15	19
6	10	8	29	18	9	4
7	13	25	1	10	8	10
8	14	22	30	6	22	14
9	26	15	18	16	28	27
10	17	30	16	8	13	29
11	20	4	3	7	27	15
12	22	11	5	23	20	1
13	29	6	24	12	23	6
14	6	17	19	4	2	17
15	12	13	2	27	25	21
16	7	7	21	19	30	12
17	23	5	10	9	3	8
18	9	12	14	2	19	2
19	24	20	26	30	21	11
20	3	18	9	5	24	23
21	30	19	12	1	16	16
22	8	27	20	26	26	25
23	15	16	13	29	1	7
24	2	3	7	17	11	13
25	27	14	25	15	4	24
26	1	24	15	21	17	5
27	16	2	23	3	18	22
28	4	21	8	24	6	26
29	28	23	6	25	12	20
30	11	29	28	22	5	18

Fig. 4. The sample data of transmission housing

Latin square design is an experiment technology in order to decrease the influence of the order on the experiment. Its principle is that in n-dimensional space, every coordinate interval $[x_k^{\min}, x_k^{\max}], k \in [1, m]$ divides into m sections evenly, each cell recorded as $[x_k^{i-1}, x_k^i], i \in [1, m]$. Randomly selected m points, ensure a factor's each level being researched only once and that constitute the n-dimension space. The sample points are Latin square experiment for m.

In this paper, sample collection is conducted by the Best Latin square, and the optimal Latin square is more uniform than the randomized Latin square. It can make the factors and response's fit more accurate. Figure (4) is the transmission housing's sample data.

3.3   Response Surface Methodology

The method of response surface is to use a polynomial function fitting the design space, polynomials can be a first, second, third and fourth order, the minimum number of samples construct the response surface model depends on the number of input variables and model order. One of commonly used polynomial is second-order polynomial and the basic theory of second-order polynomial is:

$$y = \beta_0 + \sum_{i=1}^{j} \beta_i x_i + \sum_{i=1}^{j} \beta_{ii} x_i^2 + \sum_{i=1,k>i}^{j} \beta_{ik} x_i x_k \qquad i = 1, 2, \cdots, k \qquad (3)$$

y is the objective function to be fitted, $x_1, x_2, \cdots, x_j$ as design variables, $\beta_0, \beta_i x_i, \beta_{ii}, \beta_{ik}$ are undetermined constants.

In this paper, according to the 30 group of sample points and in the basis of experimental data obtained by the ANSYS finite element analysis, we respectively set approximation model for the gearbox housing quality, static stiffness, strength and the first order modal, which through using the ISIGHT software and polynomial response surface method. Get their similarity of approximate model: $R_m^2 = 0.996$, $R_k^2 = 0.984$, $R_\sigma^2 = 0.905$, $R_f^2 = 0.896$. From the above values of R^2, the accuracy of the four approximation models is in line with the requirements, and it can replace the finite element model to the next evolutionary algorithm optimization.

4   Evolutionary Algorithms and Results

4.1   Evolutionary Algorithms

Evolutionary Computation is a robust method, which can adapt to different problems in different environments and in most cases can get quite satisfactory and effective solution.

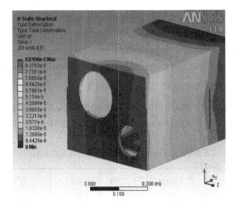

(a)   The analysis of stress

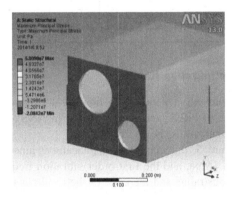

(b) The analysis of displacement

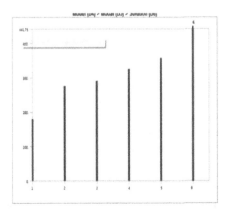

(c) The analysis of frequency

Fig. 5. Finite element analysis of the re-established model

In this paper, we use the standard genetic algorithm as a framework, which adopts binary coding, sort selection method and uniform crossover. Crossover rate is 0.5, mutation is 0.01 and the number of population is 30 [6, 7].

After 1000 iterations, the average value of 50 times to get the final results of the optimization: x_1 =4.8182mm, x_2 =14.0024mm, x_3 =3.0029mm. The quality of the gearbox is 31.9826Kg, the maximum deformation is 0.091mm, the maximum stress is 60Mpa and the first order frequency is 180Hz.As it is shown in Table 1.

Table 1. Analysis of optimization results

Model	x_1 (mm)	x_2 (mm)	x_3 (mm)	m (Kg)	K (mm)	σ (MPa)	f (Hz)
Before optimization	4	20	7	39.814	0.08161	48.62	247.77
After optimization	4.8182	14.0024	3.0029	31.348	0.091	60	180

4.2   Analysis of Optimization Results

According to the optimization data of the housing's structure, we re-establish the model by the software of PROE and then import it in ANSYS. The results obtained by the finite element analysis as shown in Figure 5.

Seen from the Figure 5, the maximum displacement amount is 0.09012mm, the maximum stress is 58.1Mpa, and the first-order vibration frequency is 178.53Hz. All the results meet the constraints and the error of the results between genetic algorithm optimized and Finite Element Analysis less than 5%, so the optimization results have high credibility. The quality of gearbox is 31.348Kg after optimization and it reduces 8.5Kg compared with the original. Lightweight extent reached 21.35%.

5   Conclusion

In this paper, we build a three-dimensional model diagram of the automatic transmission according to a car company and conduct finite element analysis on it. The premise of lightweight optimization is to ensure the performance requirements, such as strength, stiffness and vibration resistance. We create an approximate model and the accuracy of the four approximation models is in line with the requirements. Finally, we use GA to find the optimal value. The results show that this method can provide certain engineering guidance for the lightweight optimization of gear box.

Acknowledgements. This work is supported by the National Natural Science Foundation of China (No. 61202251, 60875046), the Program for Changjiang Scholars and Innovative Research Team in University (No.IRT1109), the Program for

Liaoning Innovative Research Team in University (No. LT2010005), the Natural Science Foundation of Liaoning Province (201102008), and by the Program for Liaoning Excellent Talents in University (LJQ2013133).

References

1. Zhang, Y., Li, G.Y., Wang, J.H.: Application of multi-objective genetic algorithm optimization in lightweight vehicle design. China Mechanical Engineering 4, 500–503 (2009)
2. Hu, Z.H., Cheng, A.G., Chen, S.W.: Multi-objective optimization's application in the new overall models' lightweight design. China Mechanical Engineeing 3, 404–409 (2013)
3. Zhang, S.L., Zhu, P., Chen, W.: Two methods based on robust design and its application in uncertainty lightweight body design. Shanghai Jiaotong University 5, 835–839 (2013)
4. Yan, H.S., Chen, B.S.: A calculation software based on experimental design and approximate model. Wuhan University of Technology 9, 272–274 (2010)
5. Han, D., Zheng, J.R.: Techniques of the approximate model in the engineering design optimization. East China University of Technology (Natural Science Edition) 6, 762–768 (2012)
6. Zhou, M., Sun, S.D.: Principle and application of genetic algorithms. National Defense Industry Press, Beijing (1999)
7. Fang, Q.W., Wang, P.: Substantive analysis for crossover of genetic algorithm. Beijing University of Technology 10, 1325–1336 (2010)

Design and Analysis of a Novel XY Micro-positioning Stage Used Corrugated Flexure Beams

Nianfeng Wang, Xiaohe Liang, and Xianmin Zhang

Guangdong Province Key Laboratory of Precision Equipment and Manufacturing Technology, South China University of Technology, Guangzhou, Guangdong 510640, PR China

Abstract. This paper presents the design and analysis of a novel XY micro-positioning stage which is driven by noncontact force provided by electromagnetic actuators. The stage is constructed with eight pieces of flexure structure, the corrugated flexure beam (CF beam), which have a large range of motion. Through the empirical equations, a set of optimized parameters of CF beam are decided. Then the static and modal analysis of the stage are conducted by finite element analysis(FEA) via Workbench, which validate the characteristic of large range of motion and work out its stiffness and natural frequency. The relationship between the displacement of the proposed micro-positioning stage and the exciting current is discussed.

Keywords: Micro-positioning stage, corrugated flexure beams, flexure joint.

1 Introduction

Precision positioning devices have been widely used in various areas, such as precision machine tools, surface profiler, semiconductor fabrication equipment, biological cell manipulation, optical inspections, etc. Most of the precise positioning stages are composed of compliant materials and flexure hinges [1–4]. Compared with conventional mechanical joints, flexure hinges can provide more ideal high-precision motions since they have advantages including no backlash, no friction, no need for lubrication, vacuum compatibility, ease of fabrication and basically no need for assembling [5–7].

Most of the positioning stages developed so far can only reach a few hundreds of micrometer range [8,9]. To achieve a relatively long travel range, XY stages with DC servo motor and lead screw have been proposed and popular for wafer positioning. However, it has been known that they have the nonlinear spring effect of balls and stick-slip. Such problems are difficult to eliminate; therefore, submicrometer positioning accuracy is rarely achieved by the XY stages with DC servomotor and lead screw. To achieve high precision in large motion range, many researchers have studied dual servomechanism that has a fine motion stage

X. Zhang et al. (Eds.): ICIRA 2014, Part II, LNAI 8918, pp. 586–595, 2014.

mounted on a coarse motion stage [10–12]. They used flexure joints for guidance in fine motion stages. Obviously dual servo stage makes the whole stage complex, and errors associated with coarse motion stage cause undesirable effect on the whole stage.

With the same cross section, the stiffness of flexure beam is only related to its overall length, which means that the longer the overall length the smaller the stiffness. If the spans are same in the longitudinal direction, a corrugated beam will have a longer centerline which will increase the actual length of beam to deform. Based on above observations, our previous paper [13], introduces a new type of flexure structure, the corrugated flexure beam, which has a large range of motion, low stress and simple structure. A novel XY micro-positioning stage based on CF beams is introduced in this paper with some advantages including a large range of motion, decouple character, symmetrical structure and noncontact driving force. Some of the common driving schemes such as PZT actuator, electromagnetic actuator and thermal actuator are applied to drive the high-precision positioning stage. To the proposed stage, both PZT actuator and thermal actuator can hardly meet the requirement of moving motion. As a result, a decoupled electromagnetic force actuator assembled with an armature and electromagnet is used to drive the mobile platform [14, 15].

Section 2 outlines the corrugated flexure beam. The modeling process of the micro-positioning stage is described in Section 3. The FEA is carried out in Section 4 to analyse all kinematics and dynamics modeling. Modeling and analysis of the electromagnetic actuator is presented in Section 5 and the fabrication of the novel stage is described in Section 6. Some concluding remarks are given in Section 7.

2   Corrugated Flexure Beam

In intuition, the deformation of a periodically corrugated flexible beam (Fig. 1) mentioned above is larger than a straight beam or an initially curved beam when the same load is applied. If the spans are same, CF beam will have a longer centerline which will increase the actual length of beam to deform. Thus, it can be applied to design compliant joints for its prominent feature.

CF beam is a periodical corrugated structure where each repeated unit (blue part shown in Fig. 1) consists of a semicircular and two straight segments. The CF beam shown in Fig. 1 is composed of 8 units which is similar to those applied to the proposed micro-positioning stage and all the eight CF beam are designed with identical dimensions. The cross section of CF beam is rectangle. Actually it can be circle, ellipse or other shapes. Parameters shown in Fig. 1 are the number of units N, width w and thickness t of cross section, length l of straight segment and radius R of semi-circular segment. L is the span length of CF beam. Each parameter has a definite effect on the stiffness of CF beam.

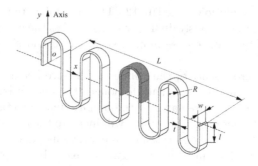

Fig. 1. Corrugated Flexure Beam

3   Design and Optimization of the Stag

3.1   Design of the Stage

The 4-Bar parallelogram with leaf springs is a typical compliant translational joint as shown in Fig. 2(a). It easily deforms in one direction but does not in other direction. Flexible translational joint composed of two 4-Bar parallelograms easily deforms in the XY-plane (Fig. 2(b)). Since the parallelogram constrains the rotation, it creates curvilinear trajectory. Thus, the two joints mentioned above cannot generate straight-line motion. To eliminate the axis drift, a symmetric configuration of plates is used. In Fig. 2(c), a symmetrical structure of the stage is designed. However, compared to the stage shown in Fig. 2(c), stage shown in Fig. 2(d) has a better compactness, which reduces the total size.

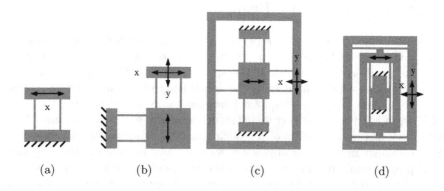

Fig. 2. Design Process of the Stage

The range of motion of a flexible segment is limited by the permissible stresses of the material. The flexible part cannot return to its original shape and size after the load is removed, providing that the elastic limit is exceeded. Notch joints and leaf spring joints can only have a small deformation when reach their elastic limit. Therefore, the stage shown in Fig. 2(d) indicates a limited displacement. In order to obtain a millimeter-scale moving range, a novel XY micro-positioning stage is proposed as shown in Fig. 3.

The micro-positioning stage contains a middle stage (Component B shown in Fig. 3) and a mobile stage (Component A shown in Fig. 3). When a force is applied to the mobile platform in the x or y-direction, it will move along one direction without axis drift. When a force is applied in y-axis direction, middle stage hardly moves, while force is applied in x-axis direction, middle stage moves along with mobile stage, so the detected object should be placed on mobile stage.

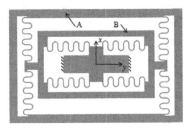

Fig. 3. XY Micro-positioning Stage used Corrugated Flexure Beams

3.2 Optimization of CF Beam

The successful design of the stage using CF beam requires knowledge of the kinematics and manipulation of the stiffness as the design variables. Each parameter has a definite effect on the stiffness of CF beam, but there is no simple yet rigorous method to calculate these stiffness values. Through searching the literature, some relevant results about axial stiffness (translational stiffness of y-axis) has been mentioned, but other stiffness components are being looked at. In this paper, both the empirical equations of x-axis translational stiffness and axial stiffness are obtained by curve-fitting. Initially the parameters of CF beam are studied through Finite Element Analysis (FEA). Then, an approximate form of the empirical equation is given. Lastly, the relationship between each two parameters is found out and the final empirical equation is determined.

By curve-fitting, x-axis translational stiffness of CF beam can be determined as:

$$K_x = \frac{Ewt^3}{Nk_{x1}} \tag{1}$$

where

$$k_{x1} = l^3 + (9.34R - 1.52)l^2 + (-2.1R^3 + 44.1R^2 - \\ 57.33R + 41.12)l + (16R^3 + 9.11R^2 - 17.4R + 7.84) \tag{2}$$

As Eqs.1 shows, there is a cubic relationship between stiffness and beam thickness, and width has only a linear effect on stiffness. There is a very complex relationship between R, l and stiffness K_x as shown in Eqs.2. The application ranges of K_x are: $N \geq 2$, $t \geq 0.1$ mm, $w \geq 2$ mm, and R and l can be any positive rational numbers. The axial stiffness (K_y) can be written as:

$$K_y = \frac{10^{-2}E(6.5wt^3 - 0.82t^3 + 0.000512)}{N^3R^2(l+3R)} \tag{3}$$

The data used in curve fitting of (K_y) are collected when parameters are in the ranges: N=6~16, R=1~7 mm, l=1~16 mm, w=2~10 mm, and t=0.1~0.4 mm.

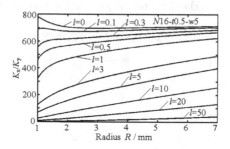

Fig. 4. Relationship between K_x/K_y and l, R

The axial stiffness (K_y) and x-axis translational stiffness (K_x) of CF beam used in this stage should be studied to work out a large K_x/K_y-value like leaf spring. Thereupon, finite element analysis (FEA) and empirical equations are carried out on CF beam to estimate the K_x/K_y-value. Results show that both length l of straight segment and radius R of semi-circular segment have a great effect on K_x/K_y as shown in Fig. 4. Increased N is required to increase the K_x/K_y-value. However, K_x/K_y is almost entirely unaffected by w and t, which can be predicted by the empirical equations. There would be no solution if all the parameters are decided as ideal values, that means N, R are very large values, l is equal to zero and so on. Therefore, some parameters should be confirmed preferentially. L is a constant as 20 mm. The thickness t is 0.6 mm for the thinner the harder to manufacture and the thicker the larger the stiffness of the stage is. The width is decided as 8 mm. The length l of straight segment can be 1 mm which is the lower limit of the empirical equations to obtain larger K_x/K_y-value. The last two structure variables are N and R, and they relational expression is:L=2RN. From Fig. 5, to maximize the K_x/K_y-value, radius R of semi-circular segment is decided as 1 mm which is the lower limit of the empirical equations and N is 10. As a result, in consideration of the moving range, a set of parameters are worked out and they are given as: N=10, R=1 mm, l=1 mm, w=8 mm and t=0.6 mm. The K_x/K_y-value is 124.6.

4   The Driving and Control System

The material used for the whole stage is aluminum alloy (AL7075-T651). The Young's modulus E assumed is 72000 MPa with Poisson's ratio of 0.33 and yield strength of 570 MPa.

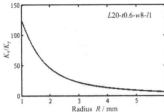

Fig. 5. Relationship between K_x/K_y and R

Fig. 6. Centre Line of the Confined and Unconfined CF Beams

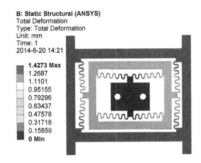

B: Static Structural (ANSYS)
Total Deformation
Type: Total Deformation
Unit: mm
Time: 1
2014-6-20 14:21

1.4273 Max
1.2687
1.1101
0.95155
0.79296
0.63437
0.47578
0.31718
0.15859
0 Min

Fig. 7. Deformation of the Stage

When force is respectively applied at two orthogonal sides of the mobile platform, the corresponding output motion of the stage are obtained as shown in Fig. 7. The x-axis and y-axis translational stiffness of the positioning stage are almost equal and they respectively are 35.09 N/mm and 35.03 N/mm. If the safe factor is assigned to be 1.8, the permissible stress is approximately equal to 317 MPa and total deformation in each direction can achieve 1.48 mm.

The first four natural frequencies evaluated by the FEA models are 116.57Hz, 136.01Hz, 145.64Hz and 147.61Hz. In addition, the first mode shape and third mode shape in Fig. 8 are translational motions in the x- and y-directions, which also reveal the mobility characteristics of the stage. The natural frequencies in other mode shapes are relatively low, so they cannot be ignored though the mobile platform works under low frequencies.

5   Analysis of the Electromagnetic Actuator

An electromagnetic force actuator assembled with an armature and electromagnet is used to drive this stage. Material of the armature is DT4C, a kind of electrical pure iron which will not be permanently magnetized. The employed electromagnets have a very simple structure which can be obtained easily. When the coil is power on, the electromagnet generates magnetic field and most of the

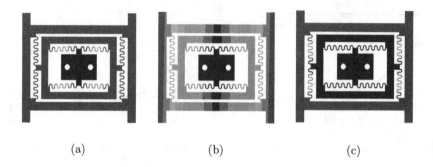

(a)   (b)   (c)

Fig. 8. First Three Mode Shapes

magnetic induction lines go through iron core, air gap and armature as shown in Fig. 9. Then, the electromagnet attracts the armature as they have opposite magnetism.

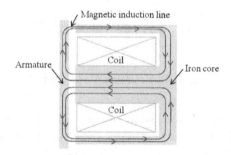

Fig. 9. Principle Scheme of the Electromagnetic Actuator

To obtain certain displacement of the stage, a certain noncontact force should be provided. The static analysis of the stage is given as:

$$F = KD \qquad (4)$$

where F is the noncontact force applied to the mobile platform, K is the translational stiffness and D indicates the movement of the stage.

For each single actuator, the electromagnetic force can be approximately calculated by the following equations. According to the Maxwell's electromagnetic theory, the driving force can be written as:

$$F = \frac{\Phi^2}{\mu_0 \cdot S} \qquad (5)$$

where Φ is the magnetic flux through the air gap, μ_0 is the vacuum permeability and S is the cross-sectional areas of the air gap.

By magnetic Ohm's law, irrespective of magnetic leakage, the magnetic flux can be obtained as:

$$\Phi = \frac{10^8 INS\mu_0}{\rho} \tag{6}$$

where I is the exciting current, N is the turns of the coil, ρ is the length of the air gap in the x or y-direction, respectively. In practice, magnetic leakage cannot be neglected even if the air gap is 2 mm, thus, a correction factor should be introduced during controlling.

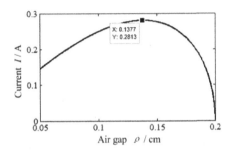

Fig. 10. Relationship between Current and the Length of Air Gap

Though F is the function of current I and air gap ρ, the current is the only variable to decide the electromagnetic force. When the air gap is 2 mm, the relationship between the length of air gap and the exciting current I is shown in Fig. 10. The rated current of the electromagnet should be larger than 0.2813 A.

6 Fabrication of the Stage

To achieve a symmetrical structure to eliminate initial magnetic force, two electromagnets are arranged opposite each other on the two sides of the stage in the x or y-direction, respectively, as shown in Fig. 11. The armatures are fixed onto the mobile platform, and the electromagnetic actuators are fixed. If the armature is attracted to the electromagnet caused by unexpected vibration, the material will exceed its elastic limit and the stage will become invalid. Thus, in order to eliminate that situation, four limited blocks are used in the positioning system as shown in Fig. 11.

After the mechanical structure is designed and the drive mode is decided, the prototype can be fabricated via wire-electrode cutting as shown in Fig. 12.

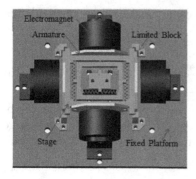

Fig. 11. Model of the Micro-positioning Stage System

Fig. 12. Prototype of the Mechanical Structure

7   Conclusion

A novel XY micro-positioning stage driven by electromagnetic actuators is proposed in this paper. The stage possesses a large range of motion by using corrugated flexible beams, symmetrical structure and noncontact driving force. Through optimization design and FEA, the stiffness of the stage system is 35 N/mm for both X and Y direction and it can reach a motion range of 1.48 mm which needs 51.8 N driving force. In our future work, DSPACE, laser displacement, dynamic signal analysis system and so on will be used to verify the whole modeling process and closed-loop control strategy will be adopted to enhance the positional accuracy and eliminate the crosstalk.

References

1. Smith, S., Chetwynd, D., Bowen, D.: Design and assessment of monolithic high precision translation mechanisms. Journal of Physics E: Scientific Instruments 20(8), 977–983 (1987)
2. Scire, F., Teague, E.: Piezodriven 50-m range stage with subnanometer resolution. Review of Scientific Instruments 49(12), 1735–1740 (1978)
3. Li, Y., Xiao, S., Xi, L., Wu, Z.: Design, modeling, control and experiment for a 2-dof compliant micro-motion stage. International Journal of Precision Engineering and Manufacturing 15(4), 735–744 (2014)
4. Choi, Y.J., Sreenivasan, S.: Large displacement flexure based nano-precision motion stage for vacuum environments. In: ASPE Proceedings, pp. 1–3 (2003)
5. Howell, L.L.: Compliant mechanisms. John Wiley & Sons (2001)
6. Lobontiu, N.: Compliant mechanisms: design of flexure hinges. CRC press (2010)
7. Trease, B.P., Moon, Y.M., Kota, S.: Design of large-displacement compliant joints. Journal of Mechanical Design 127(4), 788–798 (2005)
8. Xi, L., Li, Y., Zhao, X.: Design and analysis of a 2-dof micro-motion stage based on flexural hinges. In: 2012 12th International Conference on Control Automation Robotics & Vision (ICARCV), pp. 1335–1340. IEEE (2012)

9. Liang, Q., Zhang, D., Chi, Z., Song, Q., Ge, Y., Ge, Y.: Six-dof micro-manipulator based on compliant parallel mechanism with integrated force sensor. Robotics and Computer-Integrated Manufacturing 27(1), 124–134 (2011)
10. Leibovich, V.E., Novak, W.T.: Coarse and fine motion positioning mechanism US Patent 4,723,086 (February 2, 1988)
11. Liu, Y., Li, T., Sun, L.: Design of a control system for a macro-micro dual-drive high acceleration high precision positioning stage for ic packaging. Science in China Series E: Technological Sciences 52(7), 1858–1865 (2009)
12. Jie, D.G., Liu, Y.J., Sun, L.N., Sun, S.Y., Cai, H.G.: Modeling and control of a macro-micro dual-drive ultra-precision positioning mechanism. Optics and Precision Engineering 2, 09 (2005)
13. Wang, N., Liang, X., Zhang, X.: Pseudo-rigid-body model for corrugated cantilever beam used in compliant mechanisms. Chinese Journal of Mechanical Engineering 27(1), 122–129 (2014)
14. Xiao, S., Li, Y., Zhao, X.: Design and analysis of a novel flexure-based xy micro-positioning stage driven by electromagnetic actuators. In: 2011 International Conference on Fluid Power and Mechatronics (FPM), pp. 953–958. IEEE (2011)
15. Verma, S., Shakir, H., Kim, W.J.: Novel electromagnetic actuation scheme for multiaxis nanopositioning. IEEE Transactions on Magnetics 42(8), 2052–2062 (2006)

Author Index